D0848559

COLLOQUIA MATHEMATICA
SOCIETATIS JÁNOS BOLYAI, 8.

TOPICS IN TOPOLOGY

Edited by Á. CSÁSZÁR

NORTH-HOLLAND PUBLISHING COMPANY
AMSTERDAM - LONDON

© BOLYAI JÁNOS MATEMATIKAI TÁRSULAT

Budapest, Hungary, 1974

ISBN *North-Holland:* 0 7204 2092 X
ISBN *American Elsevier:* 0 444 10597 2

Joint edition published by

JÁNOS BOLYAI MATHEMATICAL SOCIETY

and

NORTH-HOLLAND PUBLISHING COMPANY

Amsterdam-London

Printed in Hungary

ÁFÉSZ, VÁC

Sokszorosító üzeme

QA
611
.A1
C64
1972

PREFACE

The Bolyai János Mathematical Society organized a Colloquium on Topology in Keszthely, at the Lake Balaton, from 19th to 23d June 1972. Two colloquia on the same subject, much smaller in extent, had been previously organized by the Society in the years 1956 and 1964.

The members of the Organizing Committee were the following:

Á. Császár (chairman),
P. Hamburger (secretary),
M. Bognár, E. Deák, S. Gacsályi, J. Gerlits, I. Juhász.

The importance of Topology in the development of contemporary mathematics and its many-sided applications in numerous disciplines such as Algebra or Analysis, explains the large number of mathematicians who participated and presented a communication at the Colloquium. There were 141 participants from 20 countries, among them 45 from Hungary. The number of communications was 71, among them 14 one hour lectures, and the remaining 57 were 20 minute communications (discussion included). These short communications were held in three parallel sections. One afternoon was devoted to the presentation of unsolved problems; about 40 such problems were presented in two parallel sections. Further opportunity to informal discussion was provided by an excursion by boat to Badacsony and at a dinner in one of the famous wine-cellars of this charming mountain.

The present volume contains, besides the detailed scientific programme of the Colloquium and the list of participants, the papers submitted for publication to the Organizing Committee; they vary from short summaries to detailed expositions with proofs. It also includes some papers of topologists who were not able, for various reasons, to attend the Colloquium but sent their manuscripts in order to be published in the Proceedings.*

*All papers in this volume have been refereed.

DISCARDED
WILLIAM D. McINTYRE LIBRARY
UNIVERSITY OF WISCONSIN - EAU CLAIRE
299226

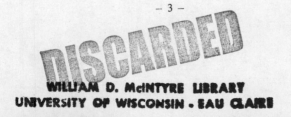

It is a pleasent duty for me to express my most sincere thanks to everybody who contributed to the success of the Colloquium by communications of high scientific level. On the other hand, I cannot fail to commemorate the irrepairable loss caused by the unexpected death of Professor J. de Groot, scarcely three months after having participated at the Colloquium in best health and having given there a lecture of great interest. The news of the death of this eminent topologist deeply shocked each member of the Organizing Committee.

The Editor

CONTENTS

SCIENTIFIC PROGRAM

June 19. Monday
Afternoon

Plenary Session

14^{30} – Opening of the Colloquium	
15^{45} – 16^{30}	M. Bognár: On generalized manifolds
16^{40} – 17^{25}	E. Deák: Theorie und Anwendung der Richtungsstrukturen

Session A.

17^{50} – 18^{10}	E.E. Moise: Cloverleaf representations of simply connected 3-manifolds
18^{10} – 18^{30}	J. Vrabec: Submanifolds of homology 3-spheres
18^{30} – 18^{50}	W. Vogel: Cohomology theories in simplicial topology

Session B.

17^{50} – 18^{10}	A. Alexiewicz: Some two norm spaces of holomorphic functions
18^{10} – 18^{30}	P.R. Andenæs: Note on metrization of compact convex sets

Session C.

17^{50} – 18^{10}	R. Frič: On a problem of J. Novák
18^{10} – 18^{30}	H.R. Fischer: On a certain class of $C(K)$ modules
18^{30} – 18^{50}	G. Helmberg: On the separability and nonmetrisability of the space of slowly increasing functions

June 20. Tuesday
Morning

Plenary Session

9^{00} – 9^{45}	J. de Groot: A generalized Schoenflies theorem for Euclidean spaces
9^{55} – 10^{40}	C.H. Dowker: Separation axioms for frames

Session A.

11^{00} – 11^{20}	R. Bennett: On some classes of noncontractible dendroids
11^{20} – 11^{40}	K.H. Hofmann – M. Mislove: Compact groups acting automorphically on compact monoids
11^{40} – 12^{00}	R.D. Anderson: Results of Chapman on Q-manifolds

Session B.

11^{00} – 11^{20}	W. Gähler: Zur Theorie der Limesräume
11^{20} – 11^{40}	E. Binz: A criterion for vector space topologies on $C(X)$
11^{40} – 12^{00}	N. Hadziivanov: On infinite dimensional Cantor manifolds
12^{00} – 12^{20}	D. Doitchinov: Groupes topologiques minimaux

Session C.

11^{00} – 11^{20}	P. Hamburger: Superextensions and preproximity
11^{20} – 11^{40}	K. Császár: On F. Riesz's separation axiom
11^{40} – 12^{00}	L. Tumarkin: On open coverings of non closed sets
12^{00} – 12^{20}	J. Langhammer: On perfect Hausdorff compactifications

June 21. Wadnesday
Morning

Plenary Session

9^{00} – 9^{45}	K. Morita: Some results on *M*-spaces
9^{55} – 10^{40}	H. Herrlich: On perfect subcategories

Session A.

11^{00} – 11^{20}	J. Rosický – M. Sekanina: Realizations of categories of topological spaces
11^{20} – 11^{40}	O.V. Lokucievskiĭ – E.V. Ščepin: On P.S. Alexandroff's axiomatic definition of dimension
11^{40} – 12^{00}	A.A. Malcev: On a Post's algebra
12^{00} – 12^{20}	M.M. Postnikov: Axiomatic homology theory and Alexander's duality theorem

Session B.

11^{00} – 11^{20}	M. Hušek: Hewitt realcompactification of products
11^{20} – 11^{40}	L. Rudolf: Minimal Hausdorff spaces are pseudo-quotient images of compact Hausdorff ones
11^{40} – 12^{00}	J. Hejcman: Uniform dimension and rings of uniformly continuous functions

Session C.

11^{00} – 11^{20}	Á. Császár: Invariant structures and transformation groups
11^{20} – 11^{40}	J. Szenthe: On transitive locally compact transformation groups
11^{40} – 12^{00}	A.J. Ostaszewski: A characterization of compact, separable, ordered spaces

June 21. Wednesday
Afternoon

Plenary Session

$15^{45} - 16^{30}$	M.E. Rudin: Using Souslin trees to construct Dowker spaces
$16^{40} - 17^{25}$	K. Kuratowski: Development of the research on indecomposable continua

Session A.

$17^{50} - 18^{10}$	H.G. Bothe: Relative Einbettbarkeit in Mannigfaltigkeiten
$18^{10} - 18^{30}$	K. Horvatić: Some sufficient conditions for embedding polyhedra in Euclidean space

Session B.

$17^{50} - 18^{10}$	R. Duda: Some problems concerning lattices in topology
$18^{10} - 18^{30}$	J. Mioduszewski: An irreducible covers
$18^{30} - 18^{50}$	K. Kuhnert: A universal property of the lattice of all topologies (Boolean-valued topologies)
$18^{50} - 19^{10}$	W. Kulpa: Remark on product of proximities

Session C.

$17^{50} - 18^{10}$	K. Alster: Cartesian products of subparacompact spaces
$18^{10} - 18^{30}$	J. Gerlits: On G_δ p-spaces
$18^{30} - 18^{50}$	T. Przymusiński: A Lindelöf space X such that X^2 is normal but not paracompact
$18^{50} - 19^{10}$	R. Telgársky: On topological properties defined by games

June 11. Thursday
Morning

Plenary Session

$9^{00} - 9^{45}$	G. Chogoshvili: On projective homology and homotopy groups
$9^{55} - 10^{40}$	S. Mardešić: Shapes for topological spaces

Session A.

$11^{00} - 11^{20}$	R. Liedl: A universal property of the lattice of all topologies
$11^{20} - 11^{40}$	M. Moszyńska: The Fox theorem in theory of shapes
$11^{40} - 12^{00}$	F. Szigeti: Differential topology in smooth Banach spaces
$12^{00} - 12^{20}$	H. Evans: Topological methods in representation of rings by shaves

Session B.

$11^{00} - 11^{20}$	J.J. Charatonik: On decomposition of continua
$11^{20} - 11^{40}$	G. Murphy: When are the isometries onto?
$11^{40} - 12^{00}$	H. Kok: On tree-like spaces and the intersection of connected subsets of a connected T_1-space
$12^{00} - 12^{20}$	F.B. Jones: Aposyndetic continua

Session C.

$11^{00} - 11^{20}$	J. van Dalen: Characterizing totally orderable spaces and their topological products
$11^{20} - 11^{40}$	M.J. Faber: A metrization theorem for linearly ordered topological spaces
$11^{40} - 12^{00}$	A.H. Siddigi: Non-Archimedean 2-metric spaces
$12^{00} - 12^{20}$	R.Z. Domiaty: Zur Topologisierung Metrischer Räume

June 22. Thursday
Afternoon

Plenary Session

$15^{45} - 16^{30}$ Ju.M. Smirnov: Extensors for metric spaces and infinite dimensional spaces

$16^{30} - 17^{25}$ V.I. Ponomarov: Problems and results in set theoretical topology

Session A.

17^{50} Problems session

Session B.

17^{50} Problems session

June 23. Friday
Morning

Plenary Session

9^{00} – 9^{45}	F.W. Bauer: Homotopy and stabilization
9^{55} – 10^{40}	A. Hajnal: Some independence results in set theoretical topology

Session A.

11^{00} – 11^{20}	O.T. Alas: A generalization of a Stone-Morita-Hanai theorem
11^{20} – 11^{40}	F.D. Tall: The countable chain condition vs. separability Applications of Martin's axiom
11^{40} – 12^{00}	I. Juhász: Remarks on cardinal functions
12^{00} – 12^{20}	L. Babai – A. Máté: Inner set mappings on locally compact spaces
12^{20}	P. Erdős: On some general properties of chromatic numbers

Session B.

11^{00} – 11^{20}	L. Babai – L. Lovász: Finite groups of spherical homeomorphisms
11^{20} – 11^{40}	V.V. Fedorčuk: On H-closed extensions of Θ-proximity spaces
11^{40} – 12^{00}	S. Negrepontis: Spaces homeomorphic to $(2^{\kappa})_{\kappa}$

LIST OF PARTICIPANTS

ALAS, OFELIA TERESA, Instituto de Matemática e Estatistica, Universidade de Sao Paulo, Brazil

ALBRYCHT, J., Poznan, ul. Grochowska 40 m. 10, Poland

ALEXIEWICZ, A., Poznan, ul. Matejki 48/49, Poland

ALÒ, R.A., Department of Mathematics, Carnegie-Mellon University, Schenley Park, Pittsburgh, Pennsylvania 15213, USA

ALSTER, K., Nowogrodzka 7/9 m. 22, Warszawa, Poland

ANDENÆS, P.R., Department of Mathematics NLH, University of Trondheim, 7000 Trondheim, Norway

ANDERSON, R.D., Louisiana State University and Agricultural and Mechanical College, Baton Rouge – Louisiana 70803, College of Arts and Sciences, USA

AQUARO, G., Istituto Matematico "Guido Castelnuovo", Università degli Studi di Roma, Città Universitaria 001000, Roma, Italy

BAUER, F.W., 6 Frankfurt am Main, Oederweg 109, German Federal Republic

BENARD, M., 2011 Milan Street, New Orleans, Louisiana 70115, USA

BENNETT, R., Department of Mathematics, Auburn University, Auburn, Alabama 36830, USA

BINZ, E., Lehrstuhl für Mathematik I., Universität Mannheim (W.H.) 68 Mannheim 1, Schloss, German Federal Republic

BOTHE, H.G., Zentralinst. für Mathematik und Mechanik der DAW, 108 Berlin, Mohrenstr. 39, German Democratic Republic

BUTZMANN, H.P., Lehrstuhl für Mathematik I. der Universität, 68 Mannheim, Schloss, German Federal Republic

CHABER, J., Warszawa 1, Litewska 10 m. 14, Poland

CHARATONIK, J.J., Wroclaw 44, skr. poczt. 1186, Ul. Zielonogorska 9 m. 3, Poland

CHOGOSHVILI, G., Institute of Mathematics, Academy of Sciences Tbilisi, Georgia, USSR

van DALEN, J., Wiskundig Seminarium der Vrije Universiteit, Amsterdam-Buitenveldert, De Boelelaan 1081, Postbus 7161, The Netherlands

DOITCHINOV, D., Sofia 26, Boul. A. Ivanov 1, Bulgaria

DOMIATY, R.Z., A-8010 Graz, Lehrkanzell und Institut für Mathematik III, Technische Hochschule in Graz, Kopernikusgasse 24, Austria

DOWKER, C.H., Department of Mathematics, Birkbeck College, Malet Street, London WCIE 7 HX, Great Britain

DUDA, R., Wroclaw 2, ul. Lukasiewicza 18 M. 1, Poland

DUNG, N.H., Mathematical Research Institute of the Hungarian Academy of Sciences, Budapest V. Reáltanoda u. 13-15.

EVANS, H., Department of Mathematics, Tulane University, College of Arts and Sciences, New Orleans, Louisiana 70118, USA

FABER, M.J., Wiskundig Seminarium der Vrije Universiteit, Amsterdam-Buitenveldert, De Boelelaan 1081, Postbus 7161, The Netherlands

FEDORČUK, V.V., Moscow State University, Dept. Geo. Topology, Moscow V-234, USSR

FISCHER, H.R., Department of Mathematics, The Commonwealth of Massachusetts University of Massachusetts, Amherst 01002, USA (nationality: Swiss)

FRIČ, R., Magurskà 15, Košice, Czechoslovakia

GÄHLER, W., Zentralinstinstitut für Mathematik und Mechanik der DAW, Berlin 108 Mohrenstr. 39, German Democratic Republic

de GROOT, J., Mathematisch Institut, Universiteit van Amsterdam, Roeterostraat 15, Amsterdam, The Netherlands

GYFTODIMOS, G., 3 rue Humbert II. 38 – Grenoble, France

HADZIIVANOV, H., Sofia 26, boul. A. Ivanov 1, Bulgaria

HEJCMAN, J., Matematický ústav ČSAV, Žitna 25, Praha 1, Czechoslovakia

HELMBERG, G., Universität Innsbruck, I. Lehrkanzel für Mathematik an der Fakultät für Bauingenieurwesen und Architektur, A-6020 Innsbruck, Technikerstr. 13, Austria

HERRLICH, H., 28 Bremen, Oberblockland 7 b, German Federal Republik

HOFMANN, K.H., Department of Mathematics, Tulane University, New Orleans, Louisiana 70118, USA

HORVATIĆ, K., Geometrijski Zavod P.P. 187, Zagreb, Yugoslavia

HUŠEK, M., Math. Institute, Praha 8. Sokolovská 83, Czechoslovakia

JACKOWSKI, S., University of Warszawa, Warszawa, Poland

JAYNE, J.E., Scuola Normale Superiore, Piazza dei Cavalieri 7, 56100 Pisa, Italy, (nationality: USA)

JONES, F.B., Department of Mathematics, University of California, California 92502, USA

KANIEWSKI, J., University of Warszawa, Warszawa, Poland

KOK, H., Wiskundig Seminarium der Vrije Universiteit, Amsterdam-Buitenveldert, De Boelelaan 1081, Postbus 7161, The Netherlands

KRASINKIEWICZ, J., Instytut Matematyczny, Polskiej Akademii Nauk, Warszawa 1 ul. Śniadeckich 8, Poland

KUHNERT, K., Mathem. Institut der Universität, A-6020 Innsbruck, Austria

KROONENBERG, N., Stichting Mathematisch Centrum, 2e Boerhaavestraat 49, Amsterdam (Oost), The Netherlands

KULPA, W., ul. Mieszka I-ego 5 m. 28, Katowice 22, Poland

KURATOWSKI, K., Instytut Matematyczny, Polskiej Akademii Nauk, Warszawa 1 ul. Śniadeckich 8, Poland

KUTZLER, K., Lehrstuhl für Mathematik I., Universität Mannheim, 368 Mannheim, Schloss, German Federal Republic

LANGHAMMER, J., 22 Greifswald Sektion Mathematik, Jahnstr. 15 a, German Democratic Republic

LEIDL, R., Institut für Mathematik der Universität Innsbruck, Innrain 52, A-6020 Innsbruck, Austria

LOKUCIEVSKIĬ, O.V., Mat. Inst. AN USSR, ul. Vavilova 28, Moskow, USSR

MALCEV, A.A., Mat. Inst. AN USSR, ul. Vavilova 28, Moscow, USSR

de MARCO, G., Istituto di Matematica Applicata, Via Belzoni 3, 35100 Padova, Italy

MARDEŠIĆ, S., Department of Mathematics, University of Zagreb, Zagreb, Yugoslavia

METELLI, C., Seminario Matematico, Via Paolotti 3, 35100 Padova, Italy

MILNER, E.C., Department of Mathematics, 16 Mill Lane, Cambridge, Great Britain

MIODUSZEWSKI, J., ul. Mieszka I-ego 5 m. 28, Katowice 22, Poland

MISLOVE, M., Tulane University, College of Arts and Sciences, New Orleans, Louisiana 70118, USA

MOISE, E.E., Department of Mathematics, Queens College (CUNY) Flushing, New York 11367, USA

MORITA, K., Department of Mathematics, Tokyo University of Education 3-29-1, Omtsuka Sonkyo-Ku, Tokyo 112, Japan

MOSZYŃSKA, MARIA, Warszawa, Sanocka 11 a m. 3, Poland

MURPHY, G., 78 Freiburg, Hebelstrasse 29, German Federal Republic, (nationality: USA)

MÜLLER, B., Lehrstuhl für Mathematik I. der Universität, 68 Mannheim, Schloss, German Federal Republic

NEGREPONTIS, S., Department of Mathematics, McGill University, Montreal, Que. Canada

NERALIĆ, L., 41000 Zagreb, Fakultet Ekonomskih Nauka, Kenedijev Trg 6

NOWIŃSKI, K., University of Warszawa, Warszawa, Poland

ORSATTI, A., Istituto di Matematico dell'Università, Via Savonarola 9, 44100 Ferrara, Italy

OSTRASZEWSKI, A.J., Department of Mathematics, University College London, Gower Street, London WC 1, Great Britain

PHILI, C., St. Niridon 67 Palaion Falirou, Athens, Greece

POL, R., University of Warszawa, Warszawa, Poland

PONOMAREV, I.V., Moscow State University, Dept. Geo. Topology, Moscow V-234, USSR

POPPE, H., 22 Greifswald, Karl Liebknecht-Ring 1, German Democratic Republic

POSTNIKOV, M.M., Mat. Inst. AN USSR, ul. Vavilova 28, Moskow, USSR

PRZYMUSIŃSKI, T., University of Warszawa, Warszawa, Poland

PUZIO, ELZBIETA, University of Warszawa, Warszawa, Poland

ROSICKÝ, J., Department of Mathematics, Purkyně Univ. Janáčkovo nám. 2 a, Brno, Czechoslovakia

RUDIN, MARY ELLEN, Department of Mathematics, University of Wisconsin, Madison 53706, Wisconsin, USA

RUDOLF, L., Gdansk Oliwa, ul. Abrahama 28 Pod Łosiem, Poland

SALCE, L., Seminario Matematico, Via Paolotti 3, 35100 – Padova, Italy

SEKANINA, M., Department of Mathematics, Purkyne Univ. Janackovo nám. 2 a, Brno, Czechoslovakia

SENNOTT, L.I., Department of Mathematics, George Mason University, Fairfax Virginia 22030, USA

SIDDIGI, A.H., Institut für Angewandte Mathematik, Universität Heidelberg, 6900 Heidelberg im Neuenheimer Feld 5, German Federal Republic, (nationality: Indian)

SMIRNOV, Yu.M., Moscow State University, Dept. Geo. Topology, Moscow V-234, USSR

SWAMINATHAN, S., Department of Mathematics, Dalhousie University, Halifax N.S., Canada, (nationality: Indian)

TALL, D.F., Department of Mathematics, University of Toronto, Toronto 181, Canada

TELGÁRSKY, R., Matematicky ustav SAV, ul. Stefanikova 41, Bratislava, Czechoslovakia

TORUŃCYZK, A.H., Warszawa, Gagarina 27 m. 42, Poland

TUMARKIN, L.A., Moscow State University, Dept. Geo. Topology, Moscow V-234. USSR

UNGAR, Š., 41001 Zagreb, PP 187, Matematički Zavod, Marulićev Trg 19, Yugoslavia

VENCEL, MARIJA, Zelena Pot 17, 61000 Ljubljana, Yugoslavia

VOGEL, W., Sektion Mathematik der Universität, Halle/Saale, Universitätsplatz 8, German Democratic Republic

VRABEC, J., Department of Mathematics, University of Ljubljana, Jadranska 19, 61000 Ljubljana, Yugoslavia

WEBER, M., 90 Karl-Marx-Stadt, Beethovenstr. 51, German Democratic Republic

WOJTKOWSKA, KATAVRYNA, University of Warszawa, Warszawa, Poland

WONG, R.Y., Math. Institute, Roeterstraat 15, Amsterdam, The Netherlands, (nationality: USA)

LIST OF HUNGARIAN PARTICIPANTS

ÁDÁM, KATALIN, Department of Mathematics, Technical University of Budapest, Faculty for Electrical Engineering, H-1111 Budapest, Stoczek u. 33.

BABAI, L., Roland Eötvös University, H-1088 Budapest, Múzeum krt. 6-8.

BÁRÁNY, I., H-1165 Budapest, Huszár u.

BÁRÁSZ, P., H-1113 Budapest, Kökörcsin u. 19.

BÉKÉSSY, P., Roland Eötvös University, H-1088 Budapest, Múzeum krt. 6-8.

BLEYER, A., Department of Mathematics, Technical University of Budapest, Faculty for Electrical Engineering, H-1111 Budapest, Stoczek u. 33.

BOGNÁR, M., Roland Eötvös University, H-1088 Budapest, Múzeum krt. 6-8.

BUZÁSI, SZVETLÁNA, Kossuth Lajos University, H-4010 Debrecen 10.

CSÁSZÁR, Á., Roland Eötvös University, H-1088 Budapest, Múzeum krt. 6-8.

CSÁSZÁR, KLÁRA, Department of Mathematics, Technical University of Budapest, Faculty for Electrical Engineering, H-1111 Budapest, Stoczek u. 33.

DEÁK, E., Mathematical Research Institute of the Hungarian Academy of Sciences, H-1053 Budapest, Reáltanoda u. 13-15.

DEÁK, J., Mathematical Research Institute of the Hungarian Academy of Sciences, H-1053 Budapest, Reáltanoda u. 13-15.

ERDŐS, P., Mathematical Research Institute of the Hungarian Academy of Sciences, H-1053 Budapest, Reáltanoda u. 13-15.

FIALA, T., Roland Eötvös University, H-1088 Budapest, Múzeum krt. 6-8.

GACSÁLYI, S., Kossuth Lajos University, H-4010 Debrecen 10.

GEHÉR, I., Computing Center of the Hungarian Academy of Sciences H-1014 Budapest, Uri u. 49.

GEHÉR, L., Center of the Hungarian Academy of Sciences, H-6720 Szeged, Somogyi Béla u. 7.

GERLITS, J., Mathematical Research Institute of the Hungarian Academy of Sciences, H-1053 Budapest, Reáltanoda u. 13-15.

GYÖNGY, I., Roland Eötvös University, H-1088 Budapest, Múzeum krt. 6-8.
HAJNAL, A., Mathematical Research Institute of the Hungarian Academy of Sciences, H-1053 Budapest, Reáltanoda u. 13-15.
HAMBURGER, P., Mathematical Research Institute of the Hungarian Academy of Sciences, H-1053 Budapest, Reáltanoda u. 13-15.
JUHÁSZ, I., Roland Eötvös University, H-1088 Budapest, Múzeum krt. 6-8.
KÁSZONYI, L., Roland Eötvös University, H-1088 Budapest, Múzeum krt. 6-8.
KERSNER, R., Computing Center of the Hungarian Academy of Sciences, H-1014 Budapest, Uri u. 49.
KOMLÓSI, S., József Attila University, H-6720 Szeged, Aradi vértanuk tere 1.
LOVÁSZ, L., Roland Eötvös University, H-1088 Budapest, Múzeum krt. 6-8.
jr. MAKAI, E., Mathematical Research Institute of the Hungarian Academy of Sciences, H-1053 Budapest, Reáltanoda u. 13-15.
MÁLYUSZ, K., Computing Center of the Hungarian Academy of Sciences, H-1014 Budapest, Uri u. 49.
MÁTÉ, A., Research Center of the Hungarian Academy of Sciences, H-6720 Szeged, Somogyi Béla u. 7.
MÁTÉ, E., József Attila University, H-6720 Szeged, Aradi vértanuk tere 1.
MÉRŐ, L., Roland Eötvös University, H-1088 Budapest, Múzeum krt. 6-8.
MICHALETZKY, GY., Roland Eötvös University, H-1088 Budapest, Múzeum krt. 6-8.
NAGY, P., József Attila University, H-6720 Szeged, Aradi vértanuk tere 1.
NAGY, ZS., Roland Eötvös University, H-1088 Budapest, Múzeum krt. 6-8.
PETRUSKA, GY., Roland Eötvös University, H-1088 Budapest, Múzeum krt. 6-8.
PINTZ, J., Roland Eötvös University, H-1088 Budapest, Múzeum krt. 6-8.
SCHMIDT, T., Mathematical Research Institute of the Hungarian Academy of Sciences, H-1053 Budapest, Reáltanoda u. 13-15.
SEBESTYÉN, Z., Roland Eötvös University, H-1088 Budapest, Múzeum krt. 6-8.

SZEMERÉDI, E., Mathematical Research Institute of the Hungarian Academy of Sciences, H-1053 Budapest, Reáltanoda u. 13-15.

SZENTHE, J., Research Center of the Hungarian Academy of Sciences, H-6720 Szeged, Somogyi Béla u. 7.

SZÉP, A., Mathematical Research Institute of the Hungarian Academy of Sciences, H-1053 Budapest, Reáltanoda u. 13-15.

SZIGETI, F., Roland Eötvös University, H-1088 Budapest, Múzeum krt. 6-8.

THIRY, M., Department of Mathematics, Technical University of Budapest, Faculty for Electrical Engineering, H-1111 Budapest, Stoczek u. 33.

THIRY, I., Department of Mathematics, Technical University of Budapest, Faculty for Electrical Engineering, H-1111 Budapest, Stoczek u. 33.

VARGA, Z., Roland Eötvös University, H-1088 Budapest, Múzeum krt. 6-8.

COLLOQUIA MATHEMATICA SOCIETATIS JÁNOS BOLYAI
8. TOPICS IN TOPOLOGY, KESZTHELY (HUNGARY), 1972.

A GENERALIZATION OF A STONE-MORITA-HANAI'S THEOREM

O.T. ALAS

Our purpose is to prove a theorem on a special type of paracompact spaces – which satisfy property (*) below – similar to that proved by Stone, Morita and Hanai ([2], [3]) for metrizable spaces. (Recalling R.H. Bing's metrization theorem, these spaces may be regarded as a generalization of metrizable spaces. A study of other properties of theirs will be published elsewhere.)

1. PRELIMINARIES

Let E be a nondiscrete Hausdorff completely regular space.

Definition 1. The index of E is the least cardinal number m for which there is a collection of cardinality m of open subsets of E whose intersection is not an open set.

If E is paracompact and its index is greater than $\aleph_0$, then for each open covering of E there is a discrete open covering which refines it.

Definition 2. Let p be an infinite cardinal number and A be a subset of E. A is compact-p if every subset of A of cardinality p has an accumulation point in A.

Definition 3. Let m be the index of E. E satisfies property $(*)$ if it has an open basis which is the union of m discrete collections of open subsets of E.

Lemma. *Let m be the index of E and suppose that $m > \aleph_0$. E satisfies the property $(*)$ if and only if E has an open basis which is the union of m discrete open coverings of E.*

If E satisfies the property $(*)$, then E is paracompact. Furthermore, if E satisfies the property $(*)$, Y is a Hausdorff space and f is a closed continuous function of E onto Y (thus, Y is paracompact), then either Y is discrete or index E = index Y.

2. MAIN THEOREM

Theorem. *Let X and Y be nondiscrete Hausdorff paracompact spaces, with indices equal to m, and f be a closed continuous function of X onto Y. Suppose that X satisfies the property $(*)$. Y satisfies the property $(*)$ if and only if the boundary of the set $f^{-1}(\{y\})$ is compact-m, for every $y \in Y$.*

Proof. If $m = \aleph_0$, then X is metrizable and the result is the Stone-Morita-Hanai's theorem.

From now onwards suppose that $m > \aleph_0$.

Let M be the set of all ordinal numbers less than the first ordinal of cardinality m. From the hypothesis and lemma above it follows that X has an open basis $\mathscr{B} = \bigcup_{i \in M} \mathscr{B}_i$, where each $\mathscr{B}_i$ is a discrete open covering of E and $\mathscr{B}_i$ refines $\mathscr{B}_j$ whenever $i > j$.

Necessity. On the contrary, let us suppose that there is $y \in Y$ such that the boundary $\mathrm{bd}(f^{-1}(\{y\}))$ is not compact-m, i.e., there exists a subset A of $\mathrm{bd}(f^{-1}(\{y\}))$, whose cardinality is m, without accumulation points. (In particular, A is closed.)

We may write $A = \{a_i \mid i \in M\}$. Let $(V_i)_{i \in M}$ be a fundamental system of neighborhoods of y, with $V_i \subset V_j$ whenever $i > j$. For each

$i \in M$ we fix an element b_i of the set $U_i \cap f^{-1}(V_i) \cap (X - f^{-1}(\{y\}))$, where $a_i \in U_i \in \mathscr{B}_i$. The set $B = \{b_i \mid i \in M\}$ is closed in X. (Indeed, suppose $d \in \mathrm{cl}(B) - B$; there is $j \in M$ such that $(U - \{d\}) \cap A = \phi$, where $d \in U \in \mathscr{B}_j$. If $k > j$ and $b_k \in U - \{d\}$, then $U_k \subset U$ and $a_k \in \in U$; it follows that $(U - \{d\}) \cap B$ has less than m elements and has no accumulation points. So d is not an accumulation point of B.)

Since f is closed, the set $\{f(b_i) \mid i \in M\}$ is closed in Y, but this is not possible because y is an accumulation point of this set and does not belong to it.

Sufficiency. (It is based on the proof which appears in [3].) For each $y \in Y$ and $i \in M$ we put $N_y^i = \cup \{T \in B_i \mid T \cap \mathrm{bd}(f^{-1}(\{y\})) \neq \phi\}$; $U_y^i = \overset{\circ}{f^{-1}}(\{y\}) \cup N_y^i$; $V_y^i = \cup \{f^{-1}(\{z\}) \mid z \in Y,\ f^{-1}(\{z\}) \subset U_y^i\}$ and $W^i(y) = \{z \in Y \mid f^{-1}(\{z\}) \subset U_y^i\}$.

1) We have that $W^i(y) = Y - f(X - U_y^i)$ and $f^{-1}(W^i(y)) = V_y^i$: it thus follows that $W^i(y)$ and V_y^i are open sets in Y and X, respectively.

2) For each $y \in Y$ the family $(W^i(y))_{i \in M}$ is a fundamental system of neighborhoods of y.

Indeed, let G be an open set to which y belongs. Then $\mathrm{bd}(f^{-1}(\{y\})) \subset f^{-1}(G)$. For each $t \in \mathrm{bd}(f^{-1}(\{y\}))$ there is $i_t \in M$ such that $S_t \subset f^{-1}(G)$, where $t \in S_t \in \mathscr{B}_{i_t}$. Since $\mathrm{bd}(f^{-1}(\{y\}))$ is compact-m, there is a set L contained in $\mathrm{bd}(f^{-1}(\{y\}))$, whose cardinality is less than m, such that

$$\cup \{S_t \mid t \in L\} \supset \mathrm{bd}(f^{-1}(\{y\}))\,.$$

Let $j \in M$ be the $\sup\{i_t \mid t \in L\}$. It follows that $U_y^j \subset f^{-1}(G)$ and $W^j(y) \subset G$.

3) We will now prove that given $i \in M$ and $y \in Y$, there is $j \in M$, $j \geqslant i$, such that for any $z \in Y$, $W^j(z) \cap W^j(y) \neq \phi$ implies $W^j(z) \subset \subset W^i(y)$.

Indeed, given $i \in M$ and $y \in Y$, if $\overset{\circ}{f^{-1}}(\{y\})$ is not empty we fix an element a_y of this set. There is $j \in M$, $j \geqslant i$, such that

I) if $\mathrm{bd}(f^{-1}(\{y\})) \neq \phi$, then $N_y^j \subset V_y^i$;

II) if $\overset{\circ}{f^{-1}}(\{y\})) \neq \phi$, then $T \subset \overset{\circ}{f^{-1}}(\{y\})$, where $a_y \in T \in \in \mathscr{B}_j$.

Suppose that $W^j(z) \cap W^j(y) \neq \phi$, for some $z \in Y - \{y\}$. Fix an element b of this intersection. We will prove that $f^{-1}(\{z\}) \subset V_y^j$.

Take $w \in f^{-1}(\{b\})$ and recall that $f^{-1}(\{b\}) \subset V_z^j \cap V_y^j \subset U_z^j \cap \cap U_y^j$. Since $U_z^j = \overset{\circ}{f^{-1}}(\{z\}) \cup N_z^j$ and $U_y^j = \overset{\circ}{f^{-1}}(\{y\}) \cup N_y^j$, we have three cases to consider:

α) $w \in \overset{\circ}{f^{-1}}(\{z\})$.

Since $w \in V_y^j$, we have $f^{-1}(\{z\}) \cap V_y^j \neq \phi$ and $f^{-1}(\{z\}) \subset V_y^j$, as we wish to prove.

β) $w \in \overset{\circ}{f^{-1}}(\{y\})$.

Then a_y exists and $f^{-1}(\{y\}) \cap V_z^j \neq \phi$, so $f^{-1}(\{y\}) \subset V_z^j$ and $a_y \in V_z^j$ which contradicts condition II).

γ) $w \in N_y^j \cap N_z^j$.

There are points $u \in \mathrm{bd}(f^{-1}(\{y\}))$ and $v \in \mathrm{bd}(f^{-1}(\{z\}))$ and elements T_1 and T_2 belonging to $\mathscr{B}_j$ such that

$\{w, u\} \subset T_1$ and $\{w, v\} \subset T_2$. Since $\mathscr{B}_j$ is discrete, we have $T_1 = T_2$. By virtue of condition I), $v \in T_2 \subset V_y^j$, so $f^{-1}(\{z\})$ inter V_y^j is not empty and $f^{-1}(\ z\}) \subset V_y^j$.

We will next prove that U_z^j is contained in U_y^j. We have already seen that $f^{-1}(\{z\}) \subset \overset{\circ}{f^{-1}}(\{y\}) \cup N_y^j$. Thus $f^{-1}(\{z\}) \subset N_y^j$. Let $u \in N_z^j$, there is $v \in \mathrm{bd}(f^{-1}(\{z\})$ and $T_1 \in \mathscr{B}_j$ such that $\{u, v\} \subset T_1$. But v belongs to N_y^j, thus there is q belonging to $\mathrm{bd}(f^{-1}(\{y\}))$ and T_2 belonging to $\mathscr{B}_j$, verifying $\{v, q\} \subset T_2$. Since $\mathscr{B}_j$ is discrete, T_1 is equal to T_2 and $\{u, q\} \subset T_1$ and $u \in N_y^j \subset U_y^j$.

Finally we have $V_z^j \subset V_y^i$ and $W^j(z) \subset W^i(y)$.

4. For each $i \in M$, let $\mathscr{D}_i$ be a discrete open covering of Y

which refines $\{W_i(y) \mid y \in Y\}$. The set $\mathscr{D} = \bigcup_{i \in M} \mathscr{D}_i$ is an open basis in Y. The proof is completed.

3. EXAMPLE

We will next provide an example of two nondiscrete paracompact spaces X and Y, with equal indices, such that X has property (*) and Y does not have this property although there is a closed continuous function of X onto Y.

Let m be an infinite regular cardinal number and M be a set of cardinality m. Put $X = M \times M$ and fix an element $w \in M$. In X we consider the following topology: for every $y \in M$.

1) (y, z) is open, $\forall z \in M - \{w\}$;

2) a subset V of X is a neighborhood of the point (y, w) if and only if $(y, w) \in V$ and the cardinality of the set $\{t \in M \mid (y, t) \notin V\}$ is less than m.

With this topology X is a paracompact space and its index is m.

Let Y be the topological space obtained from X by identifying the points of $M \times \{w\}$ to a unique point. Let f be the canonical function of X onto Y; f is continuous. Furthermore, f is closed. (Indeed, let F be a closed subset of X. If $F \cap (M \times \{w\}) = \phi$, then $f^{-1}(f(F)) = F$; if $F \cap (M \times \{w\}) \neq \phi$, then $f^{-1}(f(F)) = F \cup (M \times \{w\})$; in both cases $f(F)$ is closed.)

With this quotient topology Y is a nondiscrete paracompact space, with index equal to m. But the point $f((w, w))$ has not a fundamental system W of neighborhoods, whose cardinality $|W|$ is not greater than m. (On the contrary, let us suppose that $(V_i)_{i \in M}$ is a fundamental system of open neighborhoods of $f((w, w))$. It then follows that, for every $i \in M$, $f^{-1}(V_i)$ is an open set which contains $M \times \{w\}$; we fix an element (i, a_i) belonging to $f^{-1}(V_i)$, with $a_i \neq w$. The set $f\big(X - \{(i, a_i) \mid i \in M\}\big)$ is a neighborhood of $f((w, w))$, but no V_i is contained in it.)

In the preceding example we have that $f^{-1}\big(\{f((w, w))\}\big)$ is a noncompact-m subset of X and, for each $y \in Y - \{f((w, w))\}$, $f^{-1}(\{y\})$ is a nonempty compact open subset of X. This is an extreme situation in the sense that we can prove (with essentially the same type of arguments as in [1]) the following analogon of Lašnev's theorem.

Theorem. *Let* X *be a nondiscrete Hausdorff regular space,* m *be the index of* X, Y *be a Hausdorff space and* f *be a closed continuous function of* X *onto* Y. *Then, if* X *has the property* (*), $Y = Z \cup \bigcup_{i \in M} Y_i$, *where* M *is a set of cardinality* m, *each* Y_i *is a closed discrete subspace of* Y *and, for each* $y \in Z$, $f^{-1}(\{y\})$ *is a compact-m subset of* X *with empty interior.*

Finally it is interesting to notice that if Y is a nondiscrete Hausdorff space such that the weight at each point does not exceed the index of Y, then Y is the open continuous image of a space X (with index X = = index Y) verifying the property (*). (For $m = \aleph_0$ this is the well known result due to V.I. Ponomarev.)

REFERENCES

[1] N.N. Lašnev, Continuous decompositions and closed mappings of metric spaces, *Soviet Math.*, 6 (1965), 1504-1506.

[2] K. Morita – S. Hanai, Closed mappings and metric spaces, *Proc. Japan Acad.*, 32 (1956), 10-14.

[3] A.H. Stone, Metrizability of decomposition spaces, *Proc. Amer. Math. Soc.*, 7 (1956), 690-700.

COLLOQUIA MATHEMATICA SOCIETATIS JÁNOS BOLYAI
8. TOPICS IN TOPOLOGY, KESZTHELY (HUNGARY), 1972.

RESULTS RELATED TO *P*-EMBEDDING

RICHARD A. ALÒ

Recently one of the most difficult problems facing set theoretical topology has been resolved by M.E. Rudin in [17]. There she exhibits a collectionwise normal (and therefore normal) Hausdorff space which is not countably paracompact. This negatively resolved Dowker's conjecture whether every normal Hausdorff space is countably paracompact.

Now each of these concepts has proven to be worthwhile in furthering the study of general topology itself and in its applications to analysis. Beginning with the class of paracompact spaces as introduced by Dieudonné in [9], we had a class of spaces sufficiently abstract to include the compact spaces and metric spaces as special cases, at the same time being sufficiently concrete to permit substantial theories to be developed.

As a generalization, the class of countably paracompact spaces as introduced in [10], were sufficiently useful in homotopy theory. Here the additional axiom of collectionwise normality or normality is appropriate. In discussing metrization theory (see [8], [11]) one finds that the concept of collectionwise normality and associated concepts are very useful. Of course, not much needs to be said about the class of normal spaces.

In this article I would like to relate these concepts to various kinds of embeddings of a subspace. These may take the form of the extension of continuous functions, the extension of pseudometrics or the extension of entourage uniformities. Much has been done in this area. Relating this work with some new ideas given here, we hope further work will be encouraged in unifying a very interesting area of research. Any of the terminology not defined here may be found in [14] and [21]. The latter offers a survey of the work in the area until 1968.

A subset S of a topological space X is said to be *p-embedded* in X if every continuous pseudometric on S extends to a continuous pseudometric on X. More generally, if γ is an infinite cardinal number, then a pseudometric d on S is *γ-separable* if there is a subset A of S of cardinality at most γ and if A is dense in S with respect to the pseudometric topology $\mathscr{T}_d$ of d. Also S is *P^γ-embedded* in X is every γ-separable continuous pseudometric on S extends to a γ-separable continuous pseudometric on X. The concept of a P-embedded subspace is definitely stronger than the subspace being C-embedded. In fact S is C-embedded in X if and only if S is $P^{\aleph_0}$-embedded in X. A final concept that will be utilized is that of a *hemicompact* space, that is, a space which is the countable union of compact spaces $\{K_i\}_{i \in N}$ and such that each compact subset of X is contained in some K_i.

In [5], we showed the following result.

Theorem 1. *If S is a non empty subset of a topological space X, then for the following statements* 1) *and* 2) *are equivalent while* 3) *and* 4) *are equivalent.*

1) *The subspace S is countably paracompact, normal and C-embedded in X.*

2) *Every countable open cover of S has a refinement that can be extended to a locally finite cozero set cover of X (subsets satisfying this condition are said to be strongly C-embedded in X).*

3) *The subspace S is countably paracompact, normal and P-embedded in X.*

4) *Every* σ-*locally finite open cover of* S *has a refinement that can be extended to a locally finite cozero set cover of* X *(subsets satisfying this condition are said to be strongly* P-*embedded in* X*).*

Closed subsets of normal spaces as well as compact subsets of Tychonoff spaces are always C-embedded. In fact even more may be said but for the present, this is sufficient. Also let us recall that a closed subset being P-embedded characterizes collectionwise normal spaces (see [19]). Theorem 1, then, gives us an "embedding type" characterization of countably paracompact, normal, C-embedded subspaces and countably paracompact, normal P-embedded subspaces. It follows directly from this that closed subsets being strongly C-embedded characterize countably paracompact normal spaces while closed subsets being strongly P-embedded characterize countably paracompact collectionwise normal spaces (or *strongly normal* spaces as labelled by Katětov.)

Moreover, it follows that T_1 spaces X are countably paracompact and normal if and only if they are strongly C-embedded in their Hewitt realcompactifications υX. However if their cardinality is non-measurable then the equivalence should read "if and only if they are strongly P-embedded in υX".

Using results of Dowker in [10] and [11] one has the following:

a) If there is a space X such that S is strongly C-embedded in X, then the product of S with a compact metric space is normal

b) If there is a space X such that S is strongly P-embedded in X, then the product of S with a compact metric space is collectionwise normal.

Since Tychonoff spaces, X, have admissible uniformities, let us consider the following concepts. As we know, any collection of pseudometrics $\mathscr{D}$ gives rise to an entourage uniformity. The sets

$$U(d,\epsilon) = \{(x,y) \in X \times X\colon d(x,y) < \epsilon\}$$

for all d in $\mathscr{D}$ and all $\epsilon > 0$ form a subbase for some uniformity. A subcollection $\mathscr{P}$ of $X \times X$ *generates* a uniformity $\mathscr{U}$ on X if $\mathscr{P}$ is a subbase for $\mathscr{U}$. To every real valued function f on X is associated a pseudometric ψ_f on X defined by

$$\psi_f(x, y) = |f(x) - f(y)| \quad (x, y \in X) .$$

A family of *functions* $\mathscr{G}$ *generates* a uniformity $\mathscr{U}$ if $\{U(\psi_f, \epsilon): f \in \mathscr{G}$ and $\epsilon > 0\}$ generates $\mathscr{U}$. The pseudometric topology associated with a pseudometric d is denoted by $\mathscr{T}_d$. A pseudometric d is said to be $\aleph_0$*-separable* in case the pseudometric space $(X, \mathscr{T}_d)$ has a countable dense subset. The pseudometric is *totally bounded* if for every $\epsilon > 0$ there is a finite subset F of X such that X is the union of the d-spheres of radius ϵ centered at the points of F. A *precompact uniformity* is a uniformity $\mathscr{U}$ on X that is generated by a collection of bounded continuous real valued functions on X. If the uniformity is generated by a collection of continuous real valued functions then the uniformity is said to be *prerealcompact.*

Every uniformity $\mathscr{U}$ on a non-empty set X yields a unique topology $\mathscr{T}(\mathscr{U})$. This topology is obtained by taking as a base for the open sets the collection of sets $U[x]$ for all U in $\mathscr{U}$ and all x in X. If $\mathscr{T}(\mathscr{U})$ is a subcollection of the original topology $\mathscr{T}$ on X then we say that $\mathscr{U}$ is a *continuous uniformity.* If $\mathscr{T}(\mathscr{U})$ agrees with $\mathscr{T}$ then $\mathscr{U}$ is called an *admissible uniformity.* The admissible uniformity on X generated by the collection of all continuous pseudometrics (respectively all continuous real valued functions, all bounded continuous real valued functions) on X is denoted by $\mathscr{U}_0(X)$ (respectively, $\mathscr{C}(X)$, $\mathscr{C}^*(X)$).

If S is a subset of X and if $\mathscr{G}$ is a collection of subsets of X, then by $\mathscr{G}|S$ is meant the collection $\{G \cap S: G \in \mathscr{G}\}$. A uniformity $\mathscr{U}^*$ on X is an extension of a uniformity $\mathscr{U}$ on S in case $\mathscr{U}^* | S \times S =$ $= \mathscr{U}$. The subset S is *uniformly embedded in* X in case every admissible uniformity $\mathscr{U}$ on S has an extension that is a continuous uniformity on X. It is said to be *prerealcompact uniformly embedded in* X in case, every prerealcompact admissible uniformity $\mathscr{U}$ on X has an extension that is a continuous uniformity on X. The subset S is *precompact uniformly embedded in* X if every precompact admissible uniformity $\mathscr{U}$ on X can be extended to a continuous uniformity on X. The reader is referred to [4] for motivation and discussion of these concepts.

In the definitions of a prerealcompact uniformity and precompact uniformity, the extensions need not be just continuous uniformities as required but may be taken to be prerealcompact or precompact accordingly (see Theorem 1 of [4]).

Using these concepts, we are able to show the following results.

Theorem 2. *Let S be a non-empty subset of a topological space X.*

1) *The subset S is P-embedded in X if and only if S is uniformly embedded in X.*

2) *The subset S is C-embedded in X if and only if S is prerealcompact uniformly embedded in X.*

3) *The subset S is C^*-embedded in X if and only if S is precompact uniformly embedded in X.*

Proof. Let us just look at the proof of 1). To prove sufficiency, it is necessary to recall that S is P-embedded in X if and only if $\mathscr{U}_0(S) = \mathscr{U}_0(X) | S \times S$ (see [13], Theorem 7.5) and that for any subspace S of X, $\mathscr{U}_0(X) | S \times S$ is always contained in $\mathscr{U}_0(S)$. Since $\mathscr{U}_0(S)$ is an admissible uniformity, S uniformly embedded in X implies that there is a continuous uniformity $\mathscr{V}$ on X such that $\mathscr{V} | S \times S = \mathscr{U}_0(S)$. For any $U \in \mathscr{U}_0(S)$ there is a $V \in \mathscr{V}$ so that $V \cap (S \times S) = U$. Consequently U is a member of $\mathscr{U}_0(X) | S \times S$ and it follows that S is P-embedded in X.

For the necessity of the condition, let $\mathscr{U}$ be an admissible uniformity on S, let $\mathscr{P}$ be the set of all continuous pseudometrics on $(X, \mathscr{T})$ satisfying:

$$\text{if } d \in \mathscr{P} \text{ then } d | S \times S \text{ is uniformly continuous on } S \text{ relative to } \mathscr{U},$$

and let $\mathscr{U}^*$ be the uniformity on X generated by $\mathscr{P}$. The uniformity $\mathscr{U}^*$ is a continuous extension of $\mathscr{U}$. In fact the subbasic elements of $\mathscr{T}(\mathscr{U}^*)$ are the d-spheres about each $x \in X$ for $d \in \mathscr{P}$ and $\epsilon > 0$. The continuity of each $d \in \mathscr{P}$ relative to $\mathscr{T}$ implies that $\mathscr{T}_d \subset \mathscr{T}$ and hence $\mathscr{T}(\mathscr{U}^*) \subset \mathscr{T}$. Moreover, $\mathscr{U}^* | S \times S = \mathscr{U}$. If U is any member of

$\mathscr{U}$ then there is a continuous pseudometric d on S and an $\epsilon > 0$ so that

$$W = \{(x, y) \in S \times S: d(x, y) < \epsilon\} \subset U .$$

Since S is P-embedded in X, d has a continuous pseudometric extension d^* in $\mathscr{P}$. Let

$$W^* = \{(x, y) \in X \times X: d^*(x, y) < \epsilon\} .$$

Then $W^* \in \mathscr{U}^*$ and

$$W^* \cap (S \times S) = W \subset U \in \mathscr{U}^* | S \times S .$$

Hence $\mathscr{U} \subset \mathscr{U}^* | S \times S$. Conversely, if $U \in \mathscr{U}^* | S \times S$, then there is $U^* \in \in \mathscr{U}^*$ such that $U^* \cap (S \times S) = U$. Hence there is a $d \in \mathscr{P}$ and $\epsilon >.0$ for which

$$V = \{(x, y) \in X \times X: d(x, y) < \epsilon\} \subset U^* .$$

Thus $V \cap (S \times S) \subset U^* \cap (S \times S)$ and since $d | S \times S$ is uniformly continuous on S relative to $\mathscr{U}$, $V | S \times S \in \mathscr{U}$ and hence $U \in \mathscr{U}$. This completes the proof.

Of course this then says that a completely regular space is collectionwise normal if and only if every closed subset is uniformly embedded in X. Analogous statements may be made for completely regular spaces which are normal. In addition characterizations of the Čech – Stone compactification of a space and the Hewitt – Nachbin realcompactification of a space may be given utilizing these uniform concepts (see [4]).

With these additional characterizations of P-embedding and C-embedding it would now be good to look at analogous "uniform" characterizations for countably paracompact spaces. That is, what are the analogous "uniform" formulations for "strongly P-embedded" and "strongly C-embedded"?

Let us look at some further ideas and characterizations. If one considers the extension of continuous functions whose functional values lie in a linear topological space, then a very nice relationship to P- (and C-) embedding may be obtained.

Let X be a topological space, let γ be an infinite cardinal, and let f be a function from X to a locally convex topological vector space (abbreviated *LCTV* space). We call f an *M-valued function* if the image of X under f is contained in a complete convex metrizable subset M of L (abbreviated *CCM* subspace). The function f is a *(γ, M)-valued function* if it is an M-valued function and if the image of X under f is a γ-separable subset of M. Recall that a *Fréchet space* is a complete, metrizable *LCTV* space. The set of all bounded real-valued continuous functions on X is a Banach space under the sup norm, i.e., $\|f\| = \sup_{x \in X} |f(x)|$, and will be denoted by $C^*(X)$.

Theorem 3. *Let S be a non-empty subspace of the topological space X, let γ be an infinite cardinal number and let A be a discrete space such that* $\operatorname{card} A \geqslant \operatorname{card} S$. *The following statements are equivalent:*

1) *The subspace S is P^γ-embedded in X.*

2) *Given a CCM subspace M of a LCTV space L, every continuous (γ, M)-valued function on S extends continuously to X relative to M.*

3) *Given a CCM subspace M of a LCTV space L, every continuous (γ, M)-valued function on S extends to a continuous function from X to L.*

4) *Every continuous function from S to a Fréchet space such that the image of S is γ-separable extends to a continuous function on X.*

5) *Every continuous function from S into $C^*(S)$ such that the image of S is γ-separable extends to a continuous function on X.*

6) *Every continuous function from S into $C^*(A)$ such that the image of S is γ-separable extends to a continuous function on X.*

Furthermore, the above conditions are also equivalent to the conditions obtained from 2) *through* 6) *by requiring the image of S to be a bounded subset of the locally convex space in question.*

Proof. The proof of the implications 2) *implies* 3) *implies* 4) *implies* 5) are immediate. In showing that 1) *implies* 2), one easily verifies that the composition $d = m \circ (f \times f)$, where m is a complete metric for M and where f is a continuous (γ, M)-valued function from S to the $LCTV$ space L, is a γ-separable continuous pseudometric. A theorem of Dugundji's (see [12]) states that a continuous function from a closed subset of a metric space into a $LCTV$ space can be extended to a continuous function on the metric space with values in the convex hull of the image of the subset. It is easily seen that his proof also applies to pseudometric spaces. Therefore, the function f extends to a continuous function g from (X, d^*) into M, since M is convex. The function g is continuous with respect to the topology generated by d^*, and since this topology is contained in the original topology of X, the mapping g is the continuous extension of f relative to M that we seek. For the last implication the reader is referred to [2].

This topic was first studied by Richard Arens in [7]. There he showed the equivalence of 1) and 2) above for closed subsets of a topological space.

Now a subset S is P-embedded in X if and only if it is P^{γ}-embedded in X for each cardinal γ. Also a subset S is C-embedded in X if and only if it is $P^{\aleph_0}$-embedded in X. Thus Theorem 3 gives characterizations of P-embedded and C-embedded subsets and therefore, as before, characterizations of normal spaces and collectionwise normal spaces, as well as the Čech – Stone compactification and Hewitt – Nachbin realcompactification of Tychonoff spaces. In [2], one can also find an interesting analogue for C^*-embedded subsets (see also [23]).

Since a pseudometric on a space X is a function on the product set $X \times X$, it is of interest to relate the extension of pseudometrics to the extension of functions on $X \times X$ (without the triangle inequality). In this direction we have the following result.

Theorem 4. *Let S be a subspace of a completely regular T_1 space X. The following are equivalent:*

1) *The subspace S is P-embedded in X.*

2) *For all locally compact, hemicompact Hausdorff spaces A, the product set $S \times A$ is P-embedded in the product space $X \times A$.*

3) *For all locally compact, hemicompact Hausdorff spaces* A, *the product set* $S \times A$ *is* C*-embedded in the product space* $X \times A$.

4) *For all locally compact, hemicompact Hausdorff spaces* A, *the product set* $S \times A$ *is* C^**-embedded in the product space* $X \times A$.

5) *The product set* $S \times \beta S$ *is* P*-embedded in the product space* $X \times \beta S$.

6) *The product set* $S \times \beta S$ *is* C*-embedded in the product space* $X \times \beta S$.

7) *The product set* $S \times \beta S$ *is* C^**-embedded in the product space* $X \times \beta S$.

In the proof of this, one sees that the implications 2) *implies* 3) *implies* 4) *implies* 7) and 2) *implies* 5) *implies* 6) *implies* 7) are immediate. To show 7) *implies* 1) and 1) *implies* 2) requires some very nice applications of Theorem 3. For example in showing 1) *implies* 2), by Theorem 3 it is sufficient to prove that if f is a continuous function from the product set $S \times A$ into a Fréchet space B, then f extends to a continuous function on $X \times A$. Let f be a continuous function from $S \times A$ into a Fréchet space B, and define a map φ from S into $C(A, B)$ by $(\varphi(x))(a) = f(x, a)$ for all a in A and all x in S. Then φ is continuous and $C(A, B)$ is a Fréchet space. By assumption S is P-embedded in X. Hence by Theorem 3 the map φ extends to a continuous function φ^* from X into $C(A, B)$. Define a map f^* from $X \times A$ into B by $f^*(x, a) = (\varphi^*(x))(a)$ for all (x, a) in $X \times A$. The function f^* is an extension of f and is continuous.

One interesting corollary (in a different spirit from the ones above) is that for Tychonoff spaces X, the space X is P-embedded in υX if and only if $\upsilon(X \times \beta X) = \upsilon X \times \beta X$.

The reader can formulate corollaries to Theorem 4 for collectionwise normal and normal spaces as we have done previously. These results will appear in [3], where additional results and ideas are given.

From the point of view of analysis the following ideas seem to be very useful. Motivated by considerations of function space topologies and resulting topological properties on function spaces, the following definitions come forth. Let $\mathscr{F} = (f_a)_{a \in I}$ be a family of real-valued continuous functions on a topological space X. The family $\mathscr{F}$ is *pointwise bounded* if for each x in X there is a constant $M(x)$ such that $|f_a(x)| \leqslant M(x)$ for all $a \in I$. The family is said to be *equicontinuous* on X if for $\epsilon > 0$ and for $x \in X$ there is an open neighborhood $N(x, \epsilon)$ of x such that $y \in N(x, \epsilon)$ implies that

$$|f_a(x) - f_a(y)| \leqslant \epsilon$$

for all $a \in I$. If $(X, \mathscr{D})$ is a uniform space (with pseudometric uniform structure $\mathscr{D}$) then $\mathscr{F}$ is *uniformly equicontinuous with respect to* $\mathscr{D}$ if for $\epsilon > 0$ there is a d in $\mathscr{D}$ and a $\delta > 0$ such that $d(x, y) \leqslant \delta$ implies $|f_a(x) - f_a(y)| \leqslant \epsilon$ for all $a \in I$.

When considering paracompactness and its application to analysis, the concept of "a partition of unity" is very useful. Thus we have another motivation for wanting to consider "families of functions and their extensions" above. But the local finiteness of the cozero sets obtained from the partition is also very useful. These should serve as motivation for the following definitions (see [3]).

The subset S of X is *locally finitely embedded* in X if given a locally finite family $(f_a)_{a \in I}$ of real-valued continuous functions on S, there is a locally finite family $(g_a)_{a \in I}$ of real-valued continuous on functions X such that $g_a | S = f_a$ for all $a \in I$. The subset S is *equicontinuously embedded* in X if given an equicontinuous family $(f_a)_{a \in I}$ of real valued functions on S that is pointwise bounded, there is an equicontinuous family $(g_a)_{a \in I}$ of real-valued functions on X such that $g_a | S = f_a$ for all $a \in I$.

What is striking about these definitions is the following theorem.

Theorem 5. *Let S be a non-empty subspace of a topological space X. The following are equivalent:*

1) *The subspace S is P-embedded in X.*

2) *The subspace S is equicontinuously embedded in X.*

3) *The subspace S is locally finitely embedded in X.*

Each direction of proof in this theorem again makes use of Theorem 3. The details of this proof will be given in [18].

This theorem opens further investigations for C and C^*-embedding as well as characterizations of paracompactness and countable paracompactness in conjunction with some separation axioms as we have both alluded to and explicitly stated above. Thus we have brought forth many ways of looking at P-embedding, as well as its weaker and stronger versions as given here.

REFERENCES

[1] Richard A. Alò, Some Tietze type extension theorems, *Proc. Parague Topology Symposium,* 1971, (to appear).

[2] Richard A. Alò,– Linnea I. Sennott, Extending linear space valued functions, *Math. Ann.,* 191 (1971), 79-86.

[3] Richard A. Alò – Linnea I. Sennott, Collectionwise normality and the extension of functions on product spaces, *Fund. Math.,* (to appear).

[4] Richard A. Alò – Harvey L. Shapiro, Continuous Uniformities, *Math. Ann.,* 185 (1970), 322-328.

[5] Richard A. Alò – Harvey L. Shapiro, Countably paracompact, normal and collectionwise normal spaces, *Indag. Math.,* (to appear).

[6] Richard Arnes, Extensions of functions on fully normal spaces, *Pac. J. Math.,* 2 (1952), 11-22.

[7] Richard Arnes, Extensions of coverings, of pseudometrics, and of linear space-valued mappings, *Canad. J. Math.,* 5 (1953), 211-215.

[8] R.H. Bing, Metrization of topological spaces, *Canad. J. Math.,* 3 (1951), 175-186.

[9] J. Dieudonné, Une généralisation des espaces compacts, *J. Math. Pures Appl.,* 23 (1944), 65-76.

[10] C.H. Dowker, On countably paracompact spaces, *Canad. J. Math.*, 3 (1951), 219-224.

[11] C.H. Dowker, Homotopy extension theorems, *Proc. London Math. Soc.*, 6 (1956), 110-116.

[12] J. Dugundjii, An extension of Tietze's theorem, *Pac. J. Math.*, 1 (1951), 353-367.

[13] T.E. Gantner, Extensions of uniform structures, *Fund. Math.*, (to appear).

[14] L. Gillman – M. Jerison, *Rings of Continuous Functions*, D. Van Nostrand and Company, New York, 1960.

[15] K. Morita, Products of normal spaces with metric spaces, *Math. Ann.*, 154 (1964), 365-382.

[16] J. Nagata, *Modern General Topology*, North Holland Publishing Company, Amsterdam, 1968.

[17] M.E. Rudin, A Dowker spaces, *Bull. Amer. Math. Soc.*, 77 (1971), 246.

[18] L.I. Sennott, Equicontinuous and locally finite embeddings, (in this volume), Keszthely, 1972.

[19] H.L. Shapiro, Extensions of pseudometrics, *Canad. J. Math.*, 19 (1966), 981-998.

[20] A.H. Stone, Paracompactness and product spaces, *Bull. Amer. Math. Soc.*, 54 (1948), 477-482.

[21] Richard A. Alò, Uniformities and Embeddings, *Proc. of Kanpur Topological Conference*, 1968. *General Topology and Its Relations to Modern Analysis and Algebra.* Academia Publishing of Czechoslovakia Academy of Sciences, 1970.

[22] G. Aguaro, Ricovrimenti aperti a strutture uniformi sopra uno spazio topologico, *Ann. Mat. Pure Appl. Ser.* 4, 47 (1959), 319-390.

[23] G. Aguaro, Spazii collettivamente normali ed estensione di applicazioni continue, *Riv. Mat. Univ.*, Parma 2 (1961), 77-90.

COLLOQUIA MATHEMATICA SOCIETATIS JÁNOS BOLYAI
8. TOPICS IN TOPOLOGY, KESZTHELY (HUNGARY), 1972.

TOPOLOGIES GENERATED BY SOME VECTOR MEASURES

R.A. ALÒ — A. de KORVIN — L.B. KUNES

1. INTRODUCTION

Suppose E and F represent Banach spaces over the same scalar field $\mathscr{K}$ and suppose that $L(E, F)$ is the set of all bounded linear operators from E into F. If $\mathscr{C}$ is a ring of subsets (at least one element of which is non-empty) of a non-empty set T, let s_c be a subcollection of the collection $s = s(E, F)$ of all finitely additive set functions from $\mathscr{C}$ into $L(E, F)$. Such subcollections determine semi-norms, hence topologies, on the dual space F^* of F. Recall that a *semi-norm* p on a vector space V is a non-negative real valued function such that

$$p(x + y) \leqslant p(x) + p(y) \qquad (x, y \in V)$$

$$p(\lambda x) = |\lambda| p(x) \qquad (x \in V, \lambda \in K) .$$

In particular if $m \in s_c$, let us represent the image in $L(E, F)$ under m of $A \in \mathscr{C}$ by m_A. For $\varphi \in F^*$ one may define the measure m_φ from $\mathscr{C}$ into the dual E^* of E by

$$m_\varphi(A) = \varphi \circ m_A \qquad A \in \mathscr{C} . \tag{1}$$

In a similar manner let us represent the element $m_\varphi(A)$ in E^* by $m_{\varphi,A}$. Using the variation $\bar{m}_\varphi$ of the measure m_φ and keeping $A \in \mathscr{C}$ fixed, one may now define a semi-norm $p_{m,A}$ on F^* by

$$(2) \qquad p_{m,A}(\varphi) = \bar{m}_\varphi(A) \qquad (\varphi \in F^*) .$$

Of interest in this paper will be similar semi-norms (and the topologies they naturally generate) as determined by various subcollections s_c of s. For such a $p_{m,A}$ let $\mathscr{P}(m, A)$ be the topology *generated* (that is, the smallest topology making $p_{m,A}$ continuous) by this one semi-norm, let $\mathscr{P}(m)$ be the locally convex topology generated by $\mathscr{P}(m, A)$ for all $A \in \mathscr{C}$ and let $\mathscr{P}$ be the locally convex topology generated by $\mathscr{P}(m)$ for all $m \in s_c$.

When these topologies are restricted to the unit sphere F_1^* of F^* some very interesting facts result. For example in [5] the subcollection s_c under consideration was the subcollection s_s consisting of all $m \in s$ with finite semi-variation $\tilde{m}$. It was shown there that if F_1^* is compact in its relative $\mathscr{P}$ topology and if $m \in s_s$ then the following are equivalent.

a) The measure m_φ is countably additive for each $\varphi \in F^*$.

b) The measure m is *variationally semi regular,* that is, if $(A_n)_{n \in N}$ is a decreasing sequence of elements from $\mathscr{C}$ with empty intersection then the sequence $\{\tilde{m}(A_n)\}_{n \in N}$ converges to 0.

c) The measure m is norm countably additive.

In [6], operators on the space $C_0(H, E)$ (with the uniform norm) of continuous functions defined on a locally compact Hausdorff space H and vanishing at infinity are defined and studied. The compactness of an operator on $C_0(H, E)$ was characterized as the topology $\mathscr{P}(m)$ restricted to F_1^* being compact. In this case m was a unique (weakly regular) finitely additive vector measure used to represent the operator as an integral.

With this in mind it is natural to consider operators and corresponding results on $\mathscr{L}^p$ spaces. Here the q-semi variation is apparently pertinent (for example, see the representation theorems given in [2] for operators on $\mathscr{L}^p$ spaces). Consequently in defining the semi-norm (2) above we will replace the variation $\bar{m}_\varphi$ with the q-variation $(\bar{m}_\varphi)_q$, that is, we have

(3) $$p_{m,A}(\varphi) = (\bar{m}_\varphi)_q(A) \qquad (\varphi \in F^*)\,.$$

The topologies that these semi-norms generate will be labeled as previously.

In contrast to the countable additivity of $\bar{m}_\varphi$, the q-variation $(\bar{m}_\varphi)_q$ is only countably subadditive. In [5] and [6], conditions were given so that m may be countably additive. In our case countable additivity does follow from the fact that the q semi-variation is finite (for $q \neq 1$, see [2]). Of importance is the right continuity of the q semi-variation. We will give some conditions under which this will hold.

Among the main results in [6] is the characterization of compact operators on $C_0(H, E)$. An operator is shown to be compact if and only if the topology generated by $p_{m,A}$ for A in $\mathscr{C}$ is compact on σ^*. In this case m is the measure used to represent the operator as an integral. We have considered in [1] corresponding results for operators defined on $\mathscr{L}^p$ spaces. We can show for $p \neq \infty$ and $\dfrac{1}{p} + \dfrac{1}{q} \equiv 1$ that a continuous linear operator U from $\mathscr{L}^p_E(\mu)$ into F, $U \ll \mu$, is compact if and only if an appropriate topology on F_1^* is compact. We can also show that the q semi-variation is right continuous if and only if there exists some sequence of open Baire sets converging to ϕ and the integral satisfies some continuity condition on the unit ball of $\mathscr{L}^p_E(\mu)$ (for $\dfrac{1}{p} + \dfrac{1}{q} = 1$, and $p \neq \infty$). If U is a continuous and compact operator from $\mathscr{L}^\infty_E(\mu)$ into F with U absolutely continuous with respect to μ, it is then shown that the representative measure of U is countably additive. Finally if U is a continuous operator from $\mathscr{L}^p_E(\mu)$ into F $(p \neq \infty)$ and if

$$\langle U, \varphi \rangle (f) = \langle U(f), \varphi \rangle$$

for $f \in \mathscr{L}^p_E(\mu)$ then it is shown that whenever $\| \langle U, \varphi \rangle_A \|_p$, for $\varphi \in F_1^*$, satisfies a Fatou condition and is dominated by a set function having the 0_μ property (see [7]), the representative measure of U has a right continuous q semi-variation.

The question of norm countable additivity equivalent to variation semi-regular (defined below) has been posed by Lewis. In [5], it was shown that much information about this question was obtained by studying locally convex topologies induced by s on F^* as we have described above. In fact the above regularity condition was shown to be related to the notion of regularity for Baire and Borel measures as defined by Dinculeanu and Kluvanek. Again in [5], these topologies yielded a nice characterization of finite dimensional Banach spaces. Our remarks in the previous paragraph further demonstrate the value of these topologies.

In [7] Orlicz studied the properties of weakly absolutely continuous subadditive set functions. Some of those results are applicable to the present situation when F_1^* fails to be compact in contrast to the situation in [6] where compactness is always used.

The book [2] by N. Dinculeanu on *Vector Measures* has generated much interest in this area of research. Frequent reference to it will be made throughout the paper. By $\mathscr{L}_E^p(\mu)$ we will denote all E valued functions that are p-integrable with respect to a positive finite measure μ (in the sense of [2]). If f belongs to $\mathscr{L}_E^p(\mu)$, then $N_p(f)$ will denotee the p-norm of f. If U is a linear operator defined on $\mathscr{L}_E^p(\mu)$ we will write $U \ll \mu$ if $\| U_A \|_p = 0$ whenever $\mu(A) = 0$ (see [2]). The letter m will denote always a measure from $\mathscr{C}$ into $L(E, F)$.

As in [2], for $1 \leqslant q \leqslant \infty$, $\frac{1}{p} + \frac{1}{q} = 1$ the *q-semi variation* of the measure m is defined for A is defined for A in $\mathscr{C}$ by

$$\tilde{m}_q(A) = \sup | \sum m(A_i) x_i |$$

where the supremum is taken over all disjoint sets A_i in $\mathscr{C}$ and x_i in E for i in a finite indexing set I and for which $N_p\left(\sum_{i \in I} \chi_{A_i} x_i\right) \leqslant 1$. For A in $\mathscr{C}$, χ_A represents the characteristic functions of A.

The *q-variation* $\overline{m_q}$ of the measure m is defined for A in $\mathscr{C}$ by

$$m_q(A) = \sup \sum | m(A_i) | \, | x_i |$$

where the supremum is taken in the same manner as the q-semi variation.

Two important properties of these definitions are

d) $\tilde{m}_q(A) = \sup \{ (\overline{m_\varphi})_q(A) \colon \varphi \text{ in } F_1^* \}$.

e) $\overline{m_q} = \tilde{m}_q$ if F is the field of scalars.

For the set $s(E, F)$ defined above, we will let s_q represent that subcollection of set functions m in $s(E, F)$ whose q-variation, $\overline{m}_q$, is finite on $\mathscr{C}$. If $q \neq 1$, it is known that m is countably additive.

A sequence $\{A_n\}_{n \in N}$ of sets in $\mathscr{C}$ is said to be *decreasing monotonically* to ϕ if $\bigcap_{n=1}^{\infty} A_n = \phi$. In this case we will write $\{A_n\}_{n \in N}$ d.m. ϕ.

A scalar valued set function $\mathscr{N}$ on $\mathscr{C}$ is said to be *right continuous* at the sequence $\{A_n\}_{n \in N}$ of sets in $\mathscr{C}$ if A_n d.m. ϕ implies that the sequence $\{\mathscr{N}(A_n)\}_{n \in N}$ converges to 0. The function $\mathscr{N}$ satisfies the 0_μ *property* (as in [7]) if for every sequence $\{B_n\}_{n \in N}$ of disjoint sets in $\mathscr{C}$, the sequence $\{\mathscr{N}(B_n)\}_{n \in N}$ converges to 0 (some authors have referred to this property as "strongly bounded").

It is shown in [7] that while every function of finite variation satisfies the 0_μ condition, the converse need not be true.

The scalar valued function η is said to satisfy the *Fatou property* if η is real valued and if $\liminf \eta(E_n) \geqslant \eta(E)$ whenever $E_n \subset E$ and the sequence $\{\mu(E - E_n)\}_{n \in N}$ converges to 0.

We finally recall that

f) $(m_q)(A) = \sup \left[\sum \frac{|m(A_i)|^q}{\mu(A_i)^{q-1}} \right]^{1/q}$ if $q \neq 1$ where the sup is taken over a finite family of disjoint sets A_i from $\mathscr{C}$ with $A_i \subset A \in \mathscr{C}$ and

g) $(\overline{m}_\infty)(A) = \sup \frac{|m(B)|}{\mu(B)} = (\tilde{m}_\infty)(A)$ where the sup is taken over $B \in \mathscr{C}$, $B \subset A \in \mathscr{C}$. The convention that $\frac{0}{0}$ is interpreted as 0 is maintained.

In general all the notations and concepts pertaining to vector measures can be found in [2]. For m in s_q and A in $\mathscr{C}$ we consider, as before, the functions $p_{m,A}$ defined on F^* by

$$p_{m,A}(\varphi) = (\overline{m}_\varphi)_q(A).$$

In the unit sphere F_1^* of F^* we consider the following two topologies. We denote by $\mathscr{P}(m)_q$ the weakest topology on F_1^* making all semi-norms (as given below) $p_{m,A}$ continuous. By $\mathscr{P}_q$ we mean the topology on F_1^* generated (in the analogous way) by all the semi-norms $p_{m,A}$ for A in $\mathscr{C}$ and m in s_q.

The following lemmas are pertinent to the ensuing results. We state them here without proof (which, if desired, may be found in [1]).

Lemma 1. *For every* $1 \leqslant q \leqslant \infty$, $p_{m,A}$ *is a semi-norm on* F^*. *Thus* F^* *is a locally convex space under the topology generated by* $p_{m,A}$ *(see* [8]).

From [5] one is motivated to define the *boundary* of s_q to be all m in s_q such that whenever A is in $\mathscr{C}$ there exists some φ in F_1^* *with* $\tilde{m}_q(A) = (\overline{m_\varphi})_q(A)$.

Lemma 2. *If* F_1^* *is a compact space in the* $\mathscr{P}(m)_q$ *topology, then the boundary of* s_q *is* s_q.

Since right continuity of $\tilde{m}_q$ will be of importance for later results, the following theorem, which outlines some basic results in that direction will be of interest.

Theorem 1. *Let* $\{A_n\}_{n \in N}$ *be a sequence of sets in* $\mathscr{C}$, *decreasing monotonically to* ϕ. *If* F_1^* *is compact in the topology generated by* p_{m,A_1}, *then there exists a sequence* $\{\varphi_n\}_{n\in N}$ *in* F_1^* *such that* $(m_{\varphi_n})_q(A_n) = \tilde{m}_q(A_n)$. *Moreover if* φ *is an accumulation point of* $\{\varphi_n\}_{n \in N}$ *in the above topology, then the following statements hold.*

a) *If* $(\overline{m_\varphi})_q$ *is right continuous at* $\{A_n\}_{n\in N}$ *then* $\tilde{m}_q$ *is right continuous at* $\{A_n\}_{n \in N}$.

b) *If* $q \neq 1$ *and* m *is in* s_q *then* m *is countably additive.*

c) *If* $(m_\varphi)_q$ *is right continuous at* $\{A_n\}_{n\in N}$ *and if* $\mu(A_n) > 0$ *then* $\tilde{m}_r$ *is continuous at* $\{A_n\}_{n\in N}$ *for all* $1 \leqslant r \leqslant q$.

d) *If* m *is in* s_1 *and if* m_φ *is countably additive for every* φ *in* F_1^* *then* $\tilde{m}_1$ *is right continuous at every sequence* $\{A_n\}_{n\in N}$ d.m. ϕ.

e) *If* $(\overline{m_\varphi})_q$ *satisfies the* 0_μ *condition and the Fatou property for each increasing sequence* $\{E_n\}_{n\in N}$ *then* $(\overline{m_\varphi})_q$ *is right continuous at every sequence* $\{A_n\}_{n\in N}$ d.m. ϕ.

If F_1^* *is not necessarily compact in the topology* $\mathscr{P}(m)_q$, *then* $\widetilde{m}_q$ *is still right continuous at every sequence* $\{A_n\}_{n\in N}$ d.m. ϕ *provided there exists some set function* λ *from* $\mathscr{C}$ *into* F *for which*

f) $(\overline{m_\varphi})_q \leqslant \lambda$ *for every* φ *in* F_1^*.

g) λ *satisfies the* 0_μ *condition.*

h) *Each* $(\overline{m_\varphi})_q$ *satisfies the Fatou condition.* (φ *in* F_1^*.)

Proof. Statement a) follows from an application of Lemma 1 while statement b) is shown in [2]. Statement c) follows from that of a) and from the inequality

$$(\mu(A))^{-\frac{1}{r}} \cdot \widetilde{m}_r(A) \leqslant \mu(A)^{-\frac{1}{q}} \widetilde{m}_q(A)$$

for A such that $\mu(A) > 0$ (see [2]). In statement d), if m is in s_1, then $m \ll \mu$. Thus $m = \widetilde{m}_1$. In [5] the property is shown for $\widetilde{m}$. Statement e) follows from Theorem 4 of [7] applied to (m_φ). The second part of the theorem follows from Theorem 7 of [7]. Here one applies the theorem to the family $M = \{(\overline{m_\varphi})_q : \varphi \in F_1^*\}$. In particular as needed there, if the sequence $\{(\overline{m_\varphi})_q(A_n)\}_{n\in N}$ converges uniformly to 0 for φ in F_1^* then the sequence $\{\widetilde{m}_q(A_n)\}_{n\in N}$ converges to 0. This completes the proof of the theorem.

Applying the results of [7] to the above family M would yield conditions under which the $(m_\varphi)_q$ are uniformly absolutely continuous with respect to μ (in the $\epsilon - \delta$ sense).

We can now obtain conditions equivalent to the space F_1^* being Hausdorff in its $\mathscr{P}_q$ topology.

Lemma 3. *The following conditions are equivalent.*

1) *The space* F_1^* *is Hausdorff in its* $\mathscr{P}_q$ *topology.*

2) *The closure of the linear span of* $\bigcup_{m\in s_q} \bigcup_{A\in\mathscr{C}} m(A)E_1$ *is* F *(for* E_1 *the unit sphere of* E*).*

3) *The topology* $\mathscr{P}_q$ *is stronger than the weak* topology.*

The following interesting result now follows.

Theorem 2. 1) *If* $(F_1^*, \mathscr{P}_q)$ *is a Hausdorff space then* $(F_1^*, \mathscr{P}_q)$ *is compact if and only if* $(F_1^*, \mathscr{P}_q) = (F_1^*, wk^*)$ *where* wk^* *represents the weak* topology for* F_1^*.

2) *If* $(F_1^*, \mathscr{P}_q)$ *and* $(F_1^*, \mathscr{P}_r)$ *are Hausdorff spaces then* $(F_1^*, \mathscr{P}_q)$ *and* $(F_1^*, \mathscr{P}_r)$ *are both compact if and only if* $\mathscr{P}_q = \mathscr{P}_r = wk^*$.

Proof. We show 1). If $(F_1^*, \mathscr{P}_q)$ is Hausdorff then the identity map from $(F_1^*, \mathscr{P}_q)$ onto (F_1^*, wk^*) is continuous by Lemma 3. Since (F_1^*, wk^*) is a Hausdorff space, the map is a homeomorphism. Of course statement 2) follows immediately from statement 1).

In contrast to the situation depicted in [5] one may have $(F_1^*, \mathscr{P}_q)$ as a non Hausdorff space. If μ is identically zero, then s_q reduced to zero. Thus statement 2) of Lemma 3 shows that $(F_1^*, \mathscr{P}_q)$ is non Hausdorff. The other extreme is to have μ purely atomic. Then $(F_1^*, \mathscr{P}_q)$ is always a Hausdorff space. In fact let $m_t(A) = 0$ when $t \notin A$ and $m_t(A) = U \in L(E, F)$ when $t \in A$. If A_i is the atom containing t, $(\mu(A_i) > 0$. If $B \subset A_i$, (then $\mu(B) = 0$ if and only if $B = \phi$), then $(\widetilde{m}_t)_q(A) = \dfrac{\| U \|}{\mu(A_i)^{\frac{q-1}{q}}}$ is finite. So m_t belongs to s_q. By statement 2) of Lemma 3, it follows that $(F_1^*, \mathscr{P}_q)$ is Hausdorff.

The preceding observations point out that there are many more countably additive measures than measures in s_q (for $q \neq 1$). In [5] some conditions were pointed out which were equivalent to the topology generated by $p_{m,A}$ (m finitely additive, fixed, and A in $\mathscr{C}$ also fixed). A brief look at the proof shows that this does not carry over to the present setting since the point mass in general is not in s_q. However we have the following result.

Theorem 3. *Assume* $(F_1^*, \mathscr{P}_q)$ *is a Hausdorff space, then the following conditions are equivalent.*

1) *The topology generated by* $p_{m,A}$ *(for* m *in* s_q *fixed and* A *in* $\mathscr{C}$ *also fixed) is Hausdorff.*

2) $s_q = \{n\colon n$ *in* s_q *for which* $(\overline{m_\varphi})_q = 0$ *implies* $(\overline{n_\varphi})_q = 0\}$.

3) *The topology generated by* $p_{m,A}$ *on* F_1^* *is stronger than the* wk* *topology of* F_1^*.

Proof. If 2) holds and 1) does not there exists a non-zero φ in F_1^* such that $(\overline{m_\varphi})_q = 0$. Thus for all n in s_q, $(\overline{n_\varphi})_q = 0$. This contradicts the fact that $(F_1^*, \mathscr{P}_q)$ is a Hausdorff space. The rest of the proof follows the pattern of [5] and will not be reproduced here.

Now let $\mathscr{C}_{\sigma,f}$ denote the σ-ring of $\mu\sigma$-finite subsets of T (see [2]). If $1 \leqslant p < \infty$, if U is a continuous linear operator from $\mathscr{L}_E^p(\mu)$ into F with $U \ll \mu$ and if $T \in \mathscr{C}_{\sigma,f}$, then there exists a unique measure m from $\mathscr{C}$ into $L(E, F)$ with $\tilde{m}_q(T)$ finite and $U(f) = \int f\, dm$. If $p = \infty$, then there exists a finitely additive set function m from $\mathscr{C}$ into $L(R, X)$ with $\tilde{m}_\infty(T) < \infty$ such that $U(f) = \int f\, dm$ for all f in $\mathscr{L}_R^\infty(\mu)$ where R denotes the scalar field (see [5]). The result alluded to in our introduction may now be given as

Theorem 4.

1) *Let* $p \neq \infty$, $\dfrac{1}{p} + \dfrac{1}{q} = 1$ *and let* $T \in \mathscr{C}_{\sigma,f}$. *If* U *is a continuous linear operator from* $\mathscr{L}_E^p(\mu)$ *into* F *such that* $U \ll \mu$, *then* U *is a compact operator if and only if* $(F_1^*, \mathscr{P}(m)_q)$ *is a compact space.*

2) *Let* $p = \infty$. *If* U *is a continuous linear operator from* $\mathscr{L}_R^\infty(\mu)$ *into* F *such that* $U \ll \mu$, *then* U *is a compact operator if and only if* $(F_1^*, \mathscr{P}(m)_1)$ *is a compact space.*

Proof. In showing 1), let us assume that U is compact and let $\{\varphi_n\}_{n\in N}$ be a sequence in F_1^*. Without loss of generality we may assume that the sequence converges to φ in the weak* topology. Thus we need to show that the sequence converges to φ in the $\delta_{m,q}$ topology. (Now the sequence $\{U^*(\varphi_n)\}_{n\in N}$ converges to $U^*(\varphi)$ in the norm (see [4]).) Note that $(\overline{m_\varphi})_q = (\tilde{m}_\varphi)_q$ since m_φ has values in a dual space. Thus for A in $\mathscr{C}$, there exists a disjoint sequence of sets A_i in $\mathscr{C}$ with $A_i \subset A$, $i \in N$, such that if $\epsilon > 0$ and if $N_p\left(\Sigma \chi_{A_i} \cdot x_i\right) \leqslant 1$ then,

$$(\overline{m_{\varphi - \varphi_n}})_q(A) \leqslant |\left\langle \Sigma\, m(A_i)x_i, \varphi - \varphi_n \right\rangle| + \epsilon\,.$$

Thus

$$(m_{\varphi - \varphi_n})_q(A) \leqslant |\langle U(\Sigma \chi_{A_i} \cdot x_i), \varphi - \varphi_n \rangle| + \epsilon \leqslant$$

$$\leqslant N_p(\Sigma \chi_{A_i} \cdot x_i) \| U^*(\varphi - \varphi_n) \| + \epsilon .$$

Consequently the sequence $\{\varphi_n\}_{n \in N}$ converges to φ in the $\mathscr{P}(m)_q$ topology. The other direction is straightforward (see [1]) and the proof of 2) is similar.

For the next theorem let $\mathscr{C}$ denote the σ-ring generated by the compact G_δ subsets of the locally compact Hausdorff space T. Again if $1 \leqslant p \leqslant \infty$, $\frac{1}{p} + \frac{1}{q} = 1$, and if $m \in s_q$ then it is shown in [2] that "the integral of $f \in \mathscr{L}^p_E(\mu)$ relative to m" is defined (and is denoted by $\int f dm$) provided that $\tilde{m}_q(A)$ is finite for all $A \in \mathscr{C}_{\sigma,f}$ (the variation of m however, need not be finite). This theorem establishes a relation between the continuity of the integral $\int f\, dm$ on the unit ball of $\mathscr{L}^p_E$ and the right continuity of $\tilde{m}_q$.

Theorem 5. *Let* $\mathscr{C}$ *be as described above and let* $p \neq \infty$, $\frac{1}{p} + \frac{1}{q} = 1$. *If* m *is a measure from* $\mathscr{C}$ *into* $L(E, F)$ *with* $\tilde{m}_q$ *finite on* $\mathscr{C}_{\sigma, f}$, *then the right continuity of* $\tilde{m}_q$ *is equivalent to the following two conditions taken simultaneously.*

1) *For every sequence* $\{A_n\}_{n \in N}$ *of sets in* $\mathscr{C}$ *decreasing monotonically to* ϕ, *there exists a sequence of open Baire sets* U_n *in* T *such that* $A_n \subset U_n$, n *in* N, *and the sequence* $\{\tilde{m}_q(U_n)\}_{n \in N}$ *converges to* 0.

2) *The sequence* $\{|\int f_n\, dm|\}_{n \in N}$ *converges uniformly to* 0 *for every sequence* $\{f_n\}_{n \in N}$ *in* $\mathscr{L}^p_E(\mu)$ *with* $N_p(f_n) \leqslant 1$ *and* $f_n(x) = 0$ *for* x *in* $T \setminus U_n$, n *in* N.

Proof. Let us assume that $\tilde{m}_q$ is right continuous. As in the proof of a similar result given in [3], it can be shown (replacing the p quasi semi variation by $\tilde{m}_q$) that for every $A \in \mathscr{C}$ and $\epsilon > 0$ there exists a compact Baire set K and an open Baire set G with $K \subset A \subset G$ and $\tilde{m}_q(G - K) < \epsilon$. Thus we obtain a sequence $\{U_n\}_{n \in N}$ of open Baire sets with the sequence $\{\tilde{m}_q(U_n)\}_{n \in N}$ converging to 0. For every $f_n \in \mathscr{L}^p_E(\mu)$

satisfying 2)

$$|\int f_n \, dm| \leqslant N_p(f_n)\widetilde{m}_q(U_n):$$

Of course the latter becomes arbitrarily small.

Conversely assume 1) and 2) hold. Let $\{A_n\}_{n\in N}$ d.m. ϕ and let U_n be as above. Going to a subsequence if necessary, let us assume $\widetilde{m}_q(A_n) > \epsilon$ for all n. Pick j large enough so that whenever the support of f is a subset of U_j and $N_p(f) \leqslant 1$ $|\int f\, dm| < \epsilon/2$. There exists some $\varphi \in F_1^*$ and some finite set of disjoint subsets B_i of A_j such that $|\langle \sum_i m(B_i)x_i, \varphi \rangle| > \epsilon$ with $N_p(\sum_i \chi_{B_i} \cdot x_i) \leqslant 1$. If $f = \sum_i \chi_{B_i} \cdot x_i$ then $|\int f\, dm| < \epsilon/2$ which contradicts $|\int f\, dm| > \epsilon$. Consequently the sequence $\{\widetilde{m}_q(A_n)\}_{n\in N}$ converges to 0.

We now study the case for $q = 1$. Also for $q \neq 1$ we may ask the question for what kind of operators on $\mathscr{L}_E^p(\mu)$ is the q semi variation of the representative measure right continuous?

If U is a continuous (in the norm of $\mathscr{L}_E^p(\mu)$, $1 \leqslant p \leqslant \infty$) operator with $U \ll \mu$, from $\mathscr{L}_E^p(\mu)$ itno F, then we introduce the operator $\langle U, \varphi\rangle$ from $\mathscr{L}_E^p(\mu)$ into the scalar field R defined by

$$\langle U, \varphi\rangle (f) = \langle U(f), \varphi\rangle \qquad (\varphi \in F_1^*).$$

Theorem 6. 1) *If U is a continuous and compact operator from $\mathscr{L}_R^\infty(\mu)$ into F with $U \ll \mu$, then the representative measure of U is countably additive.*

2) *Let U be a continuous operator from $\mathscr{L}_E^p(\mu)$ into F ($p \neq \neq \infty$) with $U \ll \mu$ and let $T \in \mathscr{C}_{\sigma,f}$. If there exists a scalar valued set function λ satisfying the 0_μ condition with $\|\langle U, \varphi\rangle_A\|_p \leqslant \lambda(A)$ for all A in $\mathscr{C}$ and with $\liminf \|\langle U, \varphi\rangle_{A_n}\|_p \geqslant \|\langle U, \varphi\rangle_A\|_p$ for every A_n and A in $\mathscr{C}$ for which the sequence $\{\mu(A_n - A)\}_{n\in N}$ converges to 0, then the representative measure of U is q-variationally semi-regular.*

Proof. For the proof of this, the reader is referred to [1].

2. CONCLUDING REMARKS

It would be interesting to further study these topological spaces associated with these measures. The topological spaces under consideration, as has been seen, need not be metrizable in fact they need not even be Hausdorff. It would be interesting to consider the requirement that $(F_1^*, \mathscr{P}(m))$ or $(F_1^*, \mathscr{P}(m)_q)$ be paracompact, metacompact or any of the other "compactness type" conditions. What is the effect of these conditions on the corresponding operator defined on $\mathscr{L}_E^p(\mu)$? The compact operators are then a subclass of the class of operators so obtained. Let us emphasize again that to go beyond the more restricted setting of compactness, we found essential the results of Orlitz in [7]. They were necessary to carry the anticipated conclusions through to fruition.

REFERENCES

[1] R.A. Alò – A. de Korvin – L.B. Kunes, Topological aspects of q-regular measures, (to appear).

[2] N. Dinculeanu, *Vector Measures,* Pergamon Press, Berlin, 1967.

[3] N. Dinculeanu – P. Lewis, Regularity of Baire Measures, Proc. *Amer. Math. Soc.,* 26 (1970), 92-94.

[4] N. Dunford – J. Schwartz, *Linear Operators* Part I, Interscience Publishers, New York, 1958.

[5] P. Lewis, Vector measures and topology, *Rev. Roum. Math. Pures et Appl.,* (to appear).

[6] P. Lewis, Characterization of a class of compact operators, (submitted).

[7] W. Orlicz, Absolute continuity of set functions with respect to a finitely subadditive measure, *Annales Soc. Math. Pol.,* Series I, 14 (1970), 101-118.

[8] H.H. Schaefer, *Topological Vector Spaces,* MacMillan Company, New York, 1966.

COLLOQUIA MATHEMATICA SOCIETATIS JÁNOS BOLYAI
8. TOPICS IN TOPOLOGY, KESZTHELY (HUNGARY), 1972.

CARTESIAN PRODUCTS OF SUBPARACOMPACT SPACES

K. ALSTER

Let us recall that X is a subparacompact space [2] if and only if every open cover of X has a σ-discrete closed refinement. Obviously every paracompact space is subparacompact.

We give an example of a paracompact space X such that X^2 is not subparacompact. The space X is obtained from a lineary ordered space in the same way as the Sorgenfrey line S from the real line. Let us notice, that in [3] it is proved that $S^{\aleph_0}$ is subparacompact.

We also prove that $N^{\aleph_1}$ is not subparacompact which is a generalization of the A.H. Stone well-known result's that $N^{\aleph_1}$ is not paracompact.

Both results are obtained with R. Engelking and will be published in [1].

REFERENCES

[1] K. Alster – R. Engelking, Cartesian products of subparacompact spaces, *Bull. Acad. Polon. Sci. Ser. Sci. Math. Astronom. Phys.*, (1972) to appear.

[2] D.K. Burke, On subparacompact spaces, *Proc. Amer. Math. Soc.*, 23 (1969), 655-663.

[3] B.J. Lutzer, Another note on Sorgenfrey line, to appear.

COLLOQUIA MATHEMATICA SOCIETATIS JÁNOS BOLYAI
8. TOPICS IN TOPOLOGY, KESZTHELY (HUNGARY), 1972.

NOTE ON METRIZATION OF COMPACT CONVEX SETS

P.R. ANDENÆS

1. In this note we give a simple metrizability criterion for compact convex subsets of a Hausdorff linear topological space. In fact, we prove that a compact convex set K is metrizable if and only if the algebraic boundary $\partial_a K$ (defined in §3) is second countable in the relative topology. This result is weakly related to a theorem of Corson [1] stating that if the extreme boundary $\partial_e K$ (i.e. the set of extreme points in K) of a compact convex subset K of a locally convex Hausdorff space is second countable and topologically complete, then K is metrizable. In this connexion it seems appropriate to point out that our method of proof does not involve local convexity. Neither is topological completeness required, (although the condition will be satisfied). On the other hand, the algebraic boundary will usually be much larger than $\partial_e K$. $\partial_a K$ and $\partial_e K$ coincide if and only if K is strictly convex. Unless explicitly stated we use the notation of [3].

2. We shall need some preliminary results.

2.1. Lemma. *Let S be a subset of a Hausdorff linear topological space E and suppose that S is second countable. Then there exists a* *sequence* $\{V_n\}$ *of neighbourhoods of* 0 *such that*

$$\bigcap_{n=1}^{\infty} (x + V_n) \cap S = \{x\}$$

for each $x \in S$. *The sequence* $\{V_n\}$ *can be chosen such that each* V_n *is circled and such that* $V_{n+1} + V_{n+1} \subset V_n$ *for each* $n \in \mathbf{Z}^+$.

Proof. S is a regular Hausdorff space with a countable base, thus S is metrizable. Let d be a compatible metric for S. For a fixed $n \in \in \mathbf{Z}^+$ we now choose for each $x \in S$ a neighbourhood V_x of 0 such that

$$(x + V_x + V_x) \cap S \subset \{y \in S: d(x, y) < 2^{-n}\}$$

S is Lindelöf, so there exists $x_{n,i} \in S$, $i = 1, 2, \ldots,$ such that $\{(x_{n,i} + + V_{x_{n,i}}) \cap S\}_i$ covers S. Suppose now that

$$y \in \bigcap_{\substack{i \geq 1 \\ n \geq 1}} (x + V_{x_{n,i}}) \cap S\,,$$

$x, y \in S$. Then for each n there exists j $(= j(n))$ such that $x \in x_{n,j} + + V_{x_{n,j}}$. In particular we see that $d(x, x_{n,j}) < 2^{-n}$. We also have

$$y \in x + V_{x_{n,j}} \subset x_{n,j} + V_{x_{n,j}} + V_{x_{n,j}}\,,$$

thus $d(x_{n,j}, y) < 2^{-n}$. It follows that $d(x, y) < 2^{-n+1}$. n was arbitrary, so we have $d(x, y) = 0$, whence $x = y$. Therefore we have

(*) $$\bigcap_{\substack{i \geq 1 \\ n \geq 1}} (x + V_{x_{n,i}}) \cap S = \{x\}$$

for all $x \in S$. We now choose a circled neighbourhood V_1 of 0 such that $V_1 \subset V_{x_{1,1}}$ and then construct the sequence $\{V_n\}$ by induction such that each V_n is circled, $V_{n+1} + V_{n+1} \subset V_n$, and $V_n \subset \bigcap_{\substack{i \leq n \\ k \leq n}} V_{x_{k,i}}$.
The intersection property (*) is then valid also for the sequence $\{V_n\}$, and the proof is complete.

We now obtain a very simple metrizability criterion for compact subsets of E.

2.2. Proposition. *A non-empty compact subset* K *of a Hausdorff linear topological space* E *is metrizable if and only if there exists a sequence* $\{V_n\}$ *of neighbourhoods of* 0 *such that*

$$(*) \qquad \bigcap_{n=1}^{\infty} (x + V_n) \cap K = \{x\}$$

for every $x \in K$ *– or equivalently, such that*

$$\bigcap_{n=1}^{\infty} V_n \cap (K - K) = \{0\}.$$

Proof. Suppose first that K is metrizable. Since K is compact the second axiom of countability is satisfied, and the previous lemma ensures the existence of the sequence $\{V_n\}$.

Conversely, let $\{V_n\}$ be a sequence of neighbourhoods of 0 such that $(*)$ is satisfied. Without loss of generality we may assume that V_n is open and circled and that $V_{n+1} + V_{n+1} \subset V_n$, $n = 1, 2, \ldots$. Now let x be an arbitrary point in K and let U be an open subset of E with $x \in U \cap K$. For all $n \in Z^+$ we have $x + \bar{V}_{n+1} \subset x + V_n$ (since $V_{n+1} + V_{n+1} \subset V_n$), therefore

$$\bigcap_{n=1}^{\infty} (x + \bar{V}_n) \cap K = \bigcap_{n=1}^{\infty} (x + V_n) \cap K = \{x\}.$$

This entails

$$\bigcap_{n=1}^{\infty} (x + \bar{V}_n) \cap K \setminus (U \cap K) = \phi.$$

Since K is compact it follows that $(x + \bar{V}_n) \cap K \subset U \cap K$ for n sufficiently large. Hence $\{(x + V_n) \cap K\}_n$ is a neighbourhood base at x for each $x \in K$. The cover $\{(x + V_n) \cap K\}_{x \in K}$ of K has a finite subcover $\{(x_{n,i} + V_n) \cap K\}_{i=1}^{N_n}$. We then claim that

$$\{(x_{n,i} + V_n) \cap K : 1 \leqslant i \leqslant N_n;\ n \in Z^+\}$$

is a base for the relative topology on K. To see this, let x be an arbitrary point in K and let U be an open subset of E with $x \in U \cap K$. We then already know that we can find n such that $(x + V_n) \cap K \subset U \cap K$.

For a suitable i, $1 \leqslant i \leqslant N_{n+1}$, we have $x \in x_{n+1,i} + V_{n+1}$. If $y \in$ $\in x_{n+1,i} + V_{n+1}$ we obtain

$$y \in x + V_{n+1} + V_{n+1} \subset x + V_n$$

since V_{n+1} is circled (symmetry would suffice). Thus we have $x \in$ $\in (x_{n+1,i} + V_{n+1}) \cap K \subset (x + V_n) \cap K$, and our claim is proved. Metrizability of K now follows from Urysohn's metrization theorem. The last equivalence in the proposition is quite trivial, and so the proof is complete.

Remark. As pointed out by the referee it should be noted that an analogue of the above result is valid for arbitrary compact Hausdorff spaces. In fact, the following result is easily proved: A compact Hausdorff space X is metrizable if and only if the diagonal $\Delta = \{(x, x) : x \in X\}$ in $X \times$ $\times X$ is a G_δ-set (See f. ex. Exer. XI, 4.4. in Dugundji's "Topology").

3. We shall need some definitions. Let K be a convex subset of a linear space E. A point $x \in K$ is called an *algebraic boundary point* of K is there exists $y \in K$ such that $y + r(x - y) \notin K$ for any $r > 1$. The set of all algebraic boundary points of K is called the *algebraic boundary* of K and we denote it $\partial_a K$. (If $E = \bigcup_{n=1}^{\infty} n(K - K)$ our terminology is in accordance with [4].) In general we may have $\partial_a K = \phi$, but if K is a non-empty compact convex subset of a linear topological space, then obviously $\partial_a K$ is also non-empty. It may also happen that $\partial_a K$ is extremely large, in fact in may occur that $\partial_a K = K$, for an example of this phenomenon cf. [4]. Finally, a convex set K is called strictly convex if $\partial_a K$ contains no proper segments.

3.1. Theorem. *Let K be a non-empty compact convex subset of a Hausdorff linear topological space. Then K is metrizable if and only if $\partial_a K$ is second countable.*

Proof. If K is metrizable, then – since it is compact – it is second countable, but then so is also $\partial_a K$.

Conversely, assume that $\partial_a K$ satisfies the second axiom of countability and – to avoid trivialities – assume also that $\partial_a K \neq K$. According to lemma 2.1 there exists a sequence $\{V_n\}$ of circled neighbourhoods of 0 with $V_{n+1} + V_{n+1} \subset V_n$, $n = 1, 2, \ldots$, such that

$$\bigcap_{n=1}^{\infty} (x + V_n) \cap \partial_a K = \{x\}$$

for each $x \in \partial_a K$. Assume now that

$$\bigcap_{n=1}^{\infty} (x + V_n) \cap K \neq \{x\}$$

for some $x \in K \setminus \partial_a K$. Then we can choose $v \in \bigcap_{n=1}^{\infty} V_n$, $v \neq 0$, such that $x + v \in K$. Using the relation $V_{n+1} + V_{n+1} \subset V_n$, $n = 1, 2, \ldots,$ we see that $kv \in \bigcap_{n=1}^{\infty} V_n$ for each $k \in Z^+$, and since $\bigcap_{n=1}^{\infty} V_n$ is circled, it follows that $rv \in \bigcap_{n=1}^{\infty} V_n$ for all $r \in R$. Since K is compact, there exists $r_1 > 0$ such that $y_1 = x + r_1 v \in \partial_a K$, i.e. $y_1 \in \bigcap_{n=1}^{\infty} (x + V_n) \cap$ $\cap\, \partial_a K$. Since $x \notin \partial_a K$ the segment $[y_1 : x]$ can be extended beyond x within K, this means that there exists $r > 0$ such that $x - rv \in K$. Using the compactness of K once more we conclude that there exists $r_2 > 0$ such that $y_2 = x - r_2 v \in \partial_a K$. In particular we note that $y_2 \neq y_1$. For each $n \in Z^+$ we now have

$$y_1 \in x + V_{n+1}\ ,\ ): x \in y_1 + V_{n+1}$$

and

$$y_2 \in x + V_{n+1}\,.$$

It follows that $y_2 \in (y_1 + V_n) \cap \partial_a K$ for each $n \in Z^+$, but this contradicts the properties of the sequence $\{V_n\}$. Thus $\bigcap_{n=1}^{\infty} (x + V_n) \cap K =$ $= \{x\}$ for each $x \in K \setminus \partial_a K$. Finally, if $y \in \partial_a K$ and $x \in \bigcap_{n=1}^{\infty} (y +$ $+ V_n) \cap K$, $x \neq y$, we necessarily have $x \in K \setminus \partial_a K$. But then we just proved that $\bigcap_{n=1}^{\infty} (x + V_n) \cap K = \{x\}$ contradicting the assumption $y \in$ $\in \bigcap_{n=1}^{\infty} (x + V_n) \cap K$. Thus we must have

$$\bigcap_{n=1}^{\infty} (x + V_n) \cap K = \{x\}$$

for each $x \in K$, and we conclude from proposition 2.2 that K is metrizable, q.e.d.

An easy argument shows that a convex set K is strictly convex if and only if $\partial_a K = \partial_e K$, i.e. each algebraic boundary point is also an extreme point (cf. [4], p. 345). Then the following corollary is immediate:

3.2. Corollary. *Let K be a strictly convex, compact subset of a Hausdorff linear topological space. Then K is metrizable if and only if $\partial_e K$ is second countable.*

We conclude with another corollary which was pointed out to us by E.M. Alfsen. It generalizes the following well-known result ([2], p. 426): A Banach space V is separable if and only if the unit ball B_1^* in V^* is metrizable in the w^*-topology.

3.3. Corollary. *Let V be a Banach space. Thus `V is separable if and only if $S_1^* = \{f \in V^* : ||f|| = 1\}$ is second countable w.r.t. the w^*--topology.*

Proof. B_1^* is convex and w^*-compact, furthermore we obviously have $\partial_a B_1^* = S_1^*$. Thus B_1^* is w^*-metrizable if and only if S_1^* is second countable w.r.t. the w^*-topology, and the result follows.

REFERENCES

[1] H.H. Corson, Metrizability of compact convex sets, *Trans. AMS,* 151 (1970), 589-596.

[2] N. Dunford – J.T. Schwartz, *Linear Operators,* Part I, Interscience, New York, 1958.

[3] J. Kelley – I. Namioka, et. al., *Linear Topological Spaces,* Van Nostrand, Prinecton, 1963.

[4] G. Köthe, *Topologische Lineare Räume* I, 2. Aufl., Springer Verlag, Berlin – Heidelberg – New York, 1966.

COLLOQUIA MATHEMATICA SOCIETATIS JÁNOS BOLYAI
8. TOPICS IN TOPOLOGY, KESZTHELY (HUNGARY), 1972.

FINITE HOMEOMORPHISM GROUPS OF THE 2-SPHERE

L. BABAI — W. IMRICH — L. LOVÁSZ

To the memory of professor G. Hajós

Some investigations concerning automorphisms of planar graphs led us to the determination of all finite groups of homeomorphisms of the (2-dimensional) sphere:

Theorem. *Each finite group of homeomorphisms of the 2-sphere is topologically equivalent to a finite congruence group of the sphere, i.e. to a subgroup of the congruence group of a regular polyhedron or of a regular prism.*

Later we have learned that Kerékjártó [8] found this in 1919. He pointed out that this also follows from a result of Brouwer [2]. However, it seems to be worthwile to write down our proof for several reasons. First, investigation of the literature as well as discussion with topologists shows that this area is not investigated now and the mentioned results are almost forgotten; obvious applications as to the automorphisms of planar graphs are not formulated at all; finally, it may have some sense to put down a proof of this interesting result in modern language and accuracy.

Graph-theoretic connexions of this work are the following. Whitney showed [16] that a 3-connected planar graph without loops and multiple edges can be uniquely embedded in the sphere. From this we can easily deduce that the automorphism group of a 3-connected planar graph embedded in the sphere can be extended to a finite homeomorphism group of the sphere. Thus we obtain a characterization of the automorphism groups of the 3-connected planar graphs:

Corollary. *Let X be a 3-connected planar graph without loops and multiple edges. Then X can be embedded in the sphere S in such a way that all automorphisms of X are induced by congruences of S.*

Special consequences of this statement have been obtained by Sabidussi [14], Fleischner [6], [7] and two of the authors [1]. It should be noted that the corollary is also a consequence of a result of Mani [11] which asserts that for any 3-connected planar graph X there exists a convex polyhedron whose 1-skeleton is isomorphic to X and whose group of congruences induces all automorphisms of X. However, in [11] homeomorphism groups of the sphere have not been investigated.

The proof of the theorem makes strong use of the characterization of periodic homeomorphisms of the plane by Kerékjártó [9] and Eilenberg [4].

Let us remark that Kerékjártó [10] has shown that, apart from trivial cases, no new groups can be obtained if we consider compact groups rather than finite ones. This theorem, easily deducible also from the results of Montgomery and Zippin, is the following:

Each infinite compact homeomorphism group of the 2-sphere is topologically equivalent to one of the following groups:

(a) *the group of all congruences of the 2-sphere,*

(b) *the group of congruences stabilizing or*

(c) *fixing a pair of antipodal points,*

(d) *the subgroups of orientation preserving elements of these three groups.*

The authors' thanks are due to M. Bognár for his valuable remarks.

DEFINITIONS AND NOTATION

Let T_1 and T_2 be topological spaces, and let $\mathcal{G}_1$, $\mathcal{G}_2$ be groups acting on T_1 and T_2, respectively. $\mathcal{G}_1$ and $\mathcal{G}_2$ are said to be *topologically equivalent* If there exists a homeomorphism μ of T_1 onto T_2 such that $\mu^{-1}\mathcal{G}_1\mu = \mathcal{G}_2$.

If R is a subset of T_1 the subgroup of $\mathcal{G}_1$ leaving R invariant as a set is called the stabilizer of R.

By a *topological disc* we mean a Hausdorff space homeomorphic to the closed unit disc.

We only consider *topological graphs* X embedded in the sphere S, i.e. a set $V(X)$ of points in the sphere, connected by nonintersecting Jordan curves, which constitute the edge set $E(X)$. The set $F(X)$ of faces consists of the closures of the complementary domains of $\bigcup E(X)$. (We admit multiple edges). If X is 2-connected the faces are topological discs.

By an *automorphism of* X we mean an incidence preserving permutation acting simulatneously on $V(X)$, $E(X)$ and $F(X)$. The group of automorphisms of X will be denoted by $\mathcal{A}(X)$.

PRELIMINARY LEMMAS

Lemma 1. *Any finite homeomorphism group* $\mathcal{G}$ *of a circle* C *is topologically equivalent to a group of congruences of a circle.*

Proof. This statement is fairly obvious. For sake of brevity we do not put down the simplest proof. We point out that lemma 1 easily follows from lemma 2: extend the homeomorphisms of the circle B onto the disc D bounded by B.

Lemma 2. *Given a finite homeomorphism group* $\mathcal{H}$ *acting on the topological disc* D. *Then there is a homeomorphism* μ_2 *of* D *onto the disc* D_1 *such that* $\mu_2^{-1}\mathcal{H}\mu_2$ *is a group of congruences of* D_1. – *Given a homeomorphism* μ_1 *of the boundary* B *of* D *onto the boundary* B_1 *of* D_1 *such that* $\mu_1^{-1}\mathcal{H}\mu_1$ *is a group of congruences of* B_1, *then* μ_2 *can be chosen so that* $\mu_2 \mid B = \mu_1$.

Proof. $\mathscr{H}$ can be easily extended to a finite group of homeomorphisms of the plane. By a result of Kerékjártó and Eilenberg [9], [4]) follows:

(1) *The orientation preserving elements of* $\mathscr{H}$ *are topologically equivalent to rotations, the orientation reversing ones to reflections of* D.

Obviously, the fixed point of an orientation preserving element lies in the interior of D and the orientation reversing ones have two fixed points on B. Consequently there exists an x in B such that $x\varphi = x$, $\varphi \in \mathscr{H}$, implies $\varphi = 1$.

Let X denote the graph, with the vertex set $x\mathscr{H}$, the edges of which are the arcs of B connecting the neighboring elements of $x\mathscr{H}$. From the above it is clear that the restriction of $\mathscr{H}$ to B induces an isomorphism between $\mathscr{H}$ and a subgroup of $\mathscr{A}(X)$, hence $\mathscr{H}$ is either a cyclic group or a dihedral group containing a cyclic subgroup $\mathscr{K}$ which consists of the orientation preserving elements of $\mathscr{H}$. ($|\mathscr{H}: \mathscr{K}| = 2$.) In the first case the statement of the lemma is obvious by (1). In the second case, let φ be a generating element of $\mathscr{K}$ with the fixed point p in the interior of D. Let $\psi \in \mathscr{H} - \mathscr{K}$, then ψ is orientation reversing and has a fixed line, say l. Since φ and ψ generate $\mathscr{H}$, $\psi\varphi = \varphi^m \psi$ for some m. Hence, $p\psi = p\varphi^m \psi = p\psi\varphi$, i.e. $p\psi = p$ (the only point fixed under φ), and therefore, $p \in l$.

Let $|\mathscr{H}| = 2n$ and let $l_1, \ldots, l_n$ denote the fixed lines of the elements of $\mathscr{H} - \mathscr{K}$. Assume that $|l_i \cap l_j| \geqslant 2$, i.e. there exists a vertex $q \neq p$ fixed by $\psi_i, \psi_j \in \mathscr{H} - \mathscr{K}$. So q is fixed also under $\psi_i\psi_j$ which is an orientation preserving element of $\mathscr{H}$, hence, by (1), $\psi_i\psi_j = 1$, $\psi_i = \psi_j$.

This means $l_i \cap l_j = \{p\}$ for $i \neq j$. It is easy to see that none of the l_i can lie in one of the halfdiscs determined by l_j. Thus, there are diameters $d_1, \ldots, d_n$ of D_1 and a homeomorphism

$$\lambda\colon B_1 \cup \bigcup d_i \to B \cup \bigcup l_i$$

such that $d_i\lambda = l_i$ and such that the d_i are the symmetry-axes of a regular n-gon. We can choose λ also in such a way that, for any $\xi \in \mathscr{H}$,

$\lambda\xi\lambda^{-1}$ is a congruent transformation of $\bigcup d_i$ and $\lambda | B = \mu_1^{-1}$ if μ_1 is given.

Let E be a sector of D_1 of angle $\frac{\pi}{n}$, determined by $\bigcup d_i$. Extend λ to a homeomorphism λ^* of E and the corresponding region of D. Then λ^* can be uniquely extended to a homeomorphism ν of D_1 and D such that $\nu \mathscr{H} \nu^{-1}$ is a group of congruences of D_1. Setting $\mu_2 = \nu^{-1}$ proves the lemma.

Lemma 3. *Let* $\mathscr{H}$ *be a finite homeomorphism group of the sphere,* D *be a cap which has common interior points with all of its images under* $\mathscr{H}$, *and suppose there is a point* q *such that* $q\mathscr{H}$ *lies in one complementary domain* C *of* $D\mathscr{H}$. *Then the boundary of* C *is a Jordan curve invariant under* $\mathscr{H}$.

Proof. The invariance of C and its boundary is clear. As $D\mathscr{H}$ is a 2-connected Peano curve, the boundary of any complementary domain is a Jordan curve (see [13], p. 168).

Lemma 4. *Let* $\mathscr{G}$ *be a finite homeomorphism group of the sphere* S *and* $\mathscr{H}$ *be the stabilizer of the point* $p \in S$. *Then* $\mathscr{H}$ *is topologically equivalent to a finite group of congruences of* S *which fixes at least two (diametrically opposite) points.*

Proof. Let D be a disc with center p. If D is sufficiently small, the conditions of Lemma 3 are satisfied. The obtained Jordan curve J divides S into two discs invariant under $\mathscr{H}$. An application of Lemma 1 to J and Lemma 2 to both of these discs proves Lemma 4.

Lemma 5. *Let* $\mathscr{G}$ *be a finite homeomorphism group of the sphere* S, X *a graph on* S *invariant under* $\mathscr{G}$ *and* X_1 *a component of* X. *Then all the other components of* $X_1\mathscr{G}$ *lie in the same complementary domain of* X_1.

Proof. Assume, indirectly, that D_1, D_2 are two complementary domains of X_1 containing $X_1\varphi$ and $X_1\psi$, respectively. Then one of $D_1\varphi, D_2\varphi$ does not intersect X_1, and thus lies in D_1. $D_1\varphi \subseteq D_1$ implies $D_1\varphi = D_1$, since φ is of finite order, but this is impossible, as the boundaries of $D_1\varphi$ and D_1 lie in different components of X. Therefore $D_2\varphi \subseteq D_1$. Similarly, $D_1\psi \subseteq D_2$, whence $D_1\psi\varphi \subseteq D_2\varphi \subseteq D_1$, which implies $D_1\psi\varphi = D_2\varphi = D_1$. As before, this is impossible.

Lemma 6. *Suppose a connected planar graph Y in the sphere has no vertices of valency 2 and no 2-sided faces. If the group $\bar{\mathscr{G}} \subseteq \mathscr{A}(Y)$ acts transitively on $E(Y)$, then $\bar{\mathscr{G}}$ acts transitively on $V(Y)$ or on $F(Y)$.*

Proof. If $\bar{\mathscr{G}}$ does not act transitively on $V(Y)$, then Y is a bipartite graph. Hence, every face of Y is bounded by at least 4 edges, and therefore $|E(Y)| \geqslant 2|F(Y)|$. By the dual argument we have $|E(Y)| \geqslant 2|V(Y)|$. This implies $|E(Y)| \geqslant |F(Y)| + |V(Y)|$, contrary to Euler's theorem.

Lemma 7. *Given a connected graph Z on the sphere S which is invariant under the finite homeomorphism group $\mathscr{G}$ of S. Assume that $\mathscr{G}$ acts transitively on the set of vertices, edges and faces of Z, and suppose $|V(Z)| \geqslant 3$. Then Z has no multiple edges and is either a circle or isomorphic to the skeleton of a Platonic solid.*

Proof. Assume the vertices p_1 and p_2 are connected by two edges e_1 and e_2. The Jordan curve $e_1 \cup e_2$ divides S into two topological discs D_1 and D_2.

We can choose p_1, p_2, e_1 and e_2 such that D_1 is minimal, i.e. D_1 does not contain two vertices connected by multiple edges. Thus D_1 is a face of Z (if D_1 contained a vertex of Z in its interior the vertex transitivity would imply that D_1 is not minimal). Hence, from the face transitivity follows $|V(Z)| = 2$, contrary to assumption.

The connectedness of Z, the transitivity of $\mathscr{A}(Z)$ and $|V(Z)| \geqslant 3$ obviously imply that Z is 2-connected. If the vertices of Z have valency 2, then Z is a circle. If they have valency $\geqslant 3$ (and there are no multiple edges), a well known argument shows that Z is the skeleton of a Platonic solid.

PROOF OF THE THEOREM

(2) *By Lemma* 3 *we can assume that there exists no point of the sphere which is fixed by all elements of the finite homeomorphism group* $\mathscr{G}$. We remark that the existence of an invariant arc implies the

existence of a fixed point.

The proof of the theorem will be based on the following proposition:

Proposition. *If the finite homeomorphism group* $\mathscr{G}$ *of the sphere* S *has no fixed point, then there exists a connected graph* Z *on* S *invariant under* $\mathscr{G}$ *such that* $\mathscr{G}$ *acts transitively on* $V(Z)$ *as well as on* $E(Z)$ *and* $F(Z)$.

Proof.

1. If there is a closed Jordan curve J on S invariant under $\mathscr{G}$ one can easily find a point p on J such that the graph Z defined by subdividing J by the vertex set $V(Z) = p\mathscr{G}$ has the property that $\mathscr{G}$ acts transitively on $E(Z)$. ($|p\mathscr{G}| \geqslant 2$, as $\mathscr{G}$ has no fixed point.) Z has 2 faces, say D_1 and D_2.

If the action of $\mathscr{G}$ were not transitive on $F(Z)$, $\mathscr{G}$ would have a fixed point in D_1 (and also in D_2) by Lemma 2, contrary to assumption. Clearly, $\mathscr{G}$ acts transitively on $V(Z)$.

Henceforth we can therefore exclude that S *contains a Jordan curve invariant under* $\mathscr{G}$.

2. We construct a connected auxiliary graph X invariant under $\mathscr{G}$. Let a be an element of maximal order in the subgroup of all orientation preserving elements of $\mathscr{G}$ and let p be a point fixed by a. (If a is the identity we can select any p in S.) By Lemma 4 the stabilizer $\mathscr{H}$ of p in $\mathscr{G}$ is topologically equivalent to a finite group of congruences of S which fixes two diametrically opposite points. We can assume without loss of generality that $\mathscr{H}$ already is such a group of congruences. Clearly a generates the orientation preserving elements of $\mathscr{H}$.

As p is not fixed by $\mathscr{G}$ the orbit $p\mathscr{G}$ consists of at least two points. Let q be an element of $p\mathscr{G}$ different from p and let a be an arc of a great circle connecting p and q. Now we determine a maximal subarc d of a with endpoint p, such that the interior of d has no point in common with its images under $\mathscr{G}$. Let s be the endpoint of d different from p. If $s\mathscr{G}$ and the interior of d are disjoint we define

X by setting $V(X) = p\mathscr{G} \cup s\mathscr{G}$ and $E(X) = d\mathscr{G}$. Clearly X is invariant under $\mathscr{G}$ and $\mathscr{G}$ acts transitively on the edges of X.

If s is an inner point of some $d\varphi$ and $d\psi$ for elements φ, ψ in G, then $s\varphi^{-1}$ is an inner point of d and $d\psi\varphi^{-1}$. By the construction of d it is clear that d and $d\psi\varphi^{-1}$ have to coincide. This implies that every point of $s\mathscr{G}$ is an inner point of exactly one element of $d\mathscr{G}$. Let r be the point of $s\mathscr{G}$ in the interior of d, b the subarc of d between p and r, and c the subarc of d between r and s. Now we define the graph X on the vertex set $p\mathscr{G} \cup r\mathscr{G}$ with the edge set $b\mathscr{G} \cup \cup c\mathscr{G}$.

3. Assume X is disconnected. Let X_1 be the component of X containing p. By Lemma 5 all the other components of X are contained in one of the complementary domains of X_1. Let D be a topological disc in the interior of this complementary domain containing all the components of X but X_1. Define $\mathscr{K}$ by

$$\mathscr{K} = \{\varphi\colon \varphi \in \mathscr{G},\ X_1\varphi = X_1\}$$

and apply Lemma 3 by taking an arbitrary point of X_1 for q. Thus, we obtain a Jordan curve J invariant under $\mathscr{K}$ which separates X_1 from the other components of X. Let D_1 be the topological disc bounded by J and containing X_1. By Lemma 2 the group of homeomorphisms of D_1 induced by $\mathscr{K}$ has a fixed point p_1. This point p_1 is different from p, because X_1 contains at least 2 elements of $p\mathscr{G}$, i.e. $|p\mathscr{K}| \geqslant \geqslant 2$. Now q (see point 2 above) belongs to $\mathscr{K}$, fixes at least 2 points p and p_1 and preserves orientation. Hence a is the identity and the orientation preserving subgroup of $\mathscr{G}$ consists only of the identity. Thus $|\mathscr{G}| = 2$, $|p\mathscr{G}| \leqslant 2$, $p\mathscr{G} \subseteq V(X_1)$, and consequently $X_1 = X$, a contradiction. This proves the connectedness of X.

4. We continue with the construction of a connected graph Y which is invariant under $\mathscr{G}$ and where $\mathscr{G}$ acts transitively on $E(Y)$.

If $s\mathscr{G}$ and the interior of d are disjoint we set $Y = X$. If this is not the case we note that every element of $r\mathscr{G}$ is incident with exactly one arc of $b\mathscr{G}$. If we consider $c\varphi$, $\varphi \in G$, as directed from $r\varphi$ to $s\varphi$, this implies that every element of $r\mathscr{G}$ is the initial vertex of exactly one arc in $c\mathscr{G}$. By the definition of r every element of $r\mathscr{G}$ is the final vertex of at least one arc in $c\mathscr{G}$, hence of exactly one (by a counting argument).

Thus, the subgraph U of X spanned by $c\mathscr{G}$ consists of disjoint circuits $U_1, \ldots, U_n$. Every U_i divides the sphere into two topological discs. As X is finite there has to exist a U_i which contains all the other circuits in one of its sides. By the transitivity of $\mathscr{G}$ on $c\mathscr{G}$ this property extends to all the U_i. If U is connected it consists of only one circuit which is invariant under $\mathscr{G}$. Since the case where $\mathscr{G}$ leaves a closed Jordan curve invariant has already been treated, we can assume that U has more than one component. By the connectedness of X every point in $p\mathscr{G}$ has to be connected with at least two components of U, and this again implies that all the arcs of $b\mathscr{G}$ incident with a U_i are contained in one of the discs defined by U_i. Henceforth, there exists a disc D_i of U_i not containing any points of X in its interior.

Let $r_1, \ldots, r_k$ be the vertices of U_1 and let $\mathscr{H}_1$ be the stabilizer of D_1. By Lemma 2 there exists a point t in D_1 which is fixed by $\mathscr{H}_1$ and a set $e_1, \ldots, e_k$ of nonintersecting arcs connecting t with the r_i, such that $\cup e_i$ is invariant under $\mathscr{H}_1$.

We can assume without loss of generality that $r_1 = r$ and set $f = b \cup e_1$. Then the graph defined on $p\mathscr{G} \cup t\mathscr{G}$ with the edge set $f\mathscr{G}$ is the desired graph Y.

5. Let $\mathfrak{Y}$ denote the class of connected graphs Y is the sphere S invariant under $\mathscr{G}$ such that $\mathscr{G}$ acts transitively on $E(Y)$. If $\mathscr{A}(Y)$ is edge transitive a result of Watkins [15] implies that Y is 2-connected, or that there exists a vertex in Y which is connected with all the other vertices of Y. In the second case as G has no fixed point $Y \in \mathfrak{Y}$ has at least two vertices with this property. Thus, all graphs in $\mathfrak{Y}$ are 2-connected and their faces are topological discs.

We have just shown that $\mathfrak{Y}$ is nonvoid. It should be noted that the edge transitivity implies that there are at most two orbits induced by $\mathscr{G}$ in each of $V(Y)$ and $F(Y)$, where $Y \in \mathfrak{Y}$.

Let $\mathfrak{Z}$ be the class of connected graphs Z on S invariant under

$\mathscr{G}$ such that $\mathscr{G}$ acts transitively on $V(Z)$, $E(Z)$ and $F(Z)$ (compare Lemma 7). We define two operations on $\mathfrak{Y}$, by means of which an arbitrary element of $\mathfrak{Y}$ can be transformed into some element of $\mathfrak{Z}$.

δ-operation. If the graph $Y_1 \in \mathfrak{Y}$ has vertices of valency two define a graph Y_2 on the vertices of valency $\neq 2$ such that $\cup E(Y_1) = \cup E(Y_2)$. This is possible as the graph Y_1 is not a circuit. Clearly, $Y_2 \in \mathfrak{Y}$, $|E(Y_2)| < |E(Y_1)|$.

ϵ-operation. Let $Y_3 \in \mathfrak{Y}$, $D \in F(Y_3)$, $Q \subseteq D \cap V(Y_3)$ and set $\cup Q\mathscr{G} = Q'$, $Q' \cap D = \{q_1, \ldots, q_k\}$. By Lemma 2 we can consider D as a disc on which the stabilizer $\mathscr{H}$ of D acts as a group of congruences. Let p be the center of D and e_i be the radius connecting p with q_i $(i = 1, \ldots, k)$. Define the graph Y_4 such that $\cup E(Y_4) = \bigcup_1^k e_i \mathscr{G}$ and such that Y_4 does not contain vertices of valency 2. We remark that we have excluded the case where $\cup E(Y_4)$ is a closed Jordan curve. Clearly Y_4 is invariant under G and $V(Y_4) \subseteq p\mathscr{G} \cup Q'$. $\mathscr{G}$ acts transitively on $F(Y_4)$, because it does so on $E(Y_3)$.

(3) We prove that $Y_4 \in \mathfrak{Z}$ if Y_3 has no vertices of valency 2, no 2-sided faces and $Y_3 \notin \mathfrak{Z}$ and we put $Q = \{q\}$. By Lemma 6 the action of $\mathscr{G}$ on Y_3 is transitive either on $V(Y_3)$ or on $F(Y_3)$, but not on both, since $Y_3 \notin \mathfrak{Z}$. In both cases Y_4 is obviously connected.

Assume there are two orbits $\mathscr{O}_1, \mathscr{O}_2$ of the faces of Y_3. Then Y_3 is a regular graph of even degree ρ. As $2 < \rho < 6$, we have $\rho = 4$. Hence, $V(Y_4) = p\mathscr{G}$ (the points belonging to $q\mathscr{G} = Q'$ would have valency 2). This means that every edge of Y_4 contains exactly one element of $q\mathscr{G}$ in its interior, whence $\mathscr{G}$ acts transitively on $E(Y_4)$.

If there are two orbits of $V(Y_3)$, the dual consideration yields the desired result with $V(Y_4) = q\mathscr{G}$.

(4) *If D is a two-sided face of Y_3, then $Y_4 \in \mathfrak{Y}$* if we put $|Q| = 2$. For, in this case $V(Y_4) = Q'$ and the edges of Y_4 correspond to faces of type $D\varphi$ $(\varphi \in G)$, hence $\mathscr{G}$ acts transitively on $E(Y_4)$ and Y_4 is connected. Note that in this case $|E(Y_4)| = |D\mathscr{G}|$, hence either $|E(Y_4)| < |E(Y_3)|$ or $D\mathscr{G} = F(Y_3)$. In the later case $|V(Y_3)| = 2$, thus – as $\mathscr{G}$ has no fixed point – $Y_3 \in \mathfrak{Z}$.

Consider an arbitrary graph $Y \in \mathfrak{Y}$. If Y has a vertex of valency 2, apply δ. If Y has a two-sided face, call it D and apply ϵ. As both of these operations reduce the number of edges, repeated application of these operations will eventually lead to a graph $Y' \in \mathfrak{Y}$ containing neither vertices of valency two, nor two-sided faces (except the later case in (4) where $Y' \in \mathfrak{Z}$). If $Y' \in \mathfrak{Z}$, we are through; if not, apply ϵ. By (3) the resulting graph belongs to $\mathfrak{Z}$. This proves the proposition.

We finish the proof of the theorem with arguments of more technical nature.

The graph Z of the proposition consists of two vertices connected by at least two edges or is one of the graphs described in Lemma 7. The faces of Z are topological discs and $\mathcal{G}$ induces a permutation group $\overline{\mathcal{G}}$ on the set

$$M(Z) = V(Z) \cup E(Z) \cup F(Z) .$$

$\overline{\mathcal{G}}$ is isomorphic to $\mathcal{G}$ by Lemma 2.

We define a graph Z^* isomorphic to Z on a sphere S^* as follows:

If Z is a circuit of length n $(n > 2)$, let Z^* be a great circle of S^* subdivided by equidistant vertices.

If $|V(Z)| = 2$ we define Z^* by connecting two diametrically opposite points of S^* with the appropriate number of equally spaced meridians.

In the other cases Z is isomorphic to the skeleton of a Platonic solid $\mathfrak{P}$ with the insphere S^*. Project the skeleton of $\mathfrak{P}$ in S^* (from its center) to obtain Z^*.

We can easily define a mapping $\kappa \colon M(Z) \to M(Z^*)$ which preserves incidence. Clearly $\kappa^{-1} \overline{\mathcal{G}} \kappa$ is a permutation group $\overline{\mathcal{G}}^*$ of $M(Z^*)$ induced by a uniquely determined group $\mathcal{G}^*$ of congruences of S^*. In this way we obtain a chain of isomorphisms

$$\varphi \to \overline{\varphi} \to \overline{\varphi}^* = \kappa^{-1} \overline{\varphi} \kappa \to \varphi^*$$

between $\mathscr{G}, \bar{\mathscr{G}}, \bar{\mathscr{G}}^*$ and $\mathscr{G}^*$.

In two steps we shall extend $\kappa \mid V(Z)$ to a homeomorphism ϑ_2 of S onto S^* such that $\mathscr{G}^* = \vartheta_2^{-1} \mathscr{G} \vartheta_2$. To this end we introduce the notation $V(X) = \Sigma_0(X)$, $E(X) = \Sigma_1(X)$ and $F(X) = \Sigma_2(X)$ for any graph X embedded in a sphere. Set $T_i(X) = \bigcup \Sigma_i(X)$ for $i = 1, 2$.

We call a homeomorphism ϑ_i of $T_i(Z)$ onto $T_i(Z^*)$ i-compatible if the following two conditions hold:

(a) $\vartheta_i^{-1}(\varphi \mid T_i(Z))\vartheta_i = \varphi^* \mid T_i(Z^*)$ for any $\varphi \in \mathscr{G}$.

(b) For any l, $0 \leqslant l \leqslant i$, $\Delta \in \Sigma_l(Z)$ the mapping $\vartheta_i \mid \Delta$ is a homeomorphism of Δ onto $\Delta\kappa$.

As $T_2(Z) = S$, $T_2(Z^*) = S^*$, it suffices to prove the existence of a 2-compatible ϑ_2.

Let $\vartheta_0 = \kappa \mid V(Z)$. ϑ_0 is clearly 0-compatible. We prove that the existence of an i-compatible ϑ_i implies the existence of an $(i+1)$-compatible $\vartheta_{i+1} (i = 0, 1)$.

We choose a Γ in Σ_{i+1} with stabilizer $\mathscr{H}$ in $\mathscr{G}$. By Lemma 2 there exists a homeomorphism μ of Γ onto $\Gamma\kappa$ such that

$$\mu \mid \Gamma \cap T_i(Z) = \vartheta_i \mid \Gamma \cap T_i(Z)$$

and such that $\mu^{-1} \mathscr{H} \mu$ is a group of congruences of $\Gamma\kappa$. Let $\mathscr{H}^*$ denote the subgroup of $\mathscr{G}^*$ corresponding to $\mathscr{H}$. Then $\mathscr{H}^*$ consists of those elements of $\mathscr{G}^*$ which leave $\Gamma\kappa$ invariant and which satisfy the following relation for any φ in $\mathscr{H}$:

$$\mu^{-1}\varphi\mu \mid \Gamma\kappa \cap T_i(Z^*) = \mu^{-1}(\varphi \mid \Gamma \cap T_i(Z))\mu = (\vartheta_i^{-1} \mid \Gamma\kappa \cap T_i(Z^*))\varphi\vartheta_i .$$

By condition (a) the rightmost term is equal to $\varphi^* \mid \Gamma\kappa \cap T_i(Z^*)$. As $\varphi^* \mid \Gamma\kappa \cap T_i(Z^*)$ fully determines the congruence $\varphi^* \mid \Gamma\kappa$, we have $\mu^{-1}\varphi\mu = \varphi^* \mid \Gamma\kappa$, or, equivalently:

$$(\varphi^{-1} \mid \Gamma)\mu\varphi^* = \mu .$$

We define a mapping $\vartheta_{i+1} \colon T_{i+1}(Z) \to T_{i+1}(Z^*)$ by setting

$$\vartheta_{i+1} \mid \Gamma\psi = (\psi^{-1} \mid \Gamma\psi)\mu\psi^* \quad \text{for any} \quad \psi \in \mathscr{G} .$$

To show that ϑ_{i+1} is well defined assume that $\Gamma\psi_1 = \Gamma\psi_2$. Then $\psi_2 = \varphi\psi_1$ for a suitable φ in $\mathscr{H}$. Thus

$$(\psi_2^{-1} \mid \Gamma\psi_2)\mu\psi_2^* = (\psi_1^{-1} \mid \Gamma\psi_1)\mu\varphi^*\psi_1^* = (\psi_1^{-1} \mid \Gamma\psi_1)\mu\psi_1^* .$$

We also have to prove the unicity of ϑ_{i+1} on the intersections of type $\Gamma\psi_3 \cap \Gamma\psi_4$. We show that $\vartheta_{i+1} \mid T_i(Z) = \vartheta_i$. Let Δ be an arbitrary element of $\Sigma_i(Z)$ and let ψ be any element of $\mathscr{G}$ such that Δ is contained in $\Gamma\psi$. Then

$$\vartheta_{i+1} \mid \Delta = (\psi^{-1} \mid \Delta)\mu\psi^* = (\psi^{-1} \mid \Delta)(\mu \mid \Gamma \cap T_i(Z))\psi^* =$$
$$= (\psi^{-1} \mid \Delta)(\vartheta_i \mid \Gamma \cap T_i(Z))\psi^* .$$

By condition (a) the rightmost term equals ϑ_i. This completes the proof that ϑ_{i+1} is a well defined homeomorphism of $T_{i+1}(Z)$ onto $T_{i+1}(Z^*)$.

It remains to be shown that ϑ_{i+1} satisfies (a) and (b). Let $\Omega \in \Sigma_{i+1}$ with $\Omega = \Gamma\psi$. Then

$$\Omega\vartheta_{i+1} = \Gamma\psi\vartheta_{i+1} = \Gamma\psi\psi^{-1}\mu\psi^* = \Gamma\mu\psi^* = (\Gamma\kappa)\psi^* =$$
$$= \Gamma\kappa\bar{\psi}^* = \Gamma\kappa(\kappa^{-1}\bar{\psi}\kappa) = \Gamma\bar{\psi}\kappa = \Gamma\psi\kappa = \Omega\kappa .$$

To show (a) we choose φ, ψ in $\mathscr{G}$ and set, as before, $\Omega = \Gamma\psi \in \Sigma_{i+1}(Z)$. Then

$$(\vartheta_{i+1}^{-1} \mid \Omega\kappa)\varphi\vartheta_{i+1} = (\vartheta_{i+1} \mid \Gamma\psi)^{-1}\varphi(\vartheta_{i+1} \mid \Gamma\psi\varphi) =$$
$$= ((\psi^{-1} \mid \Gamma\psi)\mu\psi^*)^{-1}\varphi(\psi\varphi)^{-1}\mu(\psi\varphi)^* =$$
$$= ((\psi^*)^{-1} \mid \Gamma\mu\psi^*)\mu^{-1}\psi \cdot \varphi \cdot (\psi\varphi)^{-1}\mu(\psi\varphi)^* =$$
$$= ((\psi^*)^{-1} \mid \Gamma\mu\psi^*)\psi^*\varphi^* = \varphi^* \mid \Gamma\mu\psi^* =$$
$$= \varphi^* \mid \Gamma\kappa\bar{\psi}^* = \varphi^* \mid \Gamma\bar{\psi}\kappa = \varphi^* \mid \Omega\kappa .$$

Hence, $\vartheta_{i+1}^{-1}\varphi\vartheta_{i+1} = \varphi^* \mid T_{i+1}(Z^*)$, and ϑ_{i+1} satisfies (a). This completes the proof of the theorem.

REFERENCES

[1] L. Babai – W. Imrich, On groups of polyhedral graphs, *Discrete Mathematics,* 5 (1973), 101-103.

[2] L.E.J. Brouwer, *Amst. Akad. Versl.,* XXVII. (1919), 1201-1203.

[3] H.S.M. Coxeter – W.O.J. Moser, *Generators and Relations for Discrete Groups,* Springer-Verlag, Berlin, 1965.

[4] S. Eilenberg, Sur les transformations périodiques de la surface de sphére, *Fund. Math.,* 22 (1934) 28-41.

[5] L. Fejes Tóth, *Reguläre Figuren,* Akadémiai Kiadó, Budapest, 1965.

[6] H. Fleischner, Die Struktur der Automorphismen spezieller, endlicher, ebener, dreifach-knotenzusammenhängender Graphen, *Compositio Math.,* 23 (1971) 435-444.

[7] H. Fleischner, Fixpunkteigenschaften von Automorphismen spezieller, endlicher, ebener, dreifach-knotenzusammenhängender Graphen, *Compositio Math.,* 23 (1972) 445-452.

[8] B.v. Kerékjártó, Über die endlichen topologischen Gruppen der Kugelfläche, *Amst. Akad. Versl.,* XXVIII. (1919) 555-556.

[9] B.v. Kerékjártó, Über die periodischen Transformationen der Kreisscheibe und der Kugelfläche, *Math. Ann.,* 80 (1921) 36-38.

[10] B.v. Kerékjártó, A gömbfelület topológikus leképezéseinek kompakt csoportjairól (Sur les groupes compacts de transformations topologiques de la sphére; Hungarian, with French summary), *Mat. Term. Értesítő,* LIX (1940) 805-827.

[11] P. Mani, Automorphismen von polyedrischen Graphen, *Math. Ann.,* 192 (1971) 279-303.

[12] D. Montgomery – L. Zippin, *Topological transformation groups,* New York (Interscience), 1955.

[13] M.H.A. Newman, *Elements of the topology of plane sets of points,* Cambridge University Press, Cambridge, 1951.

[14] G. Sabidussi, Automorphism groups of planar graphs, unpublished.

[15] M.E. Watkins, Connectivity of Transitive Graphs, *J. Combinatorial Th.* 8(1970), 23-29.

[16] H. Whitney, Congruent graphs and the connectivity of graphs, *Amer. J. Math.*, 54 (1932), 150-168.

COLLOQUIA MATHEMATICA SOCIETATIS JÁNOS BOLYAI
8. TOPICS IN TOPOLOGY, KESZTHELY (HUNGARY), 1972.

INNER SET MAPPINGS ON LOCALLY COMPACT SPACES

L. BABAI — A. MÁTÉ

1. INTRODUCTION

Given an ordinal η not cofinal to ω, a *regressive function* on a subset X of η is a function such that $f(a) < a$ for any $a \in X$. Taking the view usual in axiomatic set theory about ordinals, i.e. that an ordinal is the set of its predecessors, the above requirement can be reformulated as follows: $f(a) + 1 \subseteq a$, and here a is a bounded open set and $f(a) + 1$ is a compact one if we endow η with the topology induced by its natural ordering. So, instead of regressive functions, we may study a function f that is defined for certain bounded open subsets of a locally compact space S, and is such that for any U in its domain $f(U) \subseteq U$ holds, and, moreover, $f(U)$ is compact. Our investigations below show that much can be saved from the theory of regressive functions for this more general context. Incidentally, *inner set mappings*, i.e. set mappings f with $f(X) \subseteq X$ for any X in its domain, were studied by P. Erdős, G. Fodor, and A. Hajnal in [2] without a topological set-up.

2. NOTATIONS AND TERMINOLOGY

In our topological considerations below, S *always denotes a locally compact but not compact Hausdorff space.* A set $X \subseteq S$ is said to be *bounded* if it is included in a compact set. Where κ is a cardinal, X is said to be κ*-bounded* if it is included in a union of κ compact sets. Instead of saying ω-bounded, it is customary to say σ*-bounded.* $\mathbf{C}$ will denote the set of all compact subsets of S, and $\mathbf{U}$ the set of all its bounded open subsets. We stress that *the above stipulations should be included among the assumptions of lemmas and theorems below,* as they will not be repeated there.

An *inner mapping* is a function f defined on a subset of $\mathbf{U}$ and with values in $\mathbf{C}$ such that $f(U) \subseteq U$ for any U in its domain. Using the notation

$$f^{-1}(X) = \{U \in \operatorname{dom}(f)\colon f(U) \subseteq X\},$$

where $\operatorname{dom}(f)$ stands for the domain of f, f is called *divergent* if $\bigcup f^{-1}(X)$ is bounded for any bounded set $X \subseteq S$.

While talking about ordinals, we take the view usual in axiomatic set theory, i.e. that an ordinal is the set of its predecessors, and a cardinal is identified with its initial ordinal. For an ordinal α, $\operatorname{cf}(\alpha)$ denotes its *cofinality number,* i.e. the least ordinal cofinal to it. We use the usual notations of logics and set theory, although there is one point to be stressed: $\subset$ *always stands for strict inclusion,* i.e. $x \subset z \leftrightarrow x \subseteq y \;\&\; x \neq y$. As customary, 0 denotes the empty set, and $|X|$ denotes the cardinality of the set X.

In our considerations below, we shall need the following concepts and results from the theory of regressive functions: Let η be an infinite limit ordinal not cofinal to ω, and endow η with the topology induced by its natural ordering. It is clear that η so becomes a locally compact space, and a set $X \subseteq \eta$ is unbounded in the sense specified above if and only if it is cofinal to η. Call a closed unbounded set a *band,* and call a set $X \subseteq \eta$ *stationary* (*in* η) if it meets every band. Calling a function f from a subset of η into η *regressive* if $f(\alpha) < \alpha$ for every nonzero $\alpha \in \operatorname{dom}(f)$, and *divergent* if the set $\{\xi \in \operatorname{dom}(f)\colon f(\xi) < \alpha\}$ is bounded for every $\alpha < \eta$, we have the following result due to W. Neumer (see [7, Sätze 2 and 4 on p. 257] or [1, Sätze 1 and 2 on p. 46]):

Theorem 2.1. *A set $X \subseteq \eta$ is stationary if and only if it is no domain of a divergent regressive function.*

G. Fodor sharpened the nontrivial half of this assertion as follows (see [4, Satz 1 on p. 141]):

Theorem 2.2. *If f is a regressive function on a stationary set $X \subseteq \eta$, then for some $\alpha < \eta$ the set $\{\xi \in X: f(\xi) < \alpha\}$ is stationary.*

An extension of the theory of regressive functions for the case when the underlying set (which is here η) is linearly ordered but not necessarily well-ordered was given by the second author in [6]. Our results may also be viewed from an angle as a futher extension of the theory of regressive functions for cases where the underlying set (which is then a certain set of subsets of S) is a partially ordered set with certain nice properties. I. Juhász also gave a generalization in another direction of the theory of regressive functions (oral communication).

3. RESTRICTIONS ON THE SOLID POINTS OF A DIVERGENT INNER MAPPING

The main result obtained in the present section can be formulated loosely as follows: given a "large" subset G of U, and a divergent inner mapping f on G, there are "many" such compact sets C for which there is no $U \in G$ with $f(U) \subset C \subseteq U$. (Such a compact set C may be called a *solid point* of f.) This result implies Neumer's theorem (Theorem 2.1), although this is of no practical use, as the latter can be proved much more easily than the former.

Now we give a precise formulation. In it, we have to assume more than the unboundedness of S, but somewhat less than the divergence of f. We also restrict the range of f; we shall comment on the significance of this restriction later.

Theorem 3.1. *Assume S is not σ-bounded, and let f be an inner mapping defined on a set $G \subseteq U$ such that $\bigcup G$ is not σ-bounded* *and suppose that* $\bigcup f^{-1}(X)$ *is* σ*-bounded for any bounded* $X \subseteq S$.* *Assume that the range of* f *is included in a set* $\mathbf{F} \subseteq \mathbf{C}$, *and* $\mathbf{F}$ *is such that it contains any intersection of its members. Then the set*

$$A = \bigcup \{C \in \mathbf{F}\colon C \subseteq \bigcup \mathbf{G} \quad \text{and there is no} \quad U \in \mathbf{G}$$

$$\text{with} \quad f(U) \subset C \subseteq U\}$$

is not σ*-bounded.*

This theorem becomes trivial in case $\mathbf{F} = \mathbf{C}$. In fact, in this case we have

$$f^{-1}(0) \cup A \supseteq f^{-1}(0) \cup \bigcup \{X \subseteq \bigcup \mathbf{G}\colon |X| = 1$$

$$\text{and there is no} \quad U \in \mathbf{G}$$

$$\text{with} \quad X \subseteq U \quad \text{and} \quad f(U) = 0\} = \bigcup \mathbf{G}\,,$$

so at least one of the sets on the left-hand side must not be σ-bounded. In general, however, the theorem does not seem to be quite trivial because, as was mentioned above, Neumer's theorem can easily be derived from it.

Proof of the theorem. Assume that the conclusion of the theorem does not hold, i.e. that the set A is σ-bounded. Choose a σ-bounded open set V such that $A \subseteq V$; using the local compactness of S, the existence of such a V can easily be shown.

Now choose an arbitrary element U_0 of $\mathbf{G}$, and write $C_1 = f(U_0)$. In case $C_1 \not\subseteq V$ we have $C_1 \not\subseteq A$ a fortiori. As we also have $C_1 \in \mathbf{F}$ and $C_1 \subseteq \bigcup \mathbf{G}$, the definition of A implies that in this case there exists an open set $U_1 \in \mathbf{G}$ such that $f(U_1) \subset C_1 \subseteq U_1$.

Using a transfinite recursion argument, for an ordinal ξ of form $\alpha + 1$ write

$$C_\xi = f(U_\alpha)\,,$$

* Such an f might be called σ-divergent, although we shall not use this term below.

and for a limit ordinal ξ put

$$C_\xi = \bigcap \{C_\alpha :\ \alpha < \xi\}.$$

By the assumptions of the theorem we clearly have $C_\xi \in F$. In case $C_\xi \not\subseteq V$ we can choose a $U_\xi \in G$ such that

$$f(U_\xi) \subset C_\xi \subseteq U_\xi .$$

Since we have

$$(3.1) \qquad C_1 \supset C_2 \supset \ldots \supset C_\xi \supset \ldots ,$$

we cannot go on indefinitely choosing new C_ξ's; that is, for some ordinal ϑ the set $C_\vartheta - V$ will be empty. (It can easily be shown by using the finite intersection property of compact sets that this ϑ is a successor ordinal, but we shall not use this fact below.)

Now it can readily be seen that the open sets V and $U_\xi - C_{\xi+1} = U_\xi - f(U_\xi)$ $(\xi < \vartheta)$ cover U_0. Indeed, if $x \in U_0 - V$, let $\gamma \leqslant \vartheta$ be the least ordinal such that $x \notin C_\gamma$. There is such a γ, as $C_\vartheta - V$ is empty; and in view of the definition of C_ξ for a limit ordinal ξ, we must have $\gamma = \alpha + 1$ with some α. Hence we have $x \in C_\alpha - C_{\alpha+1} \subseteq U_\alpha - C_{\alpha+1}$, which confirms our assertion. So

$$C_1 = f(U_0) \subseteq U_0 \subseteq V \cup \bigcup_{\xi < \vartheta} (U_\xi - f(U_\xi)) .$$

On the right-hand side we have an open cover of the compact set on the left-hand side. So there is a finite sequence $\{\xi_k :\ 1 \leqslant k \leqslant n\}$ of ordinals $\xi_k < \vartheta$ such that

$$(3.2) \qquad C_1 = f(U_0) \subseteq V \cup \bigcup_{k=1}^{n} (U_{\xi_k} - f(U_{\xi_k})) .$$

holds. Here we may assume that all ξ_k's are different from 0, as $U_0 - f(U_0)$ clearly need not occur on the right-hand side. We may also assume that the ξ_k's are indexed in an increasing order; then, writing $\xi_0 = 0$, from (3.1) we can infer that

$$f(U_{\xi_0}) \supset f(U_{\xi_1}) \supset \ldots \supset f(U_{\xi_n}) .$$

This in conjunction with (3.2) implies that

$$f(U_{\xi_k}) \subseteq V_k \quad (0 \leq k \leq n), \tag{3.3}$$

where

$$V_k = V \cup \bigcup_{j=k+1}^{n} U_{\xi_j}.$$

Put $X_0 = V$, and for any positive integer k set

$$X_k = V \cup \bigcup \{U \in G: f(U) \subseteq X_{k-1}\}.$$

Then, using (3.3), we may show by "downward" induction on k that

$$V_k \subseteq X_{n-k} \quad (0 \leq k \leq n). \tag{3.4}$$

Indeed, in case $k = n$ we have $V_k = V = X_0$, and so there is nothing to prove. Assuming that for some $k > 0$ we have already verified (3.4), by (3.3) we obtain that $f(U_{\xi_k}) \subseteq X_{n-k}$, i.e.

$$U_{\xi_k} \subseteq X_{n-k+1}.$$

Since it can easily be shown by induction that $X_0 \subseteq X_1 \subseteq \ldots$, we can infer from here that

$$V_{k-1} = U_{\xi_k} \cup V_k \subseteq X_{n-k+1} \cup X_{n-k} = X_{n-k+1},$$

which proves (3.4). In case $k = 0$,

$$f(U_0) = f(U_{\xi_0}) \subseteq V_0 \subseteq X_n,$$

that is

$$U_0 \subseteq X_{n+1}.$$

Since U_0 was chosen as an arbitrary element of G, this inclusion implies that

$$\bigcup \{X_k: k < \omega\} \supseteq \bigcup G,$$

where the set on the right-hand side was assumed not to be σ-bounded. So, there must exist a least integer k such that X_k is not σ-bounded.

Then $k \geqslant 1$, as we assumed at the beginning of this proof that $X_0 = V$ is σ-bounded. By the minimality of k, X_{k-1} is σ-bounded. So, there is a sequence $\langle W_n : n < \omega \rangle$ of bounded open sets such that $X_{k-1} \subseteq \subseteq \bigcup \{ W_n : n < \omega \}$. Now the set

$$Y = \bigcup \{ U \in G : f(U) \subseteq \bigcup_{n < \omega} W_n \}$$

is not σ-bounded, because it includes $X_k - V$.

For any positive integer m put

$$Y_m = \bigcup \{ U \in G : f(U) \subseteq \bigcup_{n < m} W_n \} .$$

By references to the compactness of $f(U)$ and to the openness of the W_n's, it is easy to see that $Y = \bigcup \{ Y_m : m < \omega \}$. Therefore, there is at least one integer m for which Y_m is not σ-bounded. This is a contradiction, as $\bigcup \{ W_n : n < m \}$ is bounded, and so $Y_m = \bigcup f^{-1}(\{ W_n : n < < m \})$ should be σ-bounded by our assumptions. This completes the proof.

One might wonder if it is really necessary to require in the definition of A in the theorem just proved that $f(U)$ be strictly included in C. This is indeed so; to see this, consider the following example:

Let S be a discrete topological space of an uncountable cardinality. Then S is locally compact and not σ-bounded, and, furthermore, C and U coincide: both consist of all finite subsets of S. Put $F = C$ and $G = U$, and for any $U \in U$ set $f(U) = U$. Then f is clearly a divergent inner mapping, and the set

(3.5) $A' = \bigcup \{ C \in C :$ there is no $U \in U$ with $f(U) \subseteq C \subseteq U \}$

is obviously empty, because in its definition we may take $U = C$.*

Now we show how to derive the nontrivial part of Neumer's theorem from Theorem 3.1. That is, we are going to prove the "only if" part

*We can give an example with similar features but with the additional stipulation that always $f(U) \subset U$ holds. It would be interesting to know whether A' may still be σ-bounded if S is connected.

of Theorem 2.1. (The "if" part is rather obvious and has no direct connection with Theorem 3.1.)

So, given a limit ordinal η not cofinal to ω, assume that in contradiction with what we are about to prove, there is a stationary set $X \subseteq \subseteq \eta$ and a divergent regressive function g defined on X. Choose S as η with the topology induced by its natural ordering. Then S is not σ-bounded, as η is not cofinal to ω. Take $G = X$ (note that, in view of the interpretation of ordinals, we do not only have $X \subseteq S$ but also $X \subseteq \subseteq U$). X being stationary, it is cofinal to η. That is $\bigcup X = \eta$, in other words, $\bigcup G = S$; so $\bigcup G$ is not σ-bounded. Setting

$$f(a) = g(a) + 1 \qquad (a \in X),$$

and making the harmless assumption that $0 \notin X$, it is clear that f is a divergent inner mapping of G into $F = \{a + 1\colon a < \eta\}$, which is a subset of C containing any intersection of its elements.

So all the assumptions of Theorem 3.1 are satisfied. Hence we see that the set A of this theorem is not bounded in the present case, i.e. it is cofinal to η. Now it is clear that

$$A = \{a + 1 < \eta\colon \forall \xi [\xi \in X \,\&\, a < \xi \rightarrow a \leqslant g(\xi)]\}.$$

So the set

$$B = \{a < \eta\colon \forall \xi [\xi \in X \,\&\, a < \xi \rightarrow a \leqslant g(\xi)]\}$$

is also cofinal to η. B is clearly also closed, so it is a band. The set B' of all accumulation points of B is also a band, as this is obviously true for every band.

Now B' is disjoint from X. Indeed, suppose $\xi \in X$. Then the definition of B implies that there is no element different from ξ of B belonging to the set $\{a\colon g(\xi) < a \leqslant \xi\}$. This latter being an open neighbourhood of ξ, we have obtained that ξ is not an accumulation point of B. So $\xi \notin B'$, which confirms that $B' \cap X = 0$.

So B' is a band that does not meet X. This contradicts the assumption that X is stationary, completing the proof of the "only if" part of Theorem 2.1. (We stress repeatedly that it is impractical to derive Neumer's theorem in the above way; nevertheless the fact that this theorem can be set in the more general context of Theorem 3.1 may be of some interest.)

4. COUNTABLY COMPACT SPACES

A topological space is called *countably compact* if every countable open cover has a finite subcover. For a Hausdorff space, this is equivalent to saying that every infinite set has an accumulation point (cf. J.L. Kelley [5, p. 162]). For countably compact S (satisfying, of course, also the additional requirements set down in Section 2), we can prove that there is no divergent inner mapping on U. (The example of an uncountable discrete space with the identical mapping of U onto itself, as described after the proof of Theorem 3.1, shows that this conclusion is not in general true without the assumption of countable compactness.) Actually, we have a stronger result:

Theorem 4.1. *Assume that* S *is countably compact. Let* $G \subseteq U$ *be such that it contains any intersection of a finite number of its elements as well as any union of its elements that is bounded. Assume that* $\bigcup G$ *includes a closed unbounded set* D. *Then, for any inner mapping* f *on* G, *there is a bounded set* $Z \subseteq S$ *such that* $\bigcup f^{-1}(Z)$ *is not* σ*-bounded.*

Proof. As D is not compact, there is an infinite subset of D that has no complete accumulation point. Assume that X is such a set of the least possible cardinality, say κ. Then $\kappa > \omega$, as in view of the countable compactness of S, any countably infinite subset of S has a (complete) accumulation point.

Clearly, κ is also regular. In fact, assuming $\mathrm{cf}(\kappa) < \kappa$, let $\langle \kappa_a : a < \mathrm{cf}(\kappa) \rangle$ be an increasing sequence cofinal to κ of infinite cardinals $< \kappa$, and take sets X_a such that $|X| = \kappa_a$ and $X = \bigcup \{X_a : a < \mathrm{cf}(\kappa)\}$. By the assumption on the cardinality of X_a, there is a complete accumulation point of X_a, say z_a. As $\{z_a : a < \mathrm{cf}(\kappa)\} \subseteq D$ (D is closed!) also has cardinality $< \kappa$, it has a complete accumulation point, say z. Now, obviously, z is a complete accumulation point of X, which is a contradiction, showing that κ is indeed regular.

Write

$$X = \{x_a : a < \kappa\},$$

and, for any limit number $\lambda < \kappa$, let

$$X_\lambda = \{x_{a_\xi} : \xi < \mathrm{cf}(\lambda)\},$$

where $\langle a_\xi : \xi < \mathrm{cf}(\lambda)\rangle$ is an increasing sequence of ordinals $< \lambda$ tending to λ. Let y_λ be a complete accumulation point of X_λ (there is one, as $X_\lambda \subseteq D$ and $|X_\lambda| < \kappa$). Clearly, $y_\lambda \in D$ since $X \subseteq D$ and D is closed. Moreover, for any neighbourhood U of y_λ,

$$U \cap \{x_a : \gamma \leqslant a < \lambda\} \neq 0, \tag{4.1}$$

provided $\gamma < \lambda$.

Now let f be an inner mapping on G. By transfinite recursion, we construct sets $U_a \in G$, $a < \kappa$, as follows: U_0 is arbitrary, $x_a \in \in U_{a+1}$, and for a limit number $\lambda < \kappa$, $U_\lambda = \bigcup\{U'_\lambda \cap U_a : a < \lambda\}$, where $U'_\lambda \in G$ is a neighbourhood of y_λ; then, in view of the assumption on the closedness of G with respect to certain set-theoretical operations, we have $U_\lambda \in G$. We note that this construction can be carried out, since x_a, $y_\lambda \in D \subseteq \bigcup G$.

From the above construction it is clear that

$$f(U_\lambda) \subseteq U_\lambda \subseteq \bigcup\{U_a : a < \lambda\}$$

holds for any limit ordinal $\lambda < \kappa$. Since on the right-hand side we have an open cover of the compact set on the left-hand side, we can select a finite subcover. That is, we have

$$f(U_\lambda) \subseteq \bigcup\{U_{\varphi_i(\lambda)} : 0 \leqslant i \leqslant k_\lambda\},$$

where k_λ is an integer, and $\varphi_0(\lambda), \ldots, \varphi_k(\lambda)$ are ordinals $< \lambda$. As the limit ordinals $< \kappa$ form a set stationary in κ, by a repeated application of Fodor's theorem (Theorem 2.2) we obtain that there are an integer k and ordinals $a_0, \ldots, a_k < \kappa$ such that the set

$$M = \{\lambda < \kappa\colon \lambda \quad \text{is a limit ordinal,} \quad k_\lambda = k\,,$$
$$\text{and} \quad \varphi_0(\lambda) = a_0, \ldots, \varphi_k(\lambda) = a_k\}$$

is stationary in κ.

Now write $V = \cup\{U_{a_i}\colon 0 \leqslant i \leqslant k\}$. Then V is bounded, and $f(U_\lambda) \subseteq V$ for any $\lambda \in M$. So

$$\cup\{U_\lambda\colon \lambda \in M\} \subseteq \cup f^{-1}(V)\,. \tag{4.2}$$

Choosing any two ordinals γ and λ such that $\gamma < \lambda < \kappa$ and λ is a limit number, we have, by the definition of the U_a's, that

$$U_\lambda \cap \{x_a\colon \gamma \leqslant a < \lambda\} = U'_\lambda \cap \{x_a\colon \gamma \leqslant a < \lambda\} \neq 0\,.$$

Here this last inequality holds in view of (4.1), as U'_λ is a neighbourhood of y_λ. Now, M, being stationary, is cofinal to κ. So, making use of the regularity of κ, from the nonemptiness of the set on the left-hand side we can infer that

$$X' = \cup\{U_\lambda \cap X\colon \lambda \in M\}$$

has cardinality κ. A subset of cardinality κ of X has no complete accumulation point (since X itself has none), so it cannot be bounded. So any bounded subset of X must have cardinality $< \kappa$. As κ is regular, this shows that $X' \subseteq X$, having cardinality κ, cannot be represented as the union of less than κ bounded sets. So, κ being greater than ω, X' is not σ-bounded, while $X' \subseteq \cup f^{-1}(V)$ (see (4.2)), where V is bounded. This completes the proof.

5. TWO-ELEMENT SOLID POINTS OF A DIVERGENT INNER MAPPING

We mentioned earlier that Theorem 3.1 becomes trivial if we put $F = C$ there. The reason is that the assumptions then directly imply that f has many one-element solid points. (For the definition of solid point, see the beginning of Section 3.) It is much more difficult to show the existence of a two-element solid point. A three-element solid point does not necessarily exist.

The following theorem confirms that, provided S is not σ-bounded, every divergent inner function f defined on U has a two-element solid point. (As can easily be seen, this is not necessarily true if S is σ-bounded.*) A similar assertion can also be established when f is defined on a "large enough" subset of U, but we do not formulate this assertion precisely.

Theorem 5.1. *Let f be a divergent inner mapping defined on U. Then the set*

$$H = S - \bigcup \{ Z \subseteq S\colon |Z| = 2 \quad \text{and there is no} \quad U \in U \quad \text{with} \quad f(U) \subset Z \subseteq U \}$$

is σ-bounded.

Proof. Assuming that S is countably compact, there nothing to prove, as in view of Theorem 4.1 there is no divergent inner mapping on U, and so the assumptions of the present theorem cannot be satisfied.

Now suppose that S is not countably compact, and let $X \subseteq S$ be a countably infinite set having no accumulation point. Then, for any $x, y \in S$ such that $y \in H$, there is an $U \in U$ with $\{x, y\} \subseteq U$ such that either $f(U) \subseteq \{y\}$ or $f(U) \subseteq \{x\}$, i.e. either $x \in \bigcup f^{-1}(\{y\})$ or $y \in \bigcup f^{-1}(\{x\})$. Keeping y fixed, we cannot have

$$X \subseteq \bigcup f^{-1}(\{y\}),$$

as the right-hand side is bounded in view of the divergency of f, and X is not so, as it has no accumulation point. So, for some $x \in X$ we must have the second alternative, i.e. that $y \in \bigcup f^{-1}(\{x\})$. Letting y run over H, we obtain

$$H \subseteq \bigcup \{ \bigcup f^{-1}(\{x\})\colon x \in X \}.$$

By the divergency of f, $\bigcup f^{-1}(\{x\})$ is bounded here; so, X being countable, we see that H is σ-bounded. This completes the proof.

*Choose S as ω with the discrete topology, and for a $U \subseteq S$ with $1 < |U| < \omega$ put $f(U) = U - \{a\}$, where a denotes the least element of U. If $|U| \leqslant 1$, put $f(U) = U$. Then f is a divergent inner mapping on U and has no solid points with more than one element.

Remarks. 1/ Suppose S is not σ-bounded. If instead of the divergency of f we only assume in the above theorem that $\bigcup f^{-1}(X)$ is σ-bounded for any bounded X, then we cannot conclude the existence of even one two-element solid point of f, i.e. we may have $H = S$. (This contrasts with Theorem 3.1, which, even under these circumstances, ensures the existence of some solid points.)

2/ The assumption that S is not σ-bounded does not imply the existence of a solid point of f having at least three elements.

Ad 1/ Let S be equal to ω_1 with the discrete topology on it. Then $\boldsymbol{U}$ is the set of all finite subsets of S. For any nonempty $U \in \boldsymbol{U}$ put $f(U) = \{a\}$, where a denotes the largest element of U. Then our modified assumptions are satisfied, and yet f has no two-element solid point.

Ad 2/ Let S again be equal to ω_1 with the discrete topology on it. For $a < \omega_1$ let h_a be a one-to-one mapping of a into ω. For any $U \subseteq S$ of cardinality $\leqslant 2$ put $f(U) = U$. For any finite $U \subseteq S$ with more than two elements put $f(U) = \{a, \beta\}$, where a is the maximal element of U, and β is such that $h_a(\beta) > h_a(\gamma)$ for any $\gamma \in U - \{a, \beta\}$. It is easy two see that f is a divergent inner mapping on $\boldsymbol{U}$. Any solid point of f can have at most two elements, as $|f(U)| \leqslant 2$ for any $U \in \boldsymbol{U}$. (This construction uses an idea of Erdős–Hajnal [3] – cf. the proof of their Theorem 2 on p. 116-117.)

If in the above theorem we assume that S is not ω_1-bounded, then it seems likely that we may expect the existence of a solid point of f having three elements.

6. A GENERALIZATION OF FODOR'S THEOREM

In addition to the stipulations imposed on S in Section 2, *throughout this and the next section we suppose that, for some cardinal* $\kappa > \omega$,

(6.1)
$$S \text{ is } \kappa\text{-bounded, and, for every cardinal } \mu < \kappa, \text{ any } \mu\text{-bounded subset of } S \text{ is bounded.}$$

As we assumed in Section 2 that S is not bounded, it is easy to see that κ must be regular. Note also that (6.1) implies that S is countably compact, as any countable set, being bounded, must have an accumulation point. If S is equal to a limit ordinal η not cofinal to ω with the topology derived from its natural ordering, then it is easy to see that S satisfies (6.1) with $\kappa = \mathrm{cf}(\eta)$.

The aim of the present section is to prove what may be considered the generalization of Fodor's theorem (Theorem 2.2) for S satisfying (6.1). Neumer's theorem will be discussed in the next section, although we are unable to give a satisfactory analogue.

Next we define what we call stagnant sets here. We chose this word to describe the analogue of stationary sets; we avoided using the term stationary instead of stagnant, as apart from the possibility of confusion, it might have suggested that the analogy went farther than it actually did.

Definition 6.1. Call a set $G \subseteq U$ stagnant if there is no divergent inner mapping defined on G.

It is clear by Theorem 4.1 that there are stagnant sets – cf. also Theorem 7.1 below. The next two lemmas are rather obvious. Note that (6.1) must be added to their assumptions.

Lemma 6.2. *Suppose that* $G \subseteq U$ *is nonstagnant, and* $G' \subseteq G$. *Then* G' *is also nonstagnant.*

Proof. If f is a divergent inner mapping on G, then its restriction to G' is a divergent inner mapping on G'.

Lemma 6.3. *Suppose that* $G_a \subseteq U$, $a < \xi$, *are nonstagnant sets, where* $\xi < \kappa$. *Then* $\bigcup \{G_a : a < \xi\}$ *is also nonstagnant.*

Proof. By the preceding lemma, we may assume that the G_a's are mutually disjoint. Let f_a be a divergent inner mapping on G_a. Then it is easy to see by making use of (6.1) that $\bigcup \{f_a : a < \xi\}$ is a divergent inner mapping on $\bigcup \{G_a : a < \xi\}$. (Here the view usual in set theory is taken that a function is a set of ordered pairs.)

We are now in position to establish the generalization of Theorem 2.2:

Theorem 6.4. *Assume that* $G \subseteq U$ *is a stagnant set and* f *is an inner mapping on* G. *Then* $f^{-1}(X)$ *is stagnant for some bounded set* $X \subseteq S$.

Proof. Take a sequence $\langle U_a : a < \kappa \rangle$ of bounded open sets such that $\bigcup \{ U_a : a < \kappa \} = S$. This is possible, for S is locally compact and under (6.1) we assumed that it is κ-bounded. Let

$$V_a = \bigcup \{ U_\beta : \beta < a \} \qquad (a < \kappa) .$$

In view of (6.1), V_a is bounded; it is clearly open. For $a < \kappa$ take

(6.2) $$G_a = \{ U \in G : f(U) \subseteq V_a \;\&\; \neg \exists \beta < a \;\; f(U) \subseteq V_\beta \} .$$

It is clear that the G_a's are mutually disjoint.

We are going to show that

(6.3) $$G = \bigcup \{ G_a : a < \kappa \} .$$

To see this, it is enough to prove that, for any $U \in G$, $f(U) \subseteq V_\gamma$ holds for some $\gamma < \kappa$. This is true for any compact set C in place of $f(U)$. In fact, we have

$$C \subseteq S = \bigcup \{ V_a : a < \kappa \} .$$

So, C, being compact, can be covered by a finite number of the V_a's. As the V_a's are linearly ordered by inclusion, we must therefore have $C \subseteq V_\gamma$ for some $\gamma < \kappa$. Thus (6.3) follows.

Now, our theorem will follow if we show that, for some $a < \kappa$, the set G_a is stagnant. Assume the contrary, i.e. that G_a is nonstagnant for every $a < \kappa$, and let f_a be a divergent inner mapping on G_a. For $U \in G_a$ put

(6.4) $$g(U) = f_a(U) \cup f(U) .$$

Then, clearly, g is an inner mapping defined on G (cf. (6.3)).

g is also divergent. To see this, choose a bounded set X. We have to show that $\bigcup g^{-1}(X)$ is bounded. Let C be a compact set such that $X \subseteq C$, and select a $\gamma < \kappa$ such that $C \subseteq V_\gamma$. The existence of such a

γ was shown while we were verifying (6.3). Now, making use of (6.2) and the definition of g under (6.4), we obtain that

$$\bigcup g^{-1}(X) \subseteq \bigcup g^{-1}(V_\gamma) = \bigcup \{ U \in G : g(U) \subseteq V_\gamma \} =$$
$$= \bigcup_{a \leqslant \gamma} \bigcup \{ U \in G_a : f_a(U) \subseteq V_\gamma \} .$$

In view of the divergency of f_a, each set on the right-hand side is bounded. So is also their union, in view of (6.1) above. Therefore, g is indeed divergent.

We have obtained that g is a divergent inner mapping on G. This contradicts the assumption that G is stagnant, proving the theorem.

We conclude this section with the following

Problem. Can every stagnant set be split into κ (or even two) mutually disjoint stagnant sets.

For stationary sets in an ordinal η with $\kappa = \mathrm{cf}(\eta) > \omega$, the answer is known to be yes (every stationary set can be split into κ mutually disjoint stationary sets), as has been established by R. M. Solovay [8]. (For a simplified proof one may also consult [6].)

7. THE ANALOGUE OF NEUMER'S THEOREM

As mentioned in the preceding section, we cannot give a statisfactory analogue of Theorem 2.1. We can only prove the following theorem, whatever it is worth:

Theorem 7.1. *Assume that* S *satisfies* (6.1) *with a cardinal* $\kappa > \omega$, *and let* $G \subseteq U$ *be a nonstagnant set. Then there is a set* $B \subseteq U - G$ *such that* $S = \bigcup B$, *and the following holds:*

Whenever $H \subseteq B$ *is such that*

(7.1) $\bigcup H$ *is bounded, and*

(7.2) *for any finite set* $H' \subseteq H$, *there is a* $U \in B$ *such that* $\bigcup H' \subseteq U \subseteq \bigcup H$, *then* $\bigcup H \in \mathbf{B}$.

Proof. As G is not stagnant, there is a divergent inner mapping f on G. Define g on a subset of $\mathbf{U}$ as follows: for $U \in \mathbf{U}$ put

$$g(U) = f(V)$$

with some $V \in G$ such that $f(V) \subseteq U \subseteq V$. If there is no such V, then do not define $g(U)$, and put

$$\mathbf{B} = \mathbf{U} - \mathrm{dom}\,(g)\,.$$

It is clear that g is a divergent inner mapping on $\mathrm{dom}\,(g)$. As, obviously, $G \subseteq \mathrm{dom}\,(g)$, we have

$$\mathbf{B} \subseteq \mathbf{U} - G\,.$$

Next we show that $S = \bigcup \mathbf{B}$. Suppose not, and let $x \in S - \bigcup \mathbf{B}$. Put

$$G' = \{U \in \mathbf{U} \colon x \in U\}\,.$$

Then clearly $G' \subseteq \mathrm{dom}\,(g)$, and the restriction of g to G' is a divergent inner mapping on G'. This contradicts Theorem 4.1, for it is clear that $\bigcup G' = S$ and that G' contains any intersection of a finite number of its elements as well as any bounded union of its elements.

Now we establish the second assertion. To this end, assume that $H \subseteq \mathbf{B}$ satisfies (7.1) and (7.2), and yet $\bigcup H \notin \mathbf{B}$. Then, in view of the definition of $\mathbf{B}$, there is a $V \in G$ such that $f(V) \subseteq \bigcup H \subseteq V$. As $f(V)$ is compact and the elements of H are open, there is a finite subset H' of H such that $f(V) \subseteq \bigcup H'$. On account of (7.2), take a $U \in \mathbf{B}$ such that $\bigcup H' \subseteq U \subseteq \bigcup H$. Then

$$f(V) \subseteq U \subseteq V\,,$$

which contradicts $U \in \mathbf{B}$. This contradiction completes the proof.

The above theorem describes a necessary condition for a set $G \subseteq \subseteq \mathbf{U}$ to be nonstagnant. We are going to show that this condition is not sufficient, and, probably, it is far from being sufficient.

To see this, take S to be an ordinal η with $\mathrm{cf}(\eta) > \omega$ endowed with the topology derived from its natural ordering, and let B be a band in η (the elements of B can be considered as bounded open subsets of S). Set

$$G = \{ a - \{0\} : 2 \leqslant a < \eta \} .$$

Then G is stagnant (e.g. by Theorem 4.1), $G \cap B = 0$, and B satisfies the assumptions imposed on it in Theorem 7.1. This shows that the conditions in Theorem 7.1 are indeed not sufficient for G to be nonstagnant.

It may be difficult give a necessary and sufficient condition of Neumer's type for G to be nonstagnant. It may be easier to give a condition of this type that is necessary and sufficient for $G \subseteq U$ to carry a divergent inner mapping f such that $f(U) \subseteq f(V)$ whenever $U, V \in G$ and $U \subseteq V$. (It is easy to see that this additional requirement does not make any difference in Theorem 2.1.)

REFERENCES

[1] H. Bachmann, *Transfinite Zahlen,* 2nd ed., Springer Verlag, Berlin – Heidelberg – New York, 1967.

[2] P. Erdős — G. Fodor — A. Hajnal, On the structure of inner set mappings, *Acta Sci. Math.,* (Szeged) 20 (1959), 81-90.

[3] P. Erdős – A. Hajnal, On the structure of set-mappings, *Acta Math. Acad. Sci. Hungar.,* 9 (1958), 111-131.

[4] G. Fodor, Eine Bemerkung zur Theorie der regressiven Funktionen, *Acta Sci. Math.,* (Szeged) 17 (1956), 139-142.

[5] J.L. Kelley, *General Topology,* Van Nostrand, Princeton – New Jersey – Toronto – New York – London, 1955.

[6] A. Máté, Regressive and divergent functions on ordered and well-ordered sets, *Acta Sci. Math.,* (Szeged) 33 (1972), 285-293.

[7] W. Neumer, Verallgemeinung eines Satzes von Alexandroff und Urysohn, *Math. Z.*, 54 (1951), 254-261.

[8] R.M. Solovay, Real-valued measurable cardinals, *Proceedings of symposia in pure mathematics* Vol. XIII. Part 1 (Proc. 1967 U.C.L.A. Summer Institute), Amer. Math. Soc., Providence, Rhode Island, 1971; 397-428.

COLLOQUIA MATHEMATICA SOCIETATIS JÁNOS BOLYAI
8. TOPICS IN TOPOLOGY, KESZTHELY (HUNGARY), 1972.

HOMOTOPY AND STABILIZATION

F.W. BAUER

INTRODUCTION

Let X, Y be two based topological spaces and $[X, Y]$ the set of all homotopy classes of basepoint preserving maps, then we have the infinite sequence of sets and maps [1]:

$$(1) \qquad [X, Y] \xrightarrow{\Sigma_*} [\Sigma X, \Sigma Y] \to [\Sigma^2 X, \Sigma^2 Y] \to \ldots$$

where Σ is the celebrated suspension functor. It is well known that $[\Sigma^k X, \Sigma^k Y]$ carries a group structure for $k \geqslant 1$ and an abelian group structure for $k \geqslant 2$. Furthermore, if we assume X to be a finite CW-complex, the sequence (1) terminates in the following sense:

There exists an index n such that

$$[\Sigma^k X, \Sigma^k Y] \xrightarrow{\Sigma_*} [\Sigma^{k+1} X, \Sigma^{k+1} Y]$$

is an isomorphism for $k \geqslant n$. This is the content of the famous "stability theorem" [1] of homotopy theory. This theorem gives an idea what we mean by the statement that the suspension functor is "almost invertible".

There exist numerous phenomena in homotopy theory which behave "as if" Σ would be an invertible functor. We call them "stable" without making any attempt at the present moment to give a precise meaning to this concept. This is postponed until §1.

On the other hand, the suspension functor is in fact far from being invertible. In order to give an idea how far we are from the existence of a "desuspension functor" let us make the following consideration:

The suspension of an n-dimensional cell is an $n+1$-dimensional cell. A hypothetical desuspension would consequently decrease dimension by one. Since all geometrical cell complexes consist of cells of non-negative dimensions, we are very drastically confronted with the impossibility of a general desuspension.

We can persuit this idea even further: Let $\underline{K}$ be a "stable" category (which means a category where some kind of topology and homotopy theory is practicable but, in addition, with an invertible suspension functor), then a "stable cell complex" $K \in \underline{K}$ must necessarily contain cells of arbitrary (postiive and negative) dimensions. Thus, in order to feel at home in such a stable category we have to develop some kind of intuition for a cell of dimension say -15.

The search for a suitable stable category has occupied the attention of topologists for the last two decades. One expected to get a deeper understanding of all kinds of stable functors (like homology, cohomology, K-theory, cobordism theory). At present, after many fruitless attempts, there are two models for a stable category available, which fulfil all claims on such a category: 1) J.M. Boardman's category [6] and 2) The category of simplicial spectra [9]. Since the two categories produce equivalent homotopy categories, it becomes a matter of taste which category one is inclined to use.

The (historically) first stable category, the S-category of E. Spanier and J.H.C. Whitehead [1], [9] was very convenient to handle; for many purposes it was in fact sufficient (e.g. for all problems concerning only finite CW-structures). The severe disadvantage of the S-

category was the absence of classifying objects for many important functors. E.g. D. Puppe was able to prove that there are not enough Eilenberg – MacLane objects available in this category.

On the following pages we will indicate a general process of stabilization for an arbitrary operator category (definition 1.1). This stabilization (whose existence is guaranteed by theorem 3.1) allows us to stabilize all phenomena which can be expressed in functorial form; it is a 2-functor between two categories of categories. There is no need for any additional convention whenever we formulate the stabilized version of a statement.

In §4 and §5 we give two examples: the stabilization of a homotopy concept and of a cell structure for the objects (in the original category).

In §6 we compare our concept with other known stable categories and try to fit them into our theory (S-category, Boardman-category).

The objective of this paper is exclusively the investigation of the theoretical (i.e. categorical) aspects of stabilization. We do not intend to compute anything explicitely; this has been accomplished in great detail at different places in the literature.

1. STABLE CATEGORIES

The main purpose of this kind of stable homotopy theory is to find a method for a categorical stabilization of a given phenomenon, simply by applying a stabilization functor.

To this end we need some definitions:

1.1. Definition. Let $\underline{A}$ be a commutative semi-group with unit 1 and $\underline{K}$ be any category. The pair $(\underline{A}, \underline{K})$ is called an operator category if each $A \in \underline{A}$ operates as a functor $A: \underline{K} \to \underline{K}$ such that

$$(A_1 A_2)(\quad) = A_1(A_2(\quad))$$

$$1(\quad) = \text{identity}.$$

Examples. 1) The category $\underline{K} = \mathrm{Top}_0$, with the semi-group $\underline{A} = Z^+ = \{1, \Sigma, \Sigma^2 \ldots\}$.

2) The category Ens^G of G-graded sets (G = an arbitrary abelian group) with operation

$$\bar{g}\{M_g\} = \{M_{\bar{g}g}\}, \quad g, \bar{g} \in G .$$

3) Let $\underline{A}$ be any semi-group, $G = \bar{\underline{A}}$ the group theoretical completion and $\varphi: \underline{A} \to G$ the canonical homomorphism. The group G can be considered as a (discrete) category such that $(\underline{A}, G)$ becomes an operator category by setting

$$A(g) = \varphi(A) \cdot g .$$

1.2. Definition. An operator category $(\underline{A}, \underline{K})$ is *stable* if $\underline{A}$ is a group.

1.3. Definition. Let $(\underline{A}, \underline{K})$, $(\underline{A}', \underline{K}')$ be two operator categories, then an operator functor $T = (\Gamma, T, \omega)$ is a triple

$$(\Gamma, T, \omega): (\underline{A}, \underline{K}) \to (\underline{A}', \underline{K}')$$

where

1) $T: \underline{K} \to \underline{K}'$ is a covariant functor

2) $\Gamma: \underline{A} \to \underline{A}'$ is a homomorphism of semi-groups

3) $\omega_{A,X}: \Gamma(A)T(X) \to T(AX)$ is a family of natural (with respect to X) transformations such that

$$\omega_{A_1A_2,X} = \omega_{A_1,A_2X} \circ \Gamma(A_1)\omega_{A_2,X}$$

$$\omega_{1,X} = 1 .$$

We call $T = (\Gamma, T, \omega)$ *stable* if all $\omega_{A,X}$ are isomorphisms.

1.4. Lemma. *Let* $T: \underline{K} \to \underline{K}'$ *be an operator functor and* $\underline{K} = (\underline{A}, \underline{K})$ *a stable operator category, then* $T = (\Gamma, T, \omega)$ *is stable. In particular, any operator functor between stable operator categories is itself stable.*

Proof. We verify immediately that $\Gamma(A)\omega_{A^{-1},AX}$ is the inverse of $\omega_{A,X}$.

1.5. Definition. An operator transformation $\zeta\colon (\Gamma, T, \omega) \to (\Gamma, T', \omega')$ is a functor transformation $\zeta\colon T \to T'$ such that

$$\zeta_{AX} \circ \omega_{A,X} = \omega'_{A,X} \circ \Gamma(A)\zeta_X .$$

We are now ready to define the following 2-categories (of categories):

Wop. Objects = Operator categories, 1-Morphisms = Operator functors, 2-Morphisms = Operator transformations

Sop. Objects = stable operator categories, 1-Morphisms, 2-Morphisms as in ***Wop.***

There is an inclusion functor i: ***Sop*** $\subset$ ***Wop*** which is a 2-functor. Furhtermore ***Sop*** is a full subcategory of ***Wop*** by lemma 1.4.

2. EXAMPLES AND REMARKS

All the interesting functors in algebraic topology are operator functors: the homology functor $H_*\colon (\boldsymbol{Z}^+, \mathrm{Top}) \to (\boldsymbol{Z}, \boldsymbol{Ens}^{\boldsymbol{Z}})$ is stable because

$$H_n(X) \approx H_{n+1}(\Sigma X) .$$

The homotopy functor $\pi_*\colon (\boldsymbol{Z}^+, \mathrm{Top}) \to (\boldsymbol{Z}, \boldsymbol{Ens}^{\boldsymbol{Z}})$ is an operator functor which is not stable because one has a natural homomorphism

$$\pi_n(X) \to \pi_{n+1}(\Sigma X)$$

which is not necessarily an isomorphism.

Let $(\underline{A}, \underline{K}) \in$ ***Wop*** be any operator category and $(1, E, \omega)\colon (\underline{A}, G) \to (\underline{A}, \underline{K})$ be an operator functor, where $(\underline{A}, G)$ is the category of example 3) in § 1. We have the morphisms $\omega_{A,g}\colon AE(g) \to E(\varphi(A)g)$. Now let $\underline{A} = \boldsymbol{Z}^+$ and $G = \boldsymbol{Z}$, $\varphi\colon \boldsymbol{Z}^+ \subset \boldsymbol{Z}$ the inclusion. We write E_n instead of $E(n)$, $n \in \boldsymbol{Z}$ and consequently we obtain a morphism

$$\omega_{\Sigma, n}\colon \Sigma E_n \to E_{n+1} .$$

Therefore the functor $E = (1, E, \omega)$ is simply a prespectrum E_n over the category $\underline{K}$ [4].

Let $(\underline{A}, \underline{K}), (\underline{B}, \underline{L}) \in \boldsymbol{Wop}$ and $\Gamma\colon \underline{A} \to B$ be a fixed homomorphism, then we consider the category $\underline{L}^{\underline{K}}$ whose objects are operator functors $T = (\Gamma, T, \omega)\colon (\underline{A}, \underline{K}) \to (\underline{B}, \underline{L})$ with operator transformations as morphisms. This category can be equipped with the structure of an operator category in two different ways:

1) $\underline{A}$ operates on $\underline{L}^{\underline{K}}$ by

$$(AT)(\quad) = T(A\quad),$$

or

2) $\underline{B}$ operates on $\underline{L}^{\underline{K}}$ by

$$(BT)(\quad) = BT(\quad).$$

We denote these operator categories by $\underline{O}(\underline{K}, \underline{L}) = (\underline{A}, \underline{O}(\underline{K}, \underline{L}))$ and $O'(\underline{K}, \underline{L}) = (\underline{B}, O(\underline{K}, \underline{L}))$ resp. The underlying category is in both cases $\underline{L}^{\underline{K}}$.

Now let $(A, \underline{M}) \in \boldsymbol{Wop}$ be a third operator category, then we have the operator category $O(\underline{K}, O(\underline{M}, \underline{L}))$. The following theorem assures us that there exists a tensor product $\underline{M} \otimes \underline{K}$ between two operator categories (with the same semigroup $\underline{A}$):

2.1. Theorem. *There exists an operator category* $(\underline{A}, \underline{M} \otimes \underline{K})$ *such that for arbitrary* $(\underline{B}, \underline{L})$, $\Gamma\colon \underline{A} \to \underline{B}$ *the operator categories* $O(\underline{K}, O(\underline{M}, \underline{L}))$ *and* $O(\underline{M} \otimes \underline{K}, \underline{L})$ *are equivalent.*

One can of course define $\underline{M} \otimes \underline{K}$ directly by using the concept of a "bi-operator functor" which is analogous to the concept of a bilinear map in algebra [5].

3. THE MAIN THEOREM

Let $(\underline{A}, \underline{K}) \in \boldsymbol{Wop}$ be a category, then we need a canonical stabilization process which converts $\underline{K}$ into a stable category. This is accomplished by the following theorem:

3.1. Theorem. *The* 2*-functor* i: $\boldsymbol{Sop} \subset \boldsymbol{Wop}$ *is furnished with a left-2-adjoint* $\hat{\ }$: $\boldsymbol{Wop} \to \boldsymbol{Sop}$.

We did not give the explicit definition of a 2-adjoint of a 2-functor. Therefore a second formulation of theorem 3.1 seems to be necessary:

3.2. Theorem. *Let* $(\underline{A}, \underline{K}) \in \boldsymbol{Wop}$ *be any operator category, then there exists a stable operator category* $(\bar{\underline{A}}, \hat{\underline{K}})$ *as well as a functor* $\beta = (\varphi, \beta, \tau)\colon (\underline{A}, \underline{K}) \to (\bar{\underline{A}}, \hat{\underline{K}})$ *such that for any* $T \in \boldsymbol{Wop}(\underline{K}, \underline{S})$, $\underline{S} = (\underline{B}, \underline{S}) \in \boldsymbol{Sop}$ *there exists an up to isomorphism a unique* $\tilde{T} \in \boldsymbol{Sop}(\hat{\underline{K}}, \underline{S})$ *with* $\tilde{T}\beta \approx T$.

Moreover any operator transformation $\zeta\colon T \to T'$ *can be uniquely lifted to an operator transformation* $\tilde{\zeta}\colon \tilde{T} \to \tilde{T}'$ *with* $\tilde{\zeta}\beta = \zeta$.

This functor $\hat{}$ is precisely the stabilization process. We will indicate some applications in the following sections. Before we do this, one word concerning the proof of theorem 3.1 seems to be in order:

To this end let us return to the category $(\underline{A}, G)$ of example 3) in § 1. According to theorem 2.1 we can form $G \otimes \underline{K} = (\underline{A}, G \otimes \underline{K})$, which is by definition not stable. However one can prove that there is a canonical way to equip $G \otimes \underline{K}$ with the structure of a stable operator category. More precisely:

3.3. Lemma. *There exists a stable operator category* $(G, \hat{\underline{K}})$ *with the same underlying category as* $G \otimes \underline{K}$ *such that the triple*

$$(\varphi, 1, 1)\colon (\underline{A},\ G \otimes \underline{K}) \to (G, \hat{\underline{K}})$$

becomes a stable operator functor.

Now it turns out that this category $\hat{\underline{K}} = (G, \hat{\underline{K}})$ is in fact the canonical stabilization.

Roughly speaking, $\hat{\underline{K}}$ is the tensor product of $\underline{K}$ with the (discrete) category $G = (\underline{A}, G)$.

4. HOMOTOPY IN STABLE CATEGORIES

All relevant operator categories $\underline{K} = (\underline{A}, \underline{K})$ are equipped with a homotopy structure. In all known cases this homotopy is induced by a functor $\phi\colon \underline{K} \to \underline{M}$ and the homotopy category $\underline{K}_h$ is the quotient category $\underline{K}/\phi$ in the sense of [1].

Roughly speaking, one converts all morphisms $f \in \underline{K}$ into isomorphisms in $\underline{K}_h$ which become isomorphisms under ϕ. There may be many different functors ϕ yielding the same homotopy.

In case of an operator category, we require the homotopy to be compatible with the operation of $\underline{A}$. Thus we assume ϕ to be a stable operator functor and $\underline{M}$ to be a stable category.

The stabilization of the ϕ-homotopy is $\hat{\phi}$-homotopy and one can prove:

4.1. Theorem. *The $\hat{\phi}$-homotopy is independent of the functor ϕ (it depends only on the homotopy concept introduced by ϕ).* We can now transfer all statements on homotopy theory in $\underline{K}$ to the stabilized category. The key for this is the following definition:

4.2. Definition. A category with homotopy $(\underline{K}, \phi)$ is a pair where $\underline{K} \in$ ***Wop*** and ϕ are as above. A functor $T\colon (\underline{K}, \phi) \to (\underline{K}', \phi')$ between two categories with homotopy is a functor $T \in$ ***Wop***$(\underline{K}, \underline{K}')$ such that $\phi' T \approx \phi$.

The natural functor $\beta\colon \underline{K} \to \hat{\underline{K}}$ from theorem 3.1 gives rise to a stable functor $\beta_h\colon \underline{K}_h \to \hat{\underline{K}}_h$ which makes the following diagram commutative

$$\begin{array}{ccc} \underline{K} & \xrightarrow{\quad\beta\quad} & \hat{\underline{K}} \\ \downarrow & & \downarrow \\ \underline{K}_h & \xrightarrow{\quad\beta_h\quad} & \hat{\underline{K}}_h \end{array}$$

The vertical functors are the natural projections on the quotient categories [3].

The theoretical aspect of stabilization of a homotopy theory is now expressed by the following theorem [4]:

4.3. Theorem. *Let* $(\underline{K}, \phi)$ be a *category with homotopy and let* $T\colon (\underline{K}, \phi) \to (\underline{S}, \phi')$ be any functor between categories *with homotopy, where $\underline{S}$ is stable. Then there exists a unique functor $\tilde{T}_h\colon \hat{\underline{K}}_h \to \underline{S}_h$ which makes the diagram*

$$\begin{array}{ccc} \underline{K}_h & \xrightarrow{\beta_h} & \hat{\underline{K}}_h \\ & \searrow^{T_h} & \downarrow \tilde{T}_h \\ & & \underline{S}_h \end{array}$$

commutative.

4.4. Corollary. *Let* $\underline{S} \in \mathbf{Sop}$ *and* $\Pi\colon \underline{K} \to \underline{S}$ *be an operator functor which factors over the homotopy category. Then* $\hat{\Pi}$ *factors also over the stabilized homotopy category.*

If Π fulfills a "Whitehead theorem" (i.e. $f \in \underline{K}$ is a homotopy equivalence if and only if $\Pi(f)$ is an isomorphism), then $\hat{\Pi}$ fulfills a "Whitehead theorem".

There is a second way to stabilize a homotopy theory in a category $\underline{K} \in \mathbf{Wop}$:

Let $\underline{K} \underset{S}{\overset{T}{\rightleftarrows}} \underline{L}$ be a pair of adjoint functors [8], then we have natural transformations $\alpha\colon 1_{\underline{L}} \to TS$, $\gamma\colon ST \to 1_{\underline{K}}$. We can consider the quotient categories $\underline{K}/\{\gamma\}$, $\underline{L}/\{\alpha\}$ where for example $\{\alpha\}$ stands for all morphisms $\alpha_{\underline{K}}\colon ST(K) \to K$, $K \in \underline{K}$. In [7] C.M. Ringel proved the following theorem:

4.5. Theorem. a) *The categories* $\underline{K}/\{\gamma\}$, $\underline{L}/\{\alpha\}$ *are well defined (within the same universe as* $\underline{K}$ *resp.* $\underline{L}$*).*

b) *They are equivalent.*

c) *In the case of* $\underline{K}$ = *category of topological spaces,* $\underline{L}$ = *category of Kan-complexes. For* $T(X)$ = *singular complex of* X, $S(L)$ = *geometrical realization, the categories* $\underline{K}/\{\gamma\}$ *resp.* $\underline{L}/\{\alpha\}$ *are the usual homotopy categories.*

This theorem can be used to stabilize a homotopy theory which is defined by a pair of adjoint functors T, S in view of the following lemma:

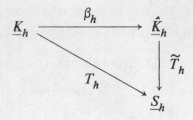

4.6. Lemma: *If* $\underline{K} \underset{S}{\overset{T}{\rightleftarrows}} \underline{L}$ *is a pair of adjoint functors in* **Wop** *)i.e. everything goes on within* **Wop***) then the functors* $\hat{T}, \hat{S}$ *are again adjoint.*

The proof is immediate from theorem 3.1.

The advantage of this method compared with the preceding one is that we obtain free of charge a proof of the following fact:

4.7. Theorem. *If* $\underline{K} \underset{S}{\overset{T}{\rightleftarrows}} \underline{L}$ *are adjoint in* **Wop**, *then the stabilized homotopy categories (induced by* $\hat{T}, \hat{S}$*) are equivalent:*

$$\hat{\underline{K}}_h \approx \hat{\underline{L}}_h \,.$$

5. CELL STRUCTURES IN STABLE CATEGORIES

The most geometrical feature of the category of *CW*-complexes is undoubtly the cell structure of its objects. The suspension raises the dimension of a cell by one and therefore a cell complex with cells of only positive dimensions can not be arbitrarily desuspended. Whenever we try to save the cell structure from the unstable category of *CW*-complexes to its stabilization, we can expect "cells" of negative dimensions. For this purpose we are forced to express the cell structure of a *CW*-complex in a functorial way because only properties of this kind admit a canonical stabilization. This problem was solved completely in a very satisfactory form by J. M. Boardman. We will indicate his ideas:

A cell-index space $(V, \dim\ )$ is a topological space V together with a function $\dim\colon V \to Z$ such that:

1) For any $x \in V$ the closure $\bar{x}$ is finite

2) A subset $A \neq \phi$ of V is closed if and only if for any $x \in A$ one has $\bar{x} \subset A$.

3) If $x \in \bar{y}$ and $x \neq y$ then $\dim x < \dim y$.

A morphism $f\colon (V, \dim\ ) \to (W, \dim\ )$ is a continuous closed embedding f of V into W such that $\dim x = \dim f(x)$.

The category $\underline{D}$ of all cell index spaces becomes an operator category whenever we define a suspension $\Sigma(V, \dim\ ) = (V, \dim\ + 1)$.

Let $\underline{C}$ be the category of CW-complexes with cellular inclusions as morphisms. Then we have an operator functor $\mathscr{X} : \underline{C} \to \underline{D}$ which assigns to each $C \in \underline{C}$ the space $\mathscr{X}(C)$ of all its cells (as points) with the natural dimension function. The topology of $\mathscr{X}(C)$ is the quotient topology (inhereted from C). The properties 1) – 3) above are now simply a restatement of the definition of a CW-complex [1]. By applying the stabilization functor $\hat{}$ to $\mathscr{X}$ we obtain a stable functor $\hat{\mathscr{X}}: \hat{\underline{C}} \to \hat{\underline{D}}$ ($\approx \underline{D}$ because $\underline{D}$ is infact a stable category). Thus every object of $\hat{\underline{C}}$ achieves the structure of a "cell complex" with cells of arbitrary dimensions.

One could now develop the entire theory of CW-spaces (in its stabilized form) by using this cell structure. Although Boardman has contributed some very promising ideas, much work remains to be done in this direction.

6. COMPARISON OF DIFFERENT STABILIZATION PROCESSES

The use of functor $\hat{}$ of theorem 3.1 is not the only way to stabilize a given operator category. The first example for a categorical stabilization was provided by the S-category of E. Spanier and J.H.C. Whitehead. In the language of operator categories, we will recognize the S-category in the following way:

Consider the category Op of operator categories with stable operator functors $T = (\Gamma, T, \omega)$ (i.e. all ω's are supposed to be isomorphisms). We have an inclusion functor i: ***Sop*** $\subset$ ***Op*** and analogously a stabilization process for ***Op***:

6.1. Theorem. *The inclusion 2-functor i:* ***Sop*** $\subset$ ***Op*** *is provided with a left 2-adjoint* $\bar{}$: ***Op*** $\to$ ***Sop***.

It is now easy to reformulate theorem 6.1 in the same way as we did with theorem 3.1, in order to obtain a more explicit description of the stabilization procedure. The disadvantage of this process is that only those functors can be stabilized which have been stable already before. There are of course other descriptions of the category $\underline{\bar{K}}$ which are more elementary [3].

The second stabilization process apart from that of theorem 3.1 considers operator categories with an additional structure: We assume that all categories $(\underline{A}, \underline{K}) = \underline{K}$ are closed under arbitrary colimits [8].

In the case of a category with inclusions as morphisms, this is equivalent to the assumption that arbitrary unions of objects are present. Consequently we have to restrict the class of our operator functors to those, compatible with colimits. Let us denote the corresponding categories by ***Wopl*** resp. ***Sopl.*** Now we have again a stabilization theorem [4]:

6.2. Theorem. *The inclusion 2-functor* i: ***Sopl*** $\subset$ ***Wopl*** *is provided with a left-2-adjoint* $\hat{}$: ***Wopl*** $\rightarrow$ ***Sopl***.

Everything which has been said about the homotopy theory of a stabilization carries over to this case: We have again a homotopy category $\underline{\hat{K}}_h$, a functor $\beta_h : \underline{K}_h \rightarrow \underline{\hat{K}}_h$ and theorem 4.3 holds with the necessary modifications (i.e. ***Wop*** resp. ***Sop*** are replaced by ***Wopl*** resp. ***Sopl***).

Let $\underline{C}$ be the category of CW-complexes with inclusions as morphisms and $\underline{C}_h$ the corresponding homotopy category, then we can prove the following assertion:

6.3. Theorem. *The category* $\underline{\hat{C}}_h$ *(in the sense of theorem* 5.2) *is equivalent to the homotopy category* $\mathfrak{B}_h$ *of* J.M. Boardman [6]. *The importance of this statement lies in the fact that* Boardman's *famous and successful approach is now presented (at least on the homotopy level) within the framework of the theory of stabilization processes.*

J.M. Boardman's original approach to his category was much more geometrical: He took the category $\underline{C}_f$ of finite CW-complexes (with cellular maps) and applied the stabilization process of theorem 6.1 to obtain the category $\underline{C}_f$. Then, in the next step, he amended the inclusion subcategory $\underline{I} \subset \underline{C}_f$ (same objects as $\underline{C}_f$ but only those morphisms, which stem from inclusions) by all colimits. His idea was that the category of CW-complexes originates from the category of finite CW-complexes by adding all direct limits (colimits) to the inclusion subcategory.

It can be proved [9] that the category $\mathfrak{S}_{p_E}$ of Kan-spectra is obtained from the category $\mathfrak{S}_E$ of simplicial sets (instead of the category of CW-complexes) by the same procedure. Therefore one has an assertion analogous to theorem 6.1 for the simplicial case.

It should be remarked that the theory of the Boardman-category can be accomplished by using prespectra (formerly called "spectra") [4]. The Boardman-category is a quotient category of the category $\mathscr{P}$ of prespectra.

We pointed out already in §2 that a prespectrum is a sequence $\{E_n\}$, $n \in Z$ of CW-complexes together with maps $\Sigma E_n \to E_{n+1}$, which are mostly assumed to be inclusions. A morphism $\{f_n\}: \{E_n\} \to \{E'_n\}$ is a family of continuous maps $f_n: E_n \to E'_n$ which fulfil a compatibility condition. The distinguished subclass F in $\mathscr{P}$ consists of all "full embeddings": A cellular inclusion $\{f_n\}: \{A_n\} \subset \{B_n\}$ in $\mathscr{P}$ is a "full embedding" if the following condition holds:

For any finite subcomplex $X \subset B_n$ there exists an index k such that $\Sigma^k X \subset A_{n+k}$.

The Boardman-category $\mathfrak{B}$ turns out to be equivalent to $\mathscr{P}/F$ [6]. This is perhaps the most adequate description of $\mathfrak{B}$.

REFERENCES

[1] F.W. Bauer, *Homotopietheorie,* BI-Taschenbücher, Bd. 475 ab Bibliogr. Institut, Mannheim (1971).

[2] F.W. Bauer, Stabile Kategorien, *Math. Zeitschr.*, 123 (1971), 139-167.

[3] F.W. Bauer, *Homotopie in stabilen Kategorien,* F. Hausdorff Gedächtnisband, VEB Deutscher Verlag d. Wissensch., Berlin (1972).

[4] F.W. Bauer, Boardman's Category and the Process of Categorical Stabilization, *Math. Z.,* 130 (1973), 95-106.

[5] F.W. Bauer, Tensor Products in Operator Categories, *Math. Ann.*, 205 (1973), 303-315.

[6] J.M. Boardman, *Stable Homotopy Theory,* Mimeographed Notes, The Johns Hopkins Univ., Baltimore (1970).

[7] C.M. Ringel, Eine Charakterisierung der Homotopiekategorie der *CW*-Komplexe *Math. Zeitshcr.*, 115 (1970), 359-365.

[8] H. Schubert, *Kategorien I, II,* Heidelberger Taschenbücher, Bd. 65, 66, Springer Verlag, Berlin – Heidelberg – New York (1970).

[9] M. Tierney, Categorical constructions in stable homotopy theory, *Lecture Notes in Mathem.*, Nr. 87, Springer Verlag, Berlin – Heidelberg – New York (1969).

COLLOQUIA MATHEMATICA SOCIETATIS JÁNOS BOLYAI
8. TOPICS IN TOPOLOGY, KESZTHELY (HUNGARY), 1972.

ON GENERALIZED MANIFOLDS

M. BOGNÁR

The notion of multi-dimensional manifolds appears in the famous treatise of Riemann [24] as a synonym of the multi-extended systems. This notion has obtained new formulations over and over again, till it got its final form. The concept, which is used today comes from Kerékjártó [14]. According to him an *n-dimensional manifold,* or rather an *n-dimensional euclidean manifold* is defined as a connected Hausdorff-space, every neighbourhood of which is homeomorphic to the n-dimensional open ball. (The neighbourhoods are here obviously the elements of a base).

The first generalization of the concept of the manifold has been introduced by Brouwer [7]. The Brouwerian *pseudomanifolds* have a pure combinatorial character, they preserve the most evident combinatorial properties of the manifolds. The definition of the n-dimensional pseudomanifold has been meanwhile somewhat modified. It has been recently defined as a strongly connected n-complex, every $(n-1)$-simplex of which is a face of precisely two n-simplexes.

In the works of Brouwer the notion of the pseudomanifolds has not obtained any independent meaning. They are used there only as instruments for the proof of the Jordan – Brouwer theorem or of the invariance of open sets [7], [8]. Only much later in the book of Aleksandrov and Hopf [1] have they got some independent function.

The first conscious step to the formation of generalized manifolds was connected to the relation between the multi-dimensional connectivity numbers of the manifolds, to the Poincaré duality. Vietoris raised in 1928 the following idea: To the holding of the Poincaré duality in a complex it is not necessary, that the union of the simplexes should be an n-dimensional manifold. This uncombinatorial condition may be substituted by other, more to the combinatorial structure connected, weaker conditions. Vietoris gave such ones and this way he was the first to define homology manifolds [26]. More general conditions were raised by Van Kampen [13], Lefschetz [16] and Pontrjagin [22].

The following step was the common generalization of the euclidean and homology manifolds. Such generalizations have been given for the first time by Lefschetz and Flexner [11], [12], [17], but their "topological manifolds" are still locally triangulable. Some years later Lefschetz [18], Čech [9], Wilder [28], Aleksandrov and Pontrjagin [2] have constructed such kinds of generalized manifolds, where the condition of local triangulability does not occur any more.

We want to show here the *generalized manifolds of Čech:*

The coefficient group is the additive group of the rational numbers or the group of rational integers mod p, where p is an arbitrary prime number.

The topological space R is said to be locally connected in the order k in its point x, if for every neighbourhood U_x of x there exists a neighbourhood V_x of x such that $V_x \subset U_x$ and every k-dimensional Čech cycle of $\bar{V}_x$ is homologous to zero in $\bar{U}_x$.

Let now R denote a topological space and S denote a subspace of R. R will be called a $V_0^n(S)$, if it satisfies the following conditions:

(a) R is compact.

(b) the open sets of R are F_σ-s.

(c) S is closed in R,

(d) the dimension of $R \setminus S$ in n,

(e) the space R is in an arbitrary point of $R \setminus S$ locally connected in the order n,

(f) if x is any point of $R \setminus S$, then $p^n(x, R) = 1$, ($p^n(x, R)$ is the n-dimensional local Betti-number of R in x),

(g) if A is a closed subset of R and x is a point of $(A \cap \cap \overline{R \setminus A}) \setminus S$, then $p^n(x, A) = 0$.

The space R is called a $V_q^n(S)$, if the following conditions are satisfied:

(i) R is a $V_0^n(S)$,

(ii) if $n - q \leqslant k \leqslant n - 1$ and if x is a point of $R \setminus S$, then the space R is in x locally connected in the order k, and furthermore the relation $p^k(x, R) = 0$ holds.

It may be observed that the generalized manifolds of Čech have a quite particular character. The generalizations of the manifolds are namely usually related to the $V_n^n(S)$, however in the paper of Čech the $V_0^n(S)$-s are also playing an important role. In fact one may regard them as a generalization of a special kind of the Brouwerian pseudomanifolds. We shall still return to this question.

A radically different point of view may be found in the works of Wilder [27], [28], [29]. He proceeds from the theorem of Schoenflies [25], which characterizes the Jordan continua in the plane by the following external properties:

If a bounded and perfect set F in the plane divides the plane in two domains, such that every point of F is accessible from each of these domains, then F is a Jordan curve.

As a generalization of this theorem, Wilder characterized the closed 2-manifolds in E^3 in the following manner [27]:

In order that a compact point set K in E^3 should be a closed 2-manifold, it is necessary and sufficient that the following conditions be *satisfied:*

(a) K *is the common boundary of two uniformly locally connected domains of* E^3,

(b) $p^1(E^3 \setminus K)$ *is finite.* (p^1 is here the 1-dimensional Betti-number mod 2).

Wilder then defined the $(n-1)$-dimensional generalized manifolds such that those embedded in E^n, may be characterized as compacta in E^n possessing analogous external properties as the 2-manifolds in E^3 or the Jordan continua in E^2 [28]. Later in his book [29] he also studied locally compact figures.

The investigations of the last fifteen years, which are concerning chiefly with generalized manifolds with boundary, have clearly demonstrated that the *gm*-s (generalized manifolds) are interesting not only because they are more general than the euclidean manifolds, and some of their basic properties are nevertheless the same, but also because, in a certain sense, they are more regular than the euclidean ones. In fact, some startling singularities of the euclidean manifolds do not occur here. I want to mention two examples for such singularities.

1°. Let X' be a topological sphere in the 3-dimensional spherical space S^3. Let us take two copies from the closure of one of the components of $S^3 \setminus X'$. Sew them along X' by the identity homeomorphism. Lininger has given in 1965 a quite simple example where the resulting figure is not homeomorphic to S^3 [19].

2°. Let the n-dimensional cell I^n be homeomorphic to the topological product of A and B ($I^n \cong A \times B$). Bing has proved that if $n \leqslant 4$, then both A and B must be cells [3]. On the other hand for the case if $n \geqslant 5$ examples of $(n-1)$-combinatorial manifolds M with boundary such that $M \times I \cong I^n$ and $M \ncong I^{n-1}$ have been given by Poénaru [21], Mazur [20], and Curtis [10].

The *gm*-s are however free of these singularities.

The results of Raymond [23] imply namely that *if* X *is a spherical* n-*gm and* X' *a spherical* $(n-1)$-*gm lying in* X, *then taking two copies from the closure of one of the components of* $X \setminus X'$ *and sewing them along* X' *we obtain again a spherical n-gm.*

On the other hand Kwun and Raymond proved in 1962 the following theorem:

Let $X \cong A \times B$. *Then* X *is a generalized n-cell if and only if* A *is a generalized k-cell and* B *is a generalized* $(n-k)$*-cell* [15].

And now we are getting to our own investigations.

We shall study mathematical objects which are near to the Brouwerian pseudomanifolds and to the $V_0^n(S)$ of Čech. In this way we can not expect the preservation of the Poincaré duality, but these objects are very suitable to the study of orientability, of the preservation and changing of banks and for other similar questions.

At first we shall define an object called *k-manifold* by general topological methods.

Let R be a Hausdorff space and let (X, A) be a compact pair in R.

A domain (a connected nonvoid open set) V in R is said to be *k-regular* mod (X, A) if the following conditions hold:

(a) $V \cap A = \phi$,

(b) $V \cap X$ is a domain in X,

(c) $V \setminus X$ consists of two components,

(d) the closure of each component of $V \setminus X$ contains $V \cap X$.

The compact pair (X, A) itself is called a *k-manifold* in R, if it satisfies the following two conditions:

(i) $X \setminus A$ is a nonvoid connected space satisfying the second axiom of countability,

(ii) for every $q \in X \setminus A$, the k-regular domains which contain the point q form a base for the neighbourhood system of the point q in R.

And now we shall give some further basic definitions concerning k-manifolds.

Let (X, A) be a k-manifold in the Hausdorff space R.

Let $\Sigma^k_{(X,A)}$ denote the family of the k-regular domains mod(X, A). $\Sigma^k_{(X,A)}$ is the *k-regular exterior base* of (X, A) in R.

Let V be a k-regular domain mod(X, A). We call the components of $V \setminus X$ the *banks* of V, or more precisely the *k-banks* of V, and design them by $P^1(V)$ and $P^2(V)$. The numeration is obviously fully arbitrary.

Let V and V' be two k-regular domains mod(X, A), and suppose that $V \subset V'$. Let us take the banks $P^1(V)$ and $P^2(V)$ of V. Then $P^1(V)$ is contained in one of the banks of V' and $P^2(V)$ in the other. That means: *the banks of V are the intersections of the banks of V' by V.*

Let now Σ be a family of sets. A sequence $g = [V_1, V_2, \ldots, V_n]$ is a *Σ-chain* if, for $i = 1, 2, \ldots, n$, $V_i \in \Sigma$ and, for $i = 1, 2, \ldots, n-1$, either $V_i \subset V_{i+1}$ or $V_{i+1} \subset V_i$. The Σ-chain g is *closed* if $V_n = V_1$.

Let us consider a $\Sigma^k_{(X,A)}$-chain or shortly a *k-chain* $g = [V_1, V_2, \ldots, V_n]$. Let us indicate by $P^1(V_1)$ and $P^2(V_1)$ the banks of V_1. Then there exists a numeration P^1_i and P^2_i of the banks of V_i such that:

(a) $P^1_1 = P^1(V_1)$, $P^2_1 = P^2(V_1)$

(b) $P^1_i \cap P^1_{i+1} \neq \phi$ and $P^2_i \cap P^2_{i+1} \neq \phi$ $(i = 1, 2, \ldots, n-1)$,

and this numeration is unique. We shall call this numeration of the banks a *regular numeration.*

Hence two chains $g_P(1) = [P^1_1, P^1_2, \ldots, P^1_n]$ and $g_P(2) = [P^2_1, P^2_2, \ldots, P^2_n]$ of the banks belong to the k-chain g. Here is, for $j = 1, 2$, $P^j_i \subset P^j_{i+1}$ in the case $V_i \subset V_{i+1}$ and $P^j_{i+1} \subset P^j_i$ in the case $V_{i+1} \subset V_i$. The chains $g_P(1)$ and $g_P(2)$ are called the *chains of banks, associated with the chain g.*

If g is closed, i.e. if $V_n = V_1$, then two cases are possible:

(1) $P_n^1 = P_1^1$ and $P_n^2 = P_1^2$,

(2) $P_n^1 = P_1^2$ and $P_n^2 = P_1^1$.

We say in the first case that g *preserves its banks,* and in the second one, that g *changes its banks.*

Let now U be a domain in $X \setminus A$. We say about a k-chain $g = [V_1, V_2, \ldots, V_n]$ that *it is leaning on* U if, for $i = 1, \ldots, n$, $V_i \cap X \subset U$.

U is called a *two-sided domain* if every closed k-chain, leaning on U, preserves its banks. If there exists a closed k-chain g, leaning on U and changing its banks, then U is called a *one-sided domain.*

The k-manifold (X, A) is *two-sided (one-sided)* if the domain $X \setminus A$ is two sided (one-sided, respectively).

And now we can formulate the following theorems:

Theorem 1. *Let* (X, A) *be a* k*-manifold in* R *and* V *a* k*-regular domain* $\operatorname{mod}(X, A)$. *Then* $V \cap X$ *is a two-sided domain of* (X, A).

Theorem 2. *Let us suppose the metrizability of the space* R *and let* U *be a two-sided domain of the* k*-manifold* (X, A) *in* R. *Then there exists a* k*-regular domain* V $\operatorname{mod}(X, A)$, *such that* $V \cap X = U$.

And here we can set immediately the following unsolved problem:

Is Theorem 2 correct if we do not suppose the metrizability of R?

And now a few words about paths and associated chains.

Let Y be a topological space. A continuous map $f\colon [a, b] \to Y$, where $[a, b]$ is a closed interval in the space of the real numbers $(a < b)$ is said to be a *(continuous) path in* Y. $f(a)$ is the *beginning,* $f(b)$ the *end* and $f([a, b])$ the *body* of the path f. For $M \subset Y$ f *is lying in* M if the body of f lies in M. f is a *closed path,* if $f(a) = f(b)$. f is a *closed Jordan path* and its body *is a Jordan curve,* if the following conditions holds: $f(x) = f(y)$ if and only if $x = y$, or $x = a$, $y = b$, or $x = b$, $y = a$.

Let $f\colon [a, b] \to Y$ be a closed path and $g = [V_1, V_2, \ldots, V_n = V_1]$ a closed Σ-chain where Σ is a family of subsets (of Y or) of a space R containing Y as its subspace. We say that g *is associated with* f, if there exists a subdivision $a = a_1 \leqslant a_2 \leqslant \ldots \leqslant a_n = b$ of the interval $[a, b]$ such that, for $i = 1, 2, \ldots, n$, $f(a_i) \in V_i$ and, for $i = 1, 2, \ldots, n-1$, $f([a_i, a_{i+1}]) \subset (V_i \cup V_{i+1})$, i.e. either $f([a_i, a_{i+1}]) \subset V_i$ or $f([a_i, a_{i+1}]) \subset V_{i+1}$.

And now we obtain the following theorems:

Let (X, A) be a k-manifold in the Hausdorff space R.

Theorem 3. *Let f be a closed continuous path in $X \setminus A$. Then there exists a closed $\Sigma^k_{(X,A)}$-chain associated with f.* Moreover *every closed k-chain associated with f either preserves its banks or changes them.* In the first case, we say that f *preserves its banks,* and in the second that *it changes them.*

Theorem 4. *A domain U in $X \setminus A$ is a two-sided domain if and only if every continuous closed path in U preserves its banks.*

Theorem 5. *Let J be a Jordan curve in $X \setminus A$. Regarding the closed Jordan paths with the body J either each such path preserves or each path changes its banks.* In the first case we say that J *preserves,* and in the second, that *J-changes its banks.*

We can assert the following principal theorem:

Theorem 6. *Each Jordan curve in $X \setminus A$ preserving its banks may be covered by a* $\operatorname{mod}(X, A)$ *k-regular domain V.*

Now we set the problem of definition of the *closed* and *bounded* *k-manifolds.* The following assertion holds: *Let (X,A) be a k-manifold in the T_2-space R. Then either for each* $\operatorname{mod}(X, A)$ *k-regular domain V the two banks of V belong to the same component of $R \setminus X$,* or for *each $V \in \Sigma^k_{(X,A)}$ the two banks of V belong to different components of $R \setminus X$.* In the first case we say that (X, A) is a *bounded,* and in the second, that (X, A) is a *closed k-manifold.* We can state now:

Theorem 7. *Closed k-manifolds are always two-sided k-manifolds.*

Now we shall turn to another object, which seems at the first sight to be wholly different from the k-manifolds. A type of generalized manifolds will be defined by internal properties.

Let p be a prime number, Z_p the cyclic group $\bmod p$ and H the Čech homology theory with the coefficient group Z_p.

Let n be a positive integer. The compact pair (X, A) is called an *(n, p)-cell* if the following conditions are satisfied:

(a) $X \setminus A$ is a nonvoid, connected, locally connected space with a countable base,

(b) $H_n(X, A) \approx Z_p$,

(c) for the injection $i\colon (X, \phi) \subset (X, A)$, the induced homomorphism i_{*n} is trivial, i.e. $i_{*n}(H_n(X)) = 0$,

(d) in every domain V of $X \setminus A$, there exists a nonempty open subset $U \subset V$ such that, for the injection $i\colon (X, A) \subset (X, X \setminus U)$, the induced i_{*n} is monomorphic.

Hence we can state:

Theorem 8. *Let (X, A) be an (n, p)-cell. Then $A \neq \phi$.*

And now an unsolved question.

Is it true for every $(1, p)$-cell (X, A) *that* $X \setminus A$ *is homeomorphic to an open interval?* (If X is embedded in the plane, then the statement is true.)

Let now (X, A) be an (n, p)-cell and let us consider the segment

$$\widetilde{H}_{n-1}(A) \xleftarrow{\partial} H_n(X, A) \xleftarrow{i_*} H_n(X)$$

of the reduced homology sequence of (X, A). (Here is, in the case $n = 1$, $\widetilde{H}_0(A)$ the reduced 0-dimensional homology group, and, in the case $n > 1$, $\widetilde{H}_{n-1}(A) = H_{n-1}(A)$). As a consequence of (c), ∂ is monomorphic and, for $0 \neq u \in H_n(X, A)$,

$$0 \neq \partial(u) \in \widetilde{H}_{n-1}(A)\,.$$

$\partial(u)$ is called the *algebraic boundary* of (X, A) and we denote it by A_*. Obviously, A_* is, in the case $p \neq 2$, not uniquely defined, it depends from the choice of u.

Let f: $[a, b] \to X \setminus A$ be a closed continuous path in $X \setminus A$ and put $Y = f([a, b])$ and $B = \{f(a) = f(b)\}$. B is a singleton, hence it is homologically trivial, and therefore, by the exactness of the sequence

$$\widetilde{H}_0(B) \leftarrow H_1(Y, B) \overset{j_*}{\leftarrow} H_1(Y) \leftarrow H_1(B) \,,$$

j_* is an isomorphism. Put $Y_1 = [a, b]$ and $B_1 = \{a, b\}$. Let us consider the induced homomorphism f_{*1}: $H_1(Y_1, B_1) \to H_1(Y, B)$ and take a non-zero element v from the group $H_1(Y_1, B_1)$, which group is isomorphic to Z_p. Let us denote $j_*^{-1} f_{*1}(v)$ by f^*. In the case, when $p \neq 2$ and $f^* \neq O$, the $f^* \in H_1(Y)$ depends obviously from the choice of v.

Let (X, A) be an (n, p)-cell in E^{n+1} and f: $[a, b] \to X \setminus A$ a closed continuous path in $X \setminus A$. That A_* and f^* are linked, or not, does not depend from the choice of A_* and f^* (but it depends from p). If A_* and f^* are linked, then we say that f *is linked to the* (n, p)*-cell* (X, A). In the opposite case f *is not linked to* (X, A).

Thereafter we define the so called *homological* (n, p)*-pseudomanifolds.*

A compact pair (X, A) is a *homological* (n, p)*-pseudomanifold* satisfies the conditions:

(1) $X \setminus A$ is a nonvoid connected space with a countable base.

(2) There exists a base σ in $X \setminus A$ such that, for any $U \in \sigma$, the pair $(\bar{U}, \bar{U} \setminus U)$ is an (n, p)-cell.

(3) There exists a base Σ in $X \setminus A$ such that Σ be closed under intersections (i.e. from $U_1, U_2 \in \Sigma$ and $U_1 \cap U_2 \neq \phi$ follows $U_1 \cap \cap U_2 \in \Sigma$) and such that, for $U \in \Sigma$ and $q > n$, $H_q(X, X \setminus U) = 0$ holds.

The question should now be raised: *Is assumption* (3) *independent from* (1) *and* (2) *or not?* I believe the answer will be yes, but in the case X lying in E^{n+1}, (3) is a consequence of (2).

We mention that *every compact pair* $(X\ A)$, *where* $X \setminus A$ *is an euclidean* n*-dimensional manifold with a countable base, forms, for every* p, *a homological* (n, p)*-pseudomanifold.*

Let (X, A) be a homological (n, p)-pseudomanifold and U a domain in $X \setminus A$. We say that U is an s-regular domain if $H_n(X, X \setminus U) \neq \neq 0$.

Theorem 9. *From the* s*-regularity of* U *follows* $H_n(X, X \setminus U) \approx \approx Z_p$.

Theorem 10. *Let* U *and* U' *be two* s*-regular domains such that* $U' \subset U$. *Then the injection* i: $(X, X \setminus U) \subset (X, X \setminus U')$ *induces an isomorphism* $i_{*n}: H_n(X, X \setminus U) \to H_n(X, X \setminus U')$. We shall denote this isomorphism by $H(U', U)$ and its inverse by $H(U, U')$.

Let $\Sigma^s_{(X,A)}$ denote the family of the s-regular domains of (X, A).

Let $g = [U_1, U_2, \ldots, U_n = U_1]$ be a closed $\Sigma^s_{(X,A)}$-chain or shortly a *closed* s*-chain*. Put

$$g_* = H(U_n, U_{n-1}) \ldots H(U_3, U_2) H(U_2, U_1) .$$

g_* is an automorphism of the group $H_n(X, X \setminus U_1)$. If g_* is the identical automorphism, then we say that g *preserves the orientation*, in the contrary case g *changes the orientation.*

The following unsolved problem should be mentioned:

Is it true, for every orientation-changing, closed s*-chain* $g = = [U_1, U_2, \ldots, U_n = U_1]$ *that, for* $u \in H_n(X, X \setminus U_1)$, $g_*(u) = -u$ *holds?*

A domain U of $X \setminus A$ is called an *orientable domain* if every closed s-chain lying in U preserves the orientation. Thus we obtain the following main theorem:

Theorem 11. *The domain* U *of* $X \setminus A$ *is orientable if and only if it is* s*-regular.*

Let us consider again a continuous closed path f: $[a, b] \to X \setminus A$.

Theorem 12. *There exists a closed s-chain associated with f, and either every closed s-chain associated with f preserves the orientation, or every one changes it.* In the first case we say that *f preserves the orientation,* and in the second, that *f changes it.*

Theorem 13. *A domain U in* $X \setminus A$ *is orientable if and only if every closed continuous path in U preserves the orientation.*

Theorem 14. *Let J be a Jordan curve in* $X \setminus A$. *Then either every closed Jordan path with the body J preserves the orientation, or every such path changes it.* In the first case we say that *J preserves the orientation,* and in the second, that *J changes it.*

Let us mention here that the following couples of Theorems: 3. and 12., 4. and 13., 5. and 14., have a common category-theoretical background (See e.g. [4], [5], [6]).

The following theorem equally holds:

Theorem 15. *Every orientation-preserving Jordan curve J in* $X \setminus A$ *may be covered by any s-regular domain.*

And now we show some connections between the *k*-manifolds and the homological (n, p)-pseudomanifolds, embedded into euclidean spaces.

Theorem 16. *Let* (X, A) *be a compact pair lying in* E^{n+1}. *Suppose that, for any p,* (X, A) *is a homological* (n, p)*-pseudomanifold. Then* (X, A) *becomes a k-manifold in* E^{n+1}.

Theorem 17. *Let* (X, A) *be a k-manifold in* E^{n+1}. *Then it is valid, for every p, that* (X, A) *is a homological* (n, p)*-pseudomanifold.*

Here are some further theorems, concerning homological (n, p)-pseudomanifolds, lying in E^{n+1}.

Let p be a prime number and (X, A) *a homological* (n, p)*-pseudomanifold lying in* E^{n+1}. According to theorem 16, (X, A) *is also a k-manifold in* E^{n+1}.

Theorem 18. *Suppose that* $p \neq 2$, *and let U be a domain in* $X \setminus A$. *In order that U should be a two-sided domain, it is necessary and* *sufficient, that* U *be* s-*regular.*

Theorem 19. *Suppose that* $p \neq 2$, *and let* f *be a continuous closed path in* $X \setminus A$. f *preserves its banks if and only if it preserves the oriantation.*

Theorem 20. *Let us suppose that* (X, A) *is not only a* k-*manifold, but also an* (n, p)-*cell in* E^{n+1}. *Let* f *be a continuous closed path in* $X \setminus A$. *The path* f *is linked to* (X, A) *if and only if it changes its banks.*

Theorem 21. (X, A) *is a bounded* k-*manifold if and only if it satisfies the following condition:* $i_{*n}(H_n(X)) = 0$, *where* $i: (X, \phi) \subset (X, A)$ *is the injection of* X *into* (X, A).

Theorem 22. *If* $A = \phi$, *then* (X, A) *is a closed* k-*manifold in* E^{n+1}

Theorem 22. means, in reality, that from the local bipartition of E^{n+1} by a figure X follows the global bipartition of E^{n+1} by the same figure.

Theorem 23. *Let* (X, A) *be a* k-*manifold in* E^{n+1} *with a local euclidean boundary point* (that means, that there exists a point $q \in A$ and a neighbourhood W of q in X such that $(W, W \cap A)$ be homeomorphic to (F, E^{n-1}), where F is a halfspace of E^n bounded by the hyperplane E^{n-1}). *Then* (X, A) *is a bounded* k-*manifold in* E^{n+1}.

And at least let us say a few words about the connection of the homological (n, p)-pseudomanifolds with the Brouwerian pseudomanifolds.

Le The Lan, assistant professor of the teachers training college of Vinh (Vietnam) has proved the following

Theorem. *Let* (X, A) *be a triangulable compact pair. That means, there exists a geometrical complex* W *and a subcomplex* W_1 *of* W *such that* $(\widetilde{W}, \widetilde{W}_1)$ *is homeomorphic to* (X, A) ($\widetilde{W}$ is the body of W and $\widetilde{W}_1$ the same for W_1). *In order that* (X, A) *should be a homological* $(n, 2)$-*pseudomanifold, it is necessary and sufficient, that* $W \setminus W_1$ *be an* n-*dimensional pseudomanifold, every star of which is strongly connected* *(and therefore the stars are also n-dimensional pseudomanifolds).*

For $p \neq 2$, (X, A) is a homological (n, p)-pseudomanifold if and only if, besides the previous conditions, the stars are orientable n-dimensional pseudomanifolds.

The following theorem of Le The Lan concerns the connection between the homological (n, p)-pseudomanifolds and the $V_0^n(S)$ of Čech:

The triangulable compact pair (R, S) is a homological (n, p)-pseudomanifold if and only if $R \setminus S$ is connected and R is a $V_0^n(S)$ (obviously with the coefficient group Z_p).

Thus we can see now also a connection of the $V_0^n(S)$ of Čech with the Brouwerian pseudomanifolds.

As to the connection between the homological (n, p)-pseudomanifolds and the $V_0^n(S)$ in the non-triangulable case, our knowledge is far from sufficient. Up to now no final solution has been found.

REFERENCES

[1] P. Alexandroff – H. Hopf, *Topologie,* Berlin, 1935.

[2] P. Alexandroff – L. Pontrjagin, Les variétés à *n*-dimensions généralisées, *Compt. Rend. Acad. Sci.*, 202 (1936), 1327-1329.

[3] R.H. Bing, A set is a 3-cell if its Cartesian product with an arc is a 4-cell, *Proc. Am. Math. Soc.*, 12 (1961), 13-19.

[4] M. Bognár, Über Kategorien mit Involution, *Acta Math. Acad. Sci. Hung.*, 23 (1972), 147-174.

[5] M. Bognár, Die *i*-Kategorie der stetigen Wege, *Acta Math. Acad. Sci. Hung.*, 24 (1973), 155-178.

[6] M. Bognár, Über Lageeigenschaften verallgemeinerter Mannigfaltigkeiten, *Acta Math. Acad. Sci. Hung.*, 24 (1973), 179-198.

[7] L.E.J. Brouwer, Beweis der Invarianz des *n*-dimensionalen Gebiets, *Math. Ann.*, 71 (1911), 305-313.

[8] L.E.J. Brouwer, Beweis des Jordanschen Satzes für den *n*-dimensionalen Raum, *Math. Ann.*, 71 (1911), 314-319.

[9] E. Čech, Sur les nombres de Betti locaux, *Ann. of Math.*, 35 (1934), 678-701.

[10] M.L. Curtis, Cartesian products with intervals, *Proc. Am. Math. Soc.*, 12 (1961), 819-820.

[11] W.W. Flexner, On topological manifolds, *Ann. of Math.*, 32 (1931), 393-406.

[12] W.W. Flexner, The Poincaré duality theorem for topological manifolds, *Ann. of Math.*, 32 (1931), 539-548.

[13] E.R. van Kampen, *Die kombinatorische Topologie und die Dualitätssätze,* Acad. Proofschriff. Dissertation, Leiden, 1929.

[14] B.V. Kerékjártó, *Vorlesungen über Topologie I. Flächentopologie,* Berlin, 1923.

[15] K.W. Kwun – F. Raymond, Factors of cubes, *Am. J. Math.*, 84 (1962), 433-439.

[16] S. Lefschetz, *Topology,* New-York, 1930.

[17] S. Lefschetz – W.W. Flexner, On the duality theorems for the Betti numbers of topological manifolds, *Proc. Nat. Acad. Sci. U.S.*, 16 (1930), 530-533.

[18] S. Lefschetz, On generalized manifolds, *Am. J. Math.*, 55 (1933), 469-504.

[19] L.L. Lininger, Somes results on crumpled cubes. *Trans. Am. Math. Soc.*, 118 (1965), 534-549.

[20] B. Mazur, A note on some contractible 4-manifolds, *Ann. of Math.*, 73 (1961), 221-228.

[21] V. Poénaru, Les décompositions de l'hypercube en produit topologique, *Bull. Soc. Math. France,* 88 (1960), 113-129.

[22] L. Pontrjagin, Über den algebraischen Inhalt topologischer Dualitätssätze, *Math. Ann.*, 105 (1931), 165-205.

[23] F. Raymond, Separation and union theorems for generalized manifolds with boundary, *Michigan Math. J.*, 7 (1960), 1-21.

[24] B. Riemann, *Ueber die Hypothesen, welche der Geometrie zu Grunde liegen,* Habilitationsschrift, 1854.

[25] A. Schoenflies, Über einen grundlegenden Satz der Analysis Situs, *Nachr. Akad. Wiss. Göttingen,* (1902), 185-192.

[26] L. Vietoris, Über die Symmetrie in den Zusammenhangszahlen kombinatorischer Mannigfaltigkeiten, *Monatsh. für Math. u. Phys.*, 35 (1928), 165-174.

[27] R.L. Wilder, On the properties of domains and their boundaries in E_n, *Math. Ann.*, 109 (1934), 273-306.

[28] R.L. Wilder, Generalized closed manifolds in n space, *Ann. of Math.*, 35 (1934), 876-903.

[29] R.L. Wilder, *Topology of manifolds,* New York, 1949.

COLLOQUIA MATHEMATICA SOCIETATIS JÁNOS BOLYAI
8. TOPICS IN TOPOLOGY, KESZTHELY (HUNGARY), 1972.

RELATIVE EMBEDDABILITY OF COMPACTA IN EUCLIDEAN SPACES

H.G. BOTHE

Let X be an n-dimensional metrizable spearable space and I^m the m-dimensional unit cube in the euclidean m-space E^m. Then by the well-known Menger – Nöbeling – Hurewicz embedding theorem there is an embedding of X into I^{2n+1}. Here we shall be concerned with a relative version of this theorem, i.e. we shall ask the following question: Let X and I^m $(m \geq 2n+1)$ be as above, X_0 a closed subset of X and $f_0\colon X_0 \to I^m$ an embedding. Is it possible to extend f_0 to an embedding $f\colon X \to I^m$?

In general the answer is "no" even in the case when $n = 0$ and m is arbitrarily high (example 1) or when X is compact, $n = 2$ and m is arbitrarily high (example 2). But if X_0 is compact the following theorems give sufficient conditions for an embedding of X_0 to be extendable to an embedding of X. In a certain sense these conditions are also necessary (see theorem 3 at the end of the paper).

Theorem 1. *Let X_0 be a compact subspace of the n-dimensional metric separable space X and $f_0\colon X_0 \to I^m$ an embedding. Then, if*

$n = 0$ *or* $n = 1$ and $m \geqslant 2n + 1$, *this embedding can be extended to an embedding of* X *into* I^m.

Theorem 2. *Let again* X_0 *be a compact subset of the* n*-dimensional metric separable space* X *and* $f_0\colon X_0 \to I^m$ *an embedding. Then, if* $m \geqslant 2n + 1$ *and* $E^m - f_0(X_0)$ *is* $1-$ULC, *the embedding can be extended to an embedding of* X *into* I^m.

(A metric space T is said to be $1-$ULC if for each $\epsilon > 0$ there is a $\delta > 0$ such that each mapping k of the unit circle S^1 into T, for which $k(S^1)$ has diameter less than δ, is homotopic to a constant map in a subset of T with diameter less than ϵ.)

Remark 1. We shall show that under the hypotheses of the theorems the set of all embeddings of X into I^m which are extensions of f_0 is dense in the set of all mappings $g\colon X \to I^m$ which are extensions of f_0 such that each mapping of this kind can be approximated by an embedding.

Remark 2. For a theorem concerning the extension of an embedding $f_0\colon X_0 \to I^m$ to an embedding $f\colon X \to I^{m+1}$ see [2, p. 179].

Remark 3. The proof of theorem 2 is an application of Stan'ko's results about embeddings of compacta in euclidian spaces.

Before proving the theorems we give the two examples which were announced above.

Example 1. Let X_0 be the set of all non negative rationals and $X = X_0 \cup \{-1, -\frac{1}{2}, -\frac{1}{3}, \ldots\}$. There is an embedding $f_0\colon X_0 \to I^m$ such that $f_0(X_0)$ is the set of all rational points in I^m and $f_0(0) = \left(\frac{1}{2}, \frac{1}{2}, \ldots, \frac{1}{2}\right)$. Since every point near $\left(\frac{1}{2}, \frac{1}{2}, \ldots, \frac{1}{2}\right)$ is a limit point of $f_0(X_0)$ the extension of f_0 to an embedding of X is impossible.

Example 2. Let X be the disjoint union of a disk D and a Cantor set C while X_0 is the union of C and the boundary curve S^1 of D. For $m \geqslant 3$ there is an embedding g of C into I^m such that $I^m - g(C)$ is not simply connected [1]. We define the embedding $f_0\colon X_0 \to I^m$ such that $f_0|_C = g$ and $f_0(S^1)$ is not contractible in $E^m -$

$-f_0(C)$. Then f_0 obviously can not be extended to an embedding $f: X \to I^m$.

We now state a lemma which we shall use in the proof of the theorems. Since simplicial complexes are needed in this connection we repeat the following definition: A *simplicial complex* $\mathfrak{K}$ in E^m is a family of euclidian simplexes s in E^m such that (1) each face of a simplex of $\mathfrak{K}$ belongs to $\mathfrak{K}$, (2) the intersection of two simplexes of $\mathfrak{K}$ is empty or a common face of these simplexes and (3) if p is a point in the *polyhedron* $|\mathfrak{K}| = \bigcup_{s \in \mathfrak{K}} s$ of $\mathfrak{K}$, then there is a neighborhood of p in E^m which intersects only a finite number of simplexes of $\mathfrak{K}$. By (3) a complex $\mathfrak{K}$ can not contain more than countably many simplexes, and the polyhedron $|\mathfrak{K}|$ of $\mathfrak{K}$ is compact if and only if the complex is finite. We say $\dim \mathfrak{K} \leqslant n$ if all simplexes of $\mathfrak{K}$ are at most n-dimensional.

Lemma. *Let A be an at most n-dimensional compact subset of I^m and $\mathfrak{K}$ an at most n-dimensional simplicial complex in I^m, $m \geqslant 2n+1$. We assume that one of the following conditions is satisfied:*

(a) $n = 0$ *or* $n = 1$.

(b) $n > 1$ *and* $E^m - A$ *is* $1-$ULC.

Then, given a positive number a and two finite subcomplexes $\mathfrak{K}_1$ and $\mathfrak{K}_2$ of $\mathfrak{K}$ such that $|\mathfrak{K}_1| \subseteq \operatorname{Int} |\mathfrak{K}_2|$ (in the topology of $|\mathfrak{K}|$), there is a topological embedding $\varphi: |\mathfrak{K}| \to I^m$ such that

(1) $d(x, \varphi(x)) < a$ *for all* $x \in |\mathfrak{K}|$,

(2) φ *is the identity on* $|\mathfrak{K}| - |\mathfrak{K}_2|$,

(3) $\varphi(|\mathfrak{K}_1|) \cap A = \phi$.

To prove the lemma in the case (a) we make the following three remarks.

(A) Let $\mathfrak{K}$ be a simplicial complex (here at most 1-dimensional) in I^m and let $p_1, \ldots, p_r$ be vertices of $\mathfrak{K}$. Then there are arbitrarily small positive numbers γ such that the following holds: If $p'_1, \ldots, p'_r$ are points in I^m such that $d(p_i, p'_i) < \gamma$ $(i = 1, \ldots, r)$, then, replacing in $\mathfrak{K}$ the vertices p_i by p'_i we get a new simplicial complex $\mathfrak{K}'$ in I^m, and $|\mathfrak{K}|$ can be mapped onto $|\mathfrak{K}'|$ by a homeomorphism which does not move any point farther than γ.

(B) Let $\mathfrak{K}$ be a 1-dimensional simplicial complex in I^m and let $s_1, \ldots, s_r$ be 1-simplexes of $\mathfrak{K}$ with end points p_i, q_i $(i = 1, \ldots, r)$. Then there are mutually disjoint neighborhoods U_i of $s_i - \{p_i, q_i\}$ in E^m $(i = 1, \ldots, r)$ such that $U_i \cap |\mathfrak{K}| = s_i - \{p_i, q_i\}$.

(C) Let $s \subseteq I^m$ be a 1-simplex with end points p, q and A an at most $(m-2)$-dimensional compact subset of I^m which contains neither p nor q. Then, if U is a neighborhood of $s - \{p, q\}$ in E^m and $\gamma > 0$, there is an embedding $\psi \colon s \to I^m$ such that $\psi(p) = p$, $\psi(q) = q$, $\psi(s - \{p, q\}) \subseteq U - A$, and $d(x, \psi(x)) < \gamma$ for each $x \in s$.

(C) follows easily from the fact that an at most $(m-2)$-dimensional compact set can not locally separate I^m.

In the case $n = 0$ the lemma follows easily from (A) and the fact that A is nowhere dense in I^m. In the proof of the lemma for $n = 1$ by (A) we may assume that the vertices of $\mathfrak{K}_1$ are all in $I^m - A$. Then using (B) and (C) we get the homeomorphism ψ.

The proof of the lemma in case (b) is a consequence of the following result of M. Stan'ko which was proved for $m \geqslant 6$ in [5] and was announced for $m = 5$ in [4].

Let A be an n-dimensional compact subset in E^m for which $E^m - A$ is $1 - \mathrm{ULC}$. Then, given $\gamma > 0$ and an at most $(m - n - 1)$-dimensional compact polyhedron P in E^m, there is a homeomorphism ψ of E^m onto E^m such that $d(x, \psi(x)) < \gamma$ for each $x \in E^m$, $\psi(x) = x$ for each $x \in E^m$ not in the γ-neighborhood of $P \cap A$ and $A \cap \psi(P) = \phi$.

Now we come to the proof of the theorems 1 and 2. Let X, X_0 and f_0 be as in these theorems. Since every n-dimensional metrizable separable space can be embedded in an n-dimensional compact metrizable space [3, p. 65] we may assume that X is compact.

By $\mathscr{G}$ we denote the set of all mappings $f \colon X \to I^m$. Then

$$d(f, g) = \sup_{x \in X} d(f(x), g(x))$$

defines a metric in $\mathscr{G}$ and $\mathscr{G}$ becomes a complete metric space [3, V.2.]. For $\epsilon > 0$ let $\mathscr{G}_\epsilon$ denote the set of all ϵ-mappings belonging to $\mathscr{G}$, i.e. all mappings $f \in \mathscr{G}$ such that $\operatorname{diam} f^{-1}(p) < \epsilon$ for each $p \in f(X)$. (Here and later on we assume that X is supplied with a fixed suitable metric.) As shown in [3, V.3.] $\mathscr{G}_\epsilon$ is an open subset of $\mathscr{G}$.

We define $\mathscr{F}$ to be the set of all mappings $f \in \mathscr{G}$ which are extensions of f_0, i.e. for which $f|_{X_0} = f_0$. Since $\mathscr{F}$ is closed in $\mathscr{G}$ it is also a complete metric space, and the set $\mathscr{F}_\epsilon = \mathscr{G}_\epsilon \cap \mathscr{F}$ of all ϵ-mappings in $\mathscr{F}$ is an open subset of $\mathscr{F}$. By Tietze's extension theorem [3, p. 82] $\mathscr{F}$ is not empty.

Similarly as in the proof of the Menger – Nöbeling embedding theorem given in [3, V.3.] we proceed as follows: We prove that for each $\epsilon > 0$ the set $\mathscr{F}_\epsilon$ is dense in $\mathscr{F}$. Then by Baire's theorem (the intersection of a countable family of dense open sets in a complete metric space is dense in this space) $\mathscr{F}_0 = \bigcap_{i=1}^{\infty} \mathscr{F}_{\frac{1}{i}}$ is dense in $\mathscr{F}$ and hence $\mathscr{F}_0 \neq \phi$. But $\mathscr{F}_0$ is just the set of all embeddings of X into I^m which extend f_0.

By a *canonical covering* of $X - X_0$ we mean a locally finite and therefore at most countable open covering $\mathfrak{U} = \{U_1, U_2, \ldots\}$ of $X - X_0$ such that

$$(*) \qquad \operatorname{diam} U_i < d(U_i, X_0)\,.$$

Since $\dim (X - X_0) \leqslant n$ there is a canonical covering of $X - X_0$ with order $\leqslant n$ (each point lies in at most $n + 1$ sets of this covering). If $\delta > 0$ this covering may be chosen to be a δ-covering (each of its sets has diameter less than δ). For proofs in this connection see [2, III.1.].

Now we prove that $\mathscr{F}_\epsilon$ is dense in $\mathscr{F}$. Let $f \in \mathscr{F}$ and $\eta > 0$. We have to find a mapping $f' \in \mathscr{F}_\epsilon$ such that $d(f, f') < \eta$. Since X is compact there is a positive number δ such that $\delta < \epsilon$ and $d(f(x), f(y)) < \frac{\eta}{3}$ whenever $d(x, y) < \delta$. Let $\mathfrak{U} = \{U_1, U_2, \ldots\}$ be a canonical δ-covering of $X - X_0$ with order $\leqslant n$. We select points $p_1, p_2, \ldots$ in $I^m - f_0(X_0)$ in general position such that

(**) $$d(p_i, f(U_i)) < \frac{\eta}{3i} \qquad (i = 1, 2, \ldots).$$

This is possible since $f_0(X_0)$ is a compact nowhere dense subset of I^m. If $\mathfrak{N}$ is the nerve of $\mathfrak{U}$ realized in I^m with the vertex p_i corresponding to U_i $(i = 1, 2, \ldots)$, then, by (*) and (**) $f_0(X_0) \cup |\mathfrak{N}|$ is compact. We consider the canonical mapping $g_1 \colon X - X_0 \to |\mathfrak{N}|$ of $X - X_0$ into the nerve of $\mathfrak{U}$ (see [3, p.58]). Then by

$$g(x) = \begin{cases} f_0(x) & \text{if} \quad x \in X_0 \\ g_1(x) & \text{if} \quad x \in X - X_0 \end{cases}$$

a mapping of $\mathscr{F}$ is defined (g is continuous by (*) and (**)). We have $d(f, g) < \frac{2\eta}{3}$ and, since $\mathfrak{U}$ is a δ-covering with $\delta < \epsilon$, the restriction of g to $X - X_0$ is an ϵ-mapping into $|\mathfrak{N}|$.

Now we use the following corollary of our lemma:

Corollary. *If $a < 0$, then there is an embedding $\varphi\colon |\mathfrak{N}| \to I^m$ for which the following conditions hold:*

(1) *$d(x, \varphi(x)) < a$ for each $x \in |\mathfrak{N}|$.*

(2) *There is a neighborhood V of X_0 in X such that φ is the identity on $g(V - X_0)$.*

(3) *If $x \in X - X_0$ is outside the α-neighborhood of X_0, then $\varphi(g(x)) \notin f_0(X_0)$.*

Since X and X_0 are compact there is a positive number a with the following property: If x, y are points in the a-neighborhood of X_0 and $d(g(x), g(y)) < a$, then $d(x, y) < \epsilon$. Let moreover be $a < \frac{\eta}{3}$ and let $\varphi\colon |\mathfrak{N}| \to I^m$ be an embedding as in the corollary. By (2)

$$f'(x) = \begin{cases} g(x) = f_0(x) & \text{if} \quad x \in X_0 \\ \varphi(g(x)) & \text{if} \quad x \in X - X_0 \end{cases}$$

defines a mapping $f' \in \mathscr{F}$. As a consequence of $d(x, \varphi(x)) < a < \frac{\eta}{3}$ and $d(f, g) < \frac{2\eta}{3}$ we have $d(f, f') < \eta$. To prove that f' is an ϵ-mapping we consider a pair of points $x, y \in X$ for which $f'(x) = f'(y)$. If $x, y \in X_0$ or $x, y \in X - X_0$, then $d(x, y) < \epsilon$; for g is one-to-one on X_0 and an ϵ-mapping on $X - X_0$, and $\varphi\colon |\mathfrak{N}| \to I^m$ is one-to-one. If x, y are not both in one of the sets X_0, $X - X_0$, say $x \in X_0$, $y \in X - X_0$, then by (3) y must be in the a-neighborhood of X_0. Using (1) we have $d(g(x), g(y)) < a$ and therefore $d(x, y) < \epsilon$ by the definition of a.

Using a construction similar to that for the second example it is possible to prove:

Theorem 3. *Let be* $n \geqslant 2$, $m \geqslant 2n + 1$, X_0 *an at most* n*-dimensional compactum and* $f_0\colon X_0 \to I^m$ *an embedding for which the following holds: if* X *is an at most* n*-dimensional compactum containing* X_0 *and* $g\colon X \to I^m$ *is a mapping extending* f_0, *then* g *can be approximated arbitrarily close by an embedding* $f\colon X \to I^m$ *which also extends* f_0. *Under these assumptions* $I^m - f_0(X_0)$ *is* $1 -$ ULC.

REFERENCES

[1] W.A. Blankinship, Generalization of a construction of Antoine, *Ann. of Math.*, 53, (1951), 276-297.

[2] K. Borsuk, *Theory of retracts,* Warszawa, 1967.

[3] W. Hurewicz – H. Wallman, Dimension theory, Princeton, 1948.

[4] M.A. Stan'ko, The embedding of compacta in euclidian space, *Dokl. Akad. Nauk SSSR,* 186, (1969) 1269-1272.

[6] M.A. Stan'ko, The embedding of compacta in euclidian space, *Mat. Sbornik,* 83 (125), (1970), 234-255.

COLLOQUIA MATHEMATICA SOCIETATIS JÁNOS BOLYAI
8. TOPICS IN TOPOLOGY, KESZTHELY (HUNGARY), 1972.

THE DIMENSION OF UNIFORM SPACES IN THE THEORY OF TRANSFORMATION GROUPS

S. BUZÁSI – BOGGYUKEVICS

J. de Groot and J. Nagata have characterized n-dimensional topological spaces (under natural conditions) by the possibility of introduction of metrics of special properties; to the 0-dimensional case there correspond the non-archimedean metrics. Following a suggestion of J. Erdős, we have investigated the corresponding problem for topological groups and for transformation groups, aiming at characterizations with the help of *invariant* metrics. One of our important results was the clarification, via the 0-dimensional case, of the decisive role of the dimension of the uniform space of a topological group (and of the uniform spaces connected with transformation groups). We have connected our investigations with the problem of the invariant metrization of factor-spaces formed with respect to subgroups of topological groups. We have also considered the introduction of invariant uniform structures instead of special invariant metrics.

The following theorems and some more pertaining results will be proved in the authors paper "The dimension of uniform spaces in the theory of continuous transformation groups (in Russian)" to appear in Vol.

19. of *Publ. Math. Debrecen.*

In the sequel G will always denote a Hausdorff topological group, and H a closed subgroup of G. By G/H we mean the factor topological space consisting of the right cosets of G according to H, and by the factor-uniformity of G/H the factor uniform structure corresponding to the right uniform structure of G. The right uniform structure of a topological group will be called for brevity'sake the uniformity of the group.

Theorem 1. *The topology of the factor space G/H can be induced by a right invariant uniform structure defined on the set G/H and having weight $\leqslant \mathfrak{m}$ (resp. also 0-dimensional[1]), if and only if in the factor space G/H, H has a neighbourhood base of cardinality $\leqslant \mathfrak{m}$ the members of which are derived from symmetrical sets (resp. from subgroups)[2].*

Theorem 2. *If the topology of the factor space G/H is induced by some right invariant uniform structure defined on the set G/H, then this coincides with the factor uniformity of G/H.*

Theorem 3. *The factor space G/H is metrizable by a right invariant metric, if and only if it contains a countable neighborhood base of H, the members of which are derived from symmetrical sets.*[2]

Theorem 4. *The factor space G/H is metrizable by a right invariant non-archimedean[3] metric, if and only if it is metrizable by a right invariant metric and the factor uniformity of G/H is 0-dimensional.*[2]

Theorem 5. *An (additively written) topological group complete from one side is metrizable by a right invariant non archimedean[3] metric, if and only if it is metrizable and the terms of each of its zero-sequences form a convergent series.*

[1]For uniform spaces we use the notion of "large" covering dimension due to J. Isbell and Yu. Smirnov.

[2]The subset S of the set G/H is said to be derived from a symmetrical set (or subgroup), if the inverse image of S with respect to the natural mapping of G onto G/H is a symmetrical subset (or subgroup) of G.

[3]The metric ρ is non-archimedean if, for any points x, y, z, $\rho(x, z) \leqslant \max(\rho(x, y), \rho(y, z))$.

Theorem 6. *For the metrizability by a right invariant non-archimedean metric of a metric group* G *complete from one side, the condition* $\dim G = 0$ *is in general not sufficient.*[4]

Theorem 7. *The class of topological groups with* 0*-dimensional*[1] *uniformity cannot be characterized (within the class of all topological groups) by topological properties of the corresponding topological spaces.*

Theorem 8. *A uniform group*[5] *is metrizable by a non-archimedean metric invariant from both sides, if and only if it is metrizable and* 0*-dimensional.*

REFERENCES

[1] J. de Groot, Non-archimedean metrics in topology, *Proc. Amer. Math. Soc.*, 7 (1956).

[2] J. Nagata, On a special metric and dimension, *Fund. Math.*, 55 (1964).

[3] Ju. M. Smirnov, On the dimension of proximity spaces, *Mat. Sbor.*, 38 (1956), (in Russian).

[4] J. Isbell, On finite-dimensional uniform spaces, *Pacific J. of Math.*, 9 (1956).

[4] $\dim G$ denotes the covering dimension of the topological space of the group G.

[5] By a uniform group we mean a group on which there is defined a uniform structure, so that the group operation considered as a function of two variables, as well as the forming of the inverse are uniformly continuous. By the metrization and by the dimension of a uniform group we mean the metrization and the dimension[1] of its uniform space.

COLLOQUIA MATHEMATICA SOCIETATIS JÁNOS BOLYAI
8. TOPICS IN TOPOLOGY, KESZTHELY (HUNGARY) 1972.

SOME RECENT RESULTS ON COMPACT HILBERT CUBE MANIFOLDS

T.A. CHAPMAN

1. INTRODUCTION

The purpose of this report is to describe some recent results concerning the point set topology of *Hilbert cube manifolds* (or *Q-manifolds*), i.e., separable metric spaces which have open covers by sets which are homeomorphic to open subsets of the Hilbert cube Q. Some obvious examples are (1) any open subset of Q, and (2) $M \times Q$, for any finite-dimensional manifold M. Some non-obvious examples of Q-manifolds are provided by the following result of J. E. West [11]: *If P is any polyhedron* (i.e., $P = |K|$, *for some countable locally-finite simplicial complex* K), *then* $P \times Q$ *is a* Q*-manifold.*

Triangulation question. *Can every Q-manifold be triangulated,* i.e., *is every Q-manifold homeomorphic to $P \times Q$, for some polyhedron P?*

Classification question. *Are there reasonable homotopy conditions which classify homeomorphic Q-manifolds?*

Strong motives for asking questions of this type can be found in the theory of *s-manifolds,* i.e., separable metric manifolds modeled on the countable infinite product of lines s. Indeed it was shown by D. W. Henderson in [8] that (1) *every s-manifold is homeomorphic to* $P \times Q$, *for some polyhedron* P, *and* (2) *any two s-manifolds of the same homotopy type are homeomorphic.* These results, along with the close technical relationship which exists between s-manifolds and Q-manifolds, make the *Triangulation Question* extremely plausible. In fact, the *Triangulation Question* has recently been answered affirmatively for all compact Q-manifolds [5]. In Section 2 we discuss some consequences of this result and its relationship to the recent triangulation results of Kirby – Siebermann concerning finite-dimensional manifolds [7]. However, it is easily seen that homotopy type is not sufficient to answer the *Classification Question* affirmatively. Just note that Q and $Q \setminus$ point are non-homeomorphic Q-manifolds which have the same homotopy type. Even in the compact case we shall see that homotopy equivalent Q-manifolds are not necessarily homeomorphic. It turns out that this question is related to the concept of simple homotopy type, and an answer involves a proof of the topological invariance of Whitehead torsion. In Section 3 we discuss this result and some of its consequences. In our discussions of both the *Triangulation Question* and the *Classification Question* we will confine our attention mainly to compact Q-manifolds. The situation involving non-compact Q-manifolds is not yet clear.

We will not have much to say about techniques of proof. The main tools are established in [3]. That paper uses a considerable amount of infinite-dimensional machinery – influenced by finite-dimensional techniques. Some of the techniques used are infinite-dimensional surgery and infinite-dimensional handle straightening. These are just infinite-dimensional versions of some finite-dimensional techniques which were used in [7] for the recent triangulation results concerning n-manifolds.

2. TRIANGULATION QUESTION

One of the key techniques that Henderson used to triangulate s-manifolds was his open embedding theorem [8]. The idea was to show that any s-manifold can be openly embedded in s, and then to show that every open subset of s can be triangulated. While it is true that any open subset of Q can be triangulated [2], Henderson's program applied to Q-manifolds does not completely succeed. The reason is that not all Q-manifolds can be openly embedded in Q (consider $S^1 \times Q$, where S^1 is the 1-sphere). However, in [2] the following weaker form of the open embedding theorem for Q-manifolds was established: *If X is any Q-manifold, then $X \times [0, 1)$ can be embedded as an open subset of Q.* This result, combined with the triangulation of open subsets of Q, gives us the following weak triangulation theorem of [2]: *If X is any Q-manifold, then $X \times [0, 1)$ can be triangulated.* Of course, this result gives us little information on the triangulation of compact Q-manifolds. We refer the reader to [1] for a list of earlier results on Q-manifolds.

More recently, the techniques that Kirby – Siebenmann used in studying the triangulation of finite-dimensional manifolds have been applied to Q-manifolds to obtain the following result [5].

Triangulation Theorem. *Every compact Q-manifold can be triangulated.*

In [11] West proved that if P_1, P_2 are compact, connected polyhedra having the same simple homotopy type, then $P_1 \times Q$ is homeomorphic ($\cong$) to $P_2 \times Q$. (Two compact polyhedra are of the same simple homotopy type if either is obtained from the other as a result of a finite number of Whitehead collapsing operations, i.e., simplicial collapses or their inverses.) Thus by using the fact that the inclusion of a compact polyhedron into its regular neighborhood in an Euclidean space is a simple homotopy equivalence, it follows that compact Q-manifolds are *combinatorially* triangulable in the following sense.

Corollary 1. *Any compact Q-manifold is homeomorphic to $M \times Q$, where M is some combinatorial manifold.*

This contrasts sharply with the corresponding finite-dimensional situation, as there exist compact finite-dimensional manifolds which cannot be combinatorially triangulated [6].

We can also use the Triangulation Theorem to obtain the following result on finiteness of homotopy types.

Corollary 2. *If* X *is any compact metric space which is locally triangulable, then* X *has the homotopy type of a finite complex.*

In particular X might be a compact n-manifold. Thus Corollary 2 strengthens the result of Kirby – Siebermann [7], where it is shown that every compact n-manifold has finite homotopy type. It also sheds some light on the more general open question concerning the finiteness of homotopy types of compact ANRs [10].

Classification question. We have already remarked that Q-manifolds of the same homotopy type are not necessarily homeomorphic. However if the manifolds are multiplied by $[0, 1)$, then the following result of [2] shows that homotopy type will alone suffice: *If* X *and* Y *are* Q*-manifolds which have the same homotopy type, then* $X \times [0, 1) \cong \cong Y \times [0, 1)$. Thus we once again see that the multiplication of a Q-manifold by $[0, 1)$ has a taming effect on the manifold. Using similar techniques from [3] the following classification result was established: If X is *a compact contratible* Q*-manifold, then* $X \cong Q$. *Until the recent* solution of the *Classification Question,* these results, and the aforementioned result of West on simple homotopy types, were the primary results known about the classification of Q-manifolds.

Our main result on the classification of Q-manifolds is stated below. It actually gives a topological characterization of simple homotopy equivalences of compact polyhedra. Details of the proof can be found in [4].

Characterization of Simple Homotopy Types. *Let* P_1, P_2 *be compact, connected polyhedra and let* $f\colon P_1 \to P_2$ *be a map. Then* f *is a simple homotopy equivalence iff the map*

$$f \times id\colon P_1 \times Q \to P_2 \times Q$$

is homotopic to a homeomorphism of $P_1 \times Q$ *onto* $P_2 \times Q$.

As an immediate corollary we get the topological invariance of Whitehead torsion. We remark that F. Waldhausen has reportedly obtained an independent and earlier proof of the topological invariance of Whitehead torsion.

Corollary 1. *Let* P_1, P_2 *be compact, connected polyhedra and* *let* $f\colon P_1 \to P_2$ *be a homeomorphism. Then* f *is a simple homotopy equivalence.*

This answer affirmatively a question which was raised by Whitehead in [2]. It also generalizes the result of Kirby – Siebenmann [7], where it is shown that any homeomorphism between compact n-manifolds is a simple homotopy equivalence.

As another application of the *Characterization of Simple Homotopy Types* we get a classification of compact connected Q-manifolds.

Corollary 2. *Let* X, Y *be compact, connected* Q*-manifolds and let*

$$X \cong P_1 \times Q\,, \qquad Y \cong P_2 \times Q$$

be triangulations of X, Y *respectively. Then* $X \cong Y$ *iff* P_1 *and* P_2 *have the same simple homotopy type.*

We can classify certain compact Q-manifolds by homotopy type. All we have to do is choose the manifolds so that their Whitehead groups vanish. In such cases all homotopy equivalences become simple homotopy equivalences. An example of this would be the following: *Let* X, Y *be compact, connected* Q*-manifolds which have the same homotopy type and for which* $\pi_1(X)$ is free abelian. Then $X \cong Y$. References showing that the Whitehead group $Wh(\pi_1(X))$ vanishes can be found in [9].

If the Q-manifolds are sufficiently bad, then it is possible for them to be homotopy equivalent, but not homeomorphic. To see this, let P_1, P_2 be compact, connected polyhedra which have the same homotopy type, but not he same simple homotopy type. Then $P_1 \times Q$ and $P_2 \times Q$ are homotopy equivalent Q-manifolds, but by Corollary 2 they are not homeomorphic. Using the comments made at the beginning of this section, it should be noted that even though $P_1 \times Q \ncong P_2 \times Q$, we do have

$$P_1 \times Q \times [0, 1) \cong P_2 \times Q \times [0, 1)\,.$$

REFERENCES

[1] T.A. Chapman, Hilbert cube manifolds, *Bull. Amer. Mat. Soc.*, 76 (1970), 1326-1330.

[2] T.A. Chapman, On the structure of Hilbert cube manifolds, *Compositio Math.*, to appear.

[3] T.A. Chapman, Surgery and handle straightening in Hilbert cube manifolds, *Pac. J. of Math.*, submitted.

[4] T.A. Chapman, Topological invariance of Whitehead torsion, *Amer. J. of Math.*, submitted.

[5] T.A. Chapman, Compact Hilbert cube manifolds and the invariance of Whitehead torsion, *Bull. Amer. Math. Soc.*, to appear.

[6] R.C. Kirby – L.C. Siebenmann, For manifolds the Hauptvermutung and the triangulation conjecture are false, *Notices Amer. Math. Soc.*, 16 (1969), 695.

[7] R.C. Kirby – L.C. Siebenmann, On the triangulation of manifolds and the Hauptvermutung, *Bull. Amer. Math. Soc.*, 75 (1969), 742-749.

[8][1] D.W. Henderson, Infinite-dimensional manifolds are open subsets of Hilbert space, *Bull. Amer. Math. Soc.*, 75 (1969), 759-762.

[9] J. Milnor, Whitehead torsion, *Bull. Amer. Math. Soc.*, 72 (1966), 358-426.

[10] L.C. Siebenmann, On the homotopy type of compact topological manifolds, *Bull. Amer. Math. Soc.*, 74 (1968), 738-742.

[11] J.E. West, Mapping cyclinders of Hilbert cube factors, *Gen. Top. and its App.*, 1 (1971), 111-125.

[12] J.H.C. Whitehead, On incident matrices, nuclei and homotopy types, *Ann, of Math.*, 42 (1941), 1197-1239.

COLLOQUIA MATHEMATICA SOCIETATIS JÁNOS BOLYAI
8. TOPICS IN TOPOLOGY, KESZTHELY (HUNGARY), 1972.

SOME PROBLEMS CONCERNING MONOTONE DECOMPOSITIONS OF CONTINUA

J.J. CHARATONIK

The bibliography on upper semi-continuous decompositions of continua is rather large. Especially a great number of papers concerns monotone decompositions of irreducible continua. E.g. Z. Janiszewski in [11], B. Knaster in [14] and [15], K. Kuratowski in [16] and [17] and also W.A. Wilson in [26] and [27] investigated such decompositions for metric continua irreducible between two points. A continuation of this topic can be found in a sequence of papers. In the last two decades several papers in this field have appeared, but from some other point of view. E.g. H.C. Miller in [21] has combined the investigations of irreducible with these of unicoherent spaces. E.S. Thomas, Jr. in his extensive study on irreducible continua [25] applied, among other things, the method of inverse limits as well as the notion of aposyndesis introduced by F.B. Jones (see [12] and [13]) to obtain some results on the structure of such continua.

Besides studies of decompositions of irreducible continua investigations of decompositions of arbitrary continua into closed sets or of monotone decompositions (i.e. into closed connected sets) were developed by a number of authors (see K. Kuratowski [18], M.E. Shanks [24]. R.W. FitzGerald and P.M. Swingle [8]). Some of Miller's results of [21] concerning decompositions of continua irreducible between two points were generalized by M.J. Russell in [23] to decompositions of continua irreducible about a finite set. Recently G.R. Gordh, Jr. considered upper semi-continuous monotone decompositions of continua with some special properties, namely of smooth [9] and nearly smooth [10] continua. Such decompositions were studied by the present author for λ-dendroids in [2] and [3], and for arbitrary continua in [6] (see also [7] which is an abstract of [6]).

A *continuum* means a compact connected metric space. It is well known that for every irreducible continuum I there exists an upper semi-continuous decomposition of I into continua (called *layers* of I, see e.g. [20], §48, IV, p. 199) with the property that the decomposition of I into layers is the finest of all linear upper semi-continuous decompositions of I into continua ([20], §48, IV. Theorem 3, p. 200; [17], Fundamental theorem, p. 259).

Let X be a continuum. A decomposition $\mathscr{D}$ of X is said to be *admissible* (see [6]) if 1° $\mathscr{D}$ is upper semi-continuous, 2° $\mathscr{D}$ is monotone, and 3° for every irreducible continuum I in X every layer T_t, $0 \leqslant t \leqslant 1$, of I is contained in some element of $\mathscr{D}$. Of course every continuum X has an admissible decomposition, namely the trivial one, i.e. such that the whole X is the only element of the decomposition.

Problem 1. (see [6]). Characterize continua X which have a non-trivial admissible decomposition.

It is known ([6], Theorem 1) that if a decomposition $\mathscr{D}$ of a continuum X is admissible, then the induced quotient space $X/\mathscr{D}$ is a hereditarily arcwise connected.

If $\mathscr{D}$ and $\mathscr{E}$ are upper semi-continuous monotone decompositions of a continuum X, then $\mathscr{D} \leqslant \mathscr{E}$ means that every element of $\mathscr{D}$ is contained in some element of $\mathscr{E}$, i.e. $\mathscr{D}$ refines $\mathscr{E}$. Thus $\leqslant$ defines a partial ordering on the family of upper semi-continuous monotone decompositions of X (see [25], p. 8; cf. [24], p. 100). It is known (see [6], theorems 2 and 3) that for every continuum X there exists an admissible decomposition of X which is minimal with respect to $\leqslant$, and that this minimal admissible decomposition is unique. The structure of elements of the minimal admissible decomposition of a continuum X can be seen from the following construction (see [6]; cf. also [2], p. 18-24).

Assign to each point $x \in X$ an increasing sequence of continua $A_a(x)$ (where a is a countable ordinal) defined by the transfinite induction. Firstly consider in X all irreducible continua I with $x \in I$, take in each of them the layer $T(x)$ to which x belongs and put $A_0(x) = \overline{\bigcup T(x)}$, where the union in the right side of the equality runs over all irreducible continua I such that $x \in I \subset X$. Secondly suppose that $A_\beta(x)$ are defined for $\beta < a$, and put

$$A_a(X) = \begin{cases} \bigcup \{ \operatorname{Ls}_{n \to \infty} A_\beta(x_n) \colon \lim_{n \to \infty} x_n \in A_\beta(x) \}, & \text{if} \quad a = \beta + 1\,, \\ \overline{\bigcup_{\beta < a} A_\beta(x)}, & \text{if} \quad a = \lim_{\beta < a} \beta\,, \end{cases}$$

where, in the case $a = \beta + 1$, the union is taken over all convergent sequences of points $x_n \in X$ with $\lim_{n \to \infty} x_n \in A_\beta(x)$. Since X, as a metric continuum, is separable, there exists a countable ordinal ξ such that if $\xi < \eta$, then $A_\xi(x) = A_\eta(x)$. Call the continuum $A_\xi(x)$ a *stratum* of the point x in the continuum X. For various points x strata of these points are either disjoint or identical. Thus the relation on X to belong to the same stratum is an equivalence. The decomposition of X into its strata is called *canonical.* It is known ([6], Theorem 4) that for every continuum X the canonical decomposition of X coincides with its minimal admissible decomposition. If the continuum X is irreducible, then its canonical decomposition coincides with the decomposition into layers, and the induced quotient space is an arc. It is well known that there are examples of irreducible continua X with the property that each layer of X is a non-trivial continuum – see e.g. Knaster's Example 5 in [20], §48, I, p. 191. In the above context of ideas one can ask the following problem, which is due to J. Krasinkiewicz.

Problem 2. Let Y be a given hereditarily arcwise connected continuum. Does there exist a continuum X every stratum of which is non-trivial and such that Y is the decomposition space of the canonical decomposition of X?

Recall that a decomposition $\mathscr{D}$ of a metric space X is said to be *continuous* at its element $D \in \mathscr{D}$ (in other words, D is called an element *of continuity* of $\mathscr{D}$) provided that, if $\{D_n\}$ is a sequence of elements in $\mathscr{D}$ and there exists a point $x_n \in D_n$ for $n = 1, 2, \ldots,$ such that the sequence $\{x_n\}$ converges to a point of D, then $\operatorname{Ls}_{n\to\infty} D_n = D$. The decomposition $\mathscr{D}$ is said to be continuous provided it is continuous at each of its elements (see [19], §19, II, p. 185 and [20], §43, V. p. 67; cf. also [25], p. 59). It is known that the family of all elements of continuity of $\mathscr{D}$ is a dense G_δ-set in the quotient space $X/\mathscr{D}$ (see [20], §43, VII, p. 73; cf. also [25], Theorem 1, p. 60). B. Knaster has proved ([15], section 5, p. 574-577) that there exists an irreducible continuum I (of type λ) such that each layer of I is a non-trivial layer of continuity. In view of these it is natural to ask the following

Problem 3. Given a hereditarily arcwise connected continuum Y, does there exist a continuum X every stratum of which is a non-trivial stratum of continuity and such that Y is the decomposition space of the canonical decomposition of X?

Of course the positive answer to Problem 3 yields one to Problem 2.

A hereditarily decomposable and hereditarily unicoherent contin--uum is called a λ-*dendroid.* If a λ-dendroid is arcwise connected, then it is a *dendroid.* It is known ([2], Corollary 1, p. 27) that the decomposition space of a canonical decomposition of a λ-dendroid is a dendroid. Therefore it is natural to ask the following modification of Problem 3.

Problem 4. Given a dendroid Y, does there exist a λ-dendroid X which has properties formulated in Problem 3?

Let a mapping f of a continuum X be monotone. The mapping f is said to belong to the class Φ if for any point $y \in f(X)$, for any point $x \in X$ and for any irreducible continuum I in X it is true that if $x \in f^{-1}(y) \cap I$, then the layer $T(x)$ of x in I is contained in

$f^{-1}(y)$. In other words $f \in \Phi$ if it takes each layer of each irreducible continuum into a point. It follows that a monotone mapping f of a continuum X is in Φ if and only if the induced decomposition of X into continua $f^{-1}(y)$, $y \in f(X)$, is admissible. For some properties of mappings belonging to Φ see [6], section 5. A continuum X is said to belong to the class $\mathscr{K}$ if every monotone mapping of X onto a hereditarily arcwise connected continuum is in Φ. It can be seen that the class $\mathscr{K}$ contains e.g. all λ-dendroids, all irreducible (hence all indecomposable) continua and also all hereditarily arcwise connected continua ([6], corollary 8). The union of a disk and of an arc which has its end point as the only common point with the disk is an example of a continuum having a nontrivial admissible decomposition but not being in $\mathscr{K}$. The known characterizations of continua belonging to $\mathscr{K}$ (see [6], Theorem 5 and Corollary 9; cf. also [7]) are rather external, expressed by transformations into another space, and in fact not too far from the definition. For example one of these characterizations says that a continuum X is in $\mathscr{K}$ if and only if for every mapping f of X onto a hereditarily arcwise connected continuum there exists one and only one mapping g of $\varphi(X)$ onto $f(X)$ such that $f(x) = g(\varphi(x))$ for each $x \in X$, where φ denotes the canonical mapping, i.e., the quotient mapping induced by the canonical decomposition of X. Since the class $\mathscr{K}$ of continua has some nice properties and it seems to be interesting enough for further investigations, the following question is very natural.

Problem 5. Give an internal characterization of continua belonging to the class $\mathscr{K}$.

A continuum X is said to be *monostratic* if it consists of only one stratum, i.e. if the canonical mapping is the trivial one of X into a point (see [6]). Each indecomposable continuum is monostratic. A monostratiform λ-dendroid (see [3]) is an example of a hereditarily decomposable monostratic continuum. An n-dimensional cube, where $n > 1$, is an example of a monostratic continuum which does not belong to $\mathscr{K}$. It is known (see [6], Proposition 11) that a continuum $X \in \mathscr{K}$ is monostratic if and only if every monotone mapping of X onto a hereditarily arcwise connected continuum is trivial.

Relatively little information concerning the inner structure of monostratic continua, in particular of monostratic λ-dendroids, has appeared in the literature. Thus the following modification of Problem 1 is open.

Problem 6. Give an internal characterization of monostratic continua.

Problem 7. Give an internal characterization of monostratic continua belonging to the class $\mathscr{K}$.

Problem 8. Give an internal characterization of monostratic λ-dendroids.

A point p of a continuum X is said to be a *terminal point* of X if every irreducible continuum in X which contains p is irreducible from p to some point ([21], p. 190). It is known that if a λ-dendroid X is monostratic, then every irreducible subcontinuum in X has empty interior ([5], p. 365); this implies that every such X has uncountably many terminal points ([5], p. 367). The known examples show that the set of all terminal points is dense in such X. So we have (see [5], p. 367).

Problem 9. Let a λ-dendroid X be monostratic. Does it follow that the set of all terminal points of X is dense in X?

It is known (see [6], Proposition 19) that monostraticity of continua belonging to $\mathscr{K}$ is an invariant under monotone mappings. The question is asked in [6] whether it is an invariant under open mappings. The answer is negative, as it can be seen from an example given in [1], p. 216: van Dantzig's solenoid is an indecomposable (thus belonging to $\mathscr{K}$) continuum which admits an open mapping onto a circle. But if we assume that X is a λ-dendroid, the answer is unknown. Hence the following two problems, due to J. B. Fugate (see [4], p. 340) are still unanswered.

Problem 10. Is monostraticity of λ-dendroids an invariant under open mappings?

Problem 11. Is monostraticity of λ-dendroids an invariant under confluent mappings?

(a mapping of X onto Y is *confluent* if for every continuum Q in Y

every component of the inverse image of Q is mapped onto the whole Q; see [1], p. 213). Since each open mapping is confluent, the positive answer to Problem 11 implies the positive answer to Problem 10. But even under a more narrow class of mappings the question is open. Namely a mapping from a topological space X to a topological space Y is said to be a *local homeomorphism* if for every point $x \in X$ there exists a neighbourhood U of x such that $f(U)$ is a neighbourhood of $f(x)$ and such that f restricted to U is a homeomorphism between U and $f(U)$. If f is a local homeomorphism, then it is an open mapping.

Problem 12. Is monostraticity of continua an invariant under local homeomorphisms?

A continuum is said to be *stratified* if it has a non-trivial stratification, i.e. if it consists of more than one stratum. In other words, a continuum is stratified if it is not monostratic. It is immediately seen that a continuum is stratified if and only if it has a non-trivial admissible decomposition. A continuum each subcontinuum of which is stratified is called *hereditarily stratified.* It is easy to observe that a continuum is hereditarily stratified if and only if it has singletons as the only monostratic subcontinua. An example of a hereditarily stratified λ-dendroid is given in [4], p. 340, which has a monotone mapping onto a monostratic λ-dendroid. Thus the hereditary stratification of λ-dendroids is not an invariant under monotone mappings.

Problem 13. Is the hereditary stratification of continua an invariant under local homeomorphisms?

Problem 14. Is the hereditary stratification of continua an invariant under open mappings?

A continuum X is said to belong to the *class* $\mathscr{L}$ if it admits a nontrivial admissible decomposition each of whose elements has void interior. In other words, $X \in \mathscr{L}$ if and only if each stratum of X has void interior. The class $\mathscr{L}$ contains by definition all irreducible continua of type λ (see [20], p. 197, the footnote; cf. also [25], p. 13 – continua of type A'), all hereditarily arcwise connected continua, and also all smooth continua (see [9], Theorem 5.2, p. 58). There are examples of continua described in [6] which show that neither $\mathscr{K} \setminus \mathscr{L}$ nor $\mathscr{L} \setminus \mathscr{K}$ is empty. It is known (see [6], Proposition 24) that if a continuum X is in the class $\mathscr{L}$, then every monostratic subcontinuum of X has void interior. This property does not characterize continua of the class $\mathscr{L}$ – there is a continuum K each monostratic subcontinuum of which is a singleton, and which is not in $\mathscr{L}$ (see [6]). So we have

Problem 15. Characterize continua belonging to the class $\mathscr{L}$.

The continuum K mentioned above is not a λ-dendroid. The following problem (see [6]) is open.

Problem 16. Let every monostratic subcontinuum of a λ-dendroid X have void interior. Does it follow that X is in $\mathscr{L}$?

L. Mohler has proved ([22], Theorem 6, p. 73) that if an irreducible continuum X is of type λ and if f is a local homeomorphism defined on X, then $f(X)$ is an irreducible continuum of type λ. The question arises if this result can be extended to continua of the class $\mathscr{L}$ (not necessarily irreducible). So we have

Problem 17. Let a continuum X belong to $\mathscr{L}$, and let f be a local homeomorphism defined on X. Does if follow that $f(X)$ is in the class $\mathscr{L}$?

Although property of being an irreducible continuum of type λ is not preserved under open mappings (see [1], p. 216; cf. [22], p. 73), if we neglect the irreducibility, we obtain the following

Problem 18. Let a continuum X belong to $\mathscr{L}$ and let f be an open mapping defined on X. Does it follow that $f(X)$ is in the class $\mathscr{L}$?

REFERENCES

[1] J.J. Charatonik, Confluent mappings and unicoherence of continua, *Fund, Math.*, 56 (1964), 213-220.

[2] J.J. Charatonik, On decompositions of λ-dendroids, *Fund. Math.*, 67 (1970), 15-30.

[3] J.J. Charatonik, An example of a monostratiform λ-dendroid, *Fund. Math.*, 67 (1970), 75-87.

[4] J.J. Charatonik, Remarks on some class of continuous mappings of λ-dendroids, *Fund. Math.*, 67 (1970), 337-344.

[5] J.J. Charatonik, Irreducible continua in monostratiform λ-dendroids, *Bull. Acad. Polon. Sci.*, ser. sci. math., astr. et phys. 19 (1971), 365-367).

[6] J.J. Charatonik, On decompositions of continua, *Fund. Math.*, 79 (1973), 113-130.

[7] J.J. Charatonik, Monotone decompositions of continua, *Bull. Acad. Polon. Sci.*, ser. sci. math., astr. et phys., 20 (1972), 567-570.

[8] R.W. FitzGerald – P.M. Swingle, Core decompositions of continua, *Fund. Math.*, 61 (1967), 35-50.

[9] G.R. Gordh, Jr., On decompositions of smooth continua, *Fund. Math.*, 75 (1972), 51-60.

[10] G.R. Gordh, Jr., Concerning closed quasi-orders on hereditarily unicoherent continua, *Fund. Math.*, 78 (1973), 61-73.

[11] Z. Janiszewski, Sur les continus irréductibles entre deux points, *Journal de l'Ecole Polytechnique*, (2) 16 (1912), 79-170.

[12] F.B. Jones, Concerning non-aposyndetic continua, *Amer. J. Math.*, 70 (1948), 403-413.

[13] F.B. Jones, Concerning aposyndetic and non-aposyndetic continua, *Bull. Amer. Math. Soc.*, 58 (1952), 137-151.

[14] B. Knaster, Sur les ensembles connexes irréductibles entre deux points, *Fund. Math.*, 10 (1927), 276-297.

[15] B. Knaster, Un continu irréductible à décomposition continue en tranches, *Fund. Math.*, 25 (1935), 568-577.

[16] K. Kuratowski, Théorie des continus irréductibles entre deux points I, *Fund. Math.*, 3 (1922), 200-231.

[17] K. Kuratowski, Théorie des continus irréductibles entre deux points II, *Fund. Math.*, 10 (1927), 225-275.

[18] K. Kuratowski, Sur les decompositions semi-continues d'espaces metriques compacts, *Fund. Math.*, 11 (1928), 169-185.

[19] K. Kuratowski, *Topology I,* Academic Press and PWN, Warszawa 1966.

[20] K. Kuratowski, *Topology II,* Academic Press and PWN, Warszawa 1968.

[21] H.C. Miller, On unicoherent continua, *Trans. Amer. Math. Soc.*, 69 (1950), 179-194.

[22] L. Mohler, On locally homeomorphic image of irreducible continua, *Colloq. Math.*, 22 (1970), 69-73.

[23] M.J. Russell, Monotone decompositions of continua irreducible about a finite set, *Fund. Math.*, 72 (1971), 255-264.

[24] M.E. Shanks, Monotone decompositions of continua, *Amer, J. Math.*, 67 (1945), 99-108.

[25] E.S. Thomas, Jr., *Monotone decompositions of irreducible continua,* Dissert. Math., (Rozprawy Matem.) 50, Warszawa 1966.

[26] W.A. Wilson, On the structure of a continuum, limited and irreducible between two points, *Amer. J. Math.*, 48 (1926), 147-168.

[27] W.A. Wilson, On upper semi-continuous decompositions of compact continua, *Amer. J. Math.*, 54 (1932), 377-386.

COLLOQUIA MATHEMATICA SOCIETATIS JÁNOS BOLYAI
8. TOPICS IN TOPOLOGY, KESZTHELY (HUNGARY), 1972.

ON PROJECTIVE HOMOLOGY AND HOMOTOPY GROUPS

G. CHOGOSHVILI

The theory of spectra – direct and inverse systems – of groups is used in algebraic topology for the extension of a homology functor given on some category of objects to another category, e.g., from the category of complexes or polyhedra to the category of topological spaces. But the theory can also be used for transition from one homology theory to another, or from homology structures to homotopy structures.

In this paper we mainly consider projective homology groups of spaces, a generalization of Steenrod's homology groups, and obtained from the so-called projective homology groups of (infinite) complexes. The latter groups as well as the spectral homology groups of complexes are defined using the theory of spectra: the spectral group of a complex is the inverse limit of the homology groups of open finite subcomplexes of the given complex, while the projective group is the homology group of the chain complex obtained as the inverse limit of the chain complexes of the corresponding subcomplexes [4]. The technique by which the generalized Steenrod groups of a space are defined using the theory of spectra, with minor alterations, will be used to define the homotopy groups as well.

Below some results obtained recently by different authors, and dealing with an unexpected relation between Steenrod's and Kolmogoroff's homology groups [13], Steenrod's duality for homology sequences [11] and with the problem of representation of Steenrod's duality in terms of linking coefficients [9] will also be indicated.

1. Let $\mathbf{R} = \{R\}$ be a category of topological spaces, $G = \{G\}$ a category of abelian groups, $\mathbf{K} = \{K\}$ a category of locally finite simplicial complexes with the inclusions $i_{KL}: K \subset L$, and $\mathbf{O}_R = \{0\}$ some directed system of open coverings O of a space R from $\mathbf{R}$. A map f of the set of vertices of a complex K into the set of vertices of a complex V will be called simplicial modulo K', $K' \subset K$, if it maps every simplex from $K \setminus K'$ onto a simplex of V. f will be called quasisimplicial if it is simplicial modulo a finite subset K' of K. A map f of the vertices of a complex K into the points of a space R is called regular relative to $\mathbf{O}_R$ if it is a quasisimplicial map of K into the vietorisian V [4, p. 125] of every covering O from $\mathbf{O}_R$.

Let $\Omega = \{\alpha\}$ be the system of pairs $\alpha = (K, f)$, where $f: K \to R$ is a map regular relative to $\mathbf{O}_R$, ordered as follows: $\alpha < \beta$, $\beta = (L, g)$, if K is a subcomplex of the complex L, and $gi_{\alpha\beta} = f$, where $i_{\alpha\beta}$ denotes the inclusion $K \subset L$. If the category $\mathbf{K}$ together with two complexes contains their sum as well, then Ω is a directed system.

Let $H_\alpha = H_r(K, G)$, $\alpha = (K, f) \in \Omega$, be the projective r-dimensional homology groups (homology groups of infinite cycles) of the complexes K over G. These groups and their homomorphisms $i_{\alpha\beta *}$, induced be the inclusions $i_{\alpha\beta}: K \subset L$, constitute a directed system of groups $\{H_\alpha, i_{\alpha\beta *}\}$. The limit group $H_r(R, G)$ of this system will be called the r-dimensional projective homology group of R over G relative to $\mathbf{K}$ and $\mathbf{O}_R$.

If R is a compact metric space, $\mathbf{K}$ is the category of all countable simplicial complexes, and $\mathbf{O}_R$ is the system of all finite coverings of R, then $H_n(R, G)$ is isomorphic to the Steenrod homology group [17]. For other values of $\mathbf{K}$ and $\mathbf{O}_R$ and for the spectral homology theory of the complexes from $\mathbf{K}$, other homology theories of spaces are obtained, including the Alexandroff–Vietoris–Čech homology theory. Cohomology groups may be constructed analogously.

2. Let $H = \{H_n\}$ be a homology structure on $\boldsymbol{R}$, i.e. a sequence of functors H_n: $\boldsymbol{R} \to \boldsymbol{G}$, $n = 0, 1, 2, \ldots$, which satisfy the axioms of homotopy and dimension [7] and the condition: $H_n(R \vee S) = H_n(R) + H_n(S)$, $n > 0$ (the spaces of $\boldsymbol{R}$ are now considered to be spaces with base points) [3, p. 108]. Let $\boldsymbol{R}_n$ be the subcategory of $\boldsymbol{R}$ consisting of the connected spaces R which satisfy the conditions: $\pi_1(R) = 1$ and $H_i(R) = 0$, $0 < i < n$. Let $\Omega(R, n)$ be the system of pairs $a = (X, f)$, where $X \in \boldsymbol{R}_n$ and f: $X \to R$ is a map from $\boldsymbol{R}$. Let $a < \beta$, $\beta = (Y, g)$, if there exists a map $i^k_{a\beta}$. $X \to Y$ with $gi^k_{a\beta} = f$; then $\Omega(R, n)$ will be a directed system if it is assumed that $X, Y \in \boldsymbol{R} \Rightarrow X \vee Y \in \boldsymbol{R}$.. The groups $H_a = H_n(X)$ and their homomorphisms $i^k_{a\beta *}$: $H_a \to H_\beta$ induced by $i^k_{a\beta}$ constitute a directed system of groups with many homomorphisms $\{H_a, i^k_{a\beta *}\}$. The limit $\Pi_n = \Pi_n(R; H)$ of this system is, by definition, the n-dimensional homotopy group of R corresponding to the homology theory H. The limit is defined as the quotient group of the direct sum $\sum_a H_a$, which is assumed to be a set, modulo the subgroup generated by the elements $h_a - i^k_{a\beta *} h_a$, $h_a \in H_a$. This limit can also be defined by identifying in the union $\bigcup_a H_a$ the elements h_a and h_β if there exists $\gamma > a, \beta$ with $i^k_{a\gamma *} h_a = i^l_{\beta\gamma *} h_\beta$. But in this case the following condition must be satisfied: if $X, Y \in \boldsymbol{R}_n$ and Z is the mapping cylinder of the pair of maps $i^k_{a\beta}$, $i^l_{a\beta}$: $X \to Y$ (i.e. the usual mapping cylinders of $i^k_{a\beta}$ and $i^l_{a\beta}$ identified over the common subspace X [6, p. 37]), then $Z \in \boldsymbol{R}_n$, i.e. $H_i(Z) = 0$, $0 < i < n$. Under these conditions any pair of elements with a common predecessor will have a common successor [8].

If φ: $R \to S$, then $\varphi(a) = (X, \varphi f)$ is a pair of the system $\Omega(S, n)$ and this defines, in an obvious way, the induced homomorphism $\Pi_n(\varphi; H)$: $\Pi_n(R; H) \to \Pi_n(S; H)$. The functor $\Pi_n(\cdot\,; H)$: $\boldsymbol{R} \to \boldsymbol{G}$ obtained in this way satisfies the axioms of homotopy and dimension if it is assumed that $\boldsymbol{R}$, together with R contains the cylinder and the cone over R. The natural transformation $T_n(\cdot)$: $\Pi_n(\cdot\,; H) \to H_n(\cdot)$ is defined by the induced homomorphisms $H_n(f)$: $H_a \to H_n(R)$. With respect to this transformation $\Pi_n(\cdot\,; H)$ satisfies the theorem of Hurewicz: $R \in \boldsymbol{R}_n \Rightarrow \Pi_n(R; H) = H_n(R)$. Furthermore, for every element $p \in \Pi_n(R; H)$ there exists a space $S \in \boldsymbol{R}$, a map ψ: $S \to R$ and an element $q \in \Pi_n(S; H)$ such that $\Pi_n(\psi; H)\,(q) = p$ and $T_n(S)$: $\Pi_n(S; H) \to H_n(S)$ is an isomorphism. Finally, if $\bar{\Pi} = \{\bar{\Pi}_n\}$ is a sequence of functors $\bar{\Pi}_n$: $\boldsymbol{R} \to \boldsymbol{G}$, possessing all the above-mentioned properties relative to the homology structure H, i.e. these functors satisfy axioms of homotopy and dimension, the theorem of Hurewich relative to the natural transformation $\bar{T}_n: \bar{\Pi}_n \to H_n$ and the condition of isomorphism, then there exists a natural transformation $\Pi_n \to \bar{\Pi}_n$ which is an isomorphism for the spaces from R_n. These theorems are proved with the aid of the basic properties of spectra. In case of the usual conditions, $\Pi_n(\cdot\,; H)$ is isomorphic to the homotopy theory constructed by F.–W. Bauer [3] relative to the homology theory H. If H is the singular homology theory, then $\Pi_n(\cdot\,; H)$ is the usual homotopy theory. If H is the Čech theory, then $\Pi_n(\cdot\,; H)$ is the homotopy theory constructed by M. Mrowka [15]. If H is the projective homology group, then $\Pi_n(\cdot\,; H)$ yields a projective homotopy theory. If H is a generalized homology theory, $\Pi_n(\cdot\,; H)$ has all the above-mentioned properties except the dimension axiom. In a similar way a relative homotopy theory corresponding to a given relative homology theory can be constructed, as well as a structure corresponding to a given cohomology theory.

3. Recently L. Mdzinarishvili has shown [13] that Kolmogoroff's homology group [10] over a discrete group of coefficients (i.e. after neglecting its topology) is isomorphic with Steenrod's group [17]. That these groups are isomorphic when the group of coefficients is compact has long been known: using homology groups based on decompositions of space it is proved in [5] that Kolmogoroff's group is isomorphic with the Alexandroff–Čech group over a compact group of coefficients; on the other hand, the latter group under the same condition is isomorphic with Steenrod's group [17]. Using the same group based on decompositions and Milnor's uniqueness theorem [14] L. Mdzinarishvili proves the isomorphism of Kolmogoroff's and Steenrod's groups for any abelian (discrete) group of coefficients. Kolmogoroff's group is constructed in the following way. Let S be a fundamental system of decompositions $D_a = \{e_i^a\}$ of a space R, i.e. the system $\{D_a\}$, which satisfies the conditions: $D_a, D_\beta \in S \Rightarrow D_a \cap D_\beta = \{e_i^a \cap e_j^\beta\} \in S$ and for any compact subset A of X and for any of its external open finite covering $\{U_k\}$ there exists a decomposition $D_a \in S$ such that every element of D_a which intersects A is contained in some U_k. If A is a compact subset of R, then $A \cap S = \{A \cap D_a\}$, where $A \cap D_a = \{A \cap e_i^a\}$ is the induced fundamental system of A. Let $T(S)$ be the

aggregate of unions of a finite number of elements from different decompositions D_a of the system S. n-chains are skew-symmetric, additive functions $c_n(e_0, \ldots, e_n)$, $e_i \in T(S)$, $i = 0, \ldots, n$, with values from G, vanishing if $\bigcap_{i=0}^{n} \bar{e}_i = \phi$. The boundary ∂c_n of c_n on $(e_0, \ldots, e_{n-1})$ is equal to $c_n(R, e_0, \ldots, e_{n-1})$. If $\bigcap_{i=0}^{n} \overline{(e_i \cap A)} = \phi \Rightarrow c_n(e_0, \ldots, e_n) = 0$, then c_n is said to lie on A. If ∂c_n lies on A, then c_r is called a relative cycle mod A, and if, in addition, there exists a c_{n+1} such that $\partial c_{n+1} - c_n$ lies on A, then c_n is said to bound mod A. Kolmogoroff's homology group of the pair (R, A) over G is the quotient group of the group of cycles mod A modulo the subgroup of the bounding cycles. L. Mdzinarishvili proves that Kolmogoroff's homology groups, based on different countable fundamental systems, are isomorphic with each other and that they satisfy all the Eilenberg – Steenrod axioms and two additional axioms of J. Milnor [14]. From this follows, as the result of Milnor's theorem on the uniqueness of Steenrod's homology theory [14], the isomorphism of Steenrod's and Kolmogoroff's groups in the discrete case.

4. G. Laitadze has recently proved [11] Steenrod's duality theorem for homology sequences: if (X, A) is a pair of closed subsets of the n-sphere S^n, $A \subset X \subset S^n$, then the homology sequences of the pairs (X, A) and $(S^n \setminus A, S^n \setminus X)$ are isomorphic, i.e. there exists a commutative diagram

$$\begin{array}{ccccccccc}
\ldots \to & H_r(S^n \setminus A) & \to & H_r(S^n \setminus X) & \to & H_r(S^n \setminus A, S^n \setminus X) & \to & H_{r-1}(S^n \setminus A) & \to \ldots \\
& \downarrow \Phi_A & & \downarrow \Phi_X & & \downarrow \Phi_{X,A} & & \downarrow \Phi_A & \\
\ldots \to & H_r(A) & \longrightarrow & H_r(X) & \longrightarrow & H_r(X, A) & \longrightarrow & H_{r-1}(A) & \longrightarrow \ldots
\end{array}$$

Here Φ_A and Φ_X are Steenrod's dualities [17]. $H_r(X, A)$ is the homology group of the chain complex $\{C_r(X) \setminus C_r(A), \partial\}$, where $C_r(A)$ and $C_r(X)$ are the groups of regular chains of A and X respectively. This group is isomorphic to the homology group of the inclusion map i: $C_r(A) \to C_r(X)$. The relative group $H_r(S^n \setminus A, S^n \setminus X)$ is the homology group of the chain map $i_{\alpha\beta}$: $C_r(\alpha) \to C_r(\beta)$, where $C_r(\alpha)$, $C_r(\beta)$ are groups of the infinite r-chains of the triangulations α, β of $S^n \setminus A$, $S^n \setminus X$ respectively, $\beta > \alpha$ in the sense that every simplex s of β is contained in a uniquely determined simplex t of α and $(i_{\alpha\beta} c_r)(s) = 0$ if $\dim t > r$ and $(i_{\alpha\beta} c_r)(s) = c_r(t)$ if $\dim t = r$. These definitions do not depend on the choice of the triangulations. For homology groups of maps see [7]. The duality isomorphism Φ_{XA} of relative homology groups is established by a certain prismatic construction.

5. The question has long been open whether it is possible, and if so how, to express Steenrod's duality in terms of linking coefficients as is the case with Alexander's duality [1]. Recently H. Inasaridze has made considerable progress in this direction: he has substituted new groups for the groups of Steenrod's duality such that although are not defined in an intrinsic manner, their duality is expressed in terms of linkings. To do this he considers homology groups with coefficients in chain complexes [9].

Let K be a complex and let L denote a pair of divisible groups (G_0, G_1) and an epimorphism $\tau\colon G_0 \to G_1$ such that $\ker \tau$ is a given abelian group G (i.e. L is a resolution of G). Let L^* denote the pair of dual compact groups (G_0^*, G_1^*), $G_0^* \mid G_0$, $G_1^* \mid G_1$, and the monomorphism $\tau^*\colon G_1^* \to G_0^*$ conjugated with τ. Consider the chain complex $\{D_r(K, L), \Delta_r\}$ with

$$D_r(K, L) = C_r(K, G_0) + C_{r+1}(K, G_1)$$

$$\Delta_r(c_r, c_{r+1}) = (\partial_r c_r, \tau' c_r + \partial_{r+1} c_{r+1}),$$

$$\tau' c_r = \tau' \left(\sum_p g_0^p \sigma_p^r \right) = (-1)^r \sum_p \tau(g_0^p) \sigma_p^r, \; g_0^p \in G_0, \sigma_p^r \in K,$$

and the cochain complex $\{D^r(K, L^*), \nabla^r\}$ with

$$D^r(K, L^*) = C^{r+1}(K, G_1^*) + C^r(K, G_0^*),$$

$$\nabla^r(c^{r+1}, c^r) = (\delta^{r+1} c^{r+1}, \overset{*}{\tau}{}' c^{r+1} + \delta^r c^r),$$

$$\overset{*}{\tau}{}' c^{r+1} = \overset{*}{\tau}{}' \left(\sum_p \overset{*}{g}{}_1^p \sigma_p^{r+1} \right) = (-1)^{r+1} \sum_p \overset{*}{\tau}(\overset{*}{g}{}_1^p) \sigma_p^{r+1}, \; \overset{*}{g}{}_1^p \in \overset{*}{G}_1,$$

where $\{C_r(K, \cdot), \partial_r\}$ and $\{C^r(K, \cdot), \delta^r\}$ are the usual chain and cochain complexes of K over $\cdot$ respectively.

Let T_i, $i = 1, 2, \ldots,$ be the triangulations of the n-sphere S^n, where T_{i+1} is the barycentric subdivision of T_i, and let K_i be the smallest subcomplex of T_i containing a given closed subset A of S^n.

The r-dimensional projective homology group $\widetilde{H}_r(A, L)$ of the chain complex

$$\widetilde{D}_* = \{\widetilde{D}_r(A, L), \widetilde{\Delta}_r\} = \varprojlim_i \{(D_r(K_i, L), \Delta_r^{(i)}), \varphi_{i*}\},$$

where φ_i: $K_{i+1} \to K_i$ maps a vertex of K_{i+1} into a vertex of its carrier in K_i, is isomorphic to the $r+1$-dimensional Steenrod group of A over G. This follows from the isomorphism of both of these groups with an auxiliary homology group which satisfies the axioms of Eilenberg–Steenrod and Milnor and consequently is unique by Milnor's theorem mentioned above. On the other hand, the spectral homology group $\bar{H}_r(A, L) = \varprojlim_i \{H_r(K_i, L), \varphi_{i*}\}$ of $\widetilde{D}_*$, where $H_r(K_i, L)$ denotes the homology group of the chain complex $\{D_r(K_i, L), \Delta_r^{(i)}\}$, is isomorphic to the r-dimensional Čech group of A over G. There exists an epimorphism φ of $\widetilde{H}_r(A, L)$ onto $\bar{H}_r(A, L)$ with the kernel $\varprojlim_i B_r(K_i, L) / \widetilde{B}(A, L)$, where $B_r(K_i, L) = \operatorname{Im} \Delta_{r+1}^{(i)}$, $\widetilde{B}_r(A, L) = \operatorname{Im} \widetilde{\Delta}_{r+1}$, i.e. the quotient group of the group of spectral r-boundaries modulo the subgroup of projective r-boundaries. The group $\prod_i B_r(K_i, L) / \widetilde{B}_r(A, L)$ being divisible, there exists a homomorphism ω: $\widetilde{H}_r(A, L) \to \prod_i B_r(K_i, L) / \widetilde{B}_r(A, L)$ with $\omega\psi = \sigma$, where ψ and σ are monomorphisms of $\varprojlim_i B_r(K_i, L) / \widetilde{B}_r(A, L)$ in $\widetilde{H}_r(A, L)$ and $\prod_i B_r(K_i, L) / \widetilde{B}_r(A, L)$, respectively. The map μ: $\widetilde{H}_r(A, L) \to \bar{H}_r(A, L) + \prod_i B_r(K_i, L) / \widetilde{B}_r(A, L)$, $\mu = \varphi + \omega$, is a monomorphism.

Let K_i^* be the complex of the barycentric stars of the simplexes of K_i, let $\bar{K}_i^*$ be the complex of barycentric stars of the complement of K_i in T_i, and let $\widetilde{\varphi}_i$: $K_i^* \to K_{i+1}^*$, $\bar{\varphi}_i$: $\bar{K}_i^* \to \bar{K}_{i+1}^*$ be the maps induced by φ_i. The group $\bar{H}_s(S^{n+1} \setminus A, L^*) = \varinjlim_i \{H_s(\bar{K}_i^*, L^*), \bar{\varphi}_{i*}\}$, where $H_s(\bar{K}_i^*, L^*)$ is the s-dimensional homology group of the chain complex

$\{D_s(\bar{K}_i^*, L^*), \overset{*}{\Delta}_s^{(i)}\}$ and $\bar{\varphi}_{i*}$ is induced by $\bar{\varphi}_i$, is dual to the group $\bar{H}_r(A, L)$, $s = n - r$, and isomorphic to the Čech spactral homology group of $S^{n+1} \setminus A$ over the compact group $G^* = \text{coker } \overset{*}{\tau} \mid G$ [4].

Using the isomorphism B of the group of r-cochains c^r and the group of $s = n - r$ star chains $B(c^r)$ [16, p. 40], consider the group $\{B(d^r_{(i)})\}$ of $B(d^r_{(i)}) = (-1)^r B(c^{r+1}) + B(c^r)$ with the boundary $\overset{*}{\Delta} B(d^r_{(i)}) = (-1)^r \partial B(c^{r+1}) + (-1)^r \overset{*}{\tau} B(c^{r+1}) + \partial B(c^r)$, where $d^r_{(i)} = c^{r+1} + c^r \in D^r(K_i, L^*)$. The homology group of the chain complex $\{(B(d^r_{(i)})), \overset{*}{\Delta}\}$ gives the $s+1$-dimensional star homology of K_i^* over L^*. The chain complex $\varinjlim_i \{\{D_{s+1}(K_i^*, L^*), \Delta^{(i)}_{s+1}\}, \tilde{\varphi}_{i*}\}$ and the cochain complex $\varinjlim_i \{\{D^r(K_i, L^*), \nabla^r_{(i)}\}, \varphi_i^*\}$ are isomorphic, for the groups of the spectra are isomorphic, the isomorphisms being compatible with the homomorphisms of the spectra. The group B_s of s-boundaries of the above chain complex is isomorphic to the group $\tilde{B}^{r+1}$ of $(r+1)$-coboundaries of the cochain complex; the quotient group $D^r(K_i, L^*) / Z^r(K_i, L^*)$ is isomorphic to the group $B_s(K_i^*, L^*)$, where $Z^r(K_i, L^*)$ is the group of r-cocycles. These isomorphisms are compatible with the natural homomorphisms $\bar{\beta}\colon \sum_i D^r(K_i, L^*) / Z^r(K_i, L^*) \to \tilde{B}^{r+1}$ and $\overset{*}{\beta}\colon \sum_i B_s(K_i^*, L^*) \to B_s$, where $\bar{\beta}$ is conjugate to the inclusion map $\beta\colon \tilde{B}_r(A, L) \to \prod_i B_r(K_i, L)$. This implies the isomorphism of $\ker \bar{\beta}$ to $\ker \overset{*}{\beta}$ which, together with $\ker \bar{\beta} \mid \text{coker } \beta$, yields the duality $\ker \overset{*}{\beta} \mid \prod_i B_r(K_i, L) / \tilde{B}_r(A, L)$. It follows that $\tilde{H}_r(A, L)$ is dual to the quotient group of the direct sum $\bar{H}_s(S^{n+1} \setminus A, L^*) + \ker \overset{*}{\beta}$ modulo the annihilator of $\text{Im } \mu$.

We have now to express the dualities $\bar{H}_r(A, L) \mid \bar{H}_s(S^{n+1} \setminus A, L^*)$ and $\ker \overset{*}{\beta} \mid \prod_i B_r(K_i, L) / \tilde{B}_r(A, L)$ in terms of linkings. Define the scalar product $(d_r^{(i)}, d^r_{(i)})$ of a chain $d_r^{(i)} = c_r + c_{r+1} = \sum_p g_0^p \sigma_p^r + \sum_q g_1^q \sigma_q^{r+1}$ from $D_r(K_i, L)$ with a cochain $d^r_{(i)} = c^{r+1} + c^r = \sum_p \overset{*}{g}{}_1^q \sigma_q^{r+1} + \sum_p \overset{*}{g}{}_0^p \sigma_p^r$

from $D^r(K_i, L^*)$ as $\sum_p g_0^p \overset{*}{g}{}_0^p + \sum_q g_1^q \overset{*}{g}{}_1^q$. Represent a $s+1$-dimensional star chain of T_i over L^* as $B(d^r_{(i)})$ and define the index of intersection $I(d_r^{(i)}, B(d^r_{(i)}))$ as $(d_r^{(i)}, d^r_{(i)})$. Assume that $d_r^{(i)}$ and $B(d^{r+1}_{(i)})$ are bounding cycles with non-intersecting carriers. Define the linking coefficient $\mathfrak{B}(d_r^{(i)}, B(d^{r+1}_{(i)}))$ as $I(d_r^{(i)}, B(d^r_{(i)}))$, where $B(d^{r+1}_{(i)}) = \overset{*}{\Delta} B(d^r_{(i)})$. Finally, if $z_r = \{d_r^{(i)}\}$ is a representative of an element from $\bar{H}_r(A, L)$ and $z_s^{(i)} = B(d^{r+1}_{(i)})$ is a cycle from a coordinate of an element of $\bar{H}_s(S^{n+1} \setminus A, L^*)$ define $\mathfrak{B}(z_r, z_s^{(i)})$ as $\mathfrak{B}(d_r^{(i)}, B(d^{r+1}_{(i)}))$. These definitions are independent of the choice of representatives, coordinates, etc.

As for the other duality, define $\mathfrak{B}(d_r^{(i)}, \overset{*}{d}{}_s^{(i)})$, $d_r^{(i)} \in B_r(K_i, L)$, $\overset{*}{d}{}_s^{(i)} \in B_s(K_i^*, L^*)$, as $I(d_r^{(i)}, \overset{*}{d}{}^{(i)}_{s+1})$, where $\overset{*}{\Delta}{}^{(i)}_{s+1} \overset{*}{d}{}^{(i)}_{s+1} = \overset{*}{d}{}_s^{(i)}$ on K_i^* and $\overset{*}{d}{}^{(i)}_{s+1}$ is zero on $T_i \setminus K_i^*$.

6. Generalizations of Steenrod's homology group and duality in different directions have been given by D. Baladze. For references to his papers see [2].

REFERENCES

[1] P.S. Aleksandrov, On certain new results in the combinatorial topology of non-closed sets, *Fund. Math.*, 41 (1954), 68-88 (in Russian).

[2] D.O. Baladze, *K*-groups of homology and cohomology from the viewpoint of homological algebra, *Bull. of the Georgian Acad. of Sciences,* 63 (1971), no. 1, 17-20 (in Russian).

[3] F.W. Bauer, Homotopie und Homologie, *Math. Annalen,* 149 (1963), 105-130.

[4] G. Chogoshvili, *On Homology Theory of Non-closed Sets, General Topology and its Relations to Modern Analysis and Algebra,* Proc. Symp., Prague, (1961), 123-132.

[5] G.S. Chogoshvili, On the equivalence of the functional and spectral theory of homology, *Izvestija AN SSSR, Ser. Math.*, 25 (1951), 421-438.

[6] A. Dold, Halbexakte Homotopiefunktoren, *Lecture Notes in Mathematics,* 12 (1966).

[7] S. Eilenberg – N. Steenrod, *Foundations of Algebraic Topology,* Princeton, 1952.

[8] W. Hurewicz – J. Dugundji – C.H. Dowker, Continuous connectivity groups in terms of limit groups, *Ann. Math.*, 49 (1948), 391-406.

[9] H.N. Inasaridze, Exact homology and linkings for the Steenrod duality, *Dokl. AN SSSR,* 204 (1972), 29-32 (in Russian).

[10] A. Kolmogoroff, Propriétés des groupes de Betti des espaces localelement bicompacts, *C. R. Paris,* 202 (1936), 1325-1327.

[11] G.L. Laitadze, On Steenrod's duality theorem for homological sequences, *Bull. of the Georgian Academy of Sciences,* 68 (1972), no. 1, 37-40. (in Russian).

[12] S. MacLane, *Homology,* Springer, 1963.

[13] L.D. Mdzinarishvili, On the connection between the homological theories of Kolmogorov and Steenrod, *Dokl. AN SSSR,* 203 (1972), 528-531.

[14] J. Milnor, *On the Steenrod Homology Theory,* Preprint, Princeton, 1959.

[15] M. Mrowka, Über Homotopiegrupper, die der Čechschen Homologie zugeordnet sind, *Math. Annalen,* 177 (1968), 310-338.

[16] L.S. Pontrjagin, Topological duality theorems, *Uspehi. Mat. Nauk,* 2 (1947), no. 2, 21-44 (in Russian).

[17] N. Steenrod, Regular Cycles of Compact Metric Spaces, *Ann. Math.,* 41 (1940), 831-851.

COLLOQUIA MATHEMATICA SOCIETATIS JÁNOS BOLYAI
8. TOPICS IN TOPOLOGY, KESZTHELY (HUNGARY), 1972.

INVARIANT STRUCTURES AND TRANSFORMATION GROUPS

Á. CSÁSZÁR

It is a classical fact that the topology of a topological group admits a compatible left invariant uniformity and a compatible right invariant uniformity. We shall generalize this theorem in two directions. First of all, we consider transformation groups instead of topological groups, and, on the other hand, we consider more general structures instead of topologies and uniformities.

1. Let us first describe what kind of structures we shall use.

A *topology* is given on a set E if we assign, to each element $x \in E$, a filter $\mathfrak{B}(x)$ in E such that

(N_0) $\quad V \in \mathfrak{B}(x) \Rightarrow x \in V,$

and

(N_1) $\quad$ For $V \in \mathfrak{B}(x)$, there exists $V_1 \in \mathfrak{B}(x)$ such that

$$y \in V_1 \Rightarrow V \in \mathfrak{B}(y).$$

The elements of $\mathfrak{B}(x)$ are the neighbourhoods of x.

If we assign again, to each point $x \in E$, a filter $\mathfrak{B}(x)$ satisfying (N_0) without necessarily fulfilling (N_1), we obtain a *neighbourhood structure* $\mathfrak{B}$ on E. It can be shown ([5], (5.2)) that this concept is equivalent to that of a "closure operation" in the sense of E. Čech [1].

A *uniformity* on E is a filter $\mathfrak{U}$ in $E \times E$ such that

(U_0) $U \in \mathfrak{U},\ x \in E \Rightarrow (x, x) \in U$,

(U_1) For $U \in \mathfrak{U}$, there exists $U_1 \in \mathfrak{U}$ such that

$$U_1 \circ U_1 \subset U,$$

(U_2) $U \in \mathfrak{U} \Rightarrow U^{-1} \in U$.

If we suppose that the filter $\mathfrak{U}$ satisfies (U_0) and (U_1), we get a *quasi-uniformity* [7]. A filter satisfying (U_0) and (U_2) is a *semi-uniformity* [1]. Finally, let us introduce the term *pseudo-uniformity* for a filter in E satisfying (U_0) without any other restriction.

A *proximity* on E is a binary relation δ for the subsets of E fulfilling the following axioms:

(P_0) (a) $\phi \,\bar{\delta}\, E,\ E \,\bar{\delta}\, \phi$,

(b) $\{x\} \,\delta\, \{x\}$ for $x \in E$,

(c) $A \,\delta\, B,\ A \subset A',\ B \subset B' \Rightarrow A' \,\delta\, B'$,

(d) $A \,\bar{\delta}\, C,\ B \,\bar{\delta}\, C \Rightarrow (A \cup B) \,\bar{\delta}\, C$,

$C \,\bar{\delta}\, A,\ C \,\bar{\delta}\, B \Rightarrow C \,\bar{\delta}\, (A \cup B)$,

(P_1) For $A \,\bar{\delta}\, B$, there exist A', B' such that

$$A' \cup B' = E,\quad A \,\bar{\delta}\, A',\quad B' \,\bar{\delta}\, B,$$

(P_2) $A \,\delta\, B \Rightarrow B \,\delta\, A$.

Let us call *quasi-proximity, semi-proximity* ("proximity" in [1]), and *pseudo-proximity* a relation δ satisfying (P_0) and (P_1), (P_0) and (P_2), or (P_0) alone, respectively.

From a pseudo-uniformity $\mathfrak{U}$ on E, we can deduce a pseudo-proximity δ in the following manner:

$$(1.1) \qquad A \,\delta\, B \Leftrightarrow (A \times B) \cap U \neq \phi \quad \text{for each} \quad U \in \mathfrak{U}\,.$$

If $\mathfrak{U}$ fulfils (U_1) or (U_2), then δ, determined by (1.1), will fulfil (P_1) or (P_2), respectively. Conversely, if δ is a pseudo-proximity, then there exist in general several pseudo-uniformities $\mathfrak{U}$ from which δ can be deduced by (1.1); if δ satisfies (P_1), or (P_2), or (P_1) and (P_2), then $\mathfrak{U}$ may be chosen in such a manner to fulfil (U_1), or (U_2), or (U_1) and (U_2).

If δ is a pseudo-proximity, then a neighbourhood structure $\mathfrak{B}$ can be deduced from δ by

$$(1.2) \qquad V \in \mathfrak{B}(x) \Leftrightarrow \{x\} \,\bar{\delta}\, E - V\,.$$

If δ satisfies (P_1), then $\mathfrak{B}$ fulfils (N_1) (i.e. it is a topology). Conversely, each neighbourhood structure $\mathfrak{B}$ can be deduced from (in general) a lot of pseudo-proximities. If $\mathfrak{B}$ is a topology, then δ may be chosen in such a manner to satisfy (P_1). A neighbourhood structure $\mathfrak{B}$ can be deduced from a semi-proximity iff it fulfils the following separation axiom:

$$(N_2) \quad E - \{y\} \in \mathfrak{B}(x) \Rightarrow E - \{x\} \in \mathfrak{B}(y).$$

Finally, a proximity from which $\mathfrak{B}$ can be deduced exists iff $\mathfrak{B}$ is a completely regular topology.

Of course, a neighbourhood structure may also be deduced directly from a pseudo-uniformity by combining (1.1) and (1.2); then the sets

$$U(x) = \{y : (x, y) \in U\} \qquad (U \in \mathfrak{U})$$

form a base for the filter $\mathfrak{B}(x)$.

2. A *transformation group* Γ on a set E is a group consisting of bijections of E onto itself under the group operation of composition of mappings. It will be always assumed that the group Γ is transitive, i.e. that, for given $x, y \in E$, there exists $a \in \Gamma$ such that $a(x) = y$. We denote, for $p \in E$, by Γ_p the subgroup of Γ consisting of those $a \in \Gamma$ for which $a(p) = p$. The transitivity of Γ implies that Γ_p and Γ_q are isomorphic for $p, q \in E$.

A *neighbourhood structure* $\mathfrak{B}$ is said to be invariant with respect to Γ, or Γ-*invariant*, if

$$V \in \mathfrak{B}(x),\, a \in \Gamma \Rightarrow a(V) \in \mathfrak{B}(a(x))\,.$$

A *pseudo-proximity* δ is said to be Γ*-invariant* if

$$A\, \delta\, B,\, a \in \Gamma \Rightarrow a(A)\, \delta\, a(B)\,.$$

A set $U \subset E \times E$ is said to be Γ-invariant if

$$(x, y) \in U,\, a \in \Gamma \Rightarrow (a(x), a(y)) \in U\,.$$

A *pseudo-uniformity* $\mathfrak{U}$ is said to be Γ*-invariant* if it has (as a filter in $E \times E$) a base composed of Γ-invariant sets.

It is clear that a pseudo-proximity deduced from a Γ-invariant pseudo-uniformity is Γ-invariant as well. Similarly, a neighbourhood structure deduced from a Γ-invariant pseudo-proximity is Γ-invariant as well. Our main purpose is to look for converses of these statements under suitable restrictions.

3. Let Γ be a (transitive) transformation group on E, and $\mathfrak{B}$ a Γ-invariant neighbourhood structure on E.

Theorem 1. *$\mathfrak{B}$ can be deduced from a Γ-invariant pseudo-uniformity iff it satisfies the condition*

(V_0) *For $p \in E$, $V \in \mathfrak{B}(p)$, there exists $W \in \mathfrak{B}(p)$ such that $a \in \Gamma_p$ implies $a(W) \subset V$;*

if (V_0) is fulfilled, there exists a unique Γ-invariant pseudo-uniformity $\mathfrak{U}$ compatible with $\mathfrak{B}$ (i.e. from which $\mathfrak{B}$ can be deduced). *$\mathfrak{B}$ satisfies* (U_i)($i = 1, 2$) *iff $\mathfrak{B}$ satisfies (V_i) where*

(V_1) *For $p \in E$, $V \in \mathfrak{B}(p)$, there exists $W \in \mathfrak{B}(p)$ such that $a \in \Gamma$, $a(p) \in W$ implies $a(W) \subset V$;*

(V_2) *For $p \in E$, $V \in \mathfrak{B}(p)$, there exists $W \in \mathfrak{B}(p)$ such that $a \in \Gamma$, $a(p) \in W$ implies $a^{-1}(p) \in V$.*

The condition (V_0) is always fulfilled if Γ_p is finite. On the other hand, (V_1) or (V_2) implies (V_0). Moreover, (V_1) and (V_2) can be unified in

(V_3) For $p \in E$, $V \in \mathfrak{B}(p)$, there exists $W \in \mathfrak{B}(p)$ such that

$a \in \Gamma$, $p \in a(W)$ implies $a(W) \subset V$.

Hence (V_3) is necessary and sufficient for $\mathfrak{B}$ to possess a compatible uniformity.

Theorem 2. *A pseudo-proximity δ on E can be deduced from a Γ-invariant pseudo-uniformity iff it is Γ-perfect, i.e. if $A \not\delta B$ implies*

$$\cup\, a_i(A) \not\delta \cap\, a_i(B) \quad \text{and} \quad \cap\, a_i(A) \not\delta \cup\, a_i(B)$$

for each collection $\{a_i\colon i \in I\}$ of elements of Γ. If δ is Γ-perfect, then there is a unique Γ-invariant pseudo-uniformity $\mathfrak{U}$ compatible with δ. $\mathfrak{U}$ fulfils (U_i) $(i = 1, 2)$ iff δ satisfies (P_i).

A Γ-perfect pseudo-proximity is clearly Γ-invariant but the converse is not true. From Theorems 1 and 2, we easily get

Theorem 3. *A Γ-invariant neighbourhood structure $\mathfrak{B}$ is compatible with a Γ-perfect pseudo-proximity iff it satisfies (V_0), and if so, then the compatible Γ-perfect pseudo-proximity δ is uniquely determined; δ satisfies (P_i) $(i = 1, 2)$ iff $\mathfrak{B}$ fulfils (V_i).*

It is to be noted that a neighbourhood structure may admit several compatible Γ-invariant pseudo-proximities.

4. A *pseudo-metric* on E is a function $\sigma\colon E \times E \to \boldsymbol{R}$ such that

(M_0) $\sigma(x, y) \geqslant 0$, $\sigma(x, x) = 0$ for $x, y \in E$,

(M_1) $\sigma(x, z) \leqslant \sigma(x, y) + \sigma(y, z)$ for $x, y, z \in E$,

(M_2) $\sigma(x, y) = \sigma(y, x)$ for $x, y \in E$.

A *quasi-metric* is a function σ satisfying (M_0) and (M_1) only.

If Σ is a family of quasi-metrics on E, then a quasi-uniformity $\mathfrak{U}$ can be deduced from Σ; for $\mathfrak{U}$, the sets

$$U_{\sigma_1, \ldots, \sigma_n; \epsilon} = \{(x, y)\colon \sigma_i(x, y) < \epsilon \quad (i = 1, \ldots, n)\}$$

form a base where $\{\sigma_1, \ldots, \sigma_n\}$ is an arbitrary finite subset of Σ and $\epsilon > 0$. If Σ is composed of pseudo-metrics, then $\mathfrak{U}$ is a uniformity. Conversely, for each quasi-uniformity (uniformity), there exists a family of quasi-metrics (pseudo-metrics) from which it can be deduced ([2]).

A *quasi-metric* σ is said to be Γ*-invariant* if

$$\sigma(x, y) = \sigma(a(x), a(y)) \quad \text{for} \quad x, y \in E, a \in \Gamma .$$

Now Theorem 1 can be completed in the following manner:

Theorem 4. *If* $\mathfrak{U}$ *is a* Γ*-invariant quasi-uniformity (uniformity) on* E*, then there exists a family of* Γ*-invariant quasi-metrics (pseudo-metrics) from which* $\mathfrak{U}$ *can be deduced.*

Theorem 5. *If* $\mathfrak{B}$ *is a* Γ*-invariant topology satisfying* (V_1) *or* (V_3)*, then there is a family* Σ *of* Γ*-invariant quasi-metrics or pseudo-metrics, respectively, such that the sets*

$$U_{\sigma_1, \ldots, \sigma_n; \epsilon}(x) \qquad (\sigma_i \in \Sigma\,, \epsilon > 0)$$

constitute a base for $\mathfrak{B}(x)$ $(x \in E)$*. If* $\mathfrak{B}$ *is first countable then* Σ *can be chosen in such a manner that it be composed of a single quasi-metric (pseudo-metric).*

5. If E is a topological group and Γ is the group of the left or right translations, then it may be easily seen that condition (V_3) is fulfilled; then Theorem 1 implies the classical theorem on the existence of a unique left invariant and a unique right invariant compatible uniformity, while Theorem 5 furnishes the existence of a compatible family of left (right) invariant pseudo-metrics, in particular, of a single left (right) invariant compatible pseudo-metric if the group is first countable. If Γ denotes the transformation group generated by all left and right translations (i.e. if Γ consists of the mappings $a_{a,b}(x) = axb$ for $a,\ b \in E$), then (V_0) is fulfilled iff, for $V \in \mathfrak{B}(e)$ (where e is the unity of E), there exists $W \in \mathfrak{B}(e)$ such that $aWa^{-1} \subset V$ for each $a \in E$. Then (V_3) is fulfilled as well and we get the well-known criterion on the existence of a Γ-invariant uniformity i.e. on the identity of the left and right uniformities of the group. From Theorem 2, we obtain that a topological group admits a unique left perfect and a unique right perfect compatible proximity which coincide iff the left and right uniformities coincide. If $E = R$ with the Euclidian topology and the additive group structure, then a compatible translation invariant proximity is given by

$$A \,\delta\, B \Leftrightarrow \bar{A} \cap \bar{B} \neq \phi ;$$

this is distinct from the translation perfect compatible proximity deduced from the Euclidian metric and is, of course, not translation perfect.

Let again $E = R$ and Γ be the group of Euclidian translations. If the neighbourhood base of $x \in R$ consists of the sets $[x, +\infty) - F$ where F is finite, $x \notin F$, then the topology obtained does not satisfy either (V_1) or (V_2). For the topology $\mathfrak{B}^+$ given by the base

$$\{[a, b) : a < b\} \qquad (a, b \in R) ,$$

(V_1) is fulfilled but (V_2) is not. For the topology in which R and the finite subsets of R are only closed, (V_2) is fulfilled without (V_1) being so. The same holds for the topology in which the open sets are those Lebesgue measurable sets that have Lebesgue density 1 in each of their points (this topology is completely regular). Of course, in these examples the topology is invariant and (V_0) is fulfilled since Γ_p consists of a single element (the identity). On the other hand, if we consider on $E = R$ the Euclidian topology and the transformation group given by the mappings

$$a_{a,b}(x) = ax + b \qquad (a, b \in R,\ a \neq 0) ,$$

then (V_0) is not satisfied.

6. Let us consider once more the condition (V_3), necessary and sufficient in order that a given topology $\mathfrak{B}$ have a compatible Γ-invariant uniformity. It can be formulated in another form if we introduce the following terminology. Let E be a set equipped with a (transitive) transformation group Γ and a Γ-invariant topology. We shall say that Γ *separates* two sets $A, B \subset E$ if, given $p \in E$, there exists $V \in \mathfrak{B}(p)$ such that $a \in \Gamma$, $a(V) \cap A \neq \phi$ implies $a(V) \cap B = \phi$. Now (V_3) is easily seen to be equivalent to the following:

(V_3^*) If $x \in E$, $F = \bar{F} \subset E$, $x \notin F$, then Γ separates $\{x\}$ and F.

If we assume that E is a T_0-space, then (V_3^*) implies the following condition:

(V_3') If $K_1, K_2 \subset E$ is compact and $K_1 \cap K_2 = \phi$, then Γ separates K_1 and K_2.

(V_3') obviously implies

(V_3'') If $x \in E$, $K \subset E$ is compact and $x \notin K$, then Γ separates $\{x\}$ and K.

In general, the converses of these implications are not valid. E.g. if $E = R$, Γ is the group of Euclidian translations and we consider the topology $\mathfrak{B}^+$, then (V_3') holds without (V_3^*) or (V_3) being true. However, if E is compact, then (V_3'') obviously implies (V_3^*), and the same is true if E is rim compact and locally connected. It would be interesting to know whether the implication $(V_3'') \Rightarrow (V_3^*)$ or at least $(V_3') \Rightarrow (V_3^*)$ holds if E is locally compact.

The proofs will be published in another paper. They use the concept of topogeneous orders [2] and the methods developed in the papers [3], [4], [5], [6].

REFERENCES

[1] E. Čech, *Topological spaces* (Prague — London — New York — Sydney, 1966).

[2] Á. Császár, *Grundlagen der allgemeinen Topologie* (Budapest — Leipzig, 1963).

[3] Á. Császár, Syntopogene Gruppen I., *Math. Nachrichten* 39 (1969), 1-20.

[4] Á. Császár, Syntopogene Gruppen II., *Mathematica, Cluj* 13 (36) (1971), 25-50.

[5] Á. Császár, Syntopogone Gruppen III., *Ann. Univ. Budapest,* Sect. Math. 14 (1971), 23-52.

[6] Á. Császár, Syntopogene Gruppen IV., *Ann. Univ. Budapest.,* Sect. Math. 14 (1971), 53-65.

[7] M.G. Murdeshwar — S.A. Naimpally, *Quasi-uniform topological spaces* (Groningen, 1967).

COLLOQUIA MATHEMATICA SOCIETATIS JÁNOS BOLYAI
8. TOPICS IN TOPOLOGY, KESZTHELY (HUNGARY), 1972.

ON F. RIESZ' SEPARATION AXIOM

K. CSÁSZÁR

I. Starting from the ideas of F. Riesz (1906), C.E. Aull [2] recently introduced and investigated the following separation axiom:

Axiom T_R. *A topological space* E *is said to be a* T_R*-space if it is* T_1 *and, for* $A \subset E$, $p, q \in A'$, $p \neq q$, *there is* $B \subset A$ *such that*

$$p \in B', \quad q \notin B'.$$

This condition turns out to be equivalent to the following one, occuring in

Theorem 1. *A topological space* E *is* T_R *iff for* $A \subset E$, $p \in \bar{A}$, $p \neq q$, *there is* $B \subset A$ *such that*

$$p \in \bar{B}, \quad q \notin \bar{B}.$$

Proof. The condition of Theorem 1 is necessary. In fact, if $p \in \bar{A}$, $q \notin \bar{A}$, then $B = A$, if $p, q \in \bar{A}$, $p \neq q$, $p \in A$, then $B = \{p\}$ satisfies the requirement ($q \notin \bar{B}$ because E is a T_1-space); if $p \notin A$, then $p \in A'$, and supposing that q is an isolated point of A, $B = A - \{q\}$ will do ($\{q\} = \overline{\{q\}}$ and so $p \in \bar{B}$); finally if $p, q \in A'$, $p \neq q$, then for $C \subset A$ such that $p \in C'$, $q \notin C'$, $B = C - \{q\}$ will do ($\{q\}$ is closed, therefore $p \in B' \subset \bar{B}$, and as $q \notin B$, $q \notin B'$, it follows that $q \notin \bar{B}$).

The condition of Theorem 1 is sufficient. Assume $p \neq q$, $A = \{p\}$, $B \subset A$, $p \in \bar{B}$, $q \notin \bar{B}$. Evidently only $B = \{p\}$ is possible, hence E is a T_1-space. On the other hand, if $p, q \in A'$, $p \neq q$, then for $B = A - \{p, q\}$, $p, q \in B'$ so that $p, q \in \bar{B}$ and there is $C \subset B$ such that $p \in \bar{C}$, $q \notin \bar{C}$; obviously $C \subset A$ and $p \in C'$, $q \notin C'$.

We need two further separation axioms introduced and studied by N.A. Šanin, K. Morita, A.S. Davis, M.W. Lodato, B. Banaschewski, J.M. Maranda, the author [3], etc.

Axiom S_1. *A topological space* E *is said to be an* S_1*-space if* $p \notin \overline{\{q\}}$ *involves* $q \notin \overline{\{p\}}$.

Axiom S_2. *A topological space* E *is said to be an* S_2*-space if* $p \notin \overline{\{q\}}$ *involves that* p *and* q *have disjoint neighbourhoods.*

These axioms generalize the usual axioms T_1 and T_2. Namely

$$T_0 + S_1 = T_1 ,$$

$$T_0 + S_2 = T_2 .$$

We now introduce a similar generalization of T_R:

Axiom S_R. *A topological space* E *is said to be an* S_R*-space if, from* $p \notin \overline{\{q\}}$ *or* $q \notin \overline{\{p\}}$ *and* $p \in \bar{A}$ $(A \subset E)$, *it follows that there is* $B \subset A$ *such that*

$$p \in \bar{B} , \quad q \notin \bar{B} .$$

Theorem 2. E *is a* T_R*-space iff it is a* T_0*-space and an* S_R*-space* $(T_R = T_0 + S_R)$.

Theorem 3. *Axiom* S_2 *implies* S_R *and* S_R *implies* S_1.

Proof. In fact, if U and V are disjoint neighbourhoods of p and q, then, if $p \in \bar{A}$, $B = A \cap U$ satisfies the requirement $p \in \bar{B}$, $q \notin \bar{B}$ of axiom S_R. For the second part of the statement, assume $p \notin \overline{\{q\}}$, and put $A = \{p\}$. By S_R there is $B \subset A$ such that $p \in \bar{B}$, $q \notin \bar{B}$. Clearly $B = \{p\}$ so that $q \notin \overline{\{p\}}$.

From Theorem 3, we get the well-known implications

$$T_2 \Rightarrow T_R \Rightarrow T_1 .$$

It is easy to show that the condition "$p \notin \overline{\{q\}}$ or $q \notin \overline{\{p\}}$" in axiom S_R can be replaced either by the condition "$p \notin \overline{\{q\}}$" or by the condition "$q \notin \overline{\{p\}}$".

In order to show this, let us introduce

Axiom P_R. *The topological space* E *is a* P_R*-space if, from* $p \notin \overline{\{q\}}$, $p \in \bar{A}$ $(A \subset E)$, *it follows that there is* $B \subset A$ *such that*

$$p \in \bar{B}, \quad q \notin \bar{B} .$$

Axiom Q_R. *The topological space* E *is a* Q_R*-space if, from* $q \notin \overline{\{p\}}$, $p \in \bar{A}$ $(A \subset E)$, *it follows that there is* $B \subset A$ *such that*

$$p \in \bar{B}, \quad q \notin \bar{B} .$$

Theorem 4. *Axioms* S_R, P_R *and* Q_R *are equivalent.*

Proof. In fact,

$$S_R = P_R + Q_R = P_R + S_1 = S_1 + Q_R ,$$

and in order to prove the statement, we must show, that

$$P_R \Rightarrow S_1 , \quad Q_R \Rightarrow S_1 .$$

The first part here is obvious, because if axiom P_R holds, and $p \notin \overline{\{q\}}$, then, for $A = \{p\}$, $B \subset A$, $p \in \bar{B}$, $q \notin \bar{B}$, clearly $B = \{p\}$ and $q \notin \overline{\{p\}}$. For the second part, let us assume that the space E is a Q_R-space, if $q \notin \overline{\{p\}}$, assume $p \in \overline{\{q\}}$. Then, for $A = \{q\}$, no set $B \subset A$ can exist with $p \in \bar{B}$, $q \notin \bar{B}$, therefore $q \notin \overline{\{p\}}$ implies $p \notin \overline{\{q\}}$.

II. Investigating the invariance properties of axiom S_R we can formulate

Theorem 5. *If the topology of* X *is the inverse image of the topology of* Y *under a map* $f\colon X \to Y$, *and* Y *is* S_R, *then the same holds for* X. *The converse holds if* $f(X) = Y$.

Proof. Assume $p \in f^{-1}(G)$, $q \notin f^{-1}(G)$ where G is open in Y, and $p \in \bar{A}$; hence $f(p) \in G$, $f(q) \notin G$, $f(p) \in \overline{f(A)}$. Since Y is an S_R-space, for suitable $C \subset f(A)$, $f(p) \in \bar{C}$, $f(q) \notin \bar{C}$. Put $B = A \cap f^{-1}(C)$. Then $B \subset A$ and since a suitable neighbourhood of $f(q)$ does not meet C, there is a neighbourhood of q, disjoint from $f^{-1}(C)$, so that $q \notin \bar{B}$. Finally, for any open set $H \subset Y$ such that $f(p) \in H$, we have $H \cap C \neq \phi$ and for $y \in H \cap C$ there is $x \in f^{-1}(y)$, $x \in A$ $(y \in C \subset f(A))$. Then $x \in A \cap f^{-1}(C) = B$, $x \in f^{-1}(H)$; here $f^{-1}(H)$ is an arbitrary open neighbourhood of p, hence $p \in \bar{B}$.

On the other hand, if $f(X) = Y$ and the space X equipped with the inverse image topology is an S_R-space, assume $p', q' \in Y$, $p' \in \bar{C}$, $p' \notin \overline{\{q'\}}$. Let $p, q \in X$ be such that $f(p) = p'$, $f(q) = q'$, and $A = f^{-1}(C)$. Then, for any open set $G \subset Y$ with $p' \in G$, we have $G \cap C \neq \phi$, hence $f^{-1}(G) \cap f^{-1}(C) \neq \phi$ and $f^{-1}(G)$ being an arbitrary open neighbourhood of p, we have $p \in \bar{A}$. Moreover, if G is taken so that $q' \notin G$, then $q \notin f^{-1}(G)$, and since X is an S_R-space, there is $B \subset A$ with $p \in \bar{B}$, $q \notin \bar{B}$. Then $f(B) \subset C$, $p' = f(p) \in \overline{f(B)}$, and $q' \notin \overline{f(B)}$; in fact, there is an open set $H \subset Y$ such that $q \in f^{-1}(H)$, $f^{-1}(H) \cap B = \phi$, so that $q' \in H$, $H \cap f(B) = \phi$.

As a special case we get

Theorem 6. *Every subspace of an* S_R*-space is* S_R.

What about the image of an S_R-space, we can say the following:

Theorem 7. *If* $f\colon X \to Y$, $f(X) = Y$, X *is* S_R, *and* f *is strongly perfect (continuous, closed and the inverse image of every point consists of a finite number of points), then* Y *is* S_R.

Proof. Assume $p', q' \in Y$, $p' \in \overline{A'}$, $p' \notin \overline{\{q'\}}$. Hence there is an open set $G' \subset Y$ such that $p' \in G'$, $q' \notin G'$. Put $f^{-1}(A') = A$, $f^{-1}(G') = G$. There is a $p \in f^{-1}(p')$ such that $p \in \bar{A}$ $(f^{-1}(p') \cap \bar{A} = \phi$ would imply $p' \notin f(\bar{A})$ and $f(\bar{A}) \supset f(A)$ being closed, we would obtain $p' \notin \overline{f(A)} = \overline{A'})$. G is then a neighbourhood of p not containing q_i $(i = 1, \ldots, n)$ where $f^{-1}(q') = \{q_1, \ldots, q_n\}$. Since X is an S_R-space, there is a $B_1 \subset A$ such that $p \in \overline{B_1}$, $q_1 \notin \overline{B_1}$, then there is a B_2 such that $B_2 \subset B_1 \subset A$, $p \in \overline{B_2}$, $q_2 \notin \overline{B_2}$, and finally there are sets $B_n \subset \subset B_{n-1} \subset \ldots \subset B_1 \subset A$ with $p \in \overline{B_n}$, $q_i \notin \overline{B_n}$ $(i = 1, 2, \ldots, n)$. The map f being continuous $f(p) \in f(\overline{B_n})$ and $q' \notin f(\overline{B_n}) \supset \overline{f(B_n)}$, since f is closed. Therefore

$$f(B_n) \subset f(A) = A' \,, \quad p' \in \overline{f(B_n)} \,, \quad q' \notin \overline{f(B_n)} \,.$$

A similar statement was proved for S_1 under the hypothesis that f is continuous and closed, and for S_2 under the condition that f is perfect [3].

We immediately get from Theorem 7:

Theorem 8. *Let E be an S_R-space and E^* the quotient space with respect to a decomposition of E, semicontinuous from above, then E^* is S_R, if each cellule of the decomposition consists of a finite number of points.*

Moreover we can formulate

Theorem 9. *If E is an S_R-space, and E^* is the family of equivalence classes of the equivalence relation $x \sim y \Leftrightarrow \mathfrak{B}(x) = \mathfrak{B}(y)$ (where $\mathfrak{B}(x)$ denotes the neighbourhood filter of x) with the quotient topology, then E^* is a T_R-space.*

That is true by Theorem 5, for if f is the natural map, the inverse image topology on E is equivalent to the original topology of E, and moreover $f(E) = E^*$.

III. Concerning product spaces, the product of S_1- or S_2-spaces is S_1 or S_2, but the product of T_R- or S_R-spaces need not be T_R or S_R. In particular, $E \times R$ need not be T_R if E is T_R.

Consider on R the following T_R-topology, introduced by C. E. Aull [2]:

The neighbourhood base of $x \in R$, $x \neq 1$ is the same as usually, and the neighbourhood base of $x = 1$ consists of the sets of the form $R - K$ where K is countable and compact in the euclidian sense, $1 \notin K$, then we get a T_R-topology $\mathcal{T}$. (The condition in T_R obviously holds for $p, q \neq 1$; if $p = 1$, $q \neq 1$, $p \in \bar{A}$, the set A is not contained in any countable compact set not containing 1, hence a suitable neighbourhood $(q-\epsilon, q+\epsilon)$ of q can be removed from A without affecting this property, since otherwise $A-[q-1, q+1]$ would be contained in a compact, countable set K_1, $p \notin K_1$, and similarly $A \cap \left(\left[q-\frac{1}{n}, q-\frac{1}{n+1}\right] \cup \cup \left[q+\frac{1}{n+1}, q+\frac{1}{n}\right]\right)$ would be contained in a countable compact set $K_{n+1} \subset \left[q-\frac{1}{n}, q+\frac{1}{n}\right]$, $p \notin K_{n+1}$, so that $A \subset \bigcup_1^\infty K_n \cup \{q\} = K$ where K is countable, compact and $p \notin K$. If $p \neq 1$, $q = 1$ and $\{x_n\} \subset A$, $x_n \to p$, $x_n \neq 1$, then, for $B = \{x_n\}$, $B \subset A$, $p \in \bar{B}$ and $q \notin \bar{B}$, since $B \subset B \cup \{p\} = K$, $1 \notin K$ and K is compact.)

Consider, on $R \times R$, the topology $\mathscr{T} \times \mathscr{E}$ where $\mathscr{E}$ is the euclidian topology on R; if A is the diagonal, $p = (1, 0)$, then $p \in \bar{A}$, since if $1 \notin K \subset R$ and K is compact, countable, then there exists, for $\epsilon > 0$, $x \in (-\epsilon, \epsilon) - K$, thus $(x, x) \in A$, $(x, x) \in (R - K) \times (-\epsilon, \epsilon)$; on the other hand, if $B \subset A$, $p \in \bar{B}$, then there exists for every $\epsilon > 0$ an $(x, x) \in B$, such that $x \in (-\epsilon, \epsilon)$, hence $(x, x) \in (-\epsilon, \epsilon) \times (-\epsilon, \epsilon)$, so that $q = (0, 0) \in \bar{B}$, and the topology $\mathscr{T} \times \mathscr{E}$ is not a T_R-topology.

In this connection it is interesting to formulate

Theorem 10. *Let E be a topological space; $E \times E$ is S_R (T_R) iff E is S_2 (T_2).*

Proof. In fact, if E is not an S_2-space, then there is $p, q \in E$ such that $p \notin \overline{\{q\}}$, but any two neighbourhoods of p and q meet each other, thus $(p, q) \in \bar{A}$ in $E \times E$ where A is the diagonal; on the other hand, if $B \subset A$, $(p, q) \in \bar{B}$, then $(p, p) \in \bar{B}$ although (p, p) has a neighbourhood not containing (p, q).

Conversely, if E is S_2, then $E \times E$ is S_2, and $E \times E$ is S_R by Theorem 3. The statement concerning T_R and T_2 is then an easy corollary.

IV. It is of interest to ask when the one-point compactification E^* of a space E is S_R or T_R. In order to formulate a result in this direction, let us introduce the following axiom:

Axiom S_∞. *The space E is S_∞ if, for $p \in \bar{A}$ where $A \subset E$ and*

$\bar{A}$ is not compact, there is $B \subset A$ such that $\bar{B}$ is not compact, and $p \notin \bar{B}$.

Let us recall (see [1]) that a space E is a k'-space iff $x \in \bar{M}$ implies that there is a closed, compact set K such that $x \in \overline{M \cap K}$.

Theorem 11. *The one-point compactification of a (non-compact) space E is S_R (T_R) iff E is S_R (T_R), k' and S_∞.*

Proof. Denote the one-point compactification of E by E^*. First we show that the conditions of Theorem 11 are sufficient.

Assume $p, q \in E$, $A \subset E^*$, $p \in \bar{A}^*$ ($\bar{A}^*$ denotes the closure of A in E^*). E is open in E^* hence if $A \cap E = A_0$, then $p \in \bar{A}_0$ ($\bar{A}_0$ is the closure of A_0 in E). If $q \notin \overline{\{p\}}^*$, then $q \notin \overline{\{p\}}$ and, since E is an S_R-space, there is a $B \subset A_0 \subset A$ satisfying the requirement $p \in \bar{B}$, $q \notin \bar{B}$. Now for the closures in E^*

$$p \in \bar{B}^* \,,\; B \subset A \,,\; q \notin \bar{B}^* \,,\; (q \notin \bar{B} \,, q \neq \infty) \,.$$

Assume now $p \in E$, $q = \infty$, $p \in \bar{A}^*$ $(A \subset E^*)$. As above, $p \in \overline{A \cap E} = \bar{A}_0$. It was supposed that E is k', therefore we can find a compact closed set K in E such that $p \in \overline{A_0 \cap K} = \bar{B}$ $(A_0 \cap K = B \subset A_0 \subset A)$, and $q \notin \bar{B}^*$ since $E^* - K$ which is a neighbourhood of q in E^* does not meet B. Hence

$$p \in \bar{B}^* \,,\; q \notin \bar{B}^* \,.$$

In the case $p = \infty$, $q \in E$, if $p \in A$, then $B = \{p\}$ is closed and satisfies the requirement $(B \subset A,\ p \in \bar{B},\ q \notin \overline{\{p\}} = \bar{B})$. If $p \notin A$, thus $A \subset E$, and $p \in \bar{A}^*$, i.e. A is not compact, we get from axiom S_∞ that there is a $B \subset A$ such that $q \notin \bar{B}$ and $\bar{B}$ is non-compact. Hence $\bar{B}^*$ contains ∞, and so

$$p \in \bar{B}^* \,,\; q \notin \bar{B}^* \,.$$

The conditions of Theorem 11 are also necessary. If E^* is S_R, then E must be S_R by Theorem 6. Now assume $A \subset E$, $p \in \bar{A} \subset \bar{A}^*$, $q = \infty$. Since E^* is an S_R-space, there is $B \subset A$ such that $p \in \bar{B}^*$, $q \notin \bar{B}^*$. Hence B is contained in a compact, closed set of E, $\bar{B}$ is compact, the space E is a k'-space.

Finally, assume $p = \infty$, $q \in \bar{A}$ $(A \subset E)$, where $\bar{A}$ is not compact, thus $p \in \bar{A}^*$ and, since E^* is an S_R-space, and $q \notin \overline{\{p\}}^*$, there is a set $B \subset A$ with the property $p \in \bar{B}^*$, $q \notin \bar{B}^*$. Hence $q \notin \bar{B}$ and $\bar{B}$ cannot be compact. Thus E is S_∞. The statement concerning T_R is an easy consequence of what was proved and of the fact that E^* is T_0 iff E is T_0.

Now, for S_∞, we have a simple condition implying it:

Theorem 12. *A regular space is always* S_∞.

Proof. In fact, if $p \in \bar{A}$ $(A \subset E)$, and A is not compact, let $\bar{A} \subset \bigcup_{i \in I} G_i$ be an open cover of A such that no finite number of G_i's cover $\bar{A}$. Assume $p \in G_{i_0}$, and let V be an open neighbourhood of p such that $\bar{V} \subset G_{i_0}$. Then $B = A - V$ satisfies $B \subset A$, $p \notin \bar{B}$, and $\bar{B}$ is not compact since $\bar{B} = \overline{A - V} \supset \overline{A - \bar{V}} \supset \bar{A} - G_{i_0}$ and $\bar{A} - G_{i_0}$ cannot be compact. Hence E is S_∞.

Theorems 11 and 12 imply the

Corollary. *If* E *is regular* (T_3) *and* k', *then* E^* *is* S_R (T_R).

It is to note that a weaker statement was proved by C.E. Aull [2].

REFERENCES

[1] A. Arhangel'skiĭ, Mappings and Spaces, *Russian Mathematical surveys,* 21 No. 4 (1966), 115-162.

[2] C.E. Aull, A Separation Axiom of F. Riesz, *in print.*

[3] K. Császár, Untersuchungen über Trennungsaxiome, *Publ. Math.*, 14 (1967), 353-364.

COLLOQUIA MATHEMATICA SOCIETATIS JÁNOS BOLYAI
8. TOPICS IN TOPOLOGY, KESZTHELY (HUNGARY), 1972.

CHARACTERIZING TOTALLY ORDERABLE SPACES AND THEIR TOPOLOGICAL PRODUCTS

J. van DALEN

§1.

Several authors have considered the following *Orderability Problem:* Which topological spaces can be supplied with a linear ordering such that the original topology and the order topology coincide? (Such spaces are called (*totally*) *orderable*).

Far from being complete we refer to Eilenberg [1], Herlich [2] and Kok [3]. At this Colloquium E. Deák announced a general solution of the orderability problem in terms of what he calls *Richtungsstrukturen.*

In the present paper we want to give characterizations of orderable spaces and their topological products in terms of conditions on a subbase for the open sets of the space.

§2.

Throughout all spaces will be T_1 spaces consisting of more than one point. If $\mathscr{S}$ is an open subbase for a space X, we always suppose that $X \notin \mathscr{S}$ and $\phi \notin \mathscr{S}$. A subbase $\mathscr{S}$ is called *binary*, if each covering of X by elements of $\mathscr{S}$ has a subcovering consisting of exactly two elements.

By Alexander's subbase theorem, a space with a binary subbase is compact.

A subbase $\mathscr{S}$ is called *comparable* if for any three elements S_0, S_1 and S_2 of $\mathscr{S}$,

$$S_0 \cup S_1 = X = S_0 \cup S_2 \Rightarrow S_1 \subset S_2 \quad \text{or} \quad S_2 \subset S_1 .$$

A subbase $\mathscr{S}$ is called *complementary* if for all $S \in \mathscr{S}$ there exists $S' \in \mathscr{S}$ such that $S \cup S' = X$.

Example: The canonical subbase of any product T of totally ordered spaces is comparable and complementary. If, moreover, T is compact, then this subbase is also binary.

§3.

The following theorem is due to J. de Groot and P.S. Schnare [4].

Theorem 1. *A space X is homeomorphic to a product of compact totally ordered spaces if and only if X is a T_1 space which has a binary and comparable subbase.*

It must be emphasized that it is the combination of the two properties "binary" and "comparable" which does the trick, because it is well-known that every compact metric space has a binary subbase (O'Connor [5]).

Also every compact metric space has a comparable subbase (this is easily seen by embedding the space in a metrizable cube). In view of Theorem 1 not every compact metric space has a subbase which is both binary and comparable.

Crucial in the proof of Theorem 1 is the following equivalence relation which was discovered by J. de Groot.

If $\mathscr{S}$ is any collection of sets, define a relation $\sim$ on $\mathscr{S}$, by $S_1 \sim S_2$ if and only if $S_1 \subset S_2$ or $S_2 \subset S_1$, (so $\sim$ is the comparability relation on $\mathscr{S}$).

Then we have the following

Proposition 2. *Let a space X have a binary and comparable open subbase $\mathscr{S}$. Then $\sim$ is an equivalence relation on $\mathscr{S}$.*

Even the following stronger proposition holds:

Proposition 3. *If X has an open subbase $\mathscr{S}$ which is comparable and complementary, then $\sim$ is an equivalence relation on $\mathscr{S}$.*

Note, that $\mathscr{S}$ is complementary if $\mathscr{S}$ is binary.

Once De Groot and Schnare had characterized products of compact totally ordered spaces, the question arose if it was possible to characterize arbitrary ordered spaces and their products among all T_1 spaces.

The following results were obtained partially together with E. Wattel.

A collection $\mathscr{N}$ of sets is called a nest provided that for any two members N_1 and N_2 of $\mathscr{N}$ it is true that either $N_1 \subset N_2$ or $N_2 \subset \subset N_1$.

Theorem 4. *If a space X has an open subbase which is the union of two nests, then X is homeomorphic to a subspace of a totally ordered space.*

Remark. Observe that in a T_1 space a subbase can not be a nest.

Theorem 5. *A space X is totally orderable if and only if X has an open subbase $\mathscr{S}$ such that:*

(i) *$\mathscr{S}$ is the union of two nests;*

(ii) *no element S of $\mathscr{S}$ is the intersection of all members of*

$\mathscr{S}$ *which properly contain* S.

Remarks. 1. In an ordered space the subbase consisting of all sets of the form $\{x \mid x < a\}$ or $\{x \mid a < x\}$, with $a \in X$ satisfies these conditions.

2. If X is compact or connected, then in Theorem 5 condition (ii) can be omitted.

Theorem 6. *A space* X *is homeomorphic to a product of totally ordered spaces if and only if* X *has an open subbase* $\mathscr{S}$ *with the following properties:*

(i) $\mathscr{S}$ *is comparable and complementary;*

(ii) *condition* (ii) *in Theorem 5;*

(iii) *every covering* $\mathscr{C}$ *of* X *by elements of* $\mathscr{S}$ *has a subcovering* $\mathscr{C}'$ *such that for any two elements* S_1 *and* S_2 *of* $\mathscr{C}'$ *either* $S_1 \subset S_2$ *or* $S_2 \subset S_1$ *or* $S_1 \cap S_2 = \phi$ *or* $S_1 \cup S_2 = X$.

If X is compact or connected then (ii) can be dropped.

Remark. Condition (i) guarantees that $\sim$ is an equivalence relation on $\mathscr{S}$ (see Proposition 3). According to an idea of de Groot, this equivalence relation can be used to define the axes of coordinates of the product of totally ordered spaces to which X is homeomorphic. We intend to give full proofs in subsequent papers.

REFERENCES

[1] S. Eilenberg, Ordered topological spaces, *Amer. J. Math.*, 63 (1941), 39-45.

[2] H. Herrlich, *Ordnungsfähigkeit topologischer Räume,* Inaugural Diss., Berlin 1962.

[3] H. Kok, On conditions equivalent to the orderability of a connected space, *Nieuw Archief Wisk.*, (3), XVIII, 250-270, (1970).

[4] J. de Groot – P.S. Schnare, A characterization of products of compact totally ordered spaces, *to be published.*

[5] J.L. O'Connor, Supercompactness of compact metric spaces, *Indag Math.*, 32 (1970), 30-34.

COLLOQUIA MATHEMATICA SOCIETATIS JÁNOS BOLYAI
8. TOPICS IN TOPOLOGY, KESZTHELY (HUNGARY), 1972.

THEORY AND APPLICATIONS OF DIRECTIONAL STRUCTURES

E. DEÁK

I have introduced the notion of directional structure in 1964 and published since a series of papers on this subject and reported on results of the theory of directional structures at several international conferences; a comprehensive treatment, however, is still missing. Now I try to give a summarized – though by no means complete – survey of the first and greater part (belonging entirely to general topology) of my German monograph [13] (in press), which presents the latest state of the theory. Most results introduced in the present paper are not yet published.

My colleagues J. Gerlits and P. Hamburger in Budapest* contributed valuable results to this theory ([14], [15]) – starting from a collection of problems pbulished in Hungarian ([12] IV), included in the aforsaid book as a separate chapter – which are partly included in the present report.

Another part of the theory connected with the area of (topological)

*Department of Topology of the *Mathematical Institute of the Hungarian Academy of Sciences.*

linear spaces and convexity of sets, will not be dealt with at the present occasion, although several motives of the notion "directional structure" itself (as well as the reason for finding appropriate names for some of the new notions occuring also in this paper) find their origin exactly there; see the papers [8], [9], [10], [11], [21].

As to the terminology and notations of commonly used notions we follow the usage of the well-known book [22] of J. L. Kelley. "Space" means generally a topological space; "order" is to be understood as a total (linear) order; for elements a, b of an ordered set X (a, b) resp. $(a, b]$ denotes the open resp. half open interval between a and b; by A^- or $A^{-(X)}$ we denote the closure of a set A in the space X; the sign $\subset$ is used in the sense of $\subsetneq$; for any cardinal number $\mathfrak{n}$ we denote by $E_\mathfrak{n}$ the corresponding power of the space $R = E_1$ of the reals, in particular the Euclidean n-space is denoted by E_n $(n = 0, 1, 2, \ldots)$.

To introduce our fundamental concept we choose from a number of possible methods a way, which seems to be most reasonable from the heuristic point of view and which in addition makes it possible to build up the material gradually. For the sake of a better presentation, however, our subject must be organized in a manner in which the original deductive order of the theorems is not preserved.

§1. DIRECTABLE SPACES*

Our starting point will be the commonly known fact that in many cases the relative topology of a subset of an ordered space (ordered set with the interval topology) does not coincide with the topology induced by the inherited order of this subset. This question (and several others related to it) has been dealt with in plenty of papers, also in [13], where some criteria referring to this are given.

Theorem 1.1. *Let X be an ordered space and $\phi \neq X^* \subset X$. In order that the relative topology of the set X^* be the same as that induced by the inherited order of this set it is necessary and sufficient that for any pair of points*

*This notion will be defined in (1.10) and is not to be confused with the well-known notion named "directed set" (see e.g. [22] p. 65).

$$x \in X \setminus X^*, \quad x^* \in X^*,$$

where

$$(x, x^*) \cap X^* = \phi, \ \{y \in X^*: y \prec x\} \neq \phi \quad (x \prec x^*)$$

resp.

$$(x^*, x) \cap X^* = \phi, \ \{y \in X^*: y \succ x\} \neq \phi \quad (x \succ x^*)$$

there exists a last element in the set $\{y \in X^*: y \prec x\}$ *resp. a first element in the set* $\{y \in X^*: y \succ x\}$.

This theorem has been (as I suppose) stated for the first time – although incompletely, because of omitting the indispensable supposition $\{y \in X^*: y \prec x\} \neq \phi$ resp. $\{y \in X^*: y \succ x\} \neq \phi$ – by R.A. Alò and O. Frink [1]; the present version is due to P. Hamburger.

Theorem 1.2. *Let* X *be an ordered space and* $\phi \neq X^* \subset X$. *In order that the relative topology of the set* X^* *be the same as that induced by the inherited order of this set, it is in case of existing no jumps in the order of* X *necessary, and in case of existing no gaps in the order of* X *sufficient, that there be no interval of type* $[x_1, x_2)$ *or* $(x_1, x_2]$ $(x_1, x_2 \in X)$ *among the order-components of* $X \setminus X^*$ *bounded in* X.

Hereby we obtained particularly a necessary *and* sufficient condition of the coincidence of the two topologies of X^* in the case that the set is continuously ordered (that means an order without jumps and gaps). This criterion can be formulated much more simply related to the special case $X = E_1$:

Theorem 1.3. *The relative topology of a set* $\phi \neq X^* \subset E_1$ *coincides with its natural order topology if and only if there is no half-open interval among the bounded components of* $E_1 \setminus X^*$.

The negative content of this latter theorem can be partly compensated by the following

Theorem 1.4. *Let* A *be a space consisting of a non-void subset* A *of* E_1 *and the topology of* A *be induced by the inherited order of this set. Then there exist a subset* B *of* E_1 *and a mapping of* A *onto*

the subspace B *of* E_1 *which is both a homeomorphism and an order-isomorphism.*

However, the general problem proves to be even more difficult, as there are subspaces A of ordered spaces which are not orderable at all (that means not only the incompatibility of the inherited order, but also the incompatibility of any order of the set A with its subspace topology). Now the question arises how to characterize *"sub-orderable" spaces* (i.e. subspaces of orderable spaces).

This problem has many solutions; the following solutions admit a uniform treatment of several questions similar to the original one by means of notions serving at the same time as basis for a far-reaching theory.

Definitions 1.5. *A direction* of a set $X \neq \phi$ is a system $\mathscr{R}$ of ordered pairs (G, F), $G \subseteqq F \subseteqq X$, satisfying the axioms

R_1) (ϕ, ϕ), $(X, X) \in \mathscr{R}$ *(these are the trivial elements of* $\mathscr{R}$*),*

R_2) $(G_1, F_1), (G_2, F_2) \in \mathscr{R}$, $(G_1, F_1) \neq (G_2, F_2) \Rightarrow$
$\Rightarrow F_1 \subseteqq G_2 \vee F_2 \subseteqq G_1$,

R_3) $\cup \{G\colon G \in \mathscr{G}^*\} \in \mathscr{G}(\mathscr{R}) \quad (\phi \neq \mathscr{G}^* \subseteqq \mathscr{G}(\mathscr{R}))$,

R_4) $\cap \{F\colon F \in \mathscr{F}^*\} \in \mathscr{F}(\mathscr{R}) \quad (\phi \neq \mathscr{F}^* \subseteqq \mathscr{F}(\mathscr{R}))$,

where $\mathscr{G}(\mathscr{R})$ denotes the family of all the G-s and $\mathscr{F}(\mathscr{R})$ the family of all the F-s; the direction $\mathscr{R}$ is *orderly* if it satisfies the further condition

R_5) $\cup \{F \setminus G\colon (G, F) \in \mathscr{R}\} = X$.

There may exist different pairs $(G_1, F_1), (G_2, F_2) \in \mathscr{R}$ with $G_1 = G_2$ or $F_1 = F_2$.

Definitions 1.6. Let $\mathscr{R}$ be a direction of a set $X \neq \phi$. Two elements of $\mathscr{R}$ are said to be *strictly different* if both the first and the second components of them are different. $\mathscr{R}$ is a *strict* resp. *strict-orderly direction* of X, if any two different non-trivial elements of $\mathscr{R}$ are strictly different resp. if $\mathscr{R}$ is strict as well as orderly.

1.7. *Every direction* $\mathscr{R}$ *admits a natural order, namely the relation*

$$(G_1, F_1) < (G_2, F_2) \Leftrightarrow F_1 \subseteqq G_2, G_1 \subset F_2$$
$$((G_1, F_1), (G_2, F_2) \in \mathscr{R}) .$$

Definition 1.8. The topology $\mathscr{T}(\mathscr{R})$ of a set X induced by a direction $\mathscr{R}$ is that for which the family

(1.8.1) $$\mathscr{G}(\mathscr{R}) \cup \{X \setminus F\colon F \in \mathscr{F}(\mathscr{R})\}$$

forms an open subbasis, or – equivalently – the family

(1.8.2) $$\mathscr{F}(\mathscr{R}) \cup \{X \setminus G\colon G \in \mathscr{G}(\mathscr{R})\}$$

forms a closed subbasis.

In case X is the underlying set of a space $(X, \mathscr{T})$ we call $\mathscr{R}$ a *direction of the space,* provided that every element of $\mathscr{G}(\mathscr{R})$ and $\mathscr{F}(\mathscr{R})$ is open resp. closed. If so, the sets (1.8.1) are said to be the $\mathscr{R}$*-open* and the sets (1.8.2) the $\mathscr{R}$*-closed halfspaces* of X.

The topology $\mathscr{T}(\mathscr{R})$ is always coarser than the topology $\mathscr{T}$; in case $\mathscr{T}(\mathscr{R}) = \mathscr{T}$, $\mathscr{R}$ is called a *compatible direction* of the space $(X, \mathscr{T})$.

Definition 1.9. The *trivial directions* of a set X are the systems

$$\{(\phi, \phi), (X, X)\}, \ \{(\phi, \phi), (\phi, X), (X, X)\} ,$$

which are – in case a topology is defined on X – naturally also directions of the space, and called *the trivial directions of the space.* (A space is indiscrete if and only if it admits no direction but the two trivial ones).

Definition 1.10. A space is called *directable* if it admits a compatible direction. If there exist among the compatible directions even an orderly, strict or strict-orderly one, we say that the space is *orderly directable, strictly directable, strictly-orderly directable,* resp.

Now we come to the main results concerning the above-mentioned subject.

Theorem 1.11. *Both directability and orderly directability are hereditary properties of a space; but the same does not hold for strict directability.*

Theorem 1.12. *Endowed with the natural order and the corresponding interval topology every direction of a non-void set becomes a compact Hausdorff space.*

Theorem 1.13. *Let* X *be a* T_0*-space* (T_1*-space*). *Then* X *is*

a) *sub-orderable if and only if it is orderly directable (directable);*

b) *orderable if and only if it is strictly-orderly directable (strictly directable).*

Theorem 1.14. *A connected* T_0*-space* (T_1*-space*) *is orderable if and only if it is orderly directable (directable).*

(Thus for connected spaces sub-orderability and orderability are equivalent properties.)

Theorem 1.13, a) is partly based on the following embedding theorem:

Theorem 1.15. *An orderly directable* T_0*-space is topologically embeddable into any of its compatible orderly directions regarded as ordered spaces with the order of* 1.7.

From this we can obtain among other things the well-known ordered compactification of an ordered space, by taking the complete hull of the underlying ordered set. (It is natural, by 1.12, 1.13, b) and 1.15. to take the closure of the topological image of an ordered space in one of its compatible orderly directions. We shall come back to this compactification method in §6, in fact to a generalization of it, which includes all T_2 compactifications of arbitrary Tychonov spaces.)

It turns out that several important known properties of orderable and sub-orderable spaces hold true for orderly directable or even merely directable spaces as well; thus new methods arise for proving the corresponding theorems for ordered spaces.

Theorem 1.16. *Every orderly directable space is collectionwise normal (and thus hereditarily collectionwise normal).*

(It was proved by L. A. Steen [26] that every subspace of an ordered space is collectionwise normal). Another method yields a more general result:

Theorem 1.17. *Every orderly directable space is (hereditarily) completely monotonically normal.*

(The notion of monotonical normality was recently introduced by Ph.L. Zenor [27]: a T_1-space $(X, \mathscr{T})$ is called *monotonically normal (completely monotonically normal),* if there is a function D defined on the set of all pairs (H, K) of mutually exclusive closed subsets of X (the set of all pairs (H, K) of subsets of X such that $H^- \cap K = H \cap K^- = \phi$) with $D(H, K) \in \mathscr{T}$ and with the properties

1) $H \subseteqq D(H, K) \subseteqq D(H, K)^- \subseteqq X \setminus K$,

2) $H_1 \subseteqq H_2, K_2 \subseteqq K_1 \Rightarrow D(H_1, K_1) \subseteqq D(H_2, K_2)$.

Ph.L. Zenor announced there that monotonical normality implies collectionwise normality; R.W. Heath and D.J. Lutzer announced in [20] that every ordered space is completely monotonically normal.)

Theorem 1.18. *Every connected directable space is locally connected.*

The problems of sub-orderable spaces have been investigated by E. Čech [2] and others (see e.g. Lutzer [23]) in terms of GO-spaces (generalized ordered spaces). The underlying set of a GO-space is an ordered set and its topology is such that at every point x there exists a local base consisting of all open intervals and possibly other intervals containing x. A space is a subspace of an ordered space if and only if it is a GO-space; hence, according to 1.13, a), the GO-spaces are exactly the orderly directable T_0-spaces. The latter can be proved also directly by constructing a compatible orderly direction for a given GO-space. It seems to be useful to utilize these two techniques simultaneously.

Finally here are some results concerning the classical dimensions of a directable space.

Theorem 1.19. *If X is a directable space, then* $\operatorname{Ind} X \leqslant 1$.

Theorem 1.20. *If X is a directable space, then* $\dim X \leqslant 1$.

(Here $\operatorname{Ind} X$ denotes the large inductive dimension of X, and $\dim X$ the covering dimension in the sense of Lebesgue.) The analogous theorem concerning small inductive dimension is almost trivial:

Theorem 1.21. *If X is a directable space, then* $\operatorname{ind} X \leqslant 1$.

By 1.13, a) we obtain from these theorems:

Theorem 1.22. *If X is a sub-orderable space, then* $\operatorname{ind} X \leqslant 1$, $\operatorname{Ind} X \leqslant 1$, $\dim X \leqslant 1$.

2.§. THE NOTION OF DIRECTIONAL DIMENSION AND A SURVEY OF SOME PRINCIPAL TRENDS OF THE THEORY

To apply the notion of direction also outside of the domain of directable spaces, the most natural idea is to work with several directions on a space at the same time.

Definition 2.1. A non-void system $\mathfrak{R}$ of directions (orderly directions) of a set $X \neq \phi$ is said to be a *directional structure* or *DS* (an *orderly directional structure* or *ODS*) of this set.

Definition 2.2. The topology $\mathscr{T}(\mathfrak{R})$ *induced by a DS* $\mathfrak{R}$ on a set X is defined by

$$\mathscr{T}(\mathfrak{R}) = \sup \{ \mathscr{T}(\mathscr{R}) \colon \mathscr{R} \in \mathfrak{R} \} .$$

Definitions 2.3. A DS of the underlying set X of a space $(X, \mathscr{T})$ is called *a DS of the space,* provided that $\mathscr{T}(\mathfrak{R})$ is coarser than $\mathscr{T}$.

In that case the $\mathscr{R}$-open ($\mathscr{R}$-closed) halfspaces, where $\mathscr{R} \in \mathfrak{R}$, are called *$\mathfrak{R}$-open ($\mathfrak{R}$-closed) halfspaces.*

In the special case in which $\mathscr{T}(\mathfrak{R}) = \mathscr{T}$, $\mathfrak{R}$ is said to be a *compactible DS or CDS of the space.* The abbreviation of "*compatible ODS*" is *CODS*.

Definition 2.4. *There exist CDS's for every space.*

1.

What is the least cardinality of a CDS of a given space?

Definition 2.5. The *directional dimension* or *DD* of a space $X \neq \neq \phi$ is the minimum of the cardinalities of the CDS's without trivial directions. (The omission of the trivial directions is of significance only in case of finite DD, of course.) Further we define $\operatorname{Dim} \phi = 0$.

Here is an equivalent version of this definition: the DD of every indiscrete space (ϕ included) is equal to 0; the DD of a non-indiscrete space is equal to the minimum of the cardinalities of its CDS's.

Accordingly the directability of a space X is equivalent to $\operatorname{Dim} X \leqslant 1$.

Definition 2.6. A CDS of cardinality $\operatorname{Dim} X$ of a space X is called a *minimal CDS* or *MCDS* of the space.

Directional dimension turns out to be really a dimension-like cardinal function allowing interesting applications and being in some connection with the usual notions of dimension too. (See the paragraphs 3, 4, 5, 7 and 8.)

2,

Which topologies can be induced by ODS-s?

Definition 2.7. A space is called *feebly orderly,* if it has a CODS.

The class of feebly orderly spaces can be determined by usual topological terms, and the question proved to be useful also otherwise (§§6, 7).

3.

The following question is the natural combination of 1. and 2.

Which (necessarily feebly orderly) spaces admit CDS's being both minimal and orderly?

Definition 2.8. A space is called *orderly,* if it admits an orderly MCDS *(OMCDS).*

There exists no complete characterization (in terms of usual topological notions) of this class of spaces. The problem of finding one seems to be difficult, but quite interesting.

3. §. THE DIRECTIONAL DIMENSION

Theorem 3.1. *Every space has a DD.*

Theorem 3.2. *Every cardinal number is the DD of a space.*

Theorem 3.3. *Denoting the topological weight of a space X with* $\tau(X)$, $\mathrm{Dim}\, X \leqslant \tau(X)$ *holds for any space X; in case* $\tau(X) \geqslant \aleph_0$, *an arbitrary CDS contains CDS's of cardinality* $\leqslant \tau(X)$ *(occasionally none of them being even a MCDS).*

Monotony theorems 3.4. *Let* X^* *be an arbitrary subspace of a space X.*

a) $\mathrm{Dim}\, X^* \leqslant \mathrm{Dim}\, X$.

b) *If X is feebly orderly, then so is* X^*.

Sum theorems 3.5. *Let B be a set of indices,* $\bar{\bar{B}} > 1$, X_β $(\beta \in B)$ *arbitrary spaces and* $X = \Sigma\{X_\beta\colon \beta \in B\}$ *their topological sum.*

a) *If every space* X_β *is indiscrete, then* $\mathrm{Dim}\, X \leqslant 1$.

b) *If at least one of the spaces* X_β $(\beta \in B)$ *is not indiscrete, then* $\mathrm{Dim}\, X = \sup\{\mathrm{Dim}\, X_\beta\colon \beta \in B\}$.

c) *If every space* X_β $(\beta \in B)$ *is feebly orderly, then so is the space X.*

d) *If every space* X_β $(\beta \in B)$ *is orderly, then so is the space X.*

Union theorems 3.6. *Let* $\{X_\beta\colon \beta \in B\}$ *be a σ-locally finite system of open-and-closed subspaces of a space X such that* $X = \cup\{X_\beta\colon \beta \in B\}$.

a) *If at least one of the spaces* X_β *is non-indiscrete, then* $\operatorname{Dim} X = \sup\{\operatorname{Dim} X_\beta : \beta \in B\}$.

b) *If every space* X_β *is feebly orderly, then so is the space* X.

c) *If every space* X_β *is orderly, then so is the space* X.

(Parts a), b) and c) of theorem 3.6 are analogous to the parts b), c) and d) of theorem 3.5; an analogue of 3.5, a) in this sense does not exist.)

Product theorems 3.7. *Let* $\{X_\beta : \beta \in B\}$ *be an arbitrary non-void family of spaces and* $X = \Pi\{X_\beta : \beta \in B\}$ *their topological product.*

a) $\operatorname{Dim} X \leqslant \Sigma\{\operatorname{Dim} X_\beta : \beta \in B\}$.

b) *If every space* X_β *is feebly orderly, then so is the space* X.

Product theorem 3.8. of J. Gerlits. *Let* B *be an uncountable set of indices and* X_β $(\beta \in B)$ *be* T_0*-spaces with at least two different points. Then*

(3.8.1) $$\operatorname{Dim} \Pi\{X_\beta : \beta \in B\} = \Sigma\{\operatorname{Dim} X_\beta : \beta \in B\}.$$

From this follows e.g.

Theorem 3.9. *If* $\mathfrak{n}$ *is a uncountable cardinal, then*

(3.8.2) $$\operatorname{Dim}[0, 1]^{\mathfrak{n}} = \mathfrak{n}.$$

It seems to be very difficult to find non-trivial conditions under which the equality (3.8.1) holds also for a countably infinite or even for a finite set of indices. Nevertheless the special case (3.8.2) can be proved (in another way, of course) also in case of an arbitrary countable cardinal $\mathfrak{n}$.

Theorem 3.10. *The* DD *of the Hilbert space is equal to* $\aleph_0$.

Theorem 3.11. *The* DD *of the* n*-dimensional Euclidean space is equal to* n.

(By 3.4 a), theorem 3.10 is equivalent to $\operatorname{Dim}[0, 1]^{\aleph_0} = \aleph_0$ and theorem 3.11 is equivalent to $\operatorname{Dim}[0, 1]^n = n$ $(n = 0, 1, 2, \ldots)$.

Theorem 3.12. *If* X *is a separable metrizable space and* $\operatorname{Ind} X < \infty$

or $\operatorname{Dim} X < \aleph_0$, *then* $\operatorname{ind} X \leqslant \operatorname{Dim} X \leqslant 2 \cdot \operatorname{ind} X + 1$.

This "asymptotic" relation has as a consequence that, within the domain of separable metrizable spaces, "finite dimensionality" means the same, wether the DD of any one of the classical dimensions is meant.

Nevertheless, since in many cases $\operatorname{Dim} X > \operatorname{ind} X$ (e.g. in the very simple case of the closed circumference K, where $\operatorname{ind} K = 1 < 2 = \operatorname{Dim} K$), it is to be expected that the smallest Euclidean universal space of separable metrizable spaces X with $\operatorname{Dim} X \leqslant n$ $(n = 1, 2, \ldots)$ will not be E_{2n+1} but an E_m with $n \leqslant m < 2n + 1$. I have regarded it as one of the basic tasks of the theory to prove that m equals the optimal value n; we shall come back to this problem in §8.

On the other hand, just because of the deviation of the DD from the usual dimensions, one should try to find a local and inductive notion of dimension having its origin in DD, as e.g. C.H. Dowker introduced local covering dimension in 1955. The method of Dowker is not the only possible one, and in recent times a whole series of similar new notions have been introduced. The following theorem (or the method of its proof) may be a suitable starting point:

Theorem 3.13. *Let* $n \geqslant 3$ *be a natural number and* X *a space with* $\operatorname{Dim} X = n$. *Then there exists a local basis at every point* $x \in X$ *consisting of open sets whose boundaries are the union of* $n - 1$ *closed sets with* DD *not greater than* $n - 1$.

(As a matter of fact, the proof yields a stronger but less "symmetrical" version of this theorem.)

We conclude this section with some nice theorems of J. Gerlits.

Theorem 3.14. *If* X *is an arcwise connected metrizable space with uncountable weight, then*

$$\operatorname{Dim} X = \begin{cases} \aleph_0 & (\tau(X) = \aleph_1)\,, \\ \tau(X) & (\tau(X) > \aleph_1)\,. \end{cases}$$

As an application of this theorem (and by adding two trivial cases) the DD of an arbitrary Kowalsky star space $S(\mathfrak{n})$ (where $\mathfrak{n}$ denotes the cardinality of the set of the rays) can be determined:

Theorem 3.15.

$$\operatorname{Dim} S(\mathfrak{n}) = \begin{cases} 1 & (\mathfrak{n} = 1, 2), \\ 2 & (3 \leqslant \mathfrak{n} \leqslant \aleph_0), \\ \aleph_0 & (\mathfrak{n} = \aleph_1), \\ \mathfrak{n} & (\mathfrak{n} > \aleph_1). \end{cases}$$

From this and the Kowalsky metrization theorem it follows:

Theorem 3.16. *If* X *is a metrizable space and* $\tau(X) \leqslant \aleph_1$, *then* $\operatorname{Dim} X \leqslant \aleph_0$.

4.§. RELATIONS BETWEEN THE DIRECTIONAL DIMENSION AND THE CLASSICAL DIMENSIONS

By "the classical dimensions" of a space X we mean, as above, $\operatorname{ind} X$, $\operatorname{Ind} X$ and $\dim X$. Besides the already mentioned relations concerning directable spaces we succeeded in proving several inequalities of the same type, namely "a classical dimension is $\leqslant \operatorname{Dim} X$" for certain classes of spaces X with $\operatorname{Dim} X < \aleph_0$. For the most part these theorems are based upon the following two fundamental results.

Theorem 4.1. *For an arbitrary space* X *with finite DD,* $\operatorname{ind} X \leqslant \operatorname{Dim} X$.

Theorem 4.2. *If* X *is a compact space with finite DD, then* $\operatorname{Ind} X \leqslant \operatorname{Dim} X$.

Obviously it is possible to combine these two theorems with known inequalities in dimension theory, but also special considerations yield further results. We lay emphasis on the following ones:

Theorem 4.3. *If* X *is a compact* T_2*-space with finite DD, then* $\dim X \leqslant \operatorname{Dim} X$.

Theorem 4.4. *If* X *is a strongly paracompact* T_2*-space with finite DD, then* $\dim X \leqslant \operatorname{Dim} X$.

(A space is called, according to Yu. Smirnov, strongly paracompact, if every open cover of it can be refined by a star-finite open cover. This class of spaces contains all regular Lindelöf T_2-spaces, e.g. the separable metrizable spaces, especially the regular σ-compact T_2-spaces, furthermore the locally compact and paracompact T_2-spaces with finite DD and so on. In each of these cases there are proofs independent of 4.4 and of each other.)

Theorem 4.5. *If X is a totally normal, strongly paracompact T_2-space with finite DD, then* $\operatorname{Ind} X \leqslant \operatorname{Dim} X$.

(A normal space is called – according to C.H. Dowker – totally normal, if any open subspace possesses a locally finite open cover consisting of F_σ-sets.)

Theorem 4.6. *If X is a metrizable S_σ-space with finite DD, then* $\dim X \leqslant \operatorname{Dim} X$.

(A space possesses – according to K. Morita – the *S_σ-property, if* it is the union of countable many closed strongly paracompact subspaces.)

The question arises naturally, wether inequalities in the opposite direction (i.e. $\operatorname{Dim} X \leqslant$ a classical dimension of X) are valid for some interesting classes of spaces X. In this respect the following result of J. Gerlits is to be mentioned:

Theorem 4.7. *If X is a finite dimensional strongly metrizable space, then*

$$\dim X \leqslant \operatorname{Dim} X \leqslant 2 \dim X + 2 ,$$

and this upper bound is the best possible one.

(A regular T_1-space is called – according to Yu. M. Smirnov – strongly metrizable, if it possesses a σ-star-finite base.)

It is worth comparing theorems 4.7 and 3.12 in the light of the fact that every separable metrizable space is strongly metrizable.

5.§. TOPOLOGICAL WEIGHT, DENSITY AND DIRECTIONAL DIMENSION

Denoting the density of a space X with $\sigma(X)$, $\tau(X) \geqslant \sigma(X)$ holds for every space X, and it is an old effort to find upper bounds for $\tau(X)$ in terms of $\sigma(X)$. The known results referring to this are not very numerous. Perhaps the most important such expressions are

$$2^{\sigma(X)}, \ \sigma(X), \ \sigma(X)^{\aleph_0} .$$

For example:

(i) If X is a regular space, then $\tau(X) \leqslant 2^{\sigma(X)}$!

(ii) If X is a metrizable space, then $\tau(X) = \sigma(X)$.

(iii) If X is a locally compact T_2-space in which every set consisting of a single point is a G_δ, then $\tau(X) \leqslant \sigma(X)^{\aleph_0}$ (P.S. Alexandrov – P.S. Urysohn).

(Each of these theorems can be formulated somewhat more generally, and in some cases $\tau(X)$ equals the upper bounds given in (i) and (iii).)

Now we succeeded in finding upper bounds for $\tau(X)$ in terms of $\sigma(X)$ and $\operatorname{Dim} X$. Each of the new upper bounds is the product of $\operatorname{Dim} X$ with one of the expressions under (i) – (iii).

Theorem 5.1. *If X is a space with $\sigma(X) \geqslant \aleph_0$, then $\tau(X) \leqslant 2^{\sigma(X)} \cdot \operatorname{Dim} X$ and there exist spaces X with $\tau(X) = 2^{\sigma(X)} \cdot \operatorname{Dim} X$.*

Theorem 5.2. *If X is a feebly locally connected space with* $\max[\sigma(X), \operatorname{Dim} X] \geqslant \aleph_0$, *then* $\tau(X) = \sigma(X) \cdot \operatorname{Dim} X$.

(A space is called feebly locally connected, if each of its points has a connected neighbourhood; all connected, all locally connected and some other spaces belong to this class.) It seems to be noteworthy that in theorem 5.2 – as well as in (ii) – $\tau(X)$ could be determined exactly.

Theorem 5.3. *If X is a perfectly normal, Lindelöf T_0-space with $\sigma(X) \geqslant \aleph_0$, then $\tau(X) \leqslant \sigma(X)^{\aleph_0} \cdot \operatorname{Dim} X$ and $\tau(X) = \sigma(X)^{\aleph_0} \cdot \operatorname{Dim} X$ for some spaces belonging to this class.*

(There exist analogous theorems concerning the case of $\sigma(X) < \aleph_0$ too, but they are less nice and naturally less important.)

With regard to 5.3 the question arises, if there exists a prefectly normal, Lindelöf T_0-space X with $\sigma(X) \geqslant \aleph_0$ and $\operatorname{Dim} X > \sigma(X)^{\aleph_0}$ (or, equivalently, $\tau(X) > \sigma(X)^{\aleph_0}$.

6. §. FEEBLY ORDERLY SPACES

Theorem 6.1. *A space is completely regular if and only if it is feebly orderly (and thus the class of Tychonov spaces coincides with the class of feebly orderly T_0-spaces).*

From this we can obtain a new intrinsic characterization of the spaces admitting T_2-compactifications (on the basis of the Tychonov characterization of these spaces, of course).

In what follows a characterization (of the spaces admitting T_2-compactifications) will be called *completely internal,* if it is not only internally *formulated* but also *proved* without any external tools as e.g. real functions (in particular the notion of complete regularity).

A lot of internal characterizations is known, but in all probability only Smirnov's characterization by means of proximity relations was proved internally. (Of course, every characterization whose equivalence with the Smirnov characterization can be proved internally, becomes in this way completely internal, as is the characterization by means of uniform structures or the characterization by means of preproximities recently introduced by P. Hamburger [18].)

I have tried to give further completely internal characterizations and new internal proofs of known ones, by means of the theory of DS-s, independently of Smirnov's theorem. In these investigations the following notion plays a fundamental role. (The concept is substantially due to V.I. Zaitsev, see 6.3, g); the difference is only formal.)

Definition 6.2. A system $\mathscr{H}$ of subsets of a set $X \neq \phi$ is called *orderly,* if there exists a symmetrical binary relation R defined on $\mathscr{H}$, having the following properties:

a) $H_1RH_2 \Rightarrow H_1 \cap H_2 = \phi$;

b) $H_1 \in \mathscr{H},\ x \in X \setminus H_1 \Rightarrow \exists H_2 \in \mathscr{H},\ x \in H_2,\ H_1RH_2$;

c) $H_1RH_2 \Leftrightarrow \exists H'_1,\ H'_2 \in \mathscr{H},\ H'_1 \cup H'_2 = X,\ H_1RH'_2,\ H_2RH'_1$.

In order to emphasize the relation R we sometimes call the system $\mathscr{H}$ of sets *R-orderly*.

If $\mathscr{H}$ is a closed basis of a topology on X and $H_1RH_2 \Leftrightarrow H_1 \cap H_2 = \phi$. then the *R-orderlyness* of $\mathscr{H}$ is exactly what is known as regularity and normality of $\mathscr{H}$, as it was introduced – following the initiation of O. Frink [16] – by J. de Groot and J.M. Aarts [17], in order to give internal characterizations of the spaces admitting T_2-compactifications. The advantage of this more general concept is that starting from orderly systems of sets new orderly systems can be obtained by certain operations. In each such case, of course, a new convenient relation R must be chosen, but it is just the ability to vary R, which justifies the generalization.

Theorem 6.3. *Let* X *be a* T_0*-space, The following statements are equivalent:*

a) *The topology of* X *admits an orderly closed basis.*

b) *The topology of* X *admits an orderly closed subbasis.*

c) *The topology of* X *can be induced by a proximity relation defined on the set* X.

d) X *is a feebly orderly space.*

e) X *admits a* T_2*-compactification.*

The proofs are based on the equivalences b) ⇔ d) and d) ⇔ e), and it is not necessary to make use of c) to prove the equivalences between the other statements. All the proofs are internal, hence it is a case of completely internal characterizations of spaces admitting T_2-compactifications.

P. Hamburger observed, that in this area some further known internal characterizations can be integrated and in this way we obtain internal proofs, namely by proving their equivalence to b):

f) *The regularity system of the space* (that is – according to Yu.

M. Smirnov – the union of all systems Σ of pairs (F, G), where F is a closed set and G is an open neighbourhood of F, such that, if $(F_1, G_1) \in \Sigma$, there exists a pair $(F_2, G_2) \in \Sigma$ with $(G_2^-, G_1), (F_1, G_2) \in \in \Sigma$) *contains every pair* $(\{x\}^-, U)$, *where* x *is an arbitrary point of* X *and* U *is an arbitrary open neighbourhood of* x (Y u. S m i r n o v [25]).

g) *There exists a basis* $\mathscr{B}$ *of the closed sets in the space* X *and a system* $\mathfrak{z}$ *of pairs* (Φ, Γ) *such that*

$g_1)$ $\phi \in \mathscr{B}$, $X \setminus \Gamma \in \mathscr{B}$, $\Phi \subseteqq \Gamma$ $((\Phi, \Gamma) \in \mathfrak{z})$;

$g_2)$ *if* $(\Phi, \Gamma) \in \mathfrak{z}$ *then there exist sets* Γ_1, Γ_2 *with* $\Gamma_1 \cap \Gamma_2 = = \phi$, (ϕ, Γ_1), $(X \setminus \Gamma, \Gamma_2) \in \mathfrak{z}$;

$g_3)$ *if* $X \setminus \Gamma \in \mathscr{B}$ *and* $x \in \Gamma$, *there exists a set* Φ *with* $(\Phi, \Gamma) \in \in \mathfrak{z}$ *and* $x \in \Phi$ (V. J. Z a i t s e v [24], Á. C s á s z á r [3]).

h) *If* $\mathscr{B}$ *is the system of all closed sets in the space* X, *then there exists a system* $\mathfrak{z}$ *of pairs satisfying the three conditions listed under* g) (V. J. Z a i t s e v [24], Á. C s á s z á r [3]).

It is uncertain, if also the Frink – Aarts – de Groot-type characterizations of spaces admitting T_2-compactifications can be proved internally (possibly even by means of the theory of DS's).

A proof of theorem 6.3 can be found in our paper written jointly with P. H a m b u r g e r [14] but the really complete proof (including the proofs of all the necessary lemmas) is one of the subjects of the book [23].

As to the proof of d) $\Rightarrow$ e) under 6.3, as well as to the proof of 6.1 the following embedding theorem, a generalization of 1.15, plays a fundamental role:

Theorem 6.4. *If* X *is a feebly orderly* T_0*-space, then* X *is homeomorphic to a subspace of the topological product* $R = \Pi\{\mathscr{R}: \mathscr{R} \in \mathfrak{R}\}$, *where* $\mathfrak{R}$ *is an arbitrary CODS of the space and the directions* $\mathscr{R} \in \mathfrak{R}$ *are taken as ordered spaces according to* 1.7 *and* 1.12; *in fact the "evaluation map"* $\varphi: X \to R$, *where the* $\mathscr{R}$*-th coordinate of* $\varphi(x)$ *is defined to be the element* (G, F) *of* $\mathscr{R}$ *with* $x \in F \setminus G$, *is a topological embedding.*

(For a given x there exists a such pair (G, F) by the orderliness of $\mathscr{R}$, and only one according to R_2).)

Since the product space R is compact Hausdorff (according to 1.12 and the Tychonov product theorem), theorem 6.4 yields a compactification method consisting in taking the closure of the topological image $\varphi[X]$ of a space X in R. Different CODS-s $\mathfrak{R}$ of a space X can correspond to the same T_2-compactification of X.

Definition 6.5. Suppose that X is a feebly orderly T_0-space, $\mathfrak{R}$ is a *CODS* of it, $R = \Pi\{\mathscr{R}: \mathscr{R} \in \mathfrak{R}\}$, and φ is the natural map defined under 6.4. Then by the *$\mathfrak{R}$-compactification* of X we mean the space $\varphi[X]^{-(R)}$.

The following theorem shows the effectiveness of this method.

Theorem 6.6. *Suppose that X is a compact T_2-space, X^* is a dense subspace of it and $\mathfrak{R}$ is a CODS of X such that*

$$\mathscr{R}_M = \{(\phi, \phi), (\phi, M), (M, X), (X, X)\} \in \mathfrak{R} ,$$

provided that M is a non-trivial open-and-closed set in the space; then the $\mathfrak{R}^*$-compactification of X^*, where $\mathfrak{R}^* = \mathfrak{R} \mid X^*$, is homeomorphic to X.

(*The trace $\mathfrak{R} \mid X^*$ of a CODS $\mathfrak{R}$* of a space X on an arbitrary subspace X^* is a *CODS* of X^*; here by $\mathfrak{R} \mid X^*$ we mean the set of all traces

$$\mathscr{R} \mid X^* = \{(G \cap X^*, F \cap X^*): (G, F) \in \mathscr{R}\} \qquad (\mathscr{R} \in \mathfrak{R}.)$$

The condition of this theorem is really necessary, but it can be satisfied in every case by adding the possibly missing directions $\mathscr{R}_M$ to an arbitrary CODS of the space X.

To put this compactification method in its true light we explain its connection with the known fundamental fact that the class of compact T_2-spaces coincides with the class of closed subspaces of Tychonov cubes.

Definition 6.7. Let X be a space and $f: X \to [0, 1]$ be a continuous function. The system

$$\mathscr{R}^f = \{(\phi, \phi)\} \cup \{(G_t^f, F_t^f): t \in [0, 1]\} \cup \{(X, X)\} ,$$

where

$$G_t^f = \{x \in X\colon f(x) < t\},\ F_t^f = \{x \in X\colon f(x) \leqslant t\} \quad (t \in [0, 1]),$$

is called a *real direction* of the space X.

Theorem 6.8. *Each real direction of a space is orderly and – regarded as an ordered space – homeomorphic to a subspace of* E_1.

(The first statement of this theorem is trivial; the second is based on 1.4.)

Theorem 6.9. *The system of all real directions of a completely regular space is a CODS satisfying the condition of theorem 6.6.*

Now it is obvious from 6.8 and 6.9 that the $\mathfrak{R}$-compactification method is a generalization of the compactification method arising from embeddings in Tychonov cubes. (See also the next section.)

Finally it is to be mentioned, without giving any details, that not only ODS's and proximity relations but also $\mathfrak{R}$-compactifications and the Smirnov compactification theorem concerning proximity spaces are closely related. In [13] these connections are investigated in details. For this purpose we introduce the notion of an *orderly direction space* (that is the pair $(X, \mathfrak{R})$, where X is a non-void set and $\mathfrak{R}$ is an ODS of it) which constitutes – since it can be regarded as a generalization of the notion of a linear space – the proper subject of the investigations concerning the convexity theory mentioned in the beginning.

7.§. ORDERLY SPACES

Theorem 7.1. *Every perfectly normal space is orderly.*

This fundamental theorem enables us to apply the results concerning orderly spaces to metrizable spaces which are the most important ones in many respects, especially in dimension theory. The theorem admits, however, applications in its general form too (as e.g. in the proof of 5.3).

Perfect normality, however, is not at all a necessary condition for the orderliness of a space. The normality properties of an orderly space X with $\operatorname{Dim} X \leqslant 1$ (i.e. an orderly directable space) are very good (see theorems 1.16 and 1.17) though the space can fail to be perfectly normal. On the other hand e.g. the well-known non-normal Sorgenfrey space S is orderly (as it can be verified easily) and $\operatorname{Dim} S = 2$.

From 3.8 the following theorem can be deduced:

Theorem 7.2. *The topological product of an uncountable family of orderly T_0-spaces each of which has at least two different points, is an orderly space.*

From 7.2 and 7.1 follows:

Theorem 7.3. *The cube* $[0, 1]^{\mathfrak{n}}$ *is an orderly space for every cardinal number* $\mathfrak{n}$.

The following theorem holds according to 6.3, 6.4, 1.12, 7.2 and the fact that compactness is a productive property.

Theorem 7.4. *If a space X admits a T_2-compactification, then there exists an orderly compact Hausdorff space containing X as a subspace.*

The following consequence of this theorem may be interesting for itself too:

Theorem 7.5. *The class of feebly orderly T_0-spaces coincides with the class of subspaces of orderly T_2-spaces.*

In theorem 7.4 arbitrary subspaces, not only dense ones occure; therefore the following question arises;

Theorem 7.6. *Which spaces admit orderly T_2-compactifications?*

(If that is the case for every space admitting a T_2-compactification, then every compact T_2-space is orderly, and conversely.)

What we know is only the following:

Theorem 7.7. *Every orderly T_0-space X admits an orderly T_2-compactification Y with* $\operatorname{Dim} Y = \operatorname{Dim} X$.

It is obvious that the principal open problem (closely related to all the other questions) concerning orderliness reads as follows:

Theorem 7.8. *Is the orderliness of* T_0*-spaces a hereditary property, or, equivalently, is every feebly orderly* T_0*-spaces orderly?*

Now we discuss briefly the connections between our embedding theorem 6.4 and the Tychonov embedding theorem; this question leads us immediatly to the nex section.

From 6.4 – considering also 3.3 – both the Tychonov theorem and an analogous theorem concerning order topology can be deduced. If we restrict also the Tychonov theorem to the domain on which the other theorem is valid – that is the class of the orderly T_0-spaces (about which we do not know wether it contains every Tychonov space or not) – then the two theorems will have a common part, namely

Theorem 7.9. *Every orderly* T_0*-space is homeomorphic to a subspace of the topological product of a family of ordered spaces.* (See 6.4.)

In the restricted theorem of Tychonov we have the following addendum to this:

Theorem 7.9a. *For coordinate spaces real intervals can always be taken, and the cardinality of the family of coordinate spaces need not to be greater than* $\tau(X)$*; in case of* $\tau(X) > \aleph_0$ *this cardinality cannot be smaller than* $\tau(X)$.

The other theorem states in addition to the common part the following:

Theorem 7.9b. *The family of the coordinate spaces can be taken in any case of the cardinality* $\operatorname{Dim} X$ *and in no case of a smaller cardinality.*

Thus the restricted Tychonov theorem specializes the coordinate spaces at the expense of increasing the cardinality of their family. In fact the weight of an orderly T_0-space can be much greater than its *DD* (e.g. every discrete space X is orderly directable, hence $\operatorname{Dim} X \leqslant 1$, but the weight of X is equal to the cardinality of the set X). On the other hand, our theorem 7.9 – 7.9b reduces the cardinality of the family of coordinate spaces as much as possible, but without specializing the coordinate spaces at the same time.

The following theorem leading to one of the main results of the theory of DS-s, is a result of the effort to combine these two aspects.

8.§. COMPLETE CHARACTERIZATION OF THE CLASS OF SUBSPACES OF A EUCLIDEAN SPACE, IN TERMS OF DIRECTIONAL DIMENSION

Theorem 8.1. *A separable metrizable space* X *is homeomorphic to a subspace of* $E_{\mathrm{Dim}\,X}$.

From this theorem among others the results 3.10, 3.11 and 3.12 can be deduced anew. While, however, theorem 3.12 makes possible only a characterization of the subspaces of Euclidean spaces in general (as well as these spaces are characterized by the finiteness of their classical dimensions), from 8.1 we obtain a complete characterization of the subspaces of every single Euclidean n-space:

Theorem 8.2. *For a separable metrizable space* X, $\mathrm{Dim}\,X \leqslant n$ $(n = 0, 1, 2, \ldots)$ *holds if and only if* X *is homeomorphic to a subspace of* E_n.

Thus $\mathrm{Dim}\,X$ is exactly the least number n such that the separable metrizable space X is topologically embeddable in the Euclidean space E_n.

REFERENCES

[1] R.A. Alò – O. Frink, Topologies of chains, *Math. Ann.*, 171 (1867).

[2] E. Čech, *Topological spaces,* Academia (Czechoslovak Acad. Sci.), Prague 1966.

[3] Á. Császár, On the characterization of completely regular spaces, *Ann. Univ. Sci.*, Budapest. Sectio Math. 11 (1968), 78-82.

[4] E. Deák, Ein neuer topologischer Dimensionsbegriff, *Revue Roum. Math. Pure Appl.*, 10 (1965), 31-42.

[5] E. Deák, Bemerkungen zu meiner vorangehenden Arbeit "Ein neuer topologischer Dimensionsbegriff, *Revue Roum. Math. pures Appl.*, 13 (1968), 303-305.

[6] E. Deák, Eine vollständige Characterisierung der Teilräume eines euklidischen Raumes mittels der Richtungsdimension, *Publ. Math. Inst. Hung. Acad. Sci.*, 9 (1964), 437-465.

[7] E, Deák, Einige Beziehungen der Richtungsdimension zu den klassischen Dimensionsbegriffen der allgemeinen Topologie, *Math. Nachr.* 37 (1968), 247-266.

[8] E. Deák, Richtungsräume und Richtungsdimension, General Topology and its Relations to Modern Analysis and Algebra II (*Proc. Second Prague Top. Symp.*, 1966), 105-106.

[9] E. Deák, Eine Verallgemeinerung des Begriffs des linearen Raumes und der Konvexität, *Ann. Univ. Sci.*, Budapest. Sectio Math. 9 (1966), 45-59.

[10] E. Deák, Topologische Richtungsräume – eine Verallgemeinerung des Begriffs des lokalkonvexen Raumes mit der schwachen Topologie, *Studia Sci. Math. Hung.*, 1 (1966), 297-308.

[11] E. Deák, Extremalpunktsbegriffe für Richungsräume und eine Verallgemeinerung des Krein – Milmanschen Satzes für topologische Richtungsräume, *Acta Math. Acad. Sci. Hung.*, 18 (1967), 113-131.

[12] E. Deák, Dimension and Convexity I-IV (Hungarian), MTA III. *Oszt. Közl.*, 17 (1968), 185-213, 311-329, 391-419; 18 (1968), 45-81.

[13] E. Deák, *Dimension und Konvexität. Theorie und Anwendung der Richtungsstrukturen,* Akadémiai Kiadó (to appear).

[14] E. Deák – P. Hamburger, Vollständig interne Charakterisierungen der T_2-kompaktifizierbaren Räume, *Periodica Math. Hung.*

[15] J. Gerlits, On some problems concerning directional dimension, *Studia Sci. Math. Hung.*, 6 (1971), 409-417.

[16] O. Frink, Compactifications and semi-normal spaces, *Amer. Journal Math.*, 86 (1964), 602-607.

[17] J. de Groot – J.M. Aarts, Complete regularity as a separation axiom, *Canad. Journal Math.*, 21 (1969), 96-105.

[18] P. Hamburger, A general method to give internal characterizations of completely regular and Tychonov spaces, *Acta Math. Acad. Sci. Hung.*, 23 (3-4), (1972), 479-494.

[19] P. Hamburger, On *k*-compactifications and realcompactifications, *Acta Math. Acad. Sci. Hung.*, 23 (1-2), (1972), 255-262.

[20] R.W. Heath – D.J. Lutzer, A note on monotone normality, *Notices Amer. Math, Soc.*, 18, No 3 (1971), 783.

[21] A. Higarshisaka, On $\mathscr{R}$-convex sets in a topological $\mathscr{R}$-space, *Proc. Japan Acad.*, 45 (1969), 102-106.

[22] J.L. Kelley, *General topology,* New York (1955).

[23] D.J. Lutzer, *On generalized ordered spaces,* Diss. Math., 89, Warszawa 1971.

[24] V.I. Zaitsev, On the theory of Tychonov spaces (Russian), *Vestnik Mosk. Univ. Math. Mech.*, (1967), 48-57.

[25] Yu.M. Smirnov, On the theory of completely regular spaces (Russian), *Dokl. Acad. Nauk SSSR,* (N.S.) 62 (1948), 749-752.

[26] L.A. Steen, A direct proof that the interval topology is collectionwise normal, *Proc. Amer. Math. Soc.*, 24 (1970), 727-728.

[27] Ph.L. Zenor, Monotonically normal spaces, *Notices Amer. Math. Soc.*, 17, No 7, (1970), 1034.

COLLOQUIA MATHEMATICA SOCIETATIS JÁNOS BOLYAI
8. TOPICS IN TOPOLOGY, KESZTHELY (HUNGARY), 1972.

MINIMAL TOPOLOGICAL GROUPS

D. DOITCHINOV

A topological group G with a Hausdorff topology is called minimal if it is not possible to introduce a Hausdorff topology on G compatible with the group sturcutre of G and strictly coarser than the topology τ given on G (in this case the topology τ is also said to be minimal). Every compact topological group is obviously minimal. On the other hand, we describe here an example showing that there are minimal topological groups that are not compact.

The set of all subgroups of form $2^n \cdot Z$ (n is a positive integer) of the additive group Z of the integers is a filter base on Z. This generates a filter which is the neighbourhood filter of 0 in a minimal topology on Z. In this way we obtain a minimal Abelian topological group Z that is not compact. Moreover, it is easily seen that the product $\widetilde{Z} \times \widetilde{Z}$ is not a minimal topological group.

On the other hand, it is shown in [1] that the product of a minimal topological group with a compact group is always minimal.

Problem 1. Does there exist a minimal topology compatible with the additive group structure of the real line R and coarser than the usual topology on R (the latter topology being not minimal)?

Problem II. Does there exist a minimal topological group which is not precompact?

Problem III. Find necessary and sufficient conditions for the product of a family of minimal topological groups to be minimal.

REFERENCES

[1] D. Doitchinov, Produits de groupes topologiques minimaux, *Bull. Sciences Math.*, 2nd series, 96 (1972.

COLLOQUIA MATHEMATICA SOCIETATIS JÁNOS BOLYAI
8. TOPICS IN TOPOLOGY, KESZTHELY (HUNGARY), 1972.

TOPOLOGIZING METRIC SPACES, II

R.Z. DOMIATY

§1. INTRODUCTION

The starting point for the considerations of [2] was the following

Definition 1. Let (R, d) be a metric space. A topology T on R is called *compatible* with d if the closed balls $B(p, r)$ are closed sets with respect to T.

In what follows we shall denote by $V(R, d)$ the set of all topologies on R compatible with d. If a topology $T \in V(R, d)$ is chosen to correspond to (R, d) we shall briefly indicate this by writing (R, d, T).

We shall not give a motivation or a discussion of this definition here, but rather refer the reader to [1] and [2]. We shall restrict ourselves to quoting several simple results which will be needed below.

In $V(R, d)$ there exists a (uniquely determined) coarsest topology, which we shall denote by T_d^*. Thus in addition to the topology T_d induced by d an R (which, of course, belongs to $V(R, d)$) that is usually mentioned as the natural metric topology of (R, d), we obtain here a second distinguished topology which, by analogy, we shall call the $*$-induced topology on R. T_d^* satisfies the T_1 (Fréchet) separation axiom, is coarser than T_d, and the closed balls form a subbasis for the closed sets of the topological space (R, T_d^*). However, T_d^* does not necessarily satisfy the T_2 (Hausdorff) separation axiom (e.g. if the diameter $\delta(R) = \infty$).

By definition 1 to every metric space (R, d) there corresponds a family $V(R, d)$ of topologies on R. For different inner geometric questions the T_2 topologies from $V(R, d)$ are of special interest. (Since $T_d \in V(R, d)$ there is always such a topology in $V(R, d)$.) Among these the minimal ones are distinguished in a natural manner. (A T_2 topology T is called minimal whenever every T_2 topology coarser than T must coincide with T.) Especially interesting seem to be also those T_2 topologies $T \in V(R, d)$ for which

$$T_d^* \subseteq T \subseteq T_d$$

([2], Problem 2.)

Let R^n stand for the usual n-dimensional euclidean space together with the usual euclidean metric e. In the present paper we shall study several situations which occur when on (R^n, e) instead of the euclidean topology T_e a topology $T \in V(R^n, e)$ with

$$T_e^* \subseteq T \subseteq T_e$$

is considered.

§2. (R, d, T_d^*)

Here we want first of all to collect several simple facts which arise when in a metric space (R, d) the $*$-induced topology is considered.

Let

$$b_d^* = \left\{ \bigcup_{i=1}^{n} B(p_i, r_i) \mid n \in N, p_i \in R, r_i \geqslant 0 \right\}. \tag{1}$$

A simple argument shows that for every $M \subseteq R$ its closure (with respect to T_d^*) can be represented as

(2) $$\bar{M} = \bigcap_{M \subseteq B \in b_d^*} B .$$

Moreover, it follows from (2) that

(3) $$\delta(M) = \infty \Rightarrow \bar{M} = R ,$$

(4) $$\delta(M) < \infty \Rightarrow \delta(\bar{M}) < \infty .$$

([2], Lemma 1). It is trivial that if $\delta(M) = \infty$ there might exist a $p \in \bar{M}$ such that $d(p, M) > 0$, however the next example shows that this can happen even if $\delta(M) < \infty$.

Example 1. Let

$$l^2 = \left\{ w = (\omega_i)_{i \in N} \mid \omega_i \in R,\ \sum_{i=1}^{\infty} \omega_i^2 < \infty \right\}$$

be the usual Hilbert space with the metric

$$d(a, b) = \left[\sum_{i=1}^{\infty} (a_i - \beta_i)^2 \right]^{\frac{1}{2}} .$$

If we put

$$M = \left\{ z = (\delta_{ij})_{j \in N} \right\}_{i \in N} \subset l^2 ,$$

then $\delta(M) = \sqrt{2}$, $0 \in \bar{M} \setminus M$ and $d(0, M) = 1$.

However if M is a totally bounded subset of (R, d), then

(5) $$\forall p \in \bar{M} : d(p, M) = 0 .$$

§3. RECTIFIABLE MAPPINGS AND PATHS IN (R, d)

Let $I = [a, b]$ be a fixed compact interval of R and $f \colon I \to R$ be a mapping of I into the metric space (R, d). We shall denote by $e(I)$ the set of all finite subsets of I (where the elements of any such finite subset $a = \{t_1, \ldots, t_n\}$ are thought to be ordered increasingly: $t_1 < t_2 < \ldots$ $\ldots < t_n$) and for all $a = \{t_1, \ldots, t_n\} \in e(I)$ we put

$$v_f(a) = \sum_{i=1}^{n-1} d[f(t_i), f(t_{i+1})] .$$

Then the quantity

$$L(f) = \sup_{a \in e(I)} v_f(a)$$

is called the length of f. If $L(f) < \infty$ then we say that f is rectifiable.

In case a topology $T \in V(R, d)$ is considered on R, then the mapping f is called a path in (R, d, T) provided that f is continuous with respect to T. Whether or not a mapping is a path depends entirely on T, while the rectifiability of a mapping is independent of the topology under consideration on (R, d).

The following example shows that not for every path f in (R, d, T) must we have the image set $f(I)$ bounded.

Example 2. Let $(\boldsymbol{R}^1, e, T_e^*)$, $I = [0, 1]$, and for every non-negative integer n let

$$f_n : \left[\frac{n}{n+1}, \frac{n+1}{n+2}\right] \to [n, n+1]$$

be a bijective and continuous mapping of the interval $\left[\frac{n}{n+1}, \frac{n+1}{n+2}\right] \subset$ $\subset I$ onto the closed interval $[n, n+1] \subset \boldsymbol{R}$ with $f_n\left(\frac{n}{n+1}\right) = n$ and $f_n\left(\frac{n+1}{n+2}\right) = n + 1$. Now we define a mapping

$$f_z : I \to \boldsymbol{R}^1$$

by the following stipulations:

$$f_z(t) = \begin{cases} f_n(t) & \text{for} \quad t \in \left[\frac{n}{n+1}, \frac{n+1}{n+2}\right], \\ z \in \boldsymbol{R}^1 & \text{for} \quad t = 1 . \end{cases}$$

Then f is continuous with respect to T_e^* no matter which value $z \in \boldsymbol{R}^1$ is chosen. By

$$f_z(I) = \{z\} \cup [0, \infty) ,$$

$f_z(I)$ is not bounded. This example also shows that the compact subsets of $(\boldsymbol{R}, d, T_d^*)$ do not have to be bounded.

§4. $(\boldsymbol{R}^n, e, T)$

In this section we shall consider $(\boldsymbol{R}^n, e)$. If $Q \subseteq \boldsymbol{R}^n$, we shall denote by $e \mid Q$ the metric induced on Q by e, and for every topology T on $\boldsymbol{R}^n$ we denote by $T \mid Q$ the trace topology on Q.

Lemma 1. ([2], §6). *Let $Q \subseteq \boldsymbol{R}^n$ with $\delta(Q) < \infty$. Then we have*

$$T_{e|Q} = T_e^* \mid Q . \tag{6}$$

Moreover, since we have

$$T_{e|Q} = T_e \mid O ,$$

we obtain

Theorem 1. *Let $Q \subseteq \boldsymbol{R}^n$ with $\delta(Q) < \infty$. Then for every topology $T \in V(\boldsymbol{R}^n, e)$ with $T_e^* \subseteq T \subseteq T_e$ we have the equality*

$$T_{e|Q} = T \mid Q .$$

Thus all topologies T with $T_e^* \subseteq T \subseteq T_e$ generate the same trace topology on a bounded subset $Q \subseteq \boldsymbol{R}^n$. Since on one hand every bounded subset of R is totally bounded, and on the other hand every rectifiable path in $(\boldsymbol{R}^n, e, T)$ is bounded, we obtain

Lemma 2. *Let $T \in V(\boldsymbol{R}^n, e)$ and $f\colon I \to \boldsymbol{R}^n$ be a rectifiable path. Then $f(I)$ is totally bounded.*

In connection with this lemma the following open question arises: If $f\colon I \to R$ is a rectifiable path in (R, d, T_d^*), is then $f(I)$ necessarily totally bounded?

To every locally compact T_2 space there exists a coarser topology which transforms it into a compact T_2 space, hence every such space possesses a minimal T_2 topology. Since $(\boldsymbol{R}^n, e, T_e)$ is such a space we can prove

Theorem 2. ([4]). *There exists a minimal T_2 topology $T_0 \in$ $\in V(\boldsymbol{R}^n, e)$ with the following properties:*

1) $T_e^* \subseteq T_0 \subseteq T_e$,

2) $(\boldsymbol{R}^n, T_0)$ *is a compact metrizable space.*

A throughgoing investigation of this group of questions was carried out in [4].

Finally we would like to consider briefly a dimension-theoretic problem. Since in our investigations the metric on a set plays a bigger rôle than the corresponding topology, we shall use here a dimension concept which takes this standpoint into consideration. Here we have to reckon with the fact that this dimension concept is not topologically invariant.

Definition 2. ([5], 172; [6], 126). Let (R, d, T) be given and let $n \in \{-1, 0, 1, \ldots, \infty\}$. We shall denote by $A_T(R)$ the family of all closed coverings of R (with respect to T).

1. We put

$$\mu - \dim_T R \leqslant n ,$$

whenever there exists a countable sequence $\{F_i\}_{i \in N} \subset A_T(R)$ with the following properties:

a) For each $i \in N$, F_i is locally finite.

b) For each $i \in N$ we have

$$\operatorname{ord} F_i \leqslant n + 1 ,$$

c)

$$\lim_{i \to \infty} \operatorname{mesh} F_i = 0 .$$

Then $\mu - \dim_T R$ is called the metric dimension of R with respect to $T \in V(R, d)$.

2. $\mu - \dim_T \phi = -1$.

3. We put

$$\mu - \dim_T R = n ,$$

if

$$\mu - \dim_T R \leqslant n \quad \text{and} \quad \mu - \dim_T R \nleqslant n - 1 .$$

4. Finally $\mu - \dim_T R = \infty$ if for every $n \in N$ we have

$$\mu - \dim_T R \nleq n .$$

Now let us add a few remarks to this definition. It is somewhat more general than it is usually formulated, as we do not restrict ourselves to the topology T_d only, but permit all the topologies $T \in V(R, d)$. When restricting to $T = T_d$, definition 2 is equivalent to the usual μ-dimension, i.e.

$$\mu - \dim_{T_d} R = \mu - \dim R$$

(cf. [5], 172, Def. 29-1 and Prop. 29-2). Since in definition 1 the closed sets of (R, d, T) are somewhat favoured, we have chosen to provide the formulation of definition 2 for the case of closed sets.

It is well known that

(7)
$$\mu - \dim_{T_e} \boldsymbol{R}^n = n .$$

Now we shall generalize (7) as follows.

Theorem 3. *Let* $T \in V(\boldsymbol{R}^n, e)$. *If* $T_e^* \subseteq T \subseteq T_e$, then

(8)
$$\mu - \dim_T \boldsymbol{R}^n = n .$$

Proof. Let us fix an arbitrary topology T with $T_e^* \subseteq T \subseteq T_e$. We shall show (8) in two steps. First we show

(9)
$$\mu - \dim_T \boldsymbol{R}^n \nleq n - 1 .$$

Indeed, if (9) was false, then by definition 2 there should exist a family $\{F_i^*\} \subseteq A_T(\boldsymbol{R}^n)$ of coverings with the properties a) – c) with n replaced by $n - 1$. However $T \subseteq T_e$ implies immediately $A_T(\boldsymbol{R}^n) \subseteq A_{T_e}(\boldsymbol{R}^n)$, and thus results in a contradiction with (7). Hence (9) must be valid. Next we show

$$\mu - \dim_T \boldsymbol{R}^n \leqslant n .$$

By (7) and definition 2 there exists a family of coverings with the properties a) – c). As an example of such a family can be taken any family $\{W_i\}_{i\in N}$ of rectangular cell divisions, where every division W_i consists of closed cells of diameter $1/i$ (cf. [3], p. 124.). Thus it remains to show that for each $i \in N$.

$$W_i \subseteq A_T(R^n),$$

i.e. that every closed rectangular cell w (in the usual sense) is a closed set with respect to T as well. Since w is closed with respect to T_e and totally bounded, it can be written in the form

$$w = \bigcap_{w \subseteq B \in b_d^*} B$$

(cf. (1)), consequently w is a closed set with respect to T_e^*, and thus with respect to T as well. The proof of theorem 3 is completed.

Added in proof. The problem following up Lemma 2 can be answerd in positiv sense (R. Domiaty, Eine Bemerkung zu total beschränkten Mengen, *Elem. Math.*, 28 (1973), 97-98.

REFERENCES

[1] R. Domiaty, Metrische Räume mit einer Elementarlänge, *Month. f. Math.*, 76 (1972), 1-20.

[2] R. Domiaty, Topologisierung metrischer Räume, I. *J. reine u. angew. Math.*, 258 (1973), 126-132.

[3] W. Franz, Topologie I. Berlin 1960.

[4] A. Ilgaz, Minimale T_2-Topologien, Dissertation, Technische Hochschule Graz, 1972.

[5] K. Nagami, *Dimension Theory,* New York, 1970.

[6] J. Nagata, *Modern Dimension Theory,* Groningen, 1965.

COLLOQUIA MATHEMATICA SOCIETATIS JÁNOS BOLYAI
8. TOPICS IN TOPOLOGY, KESZTHELY (HUNGARY), 1972.

SEPARATION AXIOMS FOR FRAMES

C.H. DOWKER and DONA PAPERT STRAUSS

A frame is a complete lattice with the infinite distributivity law

$$a \wedge \bigvee_a b_a = \bigvee_a a \wedge b_a .$$

It follows that there is a finite distributivity

$$a \vee (b \wedge c) = (a \vee b) \wedge (a \vee c) .$$

A frame map $\mu: L \to M$ is a function from a frame L to a frame M which commutes with finite meets and with arbitrary joins. The topology tX of a space X, that is, the set of all open sets of X ordered by inclusion, is a frame. A map $f: X \to Y$ of spaces induces a frame map f^{-1}: $tY \to tX$.

Many topological theorems generalize to theorems about frames (see [2]). In this paper we show how some of the separation axioms can be defined for frames so that they reduce to the usual separation axioms when applied to topologies, and we check the appropriateness of the extended definitions by proving for frames some standard theorems of topology.

A subspace $A \subset X$ induces a surjective frame map $tX \to tA$. A quotient $\mu: L \to M$ is a representative of an equivalence class of surjective frame maps of L. Methods of selecting the representative were described in [1]. In the statements and proofs of theorems, quotients are frequently used explicitly or implicitly instead of the subspaces or subsets of a space which appear in the usual topological theorems.

Epimorphisms (in the category sense) are not necessarily surjective.

Example 1. Let R be the space of real numbers, let tR be its topology and let $\mathscr{P}R$ be the set of all subsets of R. Let $j: tR \to \mathscr{P}R$ be the inclusion.

For any frame maps $f, g: \mathscr{P}R \to M$ such that $f \circ j = g \circ j$, f and g agree on open sets, in particular on the complements of points. Hence, since f and g are frame maps, they agree on one-point sets themselves and on unions of one-point sets, that is, on arbitrary sets. That is, $f = g$. Thus j is an epimorphism, though it is not surjective.

On the other hand, all monomorphisms of frames are injective. One can see this by using a three-element frame $T = \{0, t, 1\}$. For each frame L and each element $a \in L$ there is a unique frame map $f: T \to L$ for which $f(t) = a$. It follows that, if $\mu: L \to M$ is not injective, there are distinct frame maps $f, g: T \to L$ for which $\mu \circ f = \mu \circ g$.

SEPARATION AXIOMS T_0 AND T_1

By identifying inseparable points of an arbitrary space X one obtains a quotient T_0-space Y whose topology is isomorphic to the topology of X. Thus every topology is isomorphic to the topology of a T_0-space. The condition T_0 is thus not topological, that is, it is not determined by the topology of the space. Indeed T_0- places no restriction on the structure of the topology. For topologies and, by analogy, for frames T_0 is a vacuous condition.

The other separation axioms T_n contain T_0, and are to that extent not topological. To see what T_n implies for the topology, we may replace the topology by the isomorphic topology of a T_0-space.

The condition T_1 is sometimes determined by the topology. The topology of a two-point space, one point of which is closed, is not isomorphic to the topology of any T_1-space. The topology of a Hausdorff space is not isomorphic to the topology of a T_0-space which is not T_1. In fact each T_0-space whose topology is isomorphic to the topology of a Hausdorff space must be a Hausdorff space. But the following example shows that the condition T_1 is not topological for T_0-spaces.

Example 2. Let X be the space consisting of the natural numbers and ∞, whose closed sets are finite sets of natural numbers and all X. It is not a T_1-space, for the point ∞ is not closed. The subspace Y consisting of the natural numbers with the co-finite topology is a T_1-space. The canonical map of the topology of X to the topology of the subspace Y is an isomorphism.

We introduce a condition T_1' for frames which is weaker than T_1 for topological spaces.

T_1') *If* $a \nleq b$, $\exists v$ *such that* $v \vee b \neq 1$, $v \vee a = 1$.

Proposition 1. *The following conditions on a frame* L *are equivalent:*

(i) L *is a* T_1'*-frame.*

(ii) *If* $b < a$, $\exists v$ *such that* $b \leqslant v < 1$, $v \vee a = 1$.

(iii) *If* $a \neq b$, $\exists v$ *such that* $(a \wedge b) \vee v \neq 1$, $a \vee b \vee v = 1$.

A T_1'-space is a T_0-space with a T_1'-topology.

It is a T_0-space with the property that each non-empty difference $E - F$ of closed sets contains a non-empty closed set. A space has this property if points are closed. Thus each T_1-space is a T_1'-space.

We consider the property T_1' for compact frames.

Definition. A family $\{a_\alpha\}_{\alpha \in A}$ of elements of a frame is a *cover* if $\bigvee a_\alpha = 1$.

Definition. A frame is *compact* if each cover $\{a_\alpha\}$ has a finite subcover $\{a_{\alpha_i}\}$.

Definition. A *prime* element of a frame L is an element p such that $p \neq 1$ and if $p = a \wedge b$ then $p = a$ or $p = b$.

Definition. A *maximal* element of a frame L is an element m such that $m \neq 1$ and if $m \leqslant a$ then $m = a$ or $a = 1$.

Clearly each maximal element is prime.

Proposition 2. *In a compact frame each element except* 1 *precedes a maximal element.*

Proof. Let L be compact and let $a \in L$, $a \neq 1$. Let A be the partially ordered set $\{x \colon a \leqslant x < 1\}$. For each chain (totally ordered non-void subset) C in A, the join of each finite subset F of C is the greatest element of F, hence is not 1. Hence, by the compactness of L, C does not cover. Hence, $\bigvee C$ is an upper bound of C in A. By Zorn's lemma, there is an element $m \in A$ such that if $b \in A$ and $b \geqslant m$ then $b = m$. Thus m is a maximal element with $a \leqslant m$.

Proposition 3. *In a compact* T_1'*-frame each element is a meet of primes.*

Proof. Let $a \in L$ and let $b = \bigwedge\{p \in L \colon p \text{ prime},\ a \leqslant p\}$. Then $a \leqslant b$. Suppose if possible that $a \neq b$. Then by T_1', $\exists v \neq 1$ such that $a \leqslant v$ and $v \vee b = 1$. There is a maximal element m with $v \leqslant m$. Then $a \leqslant m$, m is prime, $m \vee b = 1$ and hence $m \vee p = 1$ for each prime such that $a \leqslant p$. Hence $m \vee m = 1$ which is absurd.

Proposition 4. *Each compact* T_1'*-frame* L *is isomorphic to a topology.*

Proof. Since each element $b \in L$ is a meet of primes, if $a \nleqslant b$ there is a prime p such that $a \nleqslant p$, $b \leqslant p$. It follows that the function u_L ([1], p. 292), mapping the frame L to the topology tsL of the space sL corresponding to L, is injective as well as surjective.

Neither of the two conditions by itself implies that the frame is isomorphic to a topology. The frame of all regularly open sets of the real line is a T_1'-frame but is not isomorphic to a topology. The frames K and L of example 3 below are compact frames which are not isomorphic to topologies.

SEPARATION AXIOM T_2

The weak Hausdorff separation axiom (see [6]) can be stated for a frame L as follows, with the quotient μ playing the role of a subspace.

S_2) *If* $a \in L$ *and* $a \neq 1$, *there exists a quotient* $\mu: L \to M$ *with* $\mu(a) = 0$, $\mu(1) \neq 0$ *such that if* $a \vee b = 1$ *with* $b \neq 1$, $\exists u, v$ *with* $u \wedge v = 0$, $\mu(v) = 1$, $u \nleq b$.

The requirement $\mu(1) \neq 0$ means that μ is not the zero quotient. In terms of the semicomplement $v* = \max\{x: x \wedge v = 0\}$, S_2 can be restated as follows:

If $a \in L$ *and* $a \neq 1$, *there exists a quotient* $\mu \neq 0$ *with* $\mu(a) = 0$ *such that if* $a \vee b = 1$ *with* $b \neq 1$, $\exists v$ *with* $\mu(v) = 1$, $v* \nleq b$.

A still weaker form of the axiom is as follows:

S'_2) *If* $a \vee b = 1$ *with* $a \neq 1$ *and* $b \neq 1$, $\exists u, v$ *with* $u \wedge v = 0$, $v \nleq a$, $u \nleq b$.

Proposition 5. *The following properties of a space* X *are equivalent:*

(i) *Two points with disjoint closures have disjoint neighbourhoods.*

(ii) tX *is an* S_2*-frame.*

(iii) tX *is an* S'_2*-frame.*

Proof. (i) ⇒ (ii). Let A be open and $A \neq X$. Choose $x \in X \setminus A$. Let $\mu: tX \to \{0, 1\}$ be the frame map for which $\mu(U) = 1$ if $x \in U$. Then $\mu(A) = 0$ but $\mu(X) \neq 0$. If B is open with $A \cup B = X$ but $B \neq X$, choose $y \in X \setminus B$. Then $\bar{x} \cap \bar{y} = \phi$ and by (i) there are disjoint open sets U, V with $x \in V$, $y \in U$. Then $\mu(V) = 1$ and $U \not\subset B$.

(ii) ⇒ (iii). Clear.

(iii) ⇒ (i). Let $\bar{x} \cap \bar{y} = \phi$. Putting $A = X \setminus \bar{x}$ and $B = X \setminus \bar{y}$, A and B are open and $A \cup B = X$, $A \neq X$, $B \neq X$. By (iii), $\exists U, V$ with $U \cap V = \phi$, $V \not\subset A$, $U \not\subset B$. Then $x \in V$, $y \in U$ and $U \cap V = \phi$.

We define a T_2-frame to be one satisfying both T'_1 and S_2, and we introduce a weaker property T_2^*.

$T_2 = T_1' + S_2$

T_2^*) *If* $a \in L$ *and* $a \neq 1$, *there exists a quotient* $\mu \neq 0$ *with* $\mu(a) = 0$ *such that* $\bigvee_{\mu(r)=1} r* \geqslant a$.

Proposition 6. $T_2 \Rightarrow T_2^* \Rightarrow S_2$ *and hence* $T_1' + T_2^* = T_2$.

Proof. That $T_2 \Rightarrow T_2^*$ is seen by supposing $a \nleqslant \bigvee_{\mu(r)=1} r*$. Applying T_1', $\exists b \neq 1$ with $a \vee b = 1$ and $\bigvee_{\mu(r)=1} r* \leqslant b$. This contradicts S_2.

Assume T_2^* and let $a \vee b = 1$ with $b \neq 1$. Since $\bigvee r* \geqslant a$, therefore $b \vee \bigvee r* = 1$. Hence some $r* \nleqslant b$. Thus S_2 holds.

Proposition 7. *The following properties of a* T_0*-space* X *are equivalent:*

(i) X *is a Hausdorff space.*

(ii) tX *is a* T_2*-frame.*

(iii) tX *is a* T_2^**-frame.*

Proof. (i) $\Rightarrow$ (ii). Since X is a T_1-space, tX satisfies T_1'. To show that tX satisfies S_2, take μ to be the quotient induced by a one-point subspace,

(ii) $\Rightarrow$ (iii), by Prop. 6.

(iii) $\Rightarrow$ (i). Suppose if possible that $y \in \bar{x}$ with $y \neq x$. By T_0, $x \notin \bar{y}$. There is a quotient μ: $tX \to M$ with $\mu(X \setminus \bar{y}) = 0$ but $\mu(X) \neq 0$, such that $\bigcup \{V*: \mu V = 1\} \supset X \setminus \bar{y}$. In particular, since $x \in Y \setminus \bar{y}$, there exists V with $\mu V = 1$, and hence $V \not\subset X \setminus \bar{y}$ and hence $y \in V$, for which $x \in V*$. This contradicts $y \in \bar{x}$, so X is a T_1-space.

We have $T_2^* \Rightarrow S_2 \Rightarrow S_2'$ which implies that two distinct closed points disjoint neighbourhoods. Thus X is a Hausdorff space.

The following is a generalization of the theorem that a compact subspace of a T_2-space is closed.

Proposition 8. *For any quotient* κ: $L \to K$ *such that* L *is a* T_2-

frame and K *is compact there exists* $u \in K$ *such that* $\kappa(x) = \kappa(y)$ *iff* $u \vee x = u \vee y$.

Proof. Let $u = \max\{x: \kappa x = 0\}$. If $u \vee x = u \vee y$ then $\kappa(x) = \kappa(u \vee x) = \kappa(u \vee y) = \kappa(y)$.

To show the converse suppose on the contrary that $\kappa x = \kappa y$ but $u \vee x \neq u \vee y$. Putting $p = (x \wedge y) \vee u$ and $q = x \vee y \vee u$, we have $u \leqslant p < q$ and $\kappa p = \kappa q$.

By T_1' there exists $v \neq 1$ with $p \leqslant v$ and $v \vee q = 1$. By T_2^* there exists a quotient $\mu: L \to M$ with $\mu(v) = 0$, $\mu(1) \neq 0$ and $\bigvee_{\mu(r)=1} r* \geqslant v$. Then

$$\kappa(v) = \kappa(p \vee v) = \kappa(p) \vee \kappa(v) = \kappa(q) \vee \kappa(v) = \kappa(1) = 1\,,$$

and hence $\kappa \bigvee r* = 1$. Thus $\bigvee \kappa r* = 1$ in the compact frame K. There is a finite subcover κr_i*. Thus $\kappa \bigvee r_i* = 1$ and hence $\kappa \bigwedge r_i = 0$. By the definition of u, $\bigwedge r_i \leqslant u$. But $\mu(u) \leqslant \mu(p) \leqslant \mu(v) = 0$, whereas $\mu \bigwedge r_i = \bigwedge \mu r_i = 1$. This contradiction shows that $\kappa x = \kappa y$ implies $u \vee x = u \vee y$.

Example 3. Let X be the compact T_0-space $S \times I$ where S is a space with three points s_1, s_2, s_3 and with open sets ϕ, $\{s_3\}$, $\{s_2, s_3\}$, S, and I is the closed interval $[0, 1]$ with the usual topology. Let j_n: $I \to X$ be defined by $j_n(t) = (s_n, t)$. Let the quotient $\lambda: tX \to L$ be defined by the congruence for which $U \equiv V(\lambda)$ if $j_1^{-1}U = j_1^{-1}V$, $(j_2^{-1}U)^- = (j_2^{-1}V)^-$ and $(j_3^{-1}U)^- = (j_3^{-1}V)^-$.

L is easily seen to be compact. That it is also a T_2^*-frame can be seen as follows. Let $a \neq 1$ in L. If $a = \lambda U$ then $U \neq X$ and hence does not contain $s_1 \times I$. Choose $(s_1, t_0) \in X \setminus U$, and let $\mu: L \to \{0, 1\}$ be the quotient for which $\mu(\lambda V) = 1$ iff $(s_1, t_0) \in V$. Note that if $V \equiv W(\lambda)$ then $(s_1, t_0) \in V$ iff $(s_1, t_0) \in W$; so μ is well defined. Clearly $\mu(a) = 0$.

If we set $G_n = \{t \in I, |t - t_0| < \frac{1}{n}\}$, $H_n = \{t \in I, |t - t_0| > \frac{1}{n}\}$ and $r_n = \lambda(S \times G_n)$, we have $\mu(r_n) = 1$ and $r_n* = \lambda(S \times H_n)$. Hence

$$\bigvee_{\mu(r)=1} r* \geqslant \bigvee r_n* = \bigvee \lambda(S \times H_n) = \lambda \bigcup S \times H_n = \lambda\big(S \times \big(I \setminus \{t_0\}\big)\big)$$
$$= \lambda\big(X \setminus \{(s_1, t_0)\}\big).$$

Since $X \setminus \{(s_1, t_0)\} \supset U$, $\bigvee_{\mu(r)=1} r* \geqslant \lambda U = a$, thus L satisfies T_2^*.

Let Y be the subspace $\{s_1, s_3\} \times I$ of X and let i: $Y \to X$ be the inclusion map. Let ν: $tY \to K$ be the quotient determined by the congruence for which $U \equiv V(\nu)$ if $j_1^{-1}U = j_1^{-1}V$ and $(j_3^{-1}U)^- = (j_3^{-1}V)^-$. Then K is also compact, and $\nu \circ i^{-1} \leqslant \lambda$. Hence there exists κ: $L \to K$ for which $\nu \circ i^{-1} = \kappa \circ \lambda$. Then κ: $L \to K$ is a surjective frame map, L satisfies T_2^* and K is compact. But κ is not closed, for $\kappa(u) = 0$ only when $u = 0$. Thus in Prop. 8, T_2 cannot be replaced by T_2^*.

Example 4. Let X be the compact T_1-space defined as follows. As a set, $X = \{s_1, s_2, s_3\} \times I$ where I is its closed interval $[0, 1]$. Let j_n: $I \to X$ be defined by $j_n(t) = (s_n, t)$. A set U in X is defined to be open if (i) $j_1^{-1}U$ and $j_2^{-1}U$ are open in the usual topology of I, (ii) $j_3^{-1}U \setminus j_2^{-1}U$ has no accumulation point in $j_3^{-1}U$, that is, the difference set is finite in each compact subset of $j_3^{-1}U$, and (iii) $j_1^{-1}U \setminus j_2^{-1}U$ is nowhere dense. Then the subspaces $s_1 \times I$ and $s_3 \times I$ are homeomorphic to I, and $s_2 \times I$ is discrete. It follows from (ii) that $j_3^{-1}U \setminus j_2^{-1}U$ is nowhere dense.

Let the quotient λ: $tX \to L$ be defined by the congruence for which $U \equiv V(\lambda)$ if $(j_1^{-1}U)^- = (j_1^{-1}V)^-$, $j_2^{-1}U = j_2^{-1}V$ and $j_3^{-1}U = j_3^{-1}V$. It is easily seen that L satisfies T_1' and S_2'.

To show that L does not satisfy S_2, let $a = \lambda\big(\{s_2, s_3\} \times I\big)$. Let μ: $L \to M$ be any non-zero quotient for which $\mu(a) = 0$. Then $\mu \leqslant \varphi_a$ where φ_a: $L \to R$ is the closed quotient determined by the congruence for which $x \equiv y(\varphi_a)$ when $x \vee a = y \vee a$. We can factor μ through φ_a: $\mu = \nu \circ \varphi_a$.

$$L \xrightarrow{\varphi_a} R \xrightarrow{\nu} M.$$

Here R is a complete Boolean algebra isomorphic to the frame of regularly open sets of I.

The quotient ν of R is the closed quotient φ_r for $r = \max\{x$:

$\nu(x) = 0\}$, because all quotients of a complete Boolean algebra are of this form.

Let G, H be the regularly open sets of I corresponding to r and its complement $r' \in R$ respectively. Let $b = \lambda(\{s_1, s_2\} \times I \cup s_3 \times G)$. Then $a \vee b = 1$. Since $\mu \neq 0$, we have $r \neq I$ and hence $b \neq 1$.

For each $\lambda U \in L$ such that $\mu\lambda U = 1$, $\varphi_a \lambda U \geqslant r'$ so $j_1^{-1} U$ contains a dense subset of H. Since $j_1^{-1} U \setminus j_2^{-1} U$ is nowhere dense, $j_2^{-1} U$ also contains a dense subset of H. Let V be an open set of X for which $\lambda V = (\lambda U)^*$. Then $U \cap V = \phi$, so $j_2^{-1} V \cap H$ is disjoint from a dense subset of H. Since $j_3^{-1} V \setminus j_2^{-1} V$ is nowhere dense, $j_3^{-1} V \cap H = \phi$ and hence $j_3^{-1} V \subset G$. Thus $(\lambda U)* = \lambda V \leqslant b$. Hence $\bigvee_{\mu(u)=1} u* \leqslant b$. Therefore L is not an S_2-frame.

Let Y be the subspace $\{s_2, s_3\} \times I$ of X. Then Y is a compact T_2-space. The canonical map i^{-1} of the topology of X to the topology of the subspace factors through L: $i^{-1} = \mu \circ \lambda$.

$$tX \xrightarrow{\lambda} L \xrightarrow{\mu} tY .$$

The quotient μ is not closed, for $\mu(a) = 0$ only if $a = 0$; in fact $i^{-1} U = \phi$ only if $U = \phi$. Thus, in Prop. 8, T_2 cannot be replaced by $T_1' + S_2'$.

REGULARITY

Weak regularity can be defined for frames as follows:

S_3) *If* $a \vee b = 1$ *with* $b \neq 1$, *there exists* v *such that* $a \vee v = 1$, $v* \nleqslant b$.

The topology tX of a space X satisfies S_3 iff each point x whose closure lies in an open set B is in the interior of some closed subset of B.

T_3) If $a \nleqslant b$, $\exists v$ such that $a \vee v = 1$, $v* \nleqslant b$.

Proposition 9. *A frame* L *satisfies* T_3 *iff for each* $a \in L$,

$$a = \bigvee_{a \vee r = 1} r* .$$

Proof. Let L satisfy T_3. If $a \vee r = 1$ then $r* = (a \vee r) \wedge r* = a \wedge r*$ so $r* \leqslant a$. Hence $\bigvee_{a \vee r = 1} r* \leqslant a$. Suppose if possible that $\bigvee_{a \vee r = 1} r* < a$. By T_3 there exists v such that $a \vee v = 1$ and $v* \nleqslant \bigvee_{a \vee r = 1} r*$, which is absurd.

To show the converse, let $a \nleqslant b$. Then $\bigvee_{a \vee v = 1} v* \nleqslant b$. Hence $\exists v$ such that $a \vee v = 1$ and $v* \nleqslant b$.

This proposition 9 shows that the topology tX of a space X satisfies T_3 iff each open set A is the union of the interiors of the closed sets contained in A, that is, iff X is a regular space.

Proposition 10. $T_3 = S_3 + T_1'$.

Proof. If $a \vee b = 1$ with $b \neq 1$ then $a \nleqslant b$. Hence T_3 implies S_3. If $v* \nleqslant b$ then $v \vee b \neq 1$. Hence T_3 implies T_1'.

Assume T_1' and S_3. By T_1', if $a \nleqslant b$, $\exists v$ such that $v \vee b \neq 1$, $v \vee a = 1$, and hence $v \vee b \vee a = 1$. By S_3, $\exists w$ such that $a \vee w = 1$, $w* \nleqslant v \vee b$. Thus T_3 is satisfied.

Proposition 11. $S_3 \Rightarrow S_2$ *and* $T_3 \Rightarrow T_2$.

Proof. Assume S_3. If $a \neq 1$, let $\mu = \varphi_a$. Then $\mu(a) = 0$ but $\mu(1) \neq 0$. If $a \vee b = 1$ with $b \neq 1$, by S_3 $\exists v$ with $a \vee v = 1$, $v* \nleqslant b$. Then $1 = \mu(1) = \mu(a \vee v) = \mu(v)$. Thus S_2 follows.

Since $S_3 \Rightarrow S_2$, $S_3 + T_1' \Rightarrow S_2 + T_1'$, that is $T_3 \Rightarrow T_2$.

Definition. A family $\{a_\alpha\}_{\alpha \in A}$ of elements of a frame L is called *locally finite* if there exists a cover $\{c_\gamma\}_{\gamma \in \Gamma}$ such that for each γ, $c_\gamma \wedge a_\alpha = 0$ except for a finite number of α.

If $\{a_\alpha\}_{\alpha \in A}$ is locally finite, then each subfamily $\{a_\alpha\}_{\alpha \in B}$, where $B \subset A$, is locally finite, and each family of smaller elements $\{b_\alpha\}_{\alpha \in A}$, where $b_\alpha \leqslant a_\alpha$, is locally finite. Also, for each partition Δ of A into disjoint subsets δ, if we define $d_\delta = \bigvee_{\alpha \in \delta} a_\alpha$, the family $\{d_\delta\}_{\delta \in \Delta}$ is locally finite.

Proposition 12. *The family* $\{a_\alpha\}_{\alpha \in A}$ *is locally finite iff*

$$\bigvee_{\text{fin}\, k \subset A} \Big(\bigvee_{a \in A - k} a_a \Big)* = 1 .$$

Definition. A frame L is *paracompact* if each cover $\{a_a\}_{a\in A}$ has a locally finite refinement.

Equivalently, L is paracompact if it satisfies:

P) For each cover $\{a_a\}_{a\in A}$ there is a locally finite cover $\{b_a\}_{a\in A}$ with $b_a \leqslant a_a$ for all $a \in A$.

Proposition 13. $P + S_2 \Rightarrow S_3$.

Proof. Let $a \vee b = 1$ with $b \neq 1$. By S_2 there exists a quotient $\mu \neq 0$ with $\mu(b) = 0$ such that if $b \vee c = 1$ with $c \neq 1$, $\exists \nu$ with $\mu(\nu) = 1$, $\nu * \nleqslant c$.

Let $R = \{r \in L \colon \mu(r) = 1\}$. Writing $d = \bigvee_{r\in R} r* = \bigvee_{\mu(r)=1} r*$, we have $\mu(d) = 0$, so $d \neq 1$. Suppose if possible that $a \vee d \neq 1$. Since $b \vee a \vee d = 1$, $\exists \nu$ with $\mu(\nu) = 1$ and $\nu * \nleqslant a \vee d$, which is absurd. Therefore $a \vee \bigvee_{r\in R} r* = 1$.

By P, $\exists$ locally finite $\{u_r\}$ with $u_r \leqslant r*$ and $a \vee \bigvee_{r\in R} u_r = 1$. That is, $a \vee w = 1$ where $w = \bigvee_{r\in R} u_r$ and $w* = \bigwedge_{r\in R} u_r*$. There is a cover $\{c_\gamma\}$ such that for each γ, $c_\gamma \wedge u_r = 0$ except for r in some finite set $k_\gamma \subset R$. Then $c_\gamma \leqslant \bigwedge_{r\in R - k_\gamma} u_r*$ and hence $\mu c_\gamma \leqslant \mu \bigwedge_{r\in R-k_\gamma} u_r*$. Since $u_r* \geqslant r** \geqslant r$, $\mu(u_r*) = 1$ and $\mu \bigwedge_{r\in k_\gamma} u_r* = 1$. Therefore $\mu w* = \mu \bigwedge_{r\in R} u_r* \geqslant \mu c_\gamma$ for each γ. Hence $\mu w* \geqslant \bigvee \mu c_\gamma = \mu \bigvee c_\gamma = \mu(1) = 1$. Since $\mu b = 0$, $w* \nleqslant b$. Thus $a \vee w = 1$ and $w* \nleqslant b$.

Example 5. Let X be the compact T_0-space $S \times I$ where S is a space with three points s_1, s_2, s_3 and with open sets ϕ, $\{s_2\}$, $\{s_1, s_2\}$, $\{s_2, s_3\}$, S, and I is the closed interval $[0, 1]$ with its usual topology. Let $j_n \colon I \to X$ be defined by $j_n(t) = (s_n, t)$. Then a set U of X is open if each $j_n^{-1}U$ is open and $j_2^{-1}U \supset j_1^{-1}U \cup j_3^{-1}U$. Let the quotient $\lambda\colon tX \to L$ be defined by the congruence for which $U \equiv V(\lambda)$ if $(j_n^{-1}U)^- = (j_n^{-1}V)^-$ for $n = 1, 2, 3$. For each element $u \in L$, we choose the largest element U of the congruence class u. Then $u = \lambda U$, each

$j_n^{-1}U$ is regularly open and $j_2^{-1}U \supset j_1^{-1}U \cup j_3^{-1}U$.

The quotient frame L is a paracompact S_2'-frame. To show that it is paracompact, let $\{u_a\}$ be a cover of L. Let $u_a = \lambda U_a$ where $j_n^{-1}U_a$ is regularly open. Then $\{j_1^{-1}U_a\}$ and $\{j_3^{-1}U_a\}$ are covers of the complete Boolean algebra R of regularly open sets of I. Since each cover of a complete Boolean algebra has a discrete refinement, there is a family $\{G_\beta\}$ of mutually disjoint regularly open sets such that $\cup G_\beta$ is dense in I and each $G_\beta \subset j_1^{-1}U_a \cap j_3^{-1}U_\gamma$ for some a, γ. Then $\{\lambda(\{s_m, s_2\} \times G_\beta)\}_{m,\beta}$, where m takes the values 1 and 3, is a cover of L refining $\{u_{a'}\}$ with the property that each member of the cover meets only one other member of the cover. Thus it is a locally finite refinement of the given cover.

To show that L satisfies S_2', let $a \vee b = 1$, $a \neq 1$, $b \neq 1$. Let $a = \lambda A$ and $b = \lambda B$, where $j_n^{-1}A$ and $j_n^{-1}B$ are regularly open. Since $a \neq 1$, there exists a non-empty open interval P of I such that $s_1 \times P$ or $s_3 \times P$ does not meet A. Similarly there is a non-empty open interval Q such that $s_1 \times Q$ or $s_3 \times Q$ does not meet B. Replacing P and Q if necessary by smaller non-empty intervals, we may assume that $P \cap Q = \phi$. Then $S \times P \cap S \times Q = \phi$, $S \times P \not\subset A$, $S \times Q \not\subset B$. Hence $\lambda(S \times P) \cap \lambda(S \times Q) = 0$, $\lambda(S \times P) \nleq a$, $\lambda(S \times Q) \nleq b$.

That L does not satisfy S_2 may be shown in the same way as in example 4. Hence L does not satisfy S_3. Alternately it is easy to see directly that L does not satisfy S_3. It then follows by proposition 13 that L does not satisfy S_2.

NORMALITY

A frame L is called *normal* if it satisfies the condition:

S_4) *If* $a \vee b = 1$ *there exist* u, v *such that* $a \vee v = 1$, $u \vee b = 1$, $u \wedge v = 0$.

Proposition 14. *The following conditions on a frame* L *are equivalent:*

(i) L *satisfies* S_4.

(ii) *If* $a \vee b = 1$ *there exists a frame map* $\mu\colon tR \to L$, *where* tR *is the topology of the real line, such that* $\mu(R \setminus \{0\}) \leqslant a$, $\mu(R \setminus \{1\}) \leqslant b$.

The non-trivial implication (i) ⇒ (ii) was proved in [2].

Definition. $T_4 = T_1' + S_3$.

Proposition 15. *The following implication hold:*

$$\begin{array}{ccccccc} T_4 & \longrightarrow & T_3 & \longrightarrow & T_2 & \longrightarrow & T_1' \\ \downarrow & & \downarrow & & \downarrow & & \\ S_4 & \longrightarrow & S_3 & \longrightarrow & S_2 & & \end{array}$$

Definition. A family $\{a_\alpha\}_{\alpha\in A}$ of elements of a frame L is called *conservative* if, for each subset $B \subset A$ and each $u \in L$.

$$u \vee \bigwedge_{\alpha\in B} a_\alpha = \bigwedge_{\alpha\in B} u \vee a_\alpha .$$

If $\{a_\alpha\}_{\alpha\in A}$ is conservative, then each subfamily $\{a_\alpha\}_{\alpha\in B}$ is conservative. Also for each partition Δ of A into disjoint subsets δ, if we define $d_\delta = \bigwedge_{\alpha\in\delta} a_\alpha$, then family $\{d_\delta\}_{\delta\in\Delta}$ is conservative.

In example 2, let $E_n = \{1, 2, \ldots, n\}$. The family E_n is a closure preserving family of closed sets in Y but not in X. Thus the property of being closure preserving for a family of closed sets is not topological. The families $\{X \setminus E_n\}$ and $\{Y \setminus E_n\}$ of complementary open sets are both conservative. Indeed conservativity is a topological property of a family of open sets analogous to and following from the non-topological property of closure preserving for a family of closed sets. Closure preserving for a family of open sets is also not topological. (See example 6.) We replace it by the following topological property, which is implied by closure preserving.

Example 6. Let $X = \{\infty\} \cup N \times \{0, 1\}$, where N is the set of natural numbers. Let a subset of X be closed if it contains $\{\infty\} \cup N \times 0$ or if it is a finite set of the form $A \times 0 \cup B \times 1$, where $B \subset A$. Let Y be the subspace $N \times \{0, 1\}$. The inclusion map induces an isomorphism of the topologies of X and Y. Let $G_n = \{1, 2, \ldots, n\} \times 1$. Then $\bar{G}_n = \{1, 2, \ldots, n\} \times \{0, 1\}$. The sequence (G_n) of open sets is closure preserving in Y but not in X.

Definition. A family $\{a_a\}_{a \in A}$ of elements of a frame L is called c.p. if the family $\{a_a *\}$ of semicomplements is conservative, that is, for each subset $B \subset A$ and each $u \in L$,

$$u \vee \bigwedge_{a \in B} a_a * = \bigwedge_{a \in B} u \vee a_a * .$$

Proposition 16. *Each locally finite family* $\{a_a\}_{a \in A}$ *is c.p.*

Proof. Each subfamily $\{a_a\}_{a \in B}$ is locally finite. Hence there is a cover $\{b_\gamma\}$ with $b_\gamma \wedge a_a = 0$ except for a in some finite subset $k_\gamma \subset B$. Then

$$b_\gamma \leqslant \bigwedge_{a \in B - k_\gamma} a_a * ,$$

$$b_\gamma \vee \bigwedge_{a \in B} a_a * = b_\gamma \wedge \bigwedge_{a \in k_\gamma} a_a * ,$$

and, for each $u \in L$

$$b_\gamma \wedge u \vee \bigwedge_{a \in B} a_a * = b_\gamma \wedge u \vee \bigwedge_{a \in k_\gamma} a_a *$$

$$= b_\gamma \wedge \bigwedge_{a \in k_\gamma} u \vee a_a * = b_\gamma \wedge \bigwedge_{a \in B} u \vee a_a * .$$

Therefore, since $\bigvee b_\gamma = 1$,

$$u \vee \bigwedge_{a \in B} a_a * = \bigwedge_{a \in B} u \vee a_a * ,$$

Proposition 17. $P + S_3 \Rightarrow S_4$.

Proof. We shall prove that if each cover of an S_3-frame L has a c.p. refinement, then L is normal.

Let $a \vee b = 1$. Let $R = \{r \in L : a \vee r = 1\}$. Writing

$$d = \bigvee_{r \in R} r* = \bigvee_{a \vee r = 1} r* ,$$

suppose if possible that $b \vee d \neq 1$. Then by S_3 since $a \vee b \vee d = 1$, there exists r such that $a \vee r = 1$ and $r* \nleqslant b \vee d$, which is absurd. Hence $b \vee d = 1$, that is, $b \vee \bigvee_{r \in R} r* = 1$. Since this cover has a c.p. refinement, there is a family $\{c_r\}_{r\in R}$ with $c_r \leqslant r*$ such that $b \vee \bigvee c_r = 1$ and $\{c_r *\}$ is conservative. Since $c_r * \geqslant r** \geqslant r$, $a \vee c_r * = 1$. If $u = \bigvee_{r\in R} c_r$, then $b \vee u = 1$ and

$$a \vee u* = a \vee \bigwedge c_r* = \bigwedge a \vee c_r* = 1\,,$$

with $u \wedge u* = 0$. Thus L is normal.

COMPLETE REGULARITY

A frame L is called *weakly completely regular* if it satisfies the following:

S_{3+}) *If* $a \vee b = 1$ *with* $b \neq 1$ *there exists a frame map* μ: $tR \to L$ *such that* $\mu(R \setminus \{0\}) \leqslant a$ *and* $\mu\left(\left\{t \in R\colon t > \frac{1}{2}\right\}\right) \nleqslant b$.

A frame L is called *completely regular* if it satisfies the following:

T_{3+}) *If* $a \nleqslant b$ *there is a frame map* μ: $tR \to L$ *such that* $\mu(R \setminus \{0\}) \leqslant a$ *and* $\mu\left(\left\{t \in R\colon t > \frac{1}{2}\right\}\right) \nleqslant b$.

Proposition 18. $S_{3+} + T'_1 \Rightarrow T_{3+}$.

Proof. By T'_1, if $a \nleqslant b$, there exists v such that $v \vee b \neq 1$, $v \vee a = 1$ and hence $v \vee b \vee a = 1$. By S_{3+}, there is a frame map μ: $tR \to L$ such that $\mu(R \setminus \{0\}) \leqslant a$ and $\mu\left(\left\{t \in R\colon t > \frac{1}{2}\right\}\right) \nleqslant v \vee b$. Thus T_{3+} is satisfied.

The converse follows from the next proposition.

Proposition 19. *The following implications hold:*

$$\begin{array}{ccccc} T_4 & \longrightarrow & T_{3+} & \longrightarrow & T_3 \\ \downarrow & & \downarrow & & \downarrow \\ S_4 & \longrightarrow & S_{3+} & \longrightarrow & S_3 \end{array}$$

Proof. If $a \vee b = 1$ with $b \neq 1$, then $a \nleqslant b$. Hence T_{3+} implies S_{3+}. When μ: $tR \to L$ is a frame map such that $\mu(R \setminus \{0\}) \leqslant a$

and $\mu\left(\left\{t \in R:\ t > \frac{1}{2}\right\}\right) \nleq b$, let $v = \mu\left(\left\{t \in R:\ t < \frac{1}{2}\right\}\right)$. Then $a \vee v = \mu(R) = 1$ and $v* \geqslant \mu\left(\left\{t \in R:\ t > \frac{1}{2}\right\}\right) \nleq b$. Hence S_{3+} implies S_3 and T_{3+} implies T_3.

By proposition 14, S_4 implies S_{3+}. By proposition 18, T_4 implies T_{3+}.

COVER REGULARITY

A frame L is called *cover regular* if it satsifies the following:

$\widetilde{T}_3$) *If* $\bigvee_{\alpha \in A} u_\alpha = 1$ *and* $a \nleq b$ *there is an element* v *and an index* α *such that* $v \vee u_\alpha = 1$ *and* $a \wedge v* \nleq b$.

Proposition 20. *The following conditions are equivalent:*

(i) L *satisfies* $\widetilde{T}_3$.

(ii) *If* $\bigvee_{\alpha \in A} u_\alpha = 1$ *and* $b \neq 1$ *there is an element* v *and an index* α *such that* $v \vee u_\alpha = 1$ *and* $v* \nleq b$.

(iii) *If* $\bigvee_{\alpha \in A} u_\alpha = 1$ *there is a family* $\{v_\gamma\}_{\gamma \in \Gamma}$ *and a function* $\tau\colon \Gamma \to A$ *such that* $v_\gamma \vee u_{\tau(\gamma)} = 1$ *and* $\bigvee v_\gamma* = 1$.

Proof. (i) ⇒ (ii). Substitute $a = 1$.

(ii) ⇒ (iii). Let $b = \bigvee\{r*:\ v \vee \text{some } u_\alpha = 1\}$. Then $b = 1$, for otherwise by (ii) $\exists v$ and α with $v \vee u_\alpha = 1$ and $v* \nleq b$, which is absurd. Thus $\bigvee v* = 1$.

(iii) ⇒ (i). Let $a \nleq b$. Since $\bigvee v_\gamma* = 1$, $\bigvee a \wedge v_\gamma* = a \wedge \bigvee v_\gamma* = a \nleq b$. Hence for some γ, $a \wedge v_\gamma* \nleq b$. When $\alpha = \tau(\gamma)$, $v_\gamma \vee u_\alpha = 1$.

A space X is called cover regular if it has a cover regular topology, that is, if and only if each open cover of X has a closed refinement which has an open refinement. The property that each open cover of X has a closed refinement is not topological (see example 2), but it implies that the topology tX is a frame satisfying the following condition.

$\widetilde{T}_1$) *If* $\bigvee_{\alpha \in A} u_\alpha = 1$ *and* $a \nleq b$ *there is an element* v *and an in-*

dex a *such that* $v \vee u_a = 1$ *and* $a \nleqslant b \vee v$.

Proposition 21. *The following implications hold:*

$$\begin{array}{ccccc} T_3 & \longrightarrow & \widetilde{T}_3 & \longrightarrow & S_3 \\ \downarrow & & \downarrow & & \\ T'_1 & \longrightarrow & \widetilde{T}_1 & & \end{array}$$

Proof. $T_3 \Rightarrow \widetilde{T}_3$. Let $\bigvee u_a = 1$ and $b \neq 1$. Then some $u_a \nleqslant b$. By T_3, $\exists v$ with $u_a \cup v = 1$, $v* \nleqslant b$.

$\widetilde{T}_3 \Rightarrow S_3$. Let $a \vee b = 1$ with $b \neq 1$. By (ii) of proposition 20, $\exists v$ such that $v* \nleqslant b$ and either $v \vee a = 1$ or $v \vee b = 1$. But $v \vee b = 1$ would imply $v* \leqslant b$, so $v \vee a = 1$.

$T'_1 \Rightarrow \widetilde{T}_1$. Let $\bigvee u_a = 1$ and $a \nleqslant b$. Then $\bigvee a \wedge u_a = = a \wedge \bigvee u_a = a \nleqslant b$, so some $a \vee u_a \nleqslant b$. By T'_1, $\exists v$ with $v \vee b \neq 1$ and $v \vee (a \wedge u_a) = 1$. Thus $v \vee u_a = 1$ and $a \geqslant a \wedge u_a \nleqslant v \vee b$.

$\widetilde{T}_3 \Rightarrow \widetilde{T}_1$. If $a \leqslant b \vee v$ then $a \wedge v* \leqslant b$. Hence $a \wedge v* \nleqslant b$ implies $a \nleqslant b \vee v$.

REFERENCES

[1] C.H. Dowker – Dona Papert Strauss, Quotient frames and subspaces, *Proc. London Math. Soc.*, (3) 16 (1966), 275-296.

[2] C.H. Dowker – Dona Papert Strauss, On Urysohn's lemma, *General Topology and its Relations to Modern Analysis and Algebra* II, Prague, 1966. 111-114.

[3] M. Karlowicz – K. Kuratowski, On relations between some algebraic and topological properties of lattices, *Fund. Math.*, 58 (1966), 219-228.

[4] G. Nöbeling, *Grundlagen der analytischen Topologie*, Berlin, 1954.

[5] S. Papert, An abstract theory of topological subspaces, *Proc. Cambridge Phil. Soc.*, 60 (1964), 197-203.

[6] C.-T. Yang, On paracompact spaces, *Proc. Amer. Math. Soc.*, 5 (1954), 185-189.

COLLOQUIA MATHEMATICA SOCIETATIS JÁNOS BOLYAI
8. TOPICS IN TOPOLOGY, KESZTHELY (HUNGARY), 1972.

DENSE SUBSETS WITH RATIONAL DISTANCES IN SEPARABLE METRIC SPACES (FROM A LETTER TO P. HAMBURGER)

R. ENGELKING

When you were in Warsaw you asked the following question: Does there exist for every separable metric space X a metric σ on X and a countable dense subset $D \subset X$ such that $\sigma(x, y)$ is rational for $x, y \in D$?

Here is the answer (that you probably already know):

Theorem. *For every separable metric space X and every countable dense subset D of X there exists a metric σ on X such that $\sigma(x, y)$ is rational for $x, y \in D$.*

Proof. Let $I^{\aleph_0}$ be the Hilbert cube with the usual metric ρ and let S be the countable dense subset of $I^{\aleph_0}$ composed of all points of the form $(r_1, r_2, \ldots, r_k, 0, 0, \ldots)$ where r_i is rational and k runs over all integers. Take an embedding $f\colon X \to I^{\aleph_0}$. The set $f(D) \cup S$ is countable and dense in $I^{\aleph_0}$. By a theorem of Ford (M. K. Ford, Jr. Homogeneity of infinite products of manifolds with boundary, *Pacific Journ. of Math.*, 12 (1962), 879-884; Theorem 2) there exists a homeomorphism $h\colon I^{\aleph_0} \to I^{\aleph_0}$ such that $h(f(D) \cup S) = S$. The metric $\sigma(x, y) = \rho(hf(x), hf(y))$ is the required one.

COLLOQUIA MATHEMATICA SOCIETATIS JÁNOS BOLYAI
8. TOPICS IN TOPOLOGY, KESZTHELY (HUNGARY), 1972.

ON SOME GENERAL PROPERTIES OF CHROMATIC NUMBERS

P. ERDŐS – A. HAJNAL – S. SHELAH

§1. INTRODUCTION

In his paper [7] Taylor introduced a generalization of chromatic number of graphs and stated several interesting problems. In this note we will be interested in one of his problems we will state below. We are going to formulate several possible generalizations and quite a few related questions. Our main aim is to formulate the problems but we will write down some partial results we obtained trying to clear the problems up.

Let us start with the following remark

(1) Let $\varphi(x, \lambda)$ be any statement of set theory, A a set and $\psi(x)$ an operation such that $\forall x(\psi(x) \in A)$. Let us assume that $\sigma < \lambda$ and $\varphi(x, \lambda)$ imply $\varphi(x, \sigma)$. Then there is a λ such that for all x with $\varphi(x, \lambda)$ and for all $\sigma \geqslant \lambda$ there is a y such that $\varphi(y, \sigma)$ and $\psi(y) = \psi(x)$. To see this one defines $A' \subset A$ with the stipulation

$$A' = \{ y \in A : \exists \lambda \forall x(\psi(x) = y \Rightarrow \neg\varphi(x, \lambda))\}$$

i.e. the set of $y \in A$ for which "the λ-s of $\psi^{-1}(\{y\})$ form a bounded set," and denoting by

$$\lambda(y) = \min\{\lambda: \forall x(\psi(x) = y \Rightarrow \neg\varphi(x, \lambda))\} \quad \text{for} \quad y \in A'$$

i.e. the minimal bound for $y \in A'$, $\lambda = \sup\{\lambda(y): y \in A'\}$ obviously satisfies the requirement of (1).*

It is obvious that in general one can not hope for the determination of the λ (depending on φ, and ψ). However, Taylor observed that in many cases it is quite natural to ask for the size of λ. The simplest interesting case of Taylor's general problem arises if we choose $\varphi(x, \lambda)$ to be the statement that X is a graph of chromatic number at least λ, B the set of finite graphs with vertices in ω, $A = P(B)$, and $\psi(x)$ the set of graphs in B isomorphic to a subgraph of X. Taylor's problem for chromatic numbers of graphs is to determine the minimal λ satisfying (1) in this case or to put it into words

(2) What is the minimal λ satisfying the following condition. For every graph $\mathcal{G}$ with chromatic number $\geqslant \lambda$ and for every $\sigma \geqslant \lambda$ there is a graph $\mathcal{G}'$ with chromatic number $\geqslant \sigma$ such that $\mathcal{G}$ and $\mathcal{G}'$ have the same finite subgraphs.

Taylor pointed out that known theorems imply $\lambda \geqslant \omega_1$ and he conjectured that probably $\lambda = \omega_1$.

This problem seems to be very difficult and so instead of solving it we will formulate variants of it which are probably even more difficult. We will not consider Taylor's generalization for relational structures but we will stick to set-systems. To have a brief notation we say that $\mathcal{H}$ is a set system if it consists of sets having at least two elements. For a set-system $\mathcal{H}$ we put

$$\chi(\mathcal{H}) = \min\{\chi: \exists \text{ a function } f: \cup\mathcal{H} \to \chi \text{ such that } \forall\rho\forall x \in \mathcal{H}(x \not\subset f^{-1}(\{\rho\}))\}.$$

$\chi(\mathcal{H})$ is the chromatic number of $\mathcal{H}$ and is the minimal cardinal

*This is the same proof which gives the existence of Hanf numbers in [6].

χ for which $\cup \mathscr{H}$ is the union of χ-sets none of which contains an element of $\mathscr{H}$ as a subset. $\mathscr{H}$ is said to be uniform with $\kappa(\mathscr{H}) = \kappa$ if $|X| = \kappa$ for $X \in \mathscr{H}$. A graph is a uniform set system with $\kappa(\mathscr{H}) = 2$. For further explanation of the terminology and for elementary results see e.g. [1]. Note that two set systems, $\mathscr{H}$, $\mathscr{H}'$ are considered isomorphic if there is a one-to-one mapping f of $\cup \mathscr{H}$ onto $\cup \mathscr{H}'$ such that for $X \subset \cup \mathscr{H}$

$$X \in \mathscr{H} \quad \text{iff} \quad f(X) = \{ f(u) \colon u \in X \} \in \mathscr{H}' .$$

We denote by $\mathscr{H} \cong \mathscr{H}'$ the fact that $\mathscr{H}$ and $\mathscr{H}'$ are isomorphic.

§2. STATEMENT OF SOME RESULTS ON CLASSES OF GRAPHS ADMITTING ARBITRARILY LARGE CHROMATIC NUMBERS

Definition. Let $\tau \geqslant \omega$ be a cardinal. Put

$$B(\tau) = \{ \mathscr{G} : \mathscr{G} \subset [\tau]^2 \wedge |\mathscr{G}| < \tau \} ; \qquad A(\tau) = P(B(\tau))$$

i.e. $B(\tau)$ is the set of subgraphs of cardinality $< \tau$ of the complete graph with set of vertices τ. Obviously, if $|\mathscr{G}| < \tau$, then $\mathscr{G}$ is isomorphic to an element of $B(\tau)$.

Let $\mathscr{G}$ be a graph. We denote by $\psi(\mathscr{G}, \tau)$ the set of $\mathscr{G}' \in B(\tau)$, $\mathscr{G}'$ is isomorphic to a subgraph of $\mathscr{G}$; $(\psi(\mathscr{G}, \tau) \in A(\tau))$.

Let $S \in A(\tau)$; We denote by $\mathscr{G}(S, \tau)$ the class of graphs $\mathscr{G}$ with $\psi(\mathscr{G}, \tau) \subset S$. $S \in A(\tau)$ is said τ-unbounded if

(3) For every λ there is $\mathscr{G} \in \mathscr{G}(S, \tau)$ with $\chi(\mathscr{G}) > \lambda$. An obvious approach to Taylor's problem would be first to characterize the $S \in A(\omega)$ which are ω-unbounded and then show that $\chi(\mathscr{G}) \geqslant \omega_1$ implies that $\psi(\mathscr{G}, \omega)$ satisfies this characterization.

This again seems to be hopeless at present. It is not quite easy to give nontrivial $S \in A(\omega)$ which are ω-unbounded. We now give the definition of some of them.

Let $R, \prec$ be an ordered set $i < \omega$. We will define two sorts of graphs $\mathscr{G}^{\circ}(R, i)$, $\mathscr{G}^{1}(R, i, t)$ for $i \geqslant 2$, or $i \geqslant 3$, $1 \leqslant t < i - 1$ respectively. The set of vertices will be the set of $\prec$ increasing sequences φ of length i of elements of R in both cases.

We put

$$\mathcal{G}^\circ(R, i) = \{\{\varphi, \varphi'\}: \varphi(j+1) = \varphi'(j) \quad \text{for} \quad j < i-1\} \quad \text{for} \quad i \geq 2 \quad \text{and}$$

$$\mathcal{G}^1(R, i, t) = \{\{\varphi, \varphi'\}: \varphi(j+t) < \varphi'(j) < \varphi(j+t+1) < \varphi'(j+1)$$
$$\text{for} \quad j < i - 1 - t\} \quad \text{for} \quad i \geq 3\,.$$

We put

$$S^\circ(i) = \psi(\mathcal{G}^\circ(\omega, i), \omega) = \psi(\mathcal{G}^\circ(R, i), \omega) \quad \text{for} \quad |R| \geq \omega$$

and

$$S^1(i, t) = \psi(\mathcal{G}^1(\omega, i, t), \omega) = \psi(\mathcal{G}^1(R, i, t), \omega) \quad \text{for} \quad |R| \geq \omega\,.$$

The graphs $\mathcal{G}^1(R, i, t)$ we call Specker – Graphs, (Specker used first $\mathcal{G}^1(\omega, 3, 1)$ to show $\omega^3 \nrightarrow (\omega^3, 3)^2$) and the graphs $\mathcal{G}^\circ(R, i)$ we call with some abuse of terminology the "edge graphs" having in mind the special case $i = 2$.

The following are known about these graphs:

Old-lemmas (Erdős – Hajnal)

1/ *Let* $|R| \geq (\exp_{i-1}(\lambda))^+$; $\lambda \geq \omega$, $i \geq 2$. *Then* $\chi(\mathcal{G}^\circ(R, i)) \geq \lambda^+$. *As a corollary* $S^\circ(i)$ *is* ω*-unbounded for* $2 \leq i < \omega$.

2/ $S^\circ(i)$ *does not contain odd-circuits of length* $2j + 3$ *for* $j \leq i - 2$, $i \geq 2$,

3/ *Let* $\kappa \geq \omega$ *be a cardinal. Then* $\chi(\mathcal{G}^1(\kappa, i, t)) = \kappa$ *for* $3 \leq i < \omega$ *and* $1 \leq t < i - 1$.

4/ $S^1(2i^2 + 1, i)$ *does not contain odd circuits of length* $2j + 3$ *for* $j \leq i - 1$, $1 \leq i < \omega$.

5/ *Assume* $\lambda \geq \omega$ *is a cardinal* $R, \prec$ *an ordered set with* $|R| \leq \exp(\lambda)$ *for* $2 \leq i < \omega$. *Then* $\chi(\mathcal{G}^\circ(R, i)) \leq \lambda$.

See [2] Theorem 1 for 1/, and 5/, [4] Theorem 7 for 2/, [1] Theorem 7.4 for 3/, and 4/.

The following inclusions hold:

(i) $S^\circ(i) \subsetneqq S^\circ(i+1)$ for $2 \leqslant i < \omega$

(*) (ii) $S^1(i,t) \subsetneqq S^1(i+1,t)$ for $3 \leqslant i < \omega$

(iii) $S^\circ(i) \subset S^1(i+1,1)$ for $2 \leqslant i < \omega$.

We will give the proof of (i) on p. We see now that the sets $S^\circ(i)$ corresponding to the "edge graphs" form a decreasing sequence. The members of the sequence are all ω-unbounded. The intersection $\cap\{S^\circ(i): 2 \leqslant i < \omega\}$ however by 2/ contains only graphs with $\chi(\mathscr{G}) = 2$, hence is not ω-unbounded.

One of our main points is that the $S^\circ(i)$ are not equally good as ω-unbounded classes.

To be able to formulate our result we need the following

Definition. Let $F(\lambda) \geqslant \lambda^+$ be an operation on cardinals.

We say that $S \in A(\omega)$ is ω-unbounded with the restriction F, if for all σ there is $\lambda \geqslant \sigma$, and a $\mathscr{G}$ with $\psi(\mathscr{G}, \omega) \subset S$ such that

(4) $$\chi(\mathscr{G}) > \lambda \quad \text{and} \quad |\mathscr{G}| \leqslant F(\lambda).$$

We briefly say that S is ω-unbounded with the restriction ξ if it is ω-unbounded with the restriction

$$F_\xi \quad \text{where} \quad F_\xi(\lambda) = \kappa \quad \text{iff} \quad \lambda = \omega_a, \quad \kappa = \omega_{a+1+\xi}.$$

Theorem 1. (a) $S^1(i,t)$ *is* ω*-unbounded with the restriction* 0 *for* $3 \leqslant i < \omega$.

(β) $S^\circ(i)$ *is* ω*-unbounded with the restriction* $\exp_{i-1}(\lambda)^+$ *for* $2 \leqslant i < \omega$.

(γ) $S^\circ(i)$ *is not* ω*-unbounded with the restriction* $\exp_{i-1}(\lambda)$ *for* $2 \leqslant i < \omega$.

As a corollary if G.C.H holds then for every n there is an $S \in$

$\in A(\omega)$ which is ω-unbounded with the restriction $n+1$ but not with the restriction n.

Note that (a) follows from the old lemma 3/ (β) follows from 1/. We will prove (γ) in the next chapter. If G.C.H is assumed then $S^\circ(n+2)$ is ω-unbounded with $(\exp_{n+1}(\lambda))^+$, if $\lambda = \omega_a$, $(\exp_{n+1}(\lambda))^+ = \omega_{a+n+2}$, hence $S^\circ(n+2)$ is $n+1$ unbounded, and is not ω-bounded with $\exp_{n+1}(\lambda) = \omega_{a+n+2}$, hence is not n-unbounded. Before giving the proof of (γ) in the next chapter, it is time to state the first problem.

Problem. Does there exist an $S \in A(\omega)$ which is ω-unbounded but is not ω-unbounded with the restriction $\exp_n(\lambda)$ for every $n < \omega$?

Note that there is an obvious correlation with an old Erdős – Hajnal problem stated in 2 (Problem 1). This problem asks if there is a graph of $\chi(\mathcal{G}) > \lambda$ such that all subgraphs $\mathcal{G}'$ of $|\mathcal{G}'| \leqslant \exp_\omega(\lambda)$ have chromatic number $\leqslant \lambda$. The "edge graphs" $S^\circ(i)$ were used in [2] to establish a positive answer to the above problem when $\exp_\omega(\lambda)$ is replaced by $\exp_n(\lambda)$, $n < \omega$.

We finally mention that the definition of unboundedness with a restriction had to be done as in (4) because we have the following

Theorem 2. *Assume $S \in A(\omega)$ is ω-unbounded. Then for every σ there is $\lambda \geqslant \sigma$ and $\mathcal{G} \in \mathcal{G}(S, \omega)$ with*

$$\chi(\mathcal{G}) = |\mathcal{G}| = \lambda .$$

§3. PROOFS

First we prove Theorem 2.

Let S be ω-unbounded. For every λ choose $\mathcal{G}_\lambda \in \mathcal{G}(S, \omega)$ with $\chi(\mathcal{G}_\lambda) \geqslant \lambda$. Put $S_\lambda = \psi(\mathcal{G}_\lambda, \omega)$. Put $\hat{S}_\lambda = \{\mathcal{G} \in \psi(\mathcal{G}_\lambda, \omega)$: There are uncountably many subgraphs $\mathcal{G}' \subset \mathcal{G}_\lambda$ isomorphic to $\mathcal{G}$ with pairwise disjoint sets $\cup \mathcal{G}'\}$.

For each $\mathcal{G} \in S_\lambda - \hat{S}_\lambda$ let $\mathcal{F}(\mathcal{G})$ be a maximal system of subgraphs of $\mathcal{G}_\lambda$ satisfying the following conditions:

(i) $\mathscr{G}' \in \mathscr{F}(\mathscr{G}) \Rightarrow \mathscr{G}' \cong \mathscr{G}$

(ii) $\mathscr{G}' \neq \mathscr{G}'' \in \mathscr{F}(\mathscr{G}) \Rightarrow \cup\mathscr{G}' \cap \cup\mathscr{G}'' = \phi$.

By the definition of $\hat{S}_\lambda$ we have $|\mathscr{F}(\mathscr{G})| \leqslant \omega$ for $\mathscr{G} \in S_\lambda - \hat{S}_\lambda$. Hence $\cup\cup\mathscr{F}(\mathscr{G})$ is countable for $\mathscr{G} \in S_\lambda - \hat{S}_\lambda$. We now omit the vertices in $\cup\{\cup\cup\mathscr{F}(\mathscr{G}) \colon \mathscr{G} \in S_\lambda - \hat{S}_\lambda\} = T_\lambda$ from $\mathscr{G}_\lambda$ i.e. we consider $\hat{\mathscr{G}}_\lambda = \mathscr{G}_\lambda \cap [\cup\mathscr{G}_\lambda - T_\lambda]^2$. Then for $\lambda > \omega$ we have $\chi(\hat{\mathscr{G}}_\lambda) \geqslant \lambda$. It follows from the construction that $\psi(\hat{\mathscr{G}}_\lambda, \omega) = \hat{S}_\lambda$ and, using $|T_\lambda| \leqslant \omega$, for all $\mathscr{G}' \in \hat{S}_\lambda$ there are uncountably many $\mathscr{G}'' \subset \hat{\mathscr{G}}_\lambda$ isomorphic to $\mathscr{G}'$ with pairwise disjoint $\cup\mathscr{G}''$. It follows that $\hat{S}_\lambda$ has the following property

(5) Assume $\mathscr{G}_i \in \hat{S}_\lambda$, $i < n < \omega$ such that $\cup\mathscr{G}_i \cap \cup\mathscr{G}_j = \phi$ for $i \neq j < n$. Then $\bigcup_{i<n} \mathscr{G}_i \in \hat{S}_\lambda$.

Now it follows that there is $S_\lambda \subset S$ S_λ ω-unbounded, such that S_λ satisfies (5). On the other hand, if S_λ is ω-unbounded and satisfies (5) then $\mathscr{G}(S_\lambda, \omega)$ is closed with respect to arbitrary unions of graphs with disjoint set of vertices. Let σ be given. We can choose $\mathscr{G}_0$, with $\chi(\mathscr{G}_0) \geqslant \sigma$, and $\mathscr{G}_{n+1}$ with $\chi(\mathscr{G}_{n+1}) > |\mathscr{G}_n|$ for $n < \omega$ such that $\mathscr{G}_n \in \mathscr{G}(S_\lambda, \omega)$ and the $\cup\mathscr{G}_n$ are disjoint. Then $\tilde{\mathscr{G}} = \bigcup_{n<\omega} \mathscr{G}_n \in \mathscr{G}(S_\lambda, \omega) \subset \mathscr{G}(S, \omega)$ and $\chi(\tilde{\mathscr{G}}) = |\tilde{\mathscr{G}}| > \sigma$. This proves Theorem 2.

We now state the following.

Lemma. *Let* $\mathscr{G}$ *be a graph,* $2 \leqslant i < \omega$, $\mathscr{G}$ *is isomorphic to a subgraph of* $\mathscr{G}^\circ(R, i)$ *for some* $(R, \prec)$ *if the following conditions hold:*

Put $G = \cup\mathscr{G}$. *There is a relation* $\preceq$ *on* $G \times i$ *such that*

(α) $\preceq$ *is transitive.*

(β) $\forall u, v \in G \times i\ (u \preceq v \vee v \preceq u)$ *i.e.* $\preceq$ *is a preorder on* $G \times i$ *put* $x \prec y$ *for* $x \preceq y \wedge y \npreceq x$; $x \mapsto y$ *for* $x \preceq y \wedge y \preceq z$.

(γ) $\forall x \in G(\langle x, 0\rangle \prec \ldots \prec \langle x, i-1\rangle)$.

(δ) $\forall x, y \in G(\langle x, 0\rangle \not\mapsto \langle y, 0\rangle \vee \ldots \vee \langle x, i-1\rangle \not\mapsto \langle y, i-1\rangle \vee x = y)$.

(ϵ) $\forall x, y \in G(\{x, y\} \in \mathscr{G} \Rightarrow (\langle x, 1\rangle \mapsto \langle y, 0\rangle \wedge \ldots \wedge \langle x, i-1\rangle \mapsto \langle y, i-2\rangle) \vee (\langle y, 1\rangle \mapsto \langle x, 0\rangle \wedge \ldots \wedge \langle y, i-1\rangle = \langle x, i-2\rangle)$. The lemma is obvious.

Proof of Theorem 1. We only have to prove (γ) of Theorem 2. Let $\mathcal{G} \in \mathcal{G}(S^\circ(i), \omega)$, $2 \leqslant i < \omega$. Let $\lambda \geqslant \omega$ be arbitrary, and assume $|\mathcal{G}| \leqslant \exp_{i-1}(\lambda)$. By the assumption for every $\mathcal{G}' \subset \mathcal{G}$, $|\mathcal{G}'| < \omega$, $\mathcal{G}'$ is isomorphic to an element of $S^\circ(i)$. Hence by the lemma there is a preorder $\preceq$ satisfying the conditions $(a) \ldots (\epsilon)$ of the lemma for $\cup \mathcal{G}' \times i$. Then by the compactness theorem the same holds for $\cup \mathcal{G} \times i$. Hence by the lemma there is $R, \prec$ such that $\mathcal{G}$ is isomorphic to a subgraph of $\mathcal{G}^\circ(R, i)$. By $|\mathcal{G}| \leqslant \exp_{i-1}(\lambda)$ we may choose R with $|R| \leqslant \exp_{i-1}(\lambda)$. Then by the old lemma 5./ $\chi(\mathcal{G}) \leqslant \chi(\mathcal{G}^\circ(R, i)) \leqslant \lambda$. Thus $S^\circ(i)$ is not ω-unbounded with the restriction $\exp_{i-1}$. This proves Theorem 1.

Finally we prove (*) (i).

It is sufficient to prove $S^\circ(i+1) \subset S^\circ(i)$ for $2 \leqslant i < \omega$. Let $\mathcal{G} \in S^\circ(i+1)$. We may assume $\mathcal{G} = \mathcal{G}^\circ(n, i+1)$, $\cup \mathcal{G} = {}^{i+1}n$ for some $n < \omega$. We define a partial order $\prec'$ on $({}^{i+1}n) \times i$ by the stipulation

$$(\varphi, j) \prec' (\psi, k) \quad \text{iff} \quad \varphi(j) < \psi(k) \vee (\varphi(j) = \psi(k) \wedge \varphi(j+1) < \psi(k+1))$$

and we extend $\prec'$ to an arbitrary preorder of ${}^{i+1}n \times i$. It is easy to see that the requirements $(a) \ldots (\epsilon)$ hold for $\preceq'$ hence by the Lemma, $\mathcal{G}$ is isomorphic to an element of $S^\circ(i)$.

§4. A THEOREM OF DIFFERENT TYPE

Old result. (see Erdős – Hajnal [1] (Corollary 5.6)).

Assume $\chi(\mathcal{G}) > \omega$. Then $\mathcal{G}$ contains a complete bypartite graph $[\kappa, \omega,]$ for all $\kappa < \omega$. As a corollary if $\mathcal{G}_0$ is a fixed finite bypartite graph and $\chi(\mathcal{G}) > \omega$, then $\mathcal{G}$ contains a subgraph isomorphic to $\mathcal{G}_0$ and again in another formulation if $S \in A(\omega)$. S is ω-unbounded, then S contains all bypartite finite graphs.

On the other hand, the old lemmas show that this statement is no longer true for any fixed nonbypartite graph.

However, it is still possible to prove statements of the following type:

(i) If $\chi(\mathcal{G}) > \lambda$ then $\psi(\mathcal{G}, \omega) \cap S \neq \phi$ for some fixed $S \in A(\omega)$.

(ii) If $\chi(\mathcal{G}) > \lambda$ then there is n_0 such that $S_n \in \psi(\mathcal{G}, \omega)$ for some fixed sequence $\langle S_n;\ n < \omega\rangle$, $S_n \in B(\omega)$, $n > n_0$.

Taylor's conjecture implies that if a statement like (i) or (ii) holds for some λ then it holds for $\lambda = \omega$ as well.

The following theorem is an example of a statement of this kind. In [1] we only could prove it in case $\chi(\mathcal{G}) > \omega_1$.

Theorem 3. *Assume* $\chi(\mathcal{G}) > \omega$. *Then there is* $n < \omega$ *such that* $\mathcal{G}$ *contains odd circuits of length* $2j + 1$ *for all* $n < j < \omega$.

Proof. Let $\chi(\mathcal{G}) > \omega$. Put $\cup \mathcal{G} = G$ for the set of vertices. We may assume $\mathcal{G}$ is connected. Let x be an arbitrary vertex of $\mathcal{G}$. Put $G_i = \{y \in G$: The length of the shortest path connecting x and y in $\mathcal{G}$ is $i\}$. Then $G_0 = \{x\}$, $G = \bigcup_{i<\omega} G_i$. Put $\mathcal{G}^i = \mathcal{G} \cap [G_i]^2$. Then there is $1 \leqslant i < \omega$ such that $\chi(\mathcal{G}^i) > \omega$. Let $\mathcal{G}^{i,m} = \{\{u, v\} \in \mathcal{G}^i$: There is a path of length $2(m+1)$ in $\mathcal{G}$ connecting u and v, all whose vertices but u and v do not belong to $G_i\}$. By the definition of G_i we have

$$\mathcal{G}^i = \bigcup_{m<i} \mathcal{G}^{i,m} .$$

Considering that then $\chi(\mathcal{G}^i) \leqslant \prod_{m<i} \chi(\mathcal{G}^{i,m})$ it follows that there is $m < i$ with $\chi(\mathcal{G}^{i,m}) > \omega$. By the old result, for all j, $2 \leqslant j < \omega$, there is an edge $\{u, v\} \in \mathcal{G}^{i,m}$ contained in an odd circuit of length $2j$ of $\mathcal{G}^{i,m}$. Omitting from this circuit the edge $\{u, v\}$ and adding to it the edges of the path of length $2(m+1)$ the existence of which is required by the definition of $\mathcal{G}^{i,m}$, we get an odd circuit of length $2(m+j)+1$ contained in $\mathcal{G}$ for $m + j \geqslant m + 2 = n$. This proves Theorem 3.

We have no counterexample to

Problem 2. Let $\chi(\mathcal{G}) > \omega$. Then there is i, with $2 \leqslant i < \omega$ such that

$$S^0(i) \subset \psi(\mathcal{G}, \omega) .$$

We think that the answer is no. A positive answer would yield the solution of all the problems mentioned so far, since, by the old lemmas, it would imply that if $\chi(\mathcal{G}) > \omega$, then $\psi(\mathcal{G}, \omega)$ is ω-unbounded with the restriction $\exp_n^+$ for some $n < \omega$. (i.e. that the answer to Taylor's problem (2) is yes and the answer to Problem 1 is no.)*

§5. FURTHER SPECULATIONS

1/ Let $F(\tau) \geqslant \tau^+$ be an operation on cardinals. Choosing the property $\varphi(x, \lambda)$ appearing in (1) to be $\exists \tau(\tau^+ \geqslant \lambda \wedge x$ is a graph $\wedge$ $\wedge\, \chi(x) > \tau \wedge |x| \leqslant F(\tau))$, we see that there is a Taylor number corresponding to each restriction. Obviously we can expect results only if F in some way reasonlable, (e.g. $F(\omega) = (2^\omega)^+$, $F(\lambda) = \lambda^+$ for $x \geqslant \omega_1$, is unreasonable.) Without going into details we state the simplest problems.

Problem 3. Is it true that $\chi(\mathcal{G}) > \omega$, $|\mathcal{G}| \leqslant (\exp_i(\omega))^+$ implies that $\psi(\mathcal{G}, \omega)$ is ω-unbounded with the restriction $(\exp_i(\lambda))^+$ for $i < < \omega$?

There is no counterexample to the following stronger

Problem 4. Let $\lambda \geqslant \omega$, $i < \omega$. Assume that there is $\mathcal{G}$ with $\chi(\mathcal{G}) > \lambda$, $|\mathcal{G}| \leqslant (\exp_i(\lambda))^+$ then for every infinite τ there is $\mathcal{G}'$ with

$$\chi(\mathcal{G}') > \tau\,, \quad |\mathcal{G}'| \leqslant (\exp_i(\tau))^+\,, \quad \psi(\mathcal{G}, \omega) = \psi(\mathcal{G}', \omega)\,.$$

(We emphasize again that $\chi(\mathcal{G}) = \lambda$, $|\mathcal{G}| = \lambda$ does not imply that there is $\mathcal{G}'$ with $\chi(\mathcal{G}') = |\mathcal{G}'| = \tau$

$$\psi(\mathcal{G}', \omega) = \psi(\mathcal{G}, \omega)$$

as is shown e.g. by the fact that $\chi(\mathcal{G}^\circ(\kappa, 2)) = \kappa$ for all strong limit κ.)

2/ The problems we mentioned so far had not been studied in detail for set systems, not even in case $\kappa(\mathcal{H}) = 3$.

Let us now extend for uniform set systems, with $\kappa(\mathcal{H}) = \kappa$, in a self-explanatory way the notation $B(\tau)$, $A(\tau)$, $\psi(\mathcal{G}, \tau)$, $\mathcal{G}(S, \tau)$ intro-

*Problem 2 has already been stated in Taylor [8].

duced in §2. We will use the notation $B_\kappa(\tau)$, $A_\kappa(\tau)$, $\psi_\kappa(\mathscr{H}, \tau)$, $\mathscr{H}_\kappa(S, \tau)$ respectively. Though we can not disprove the analogue of (2) which is

(2′) Let $\kappa(\mathscr{H}) = 3$, $\chi(\mathscr{H}) > \omega$. Then for each λ there is $\mathscr{H}'$ with $\chi(\mathscr{H}') > \lambda$,

$$\psi_3(\mathscr{H}', \omega) = \psi_3(\mathscr{H}, \omega)$$

we want to point out one new phenomenon.

As we explained in §4, the old lemmas even imply the following (trivial) statement.

(6) If $\mathscr{G}_0$ is such that $\mathscr{G}_0 \in S$ for all $S \in A(\omega)$ which is ω-unbounded with the restriction λ^+ (the strongest possible restriction) then $\mathscr{G}_0 \in S$ for all unbounded $S \in A(\omega)$. (Namely the $\mathscr{G}_0$ in question are the bipartite graphs only).

A result of Erdős – Hajnal – Rothschild [5] implies that the analogue of (6) does not hold true for uniform set systems with $\kappa(\mathscr{H}) = 3$. The following is true:

(7) Let $\mathscr{H}_0$ consist of two triples having two points in common. Then

(α) $\mathscr{H}_0 \in S$ for all $S \in A_3(\omega)$ which is ω-unbounded with the restriction $\lambda^+ = \exp_0(\lambda)^+$ but

(β) There is $S \in A_3(\omega)$, $\mathscr{H}_0 \notin S$, S is ω-unbounded with the restriction $\exp_1(\lambda)^+$.

We do not know what is the natural bound to this sort of counter-examples in case $\kappa(\mathscr{H}) = 3$.

Again we do not state the problem here in general context but we formulate a rather simple Taylor type problem.

Let $C_3(\omega) = \{\mathscr{H}_0 \in B_3(\omega)\colon \mathscr{H}_0$ does not occur in some $\psi_3(\mathscr{H}, \omega)$ for which $\chi(\mathscr{H}) > \omega\}$. For each $\mathscr{H}_0 \in C_3(\omega)$ there is a minimal $\tau = \tau(\mathscr{H}_0)$ such that there is $\mathscr{H}$, $|\mathscr{H}| = \tau$ $\mathscr{H}_0 \notin \psi_3(\mathscr{H}, \omega)$,

$\chi(\mathscr{H}) > \omega$. Put $\tau(3, \omega) = \sup\{\tau(\mathscr{H}_0): \mathscr{H}_0 \in C_3(\omega)\}$.

Problem 5. $\tau(3, \omega) \leqslant (\exp_1(\omega))^+$, (Or at least $\leqslant \exp_\omega(\omega)$.)

In a forthcoming Erdős – Galvin – Hajnal paper it will be proved that if $\mathscr{H}_0$ is the system consisting of three triangles, which have an empty intersection and pairwise one point in common then $\tau(\mathscr{H}_0) \leqslant \leqslant \exp_2(\omega)^+$.

3/ Let $\mathscr{G}$ be a graph. Define a function $f_{\mathscr{G}}(n)$ for $n < \omega$ by $f_{\mathscr{G}}(n) = \max\{\chi(\mathscr{G}'): \mathscr{G}' \subset \mathscr{G} \wedge |\cup \mathscr{G}'| = n\}$ for $n < \omega$. We mention without proof that the example of the graphs $\mathscr{G}°(\xi, i)$ shows that

(8) For every $i < \omega$ there are graphs $\mathscr{G}_i$ with $\chi(\mathscr{G}_i) > \omega$, $f_{\mathscr{G}_i}(n) < C_i \log_i(n)$ for $n < \omega$. We state

Problem 6. Does there exist a $\mathscr{G}$ with $\chi(\mathscr{G}) > \omega$ and such that

$$f_{\mathscr{G}}(n) < \log_i(n) \quad \text{for} \quad n > n(i), \quad i < \omega\ ?$$

This should be compared with Problem 2.

4/ Interesting new problems arise if we investigate the case of uniform set systems with $\kappa(\mathscr{H}) = \omega$. We only mention

Problem 7. Does there exist a cardinal λ such that if $\mathscr{H}$ is a uniform set system with $\chi(\mathscr{H}) > \lambda$ and $\kappa(\mathscr{H}) = \omega$, then there always exists an $S \subset \cup \mathscr{H}$, $|S| = \omega$ for which $\chi(\mathscr{H} \cap P(S)) > 2$?

The answer to this question might turn out to be trivial, but certainly a lot of similar interesting problems could be raised.

REFERENCES

[1] P. Erdős – A. Hajnal, On chromatic numbers of graphs and set systems, *Acta Math. Acad. Sci. Hung.*, 17 (1966), 61-99.

[2] P. Erdős – A. Hajnal, On chromatic numbers of infinite graphs, *Graph theory symposium held in Tihany*, Hungary, (1966), 83-98.

[3] P. Erdős – A. Hajnal, On decomposition of graphs, *Acta Math. Acad. Sci. Hung.*, 18 (1967), 359-377.

[4] P. Erdős – A. Hajnal, Some remarks on set theory IX. *Michigan Math. Journal,* 11 (1964), 107-127.

[5] P. Erdős – A. Hajnal – B. Rothchild, On chromatic numbers of graphs and set systems, *Proceedings of the Cambridge Summer School,* 1971, (to appear).

[6] W. Hanf, Incompactness in languages, with infinitely long expressions, *Fund. Math.*, 53 (1964), 309-324.

[7] W. Taylor, Atomic compactness and elementary equivalence, *Fund. Math.*, 71 (1971), 103-112.

[8] W. Taylor, Problem 42, Combinatorial structures and their applications, *Proceedings of the Category International Conference,* 1969.

COLLOQUIA MATHEMATICA SOCIETATIS JÁNOS BOLYAI
8. TOPICS IN TOPOLOGY, KESZTHELY (HUNGARY), 1972.

METRIZABILITY OF LINEARLY ORDERED SPACES

M.J. FABER

Some classical theorems of R. H. Bing [1] state that in a regular T_1-space R the following properties are equivalent:

1. R is metrizable
2. R has a σ-discrete open base
3. R is a collectionwise normal Moore space

In this paper we want to use these results in the class of linearly ordered topological spaces and their subspaces. We establish a characterization for the metrizability, essentially based on the order structure. After that we illustrate the usefullness of this characterization by some applications to lexicographically ordered product spaces. The proofs of these applications can be found in [2].

1. TERMINOLOGY

Recall that a linearly ordered topological space (abbreviated LOTS) $(X, I_<)$ is a topological space X whose topology $I_<$ agrees with the topology induced by a linear ordering $<$. In the following a LOTS $(X, I_<)$

will mostly be denoted by X.

If X is a LOTS and Y is a subset of X, then the relative topology induced in Y by $I_<$ will be denoted by $I_<^{(Y)}$. By $<$ an ordering $<_Y$ is induced in Y. In general the spaces $(Y, I_<^{(Y)})$ and $(Y, I_{<_Y})$ do not coincide, even not if Y is a closed subset of X. However it is clear that $I_{<_Y} \subset I_<^{(Y)}$. If Y is a compact subset of X then, as can be easily verified, $I_{<_Y} = I_<^{(Y)}$.

A subset A of a linearly ordered set X is called a convex subset of X if, whenever $a, b \in A$ and $a \leqslant b$, then $\{x \in X \mid a \leqslant x \leqslant b\}$ is a subset of A. An interval of X is a convex subset of X having two endpoints in X (however the endpoints need not be points of the interval). We denote intervals by:

$$(a, b) = \{x \in X \mid a < x < b\}; [a, b) = \{x \in X \mid a \leqslant x < b\};$$

$$[a, b] = \{x \in X \mid a \leqslant x \leqslant b\} \quad \text{or} \quad (a, b] = \{x \in X \mid a < x \leqslant b\}.$$

Subsets of X of the form $(\leftarrow, a) = \{x \in X \mid x < a\}$; $(\leftarrow, a]$; $(a, \rightarrow)$ or $[a, \rightarrow)$ will be called half-lines of X. If $A \subset X$ and $a, b \in A$, $a \leqslant b$, then if necessary we distinguish between $(a, b)_X$ and $(a, b)_A$ etc. If $a, b \in X$, $a < b$, we will say that a and b are neighbours in X if $(a, b) = \phi$. In this case a (resp. b) is called a left (resp. right) neighbour of b (resp. a). A point $a \in X$ is said to be a neighbourpoint if there is a point $b \in X$ such that a and b are neighbours in X. Observe that for a dense subset Y of a LOTS $X, I_{<_Y} = I_<^{(Y)}$ if and only if for each pair of neighbours a and b in X, $a \in Y$ implies $b \in Y$.

For any ordertype τ, the inverse ordertype will be denoted by τ^*. And the initial ordinal of $\aleph_\mu$, where $\aleph_\mu$ is a cardinal number with ordinal index μ, is denoted by ω_μ. Suppose for each ordinal number a from a certain set M, we are given a linearly ordered set $X_a = (X_a, <_a)$.

Then the lexicographically ordered product, denoted by $\coprod_{a \in M} X_a$, is defined as the set of all (transfinite) sequences $x = (x_a)_{a \in M}$, $(x_a \in X_a$ for all $a \in M)$, supplied with an ordering $<$, given by

$x < y \Leftrightarrow \exists \beta \in M$: $(x_\beta <_\beta y_\beta$ and $x_a = y_a$ if $a < \beta$ and $a \in M)$

$(\prod_{a \in M} X_a, I_<)$ is called the lexicographically ordered product space of the family $(X_a)_{a \in M}$.

In particular, if X is a linearly ordered set, then X^μ, where μ is an ordinal number, is the lexicographically ordered product

$$\prod_{a<\mu} \{X_a \mid X_a = X \quad \text{for all} \quad a < \mu\};$$

and if both X and Y are linearly ordered sets, then $X \times_{\text{lex}} Y$ is the lexicographically ordered product

$$\prod_{a<2} \{X_a \mid X_0 = X \quad \text{and} \quad X_1 = Y\}.$$

Observe that if ν and μ are ordinal numbers, $\nu < \mu$, then clearly $\prod_{a<\mu} X_a$ is canonically homeomorphic to $(\prod_{a<\mu} X_a) \times_{\text{lex}} (\prod_{\nu \leq a < \mu} X_a)$.

Finally the set of real numbers, the set of integers and the set of positive integers will be denoted by R, Z and N respectively.

Most undefined terms in this paper are as in [5].

2. METRIZABILITY OF A LOTS AND ITS SUBSPACES

Let X be a LOTS and let Y be a subspace of X. We define the following two subsets of Y:

$$E(Y) = \{p \in Y \mid [p, \rightarrow)_Y \text{ or } (\leftarrow, p]_Y \text{ is an open subset of } Y\}$$

and

$$N(Y) = \{p \in Y \mid \exists q \in Y: (p \text{ and } q \text{ are neighbours in } Y \text{ and neither } p \text{ nor } q \text{ is isolated in } Y)\}.$$

If $Y = X$ then $E(X)$ is precisely the set of all neighbourpoints of X together with possible endpoints of X. Obviously $N(Y) \subset E(Y)$ and in general $N(Y) \neq E(Y)$, even in case $Y = X$.

2.1. Theorem. *The following properties of* X *are equivalent*

1) *X is metrizable*

2) *There exists a dense subset A in X, with* $E(X) \subset A$, *and such that* $A = \bigcup_{n=1}^{\infty} A_n$, *where, for all* $n \in N$:

(i) A_n *is a closed subset of X*

(ii) $(A_n, I_<^{(A_n)})$ *is a discrete space*

3) *There exists a dense subset A in X, with* $N(X) \subset A$, *and such that* $A = \bigcup_{n=1}^{\infty} A_n$, *where, for all* $n \in N$:

(i) A_n *is a closed subset of X*

(ii) $(A_n, I_<^{(A_n)})$ *is a discrete space*

Proof:

1) ⇒ 2). X is metrizable. Therefore X has a σ-discrete open base $\mathscr{B} = \bigcup_{n=1}^{\infty} \mathscr{B}_n$, where, for all $n \in N$, every $\mathscr{B}_n$ is a discrete family of open subsets of X. Let $n \in N$. From each $B \in \mathscr{B}_n$ we select one or two points as follows: if B contains one (resp. two) endpoints, in the ordering on X, then we select that (resp. both) endpoint(s) from B; and if B contains neither a left nor a right endpoint, then we choose an arbitrary point from B. Let A_n be the set of all points thus selected.

Finally, put $A = \bigcup_{n=1}^{\infty} A_n$. Then A satisfies the required conditions.

2) ⇒ 3). Obvious

3) ⇒ 1). A LOTS is hereditarily collectionwise normal [6], so it suffices to show that X is a Moore space. For all $n \in N$ and every $x \in X$ it follows from the fact that A_n is closed in X and discrete in the relative topology, that there is an open interval $I(x, n)$ in X, containing x, such that $I(x, n) \cap (A_n \setminus \{x\}) = \phi$. Without loss of generality we may assume that, for all $n \in N$, $I(x, n) \supset I(x, n + 1)$ and moreover, if x is an isolated point of X, that $I(x, n) = \{x\}$ and, if x is a non-isolated point of X having a left (resp. right) neighbour in X, that $I(x, n) = [x, p)$ for some

$p \in X$ (resp. $I(x, n) = (p, x]$ for some $p \in X$). Now, for all $n \in N$, we put $\mathcal{U}_n = \{I(x, n) \mid x \in X\}$; then it is clear that $\mathcal{U}_n$ is an open cover of X and hence the proof is complete once we have shown that, for every $x \in X$, the family $\{S(x, \mathcal{U}_n)\}_{n=1}^{\infty}$ is a local base at $x \in X$. So, choose $x \in X$ and let 0 be an arbitrary open interval in X, containing x.

a) Suppose x is isolated in X.

Since $\bar{A} = X$, there is an integer $i \in N$ such that $x \in A_i$. We claim that $S(x, \mathcal{U}_i) \subset 0$. For, if $x \in I(y, i)$ for some $y \in X$, then from $I(y, i) \cap (A_i \setminus \{y\}) = \phi$ and $x \in I(y, i) \cap A_i$ it follows that $x = y$. Hence $S(x, \mathcal{U}_i) = I(x, i) = \{x\} \subset 0$.

b) Suppose x has a left neighbour $x^- \in X$. Let x not be isolated in X.

Certainly $x^- \in A$ and so $x^- \in A_i$ for some $i \in N$. Since x is not an isolated point of X we can find integers j and $k \in N$ and points $a \in A_j \cap 0$ and $b \in A_k \cap 0$ such that $i < j < k$ and $x < b < a$. We claim that $S(x, \mathcal{U}_k) \subset 0$. For, if $x \in I(y, k)$ for some $y \in X$, then from $I(y, k) \cap (A_k \setminus \{y\}) = \phi$ and $b \in A_k$ it follows that $y \leqslant b < a$. Furthermore, since $I(y, i) \cap (A_i \setminus \{y\}) = \phi$, $x^- \in A_i$, $x \in I(y, k) \subset I(y, i)$ and $x \in I(x^-, k)$, we have that $x^- < y$. Therefore from $I(y, j) \cap (A_j \setminus \{y\}) = \phi$ and $y < a$ it follows that $a \notin I(y, j) \supset I(y, k)$ and from $I(y, i) \cap (A_i \setminus \{y\}) = \phi$ and $x^- < y$ it follows that $x^- \notin I(y, i) \supset I(y, k)$. So $I(y, k) \subset 0$. Hence $S(x, \mathcal{U}_k) \subset 0$.

The case that x has a right neighbour $x^+ \in X$ and x is not isolated in X, can be treated similarly.

c) Suppose x is an endpoint of X. Let x not be isolated in X.

The proof of this case is an easy modification of the proof of case b).

d) Suppose x has neither a left nor a right neighbour in X.

Consider the LOTS $[x, \rightarrow)$ and the LOTS $(\leftarrow, x]$. By the previous case c) we may conclude that there are integers m and $n \in N$ such that

$$S(x, \mathcal{U}_m) \cap [x, \rightarrow) \subset 0 \cap [x, \rightarrow) \quad \text{and} \quad S(x, \mathcal{U}_n) \cap (\leftarrow, x] \subset 0 \cap (\leftarrow, x] .$$

If we assume that $m \leqslant n$ then $S(x, \mathcal{U}_n) \subset S(x, \mathcal{U}_m)$ and consequently $S(x, \mathcal{U}_n) \subset 0$.

2.2. Theorem. *The following properties of* Y *are equivalent*

1) Y *is metrizable*

2) *There exists a dense subset* A *in* Y, *with* $E(Y) \subset A$, *and such that* $A = \bigcup_{n=1}^{\infty} A_n$, where, *for all* $n \in N$,

(i) A_n *is a closed subset of* Y

(ii) $(A_n, I_{<}^{(A_n)})$ *is a discrete space.*

Proof.

1) ⇒ 2). Analogous to the proof of 2.1, 1) ⇒ 2)

2) ⇒ 1). Let Y^* be a subset of the lexicographically ordered product $Y \times_{\text{lex}} Z$, defined by

$$Y^* = \left(Y \times_{\text{lex}} \{0\}\right) \cup \{(y, n) \in Y \times_{\text{lex}} Z \mid [y, \rightarrow)_Y \text{ is open in } Y, \text{ but not open in } X;\ n \leqslant 0\} \cup$$

$$\cup \{(y, n) \in Y \times_{\text{lex}} Z \mid (\leftarrow, y]_Y \text{ is open in } Y, \text{ but not open in } X;\ n \geqslant 0\}.$$

Now consider the LOTS $Y^* = (Y^*, I_{<_{Y^*}})$. Then Y is homeomorphic to the closed subset $Y \times_{\text{lex}} \{0\}$ of Y^*. Next define $f\colon Y^* \to Y$ by $f(y, n) = y$, for every $(y, n) \in Y^*$. Then f is a continuous map. Finally it is easily verified that $f^{-1}[A]$ satisfies the following conditions: $f^{-1}[A]$ is a dense subset of Y^*; $f^{-1}[A] \supset N(Y^*) = N(Y)$ and $f^{-1}[A] = \bigcup_{n=1}^{\infty} f^{-1}[A_n]$, where, for all $n \in N$, $f^{-1}[A_n]$ is a closed subset of Y^* which is discrete in its relative topology. Hence, from 2.1 it follows that Y^* is metrizable and consequently Y is metrizable.

2.3. Remark. *If* Y *is a separable subspace of* X, *then* Y *is hereditarily separable* [4]. *Therefore, in this case, a subset* A *of* Y *satisfying* $A = \bigcup_{n=1}^{\infty} A_n$ *such that for all* $n \in N$, $(A_n, I_{<}^{(A_n)})$ *is a discrete* *space, has to be countable.*

2.4. Examples. We like to illustrate these theorems by applying them to the ordinal space $W(\omega_1)$ and the Sorgenfrey-line S. (Of course the non-metrizability of these spaces is well-known and can be shown in various easy manners).

– Let $W(\omega_1)$ be the set of all ordinal numbers less than the first uncountable one ω_1. It is easy to see that no infinite subset of the LOTS $W(\omega_1)$ can be closed in $W(\omega_1)$ and at the same time discrete in its relative topology.

Since $E(W(\omega_1)) = W(\omega_1)$ we conclude now from 2.1 that $W(\omega_1)$ is not metrizable.

– Let X denote the unit square supplied with the lexicographic order topology. Let $S \subset X$ be defined by $S = \{(x, y) \in X \mid x \neq 0,1;\ y = 0\}$. S is called the Sorgenfrey-line. S is homeomorphic to $\boldsymbol{R}$ supplied with a topology generated by the family of all half-open intervals $(a, b]$, $a, b \in \boldsymbol{R}$. Since $E(S) = S$ and moreover S is separable, 2.2 together with 2.3 yield the non-metrizability of S.

3. SOME APPLICATIONS TO LEXICOGRAPHICALLY ORDERED PRODUCT SPACES

We first notice that in 2.1, the condition on A_n to be a discrete subspace of X is equivalent with the condition that for every $p \in A_n$ both half-lines $(\leftarrow, p]_{A_n}$ and $[p, \rightarrow)_{A_n}$ are open subsets of A_n. In other words, A_n is a discrete subspace of X if and only if all closed half-lines of A_n are also open in A_n.

For all ordinal numbers a less than ω_0, let X_a be a LOTS. We investigate the following results.

3.1. Suppose, for all $a < \omega_0, X_a$ has no endpoints.

Then $\coprod_{a < \omega_0} X_a$ is metrizable.

3.2. Suppose, for all $a < \omega_0, X_a$ has two endpoints.

Then $\coprod_{a<\omega_0} X_a$ is metrizable if and only if, for all $a < \omega_0$, $X_a = \bigcup_{n=1}^{\infty} A_n^a$ such that, for all $n \in N$,

(i) A_n^a is a closed subset of X_a.

(ii) both $(\leftarrow, p]_{A_n^a}$ and $[p, \rightarrow)_{A_n^a}$ are open subsets of A_n^a, for every $p \in A_n^a$.

3.3. Suppose, for all $a < \omega_0$, X_a has a right endpoint, but no left endpoint.

Then $\coprod_{a<\omega_0} X_a$ is metrizable if and only if, for all $a < \omega_0$, X_{a+1} is coinitial with ω_0^* whenever X_a contains neighbourpoints, and $X_a = \bigcup_{n=1}^{\infty} A_n^a$ such that, for all $n \in N$,

(i) A_n^a is a closed subset of X_a

(ii) $(\leftarrow, p]_{A_n^a}$ is an open subset of A_n^a, for every $p \in A_n^a$.

3.4. Examples

– Let X be a non-trivial discrete topological space. Certainly X is a LOTS [3]. From the previous results we have now:

X^{ω_0} is metrizable $\Longleftrightarrow$ One of the following conditions is satisfied

1. X does not have endpoints. (3.1)
2. X has two endpoints. (3.2)
3. X has a right endpoint and X is coinitial with ω_0^*. (3.3)
4. X has a left endpoint and X is cofinal with ω_0 (symmetric to 3.3)

In [2] we studied, among others, the compactness and connectedness of lexicographically ordered product spaces. It turns out, in case X is a non-trivial discrete space, that

$$X^{\omega_0} \text{ is compact} \Leftrightarrow |X| < \aleph_0 \Leftrightarrow X^{\omega_0}$$

is homeomorphic to the Cantorspace $\{0,1\}^{\aleph_0} = \{0,1\}^{\omega_0}$

and

$$X^{\omega_0} \text{ is connected} \Leftrightarrow X \sim N \Leftrightarrow X^{\omega_0} \text{ is homeomorphic to}$$

$$[0,1) = \{x \in R \mid 0 \leqslant x < 1\} \quad \text{(with the usual topology).}$$

Remark. $X \sim N$ means that X is similar (or order-isomorphic) to N; (i.e. there is a one to one map f of X onto N which is monotone: $x_1 < x_2 \Rightarrow f(x_1) < f(x_2)$).

– Let X be the set of all ordinal numbers less than or equal to the first uncountable ordinal ω_1, preceded by the negative integers. So X is a LOTS of ordertype $\omega_0^* + \omega_1 + 1$. Obviously X is not metrizable. However X^{ω_0} is metrizable, as follows immediatly from 3.3, since every point of X has a right neighbour in X.

REFERENCES

[1] R.H. Bing, Metrization of topological spaces, *Can. J. Math.*, 3, (1951), 175-186.

[2] M.J. Faber, *Metrizability of linearly ordered spaces with applications to lexicographically ordered product spaces,* report 24 (1971), Free University, Amsterdam.

[3] H. Herrlich, *Ordnungsfähigkeit topologischer Räume,* inaugural dissertation. Freie Universitat, Berlin, 1962.

[4] D.J. Lutzer, *On generalized ordered spaces,* Dissertation, University of Washington, 1970.

[5] J. Nagata, *Modern general topology,* North Holland Publ. Cy., Amsterdam, 1968.

[6] L.A. Steen, A direct proof that a linearly ordered space is hereditarily collectionwise normal, *Proc. Amer. Math. Soc.*, 24 (1970), 727-728. Amsterdam, Free University.

COLLOQUIA MATHEMATICA SOCIETATIS JÁNOS BOLYAI
8. TOPICS IN TOPOLOGY, KESZTHELY (HUNGARY), 1972.

COMPLETENESS AND PRODUCTS OF ϑ-UNIFORM SPACES

V.V. FEDORČUK

In this work the ϑ-uniform spaces introduced in [9] are studied. We shall point out the connections between ϑ-uniform and uniform spaces as well as ϑ-uniform and ϑ-proximity spaces. We shall introduce the notion of complete ϑ-uniform spaces and study completions of products.

We shall employ the terminology and notation of [7], [8] and [10], where ϑ-proximities are extensively studied. For the sake of brevity symbols for embedding maps will usually be omitted and extensions will be considered as superspaces. If it is clear from the context what space we consider (topological, ϑ-proximity or ϑ-uniform), the symbol for the topology, ϑ-proximity or ϑ-uniformity will be omitted.

§1. CONNECTIONS BETWEEN ϑ-UNIFORM SPACES AND UNIFORM AND ϑ-PROXIMITY SPACES*

We shall say that a system $\nu = \{V\}$ of canonical open sets of a topological space X is a ϑ-cover of locally finite type, if for any point

*In the first and second chapters of this paper we shall give a detailed exposition of the results formulated in [9].

$x \in X$ there exist finitely many elements $V_1, \ldots, V_n$ of ν such that $x \in \left\langle \bigcup_{i=1}^{n} [V_i] \right\rangle$.

The following result yields us a great number of ϑ-covers of locally finite type.

Lemma 1. *Let $f\colon X \to Y$ be a ϑ-perfect irreducible mapping, and $\nu = \{V\}$ be a cover of X by canonical open sets. Then the system $f^{\#}\nu = \{f^{\#}V \mid V \in \nu\}$ froms a ϑ-cover of finite type of the space Y.*

Proof. According to lemma 2 of [8] the system $f^{\#}\nu$ consists of canonical open sets. Now let $y \in Y$. There are a finite number of elements $V_1, \ldots, V_n \in \nu$ such that the compactum $f^{-1}y$ is contained in the union $\bigcup_{i=1}^{n} V_i$. It remains to show that $y \in \left\langle \bigcup_{i=1}^{n} [f^{\#}V_i] \right\rangle$. By lemma 1 of [8] we have $fV_i \subset [f^{\#}V_i]$, hence $f^{\#}\left(\bigcup_{i=1}^{n} V_i\right) \subset f\left(\bigcup_{i=1}^{n} V_i\right) = \bigcup_{i=1}^{n} fV_i \subset \bigcup_{i=1}^{n} [f^{\#}V_i]$.

However, the closedness of the map f implies that the set $f^{\#}\left(\bigcup_{i=1}^{n} V_i\right)$ is a neighbourhood of the point y. Consequently

$$y \in \left\langle \bigcup_{i=1}^{n} [f^{\#}V_i] \right\rangle .$$

The proof of the lemma is completed.

A family $\mathfrak{B} = \{\nu\}$ of ϑ-covers of locally finite type of a space X is called a ϑ-uniformity if it satisfies the following axioms:

$\mathrm{I_U}$. If the ϑ-cover $\nu \in \mathfrak{B}$ is refinement of the ϑ-cover w, then $w \in \mathfrak{B}$.

$\mathrm{II_U}$. For any two ϑ-covers $u, \nu \in \mathfrak{B}$ there exists a ϑ-cover $w \in \mathfrak{B}$, which star refines both u and ν.*

$\mathrm{III_U}$. If x and y are distinct points of the space X, then there are neighbourhoods G and H, respectively, of these points and a ϑ-cover $\nu \in \mathfrak{B}$ so that

*In symbols: $w \ast\!> u$ and $w \ast\!> \nu$.

$$G \cap \mathrm{st}_\nu H = \phi .$$

IV_U. For any point $x \in X$ and its arbitrary canonical open neighbourhood G there exists a neighbourhood Ox of the point x and a ϑ-cover $\nu \in \mathfrak{B}$ such that $\mathrm{st}_\nu Ox \subset G$.

Thus if in these axioms ϑ-covers of locally finite type are replaced by arbitrary covers and we do not require in axiom IV_U that the neighbourhood G be canonical open, than we obtain uniformities compatible with the given topology of X.

The pair $(X, \mathfrak{B})$, where X is a topological space and $\mathfrak{B}$ is a ϑ-uniformity on X, is called a ϑ-uniform space.

Every separated uniformity yields a ϑ-uniformity if it is identified with its basis consisting of all those covers, whose elements are canonical open sets. In the present paper, with some abuse of language, by uniformity we shall always mean a ϑ-uniformity all of whose members are covers.

We obviously have

Proposition 1. *A ϑ-uniform space $(X, \mathfrak{B})$ is uniform if and only if the ϑ-uniformity $\mathfrak{B}$ is a uniformity and the space X is semiregular.*

On an extremely disconnected space every ϑ-uniformity is a uniformity, since every ϑ-cover of locally finite type of an extremely disconnected space is a cover.

Axiom III_U implies that ϑ-uniformities can exist on Hausdorff spaces only.

As an example of a ϑ-uniformity on an arbitrary Hausdorff space X we can take the ϑ-uniformity whose basis is the system of all finite ϑ-covers of X by canonical open sets.

Let $\mathfrak{B}$ be a ϑ-uniformity on the space X, and consider the binary relation $\delta_{\mathfrak{B}}$ defined as follows:

$A \,\bar{\delta}_{\mathfrak{B}}\, B$ if and only if there are neighbourhoods G and H of the sets A and B, respectively, and a ϑ-cover $\nu \in \mathfrak{B}$ such that

$$G \cap \mathrm{st}_\nu H = \phi .$$

Proposition 2. *The relation* $\delta_{\mathfrak{B}}$ *is a* ϑ*-proximity on the space* X.

Proof. We have to show that axioms I – VI of a ϑ-proximity are fulfilled.

Axioms I and III ($A\ \delta_{\mathfrak{B}}\ B \Rightarrow B\ \delta_{\mathfrak{B}}\ A$ and $\phi\ \bar{\delta}_{\mathfrak{B}}\ X$) are obviously satisfied.

Axiom II ($A\ \delta_{\mathfrak{B}}\ B_i,\ i = 1, 2 \Leftrightarrow A\ \bar{\delta}_{\mathfrak{B}}\ (B_1 \cup B_2)$). The implication $\Rightarrow$ follows from axiom $\mathrm{II_U}$, while the other implication is obvious.

Axiom IV $(\{x\}\ \delta_{\mathfrak{B}}\ \{y\} \Leftrightarrow x = y)$. The implication $\Rightarrow$ follows from axiom $\mathrm{III_U}$, and the other implication is obvious.

Axiom V ($A\ \bar{\delta}_{\mathfrak{B}}\ B \Rightarrow$ there exists a canonical open set $C \supset A$ such that $C\ \bar{\delta}_{\mathfrak{B}}\ B$ and $A\ \bar{\delta}_{\mathfrak{B}}\ (X \setminus [C])$). By the definition of $\delta_{\mathfrak{B}}$ there are neighbourhoods G and H of the sets A and B respectively and a ϑ-cover $v \in \mathfrak{B}$ such that $G \cap \mathrm{st}_v\ H = \phi$. According to axiom $\mathrm{II_U}$ there exists a ϑ-cover $w \in \mathfrak{B}$ which doubly star refines v. Let us put $C = \langle[\mathrm{st}_w\ G]\rangle$. It is obvious that the canonical open set C contains A. Moreover, if for a set $W \in w$ we have

$$W \cap \langle[\mathrm{st}_w\ G]\rangle \neq \phi\ ,$$

then $W \cap \mathrm{st}_w\ G \neq \phi$, hence

$$W \subset \mathrm{st}_w\ \mathrm{st}_w\ G \subset \mathrm{st}_v\ G\ .$$

Consequently $\mathrm{st}_w\ C \subset \langle[\mathrm{st}_v\ G]\rangle$, and thus $H \cap \mathrm{st}_w\ C = \phi$. Therefore, we have $C\ \bar{\delta}_{\mathfrak{B}}\ B$, moreover

$$\mathrm{st}_w\ G \cap (X \setminus [\mathrm{st}_w\ G]) = \phi\ .$$

But $X \setminus [\mathrm{st}_w\ G] = X \setminus [C]$, hence $A\ \bar{\delta}_{\mathfrak{B}}\ (X \setminus [C])$.

Axiom VI (if the point x and the set A have non-intersecting neighbourhoods, then $\{x\}\ \bar{\delta}_{\mathfrak{B}}\ A$) follows easily from axiom $\mathrm{IV_U}$. Thus proposition 2 is proved.

Proposition 3. *Every* ϑ*-uniformity which generates a proximity on the space* X *must be a uniformity.*

Proof. We shall show that in this case the ϑ-uniformity $\mathfrak{B}$ consists of covers. Let $v \in \mathfrak{B}$, x be an arbitrary point, and the ϑ-cover $w \in \in \mathfrak{B}$ be a double star refinement of v, i.e. for any $W \in w$ there exists a set $V \in v$ such that $\mathrm{st}_w\, W \subset V$. Since w is a ϑ-cover of locally finite type, it has a member W such that $x \in [W]$. By the definition of the ϑ-proximity $\delta_{\mathfrak{B}}$ we have

$$W\, \bar{\delta}_{\mathfrak{B}}\, (X \setminus [\mathrm{st}_w\, W]) ,$$

hence a fortiori $W\, \bar{\delta}_{\mathfrak{B}}\, (X \setminus [V])$. Since $\delta_{\mathfrak{B}}$ is a proximity, the closures of distant sets are distant too. Consequently $[W]\, \bar{\delta}_{\mathfrak{B}}\, X \setminus V$, and therefore $[W] \cap (X \setminus V) = \phi$, i.e. $[W] \subset V$. Thus we have $x \in [W] \subset V$, and the system v is indeed a cover. Proposition 3 is thus established.

We remark that in the above proof we have not made use of the fact that w is a ϑ-cover of locally finite type, but only that the system $\{[W] \mid W \in w\}$ is a cover.

Theorem 1. *Let $f\colon Y \to X$ be a ϑ-perfect irreducible mapping of a completely regular space Y onto a space X, and let $\mathfrak{B}$ be a uniformity on the space Y. Then the family $f^{\#}\mathfrak{B} = \{f^{\#}v \mid v \in \mathfrak{B}\}$ is a ϑ-uniformity on the space X, moreover we have $\delta_{f^{\#}\mathfrak{B}} = f\delta_{\mathfrak{B}}$.**

Proof. According to lemma 1, the family $f^{\#}\mathfrak{B}$ consists of ϑ-covers of locally finite type. Let us verify the axioms of ϑ-uniformities.

Axiom $\mathrm{I_U}$. Assume that the ϑ-cover $f^{\#}v \in f^{\#}\mathfrak{B}$ is a refinement of the ϑ-cover w which is of locally finite type. Let us put $\widetilde{w} = \{\langle[f^{-1}W]\rangle \mid W \in w\}$ and let $V \in v$. There exists a set $W \in w$ such that $f^{\#}V \subset W$. Then $V = \langle[f^{-1}f^{\#}V]\rangle \subset \langle[f^{-1}W]\rangle$. Thus the cover v is a refinement of $\widetilde{w}$, and therefore $\widetilde{w} \in \mathfrak{B}$. According to lemma 9 of [8], the set $[f^{-1}W]$ is a canonical closed set and $\langle[f^{-1}W]\rangle \supset f^{-1}W$. Therefore $f^{\#}\langle[f^{-1}W]\rangle \supset W$. Analogously, we have

$$\langle[f^{-1}(X \setminus [W])]\rangle \supset f^{-1}(X \setminus [W]) ,$$

hence

$$Y \setminus [\langle f^{-1}[W]\rangle] \supset Y \setminus f^{-1}[W] ,$$

*Here $f^{\#}v = \{f^{\#}V \mid V \in v\}$, and $f\delta_{\mathfrak{B}}$ denotes the ϑ-proximity on X generated by the proximity $\delta_{\mathfrak{B}}$ and the mapping f (cf. theorem 1 of [8]).

or

$$[\langle f^{-1}[W]\rangle] \subset f^{-1}[W] .$$

But $f^{-1}W \subset \langle f^{-1}[W]\rangle$, hence $\langle [f^{-1}W]\rangle \subset f^{-1}[W]$, which implies

$$f^{\#}\langle [f^{-1}W]\rangle \subset [W] ,$$

and thus

$$f^{\#}\langle [f^{-1}W]\rangle \subset W .$$

This implies $f^{\#}\langle [f^{-1}W]\rangle = W$, and $f^{\#}\widetilde{w} = w$, consequently $w \in f^{\#}\mathfrak{B}$.

Axiom II_{U}. The proof is reduced to the trivial verification of the fact that $v *> w$ implies $f^{\#}v *> f^{*}w$.

Axiom $\mathrm{III}_{\mathrm{U}}$ follows from the fact that in a uniform space its analogon is satisfied, in which the roles of the distinct points are played by disjoint compacta.

Axiom IV_{U}. Let G be a canonical open neighbourhood of the point x. Then $\langle f^{-1}[G]\rangle$ is a neighbourhood of the compactum $f^{-1}x$. It has a neighboruhood $Of^{-1}x$ for which a cover $v \in W$ exists so that $\mathrm{st}_v\, Of^{-1}x \subset \langle f^{-1}[G]\rangle$. It is easy to see that the neighbourhood $f^{\#}Of^{-1}x$ of the point x and the ϑ-cover $f^{\#}v \in f^{\#}\mathfrak{B}$ are as required.

Now we prove the second statement of our theorem. Let $A\, \overline{\delta}_{f^{\#}\mathfrak{B}}\, B$, i.e. there exist neighbourhoods G and H of the sets A and B and a ϑ-cover $f^{\#}v \in f^{\#}\mathfrak{B}$ so that $G \cap \mathrm{st}_{f^{\#}v}\, H = \phi$. Then $[G] \cap \mathrm{st}_{f^{\#}v}\, H = \phi$ and

$$f^{-1}[G] \cap f^{-1}\mathrm{st}_{f^{\#}v}\, H = \phi .$$

But the set $f^{-1}[G]$ contains a neighbourhood $Of^{-1}A$ of the set $f^{-1}A$ and

$$f^{-1}\mathrm{st}_{f^{\#}v}\, H = \mathrm{st}_{f^{-1}f^{\#}v}\, f^{-1}H .$$

Consequently

$$f^{-1}A \cap [\mathrm{st}_{f^{-1}f^{\#}v}\, f^{-1}H] = \phi ,$$

moreover

$$[\mathrm{st}_{f^{-1}f^{\#}\nu}\, f^{-1}H] \supset \mathrm{st}_\nu\, f^{-1}H\,,$$

and thus $f^{-1}A \cap \mathrm{st}_\nu\, f^{-1}H = \phi$. This implies $f^{-1}A \cap \mathrm{st}_\nu\, f^{-1}B = \phi$, hence $f^{-1}A\ \bar{\delta}_{\mathfrak{V}}\, f^{-1}B$, i.e. $A\ \overline{f\delta_{\mathfrak{V}}}\, B$.

Now let $A\ \overline{f\delta_{\mathfrak{V}}}\, B$, i.e. $f^{-1}A\ \bar{\delta}_{\mathfrak{V}}\, f^{-1}B$. Then there are neighbourhoods $Of^{-1}A$ and $Of^{-1}B$ and a cover $\nu \in \mathfrak{V}$ so that

$$Of^{-1}A \cap \mathrm{st}_\nu\, Of^{-1}B = \phi\,.$$

Then

$$f^{\#}Of^{-1}A \cap \mathrm{st}_{f^{\#}\nu}\, f^{\#}Of^{-1}B = \phi\,,$$

which implies $A\ \bar{\delta}_{f^{\#}\mathfrak{V}}\, B$. This completes the proof of theorem 1.

The following theorem shows us that theorem 1 yields the most general example of a ϑ-uniformity.

Theorem 2. *Let* $(X, \mathfrak{V})$ *be a* ϑ*-uniform space. There exist a unique uniform space* $(X_{\mathfrak{V}}, \widetilde{\mathfrak{V}})$ *and a unique* ϑ*-perfect irreducible mapping* $\pi_{\mathfrak{V}}\colon X_{\mathfrak{V}} \to X$ *such that* $\mathfrak{V} = \pi^{\#}_{\mathfrak{V}}\widetilde{\mathfrak{V}}$.

Proof. By proposition 2, the pair $(X, \delta_{\mathfrak{V}})$ is a ϑ-proximity space. Hence, by theorem 2 of [8], there are a proximity space $(X_{\delta_{\mathfrak{V}}}, \delta)$ and a ϑ-perfect irreducible mapping $\pi_{\delta_{\mathfrak{V}}}\colon X_{\delta_{\mathfrak{V}}} \to X$, for which $\pi_{\delta_{\mathfrak{V}}}\delta = \delta_{\mathfrak{V}}$. Let us put $X_{\mathfrak{V}} = X_{\delta_{\mathfrak{V}}}$ and $\pi_{\mathfrak{V}} = \pi_{\delta_{\mathfrak{V}}}$, furthermore for any ϑ-cover $\nu \in \mathfrak{V}$ put $\tilde{\nu} = \{\widetilde{V} \mid V \in \nu\}$, where $\widetilde{V} = \langle[\pi_{\mathfrak{V}}^{-1}]\rangle$. We shall prove that the family $\widetilde{\mathfrak{V}} = \{\tilde{\nu} \mid \nu \in \mathfrak{V}\}$ yields a uniformity on the space $X_{\mathfrak{V}}$. To start with, we shall show that $\tilde{\nu}$ is a ϑ-cover of locally finite type. Let $y \in Y$ and $x = \pi_{\mathfrak{V}}y$. Then there are a finite number of elements $V_1, \ldots, V_n$ of the ϑ-cover ν such that $x \in \left\langle \bigcup_{i=1}^{n} [V_i] \right\rangle$. The set $\bigcup_{i=1}^{n} \pi_{\mathfrak{V}}^{-1}[V_i]$ contains a neighbourhood O of the compactum $\pi_{\mathfrak{V}}^{-1}x \in y$. Since the inverse image of a dense set by a closed irreducible mapping is again dense, the set $\bigcup_{i=1}^{n} \pi_{\mathfrak{V}}^{-1}V_i$ is dense in the neighbourhood O of the compactum $\pi_{\mathfrak{V}}^{-1}x$. Consequently

$$y \in O \subset \Big[\bigcup_{i=1}^{n} \pi_{\mathfrak{V}}^{-1}V_i\Big] = \bigcup_{i=1}^{n} [\pi_{\mathfrak{V}}^{-1}V_i] = \bigcup_{i=1}^{n} [\widetilde{V}_i]\,.$$

Thus the family $\widetilde{\mathfrak{V}}$ consists of ϑ-covers of locally finite type, moreover, similarly as in the proof of theorem 1, it can be shown that $\pi^{\#}_{\mathfrak{V}}\widetilde{\nu} = \nu$, i.e. $\pi^{\#}_{\mathfrak{V}}\widetilde{\mathfrak{V}} = \mathfrak{V}$.

Next we show that the relation $\delta_{\widetilde{\mathfrak{V}}}$ coincides with the proximity δ. Suppose $A \,\bar{\delta}\, B$. Then they have distant neighbourhoods G and H respectively, hence

$$\pi^{\#}_{\mathfrak{V}} G \;\bar{\delta}_{\mathfrak{V}}\; \pi^{\#}_{\mathfrak{V}} H .$$

According to the definition of the ϑ-proximity $\delta_{\mathfrak{V}}$ there exists a ϑ-cover $\nu \in \mathfrak{V}$ so that

$$\pi^{\#}_{\mathfrak{V}} G \cap \mathrm{st}_{\nu}\, \pi^{\#}_{\mathfrak{V}} H = \phi .$$

Since $\pi^{\#}_{\mathfrak{V}}\widetilde{\nu} = \nu$, this implies $G \cap \mathrm{st}_{\widetilde{\nu}}\, H = \phi$, i.e. $A \;\bar{\delta}_{\widetilde{\mathfrak{V}}}\; B$.

Now assume $A \;\bar{\delta}_{\widetilde{\mathfrak{V}}}\; B$ and let G and H be neighbourhoods of the sets A and B respectively and let $\widetilde{\nu} \in \widetilde{\mathfrak{V}}$ be a ϑ-cover so that $G \cap$ $\cap\, \mathrm{st}_{\widetilde{\nu}}\, H = \phi$. Then we have

$$\pi^{\#}_{\mathfrak{V}} G \cap \mathrm{st}_{\nu}\, \pi^{\#}_{\mathfrak{V}} H = \phi ,$$

i.e. $\pi^{\#}_{\mathfrak{V}} G \;\bar{\delta}_{\mathfrak{V}}\; \pi^{\#}_{\mathfrak{V}} H$, consequently

$$\pi^{-1}_{\mathfrak{V}}\, \pi^{\#}_{\mathfrak{V}} G \;\bar{\delta}\; \pi^{-1}_{\mathfrak{V}}\, \pi^{\#}_{\mathfrak{V}} H .$$

From this, since the sets $\pi^{-1}_{\mathfrak{V}}\, \pi^{\#}_{\mathfrak{V}} G$ and $\pi^{-1}_{\mathfrak{V}}\, \pi^{\#}_{\mathfrak{V}} H$ are dense in G and H, respectively, we obtain $G \,\bar{\delta}\, H$, hence a fortiori $A \,\bar{\delta}\, B$.

Now let us show that the family $\widetilde{\mathfrak{V}}$ satisfies the axioms of a ϑ-uniformity. Axiom $\mathrm{I_U}$ is implied by axiom $\mathrm{I_U}$ for $\mathfrak{V}$ and lemma 1. Axiom $\mathrm{II_U}$ follows from $\mathrm{II_U}$ for $\mathfrak{V}$ and the following fact which is easily verifiable: if $w \ast\!\ast\!> \nu$ then $\widetilde{w} \ast\!> \widetilde{\nu}$. Axioms $\mathrm{III_U}$ and $\mathrm{IV_U}$ follow immediately from the fact that the family $\widetilde{\mathfrak{V}}$ generates the proximity δ.

Thus $\widetilde{\mathfrak{V}}$ is a ϑ-uniformity which generates a proximity, i ence in view of proposition 3, is a uniformity.

It remains to show that the uniform space $(X_{\mathfrak{V}}, \widetilde{\mathfrak{V}})$ and the mapping $\pi_{\mathfrak{V}}$ are unique. Indeed, let $(Y, \mathfrak{W})$ be a uniform space and π: $(Y, \mathfrak{W}) \to (X, \mathfrak{V})$ be a ϑ-perfect irreducible mapping so that $\mathfrak{V} = \pi^{\#}\mathfrak{W}$.

According to theorem 1 we have $\delta_{\mathfrak{B}} = \pi\delta_{\mathfrak{W}}$. Then the uniqueness of a proximity space generating a given ϑ-proximity space (cf. theorem 3 of [8]) implies $Y = X_{\mathfrak{B}}$ and $\pi = \pi_{\mathfrak{B}}$. However, on the space $X_{\mathfrak{B}}$ obviously there is only one such uniformity $\mathfrak{W}$, for which $\pi^{\#}\mathfrak{W} = \mathfrak{B}$, namely $\mathfrak{W} = \widetilde{\mathfrak{B}}$. This completes the proof of theorem 2.

Corollary 1. *There exists a one-to-one correspondence between the set of all ϑ-uniformities on a space X and the set of all pairs $(Z, f\colon Z \to X)$, where Z is a uniform space and f is a ϑ-perfect irreducible mapping.**

Corollary 2. *The set of all ϑ-uniformities on a compactum can be put into a one-to-one correspondence with the set of all perfect irreducible mappings onto this compactum.*

§2. MAPPINGS OF ϑ-UNIFORM SPACES

Let $(X, \mathfrak{B})$ and $(Y, \mathfrak{W})$ be two ϑ-uniform spaces. A mapping $f\colon (X, \mathfrak{B}) \to (Y, \mathfrak{W})$ is called ϑ-uniformly continuous if the following two condition are satisfied:

1°. For every ϑ-cover $w \in \mathfrak{W}$ there exists a ϑ-cover $v \in \mathfrak{B}$, for which the cover $\{f[V] \mid V \in v\}$ refines the cover $\{[W] \mid W \in \mathfrak{W}\}$.

2°. If $w \in \mathfrak{W}$, then the family $\{\langle[f^{-1}W]\rangle \mid W \in w\}$ is a uniform ϑ-cover of the space $(X, \mathfrak{B})$.

Lemma 2. *Every ϑ-uniformly continuous mapping is ϑ-continuous.*

Proof. Let $f\colon (X, \mathfrak{B}) \to (Y, \mathfrak{W})$ be a ϑ-uniformly continuous mapping, $x \in X$ and Ofx be an arbitrary neighbourhood of the point fx. According to IV_{U} there are a neighbourhood U of the point fx and a ϑ-cover $w \in \mathfrak{W}$ so that $\mathrm{st}_w U \subset [Ofx]$. Let v be a ϑ-cover from $\mathfrak{B}$ satisfying condition 1° of the ϑ-uniform continuity. Then there are a finite number of elements $V_1, \ldots, V_n$ of the ϑ-cover v in such a way

*Here two pairs $(Z, f\colon Z \to X)$ and $(Z', f'\colon Z \to X)$ are identified if there exists an isomorphism $g\colon Z \to Z'$ such that $f = f'g$.

that the set $G = \left\langle \bigcup_{i=1}^{n} [V_i] \right\rangle$ forms a neighbourhood of the point x.

Moreover, we may assume that $x \in [V_i]$, $i = 1, \ldots, n$. Then $f[G] \subset [Ofx]$. Indeed, $f[G] = \bigcup_{i=1}^{n} f[V_i]$, hence $x \in [V_i]$ implies $W \subset \subset \mathrm{st}_w U$ for each $W \in w$ with $f[V_i] \subset W$. Consequently $f[G] \subset \mathrm{st}_{[w]} U \subset \subset [Ofx]$, and the lemma is proved.

Proposition 4. *Every ϑ-uniformly continuous mapping $f: (X, \mathfrak{B}) \to \to (Y, \mathfrak{W})$ is a ϑ-proximately continuous mapping of the space $(X, \delta_{\mathfrak{B}})$ into the space $(Y, \delta_{\mathfrak{W}})$.*

Proof. Let $A, B \subset Y$ and $A \, \bar{\delta}_{\mathfrak{W}} \, B$. Then there exist neighbourhoods OA, OB and a ϑ-cover $w \in \mathfrak{W}$ such that $OA \cap \mathrm{st}_w OB = \phi$. In view of the ϑ-continuity of the mapping f, the set $\langle f^{-1}[OA] \rangle$ and $\langle f^{-1}[OB] \rangle$, are neighbourhoods of the sets $f^{-1}A$ and $f^{-1}B$, respectively. According to condition 2° we have

$$\widetilde{w} = \left\{ \langle [f^{-1} W] \rangle \mid W \in w \right\} \in \mathfrak{B} \, ,$$

moreover

$$\langle f^{-1}[OA] \rangle \cap \mathrm{st}_{\widetilde{w}} \langle f^{-1}[OB] \rangle = \phi \, .$$

Indeed, if $W \in w$ and $W \cap OA = \phi$, then $f^{-1} W \cap f^{-1}[OA] = \phi$, hence $[f^{-1} W] \cap \langle f^{-1}[OA] \rangle = \phi$, and a fortiori

$$\langle [f^{-1} W] \rangle \cap \langle f^{-1}[OA] \rangle = \phi \, .$$

Thus we obtain $f^{-1}A \, \bar{\delta}_{\mathfrak{B}} \, f^{-1}B$.

In the same way we can prove $\langle f^{-1}[A] \rangle \, \bar{\delta}_{\mathfrak{B}} \, \langle f^{-1}[B] \rangle$ as well, by noticing only that $\langle f^{-1}[A] \rangle \subset \langle f^{-1}[OA] \rangle$ and $\langle f^{-1}[B] \rangle \subset \langle f^{-1}[OB] \rangle$. Thus proposition 4 is proven.

Remark. It can be shown by easy examples that the composition of two ϑ-uniformly continuous mappings need not be ϑ-uniformly continuous. A mapping which is the composition of a finite number of ϑ-uniformly continuous mappings, will be called uniformly ϑ-continuous. Now, the composition of two uniformly ϑ-continuous mappings is obviously uniformly ϑ-continuous. Therefore the ϑ-uniform spaces and the uniformly ϑ-continuous mappings constitute a category, of which the category of uniform spaces is a complete subcategory.

Theorem 3. *Let* $f\colon (X, \mathfrak{B}) \to (Y, \mathfrak{W})$ *be a continuous and* ϑ*-uniformly continuous mapping. Then there exists a uniformly continuous mapping* $\tilde{f}\colon (X_{\mathfrak{B}}, \widetilde{\mathfrak{B}}) \to (Y_{\mathfrak{W}}, \widetilde{\mathfrak{W}})$, *which makes the following diagram* (1) *commutative.*

$$(1)\qquad \begin{array}{ccc} X_{\mathfrak{B}} & \xrightarrow{\ \tilde{f}\ } & Y_{\mathfrak{W}} \\ {\scriptstyle \pi_{\mathfrak{B}}}\big\downarrow & & \big\downarrow{\scriptstyle \pi_{\mathfrak{W}}} \\ X & \xrightarrow{\ f\ } & Y \end{array}$$

Proof. According to proposition 4 the mapping $f\colon (X, \delta_{\mathfrak{B}}) \to$ $\to (Y, \delta_{\mathfrak{W}})$ is ϑ-proximately continuous. By theorem 4 of [8] there exists a proximately continuous mapping $\tilde{f}$ of the space $X_{\delta_{\mathfrak{B}}} = X_{\mathfrak{B}}$ into the space $Y_{\delta_{\mathfrak{W}}} = Y_{\mathfrak{W}}$ which makes the corresponding diagram commutative (the assumption of this theorem that f be a mapping onto Y has in fact not been employed in its proof). Therefore, to prove theorem 3 it will suffice to show that the mapping $\tilde{f}$ is uniformly continuous.

Let w be a uniform cover of the space $Y_{\mathfrak{W}}$. Then $\pi^{\#}_{\mathfrak{W}} w \in \mathfrak{W}$. According to condition 2° of the ϑ-uniform continuity we have $\widetilde{\pi^{\#}_{\mathfrak{W}} w} \in$ $\in \mathfrak{W}$. Finally there exists a uniform cover $v \in \widetilde{\mathfrak{B}}$ such that $\pi^{\#}_{\mathfrak{B}} v = \widetilde{\pi^{\#}_{\mathfrak{W}} w}$. We shall show that $v > \tilde{f}^{-1}[w]$. For an arbitrary $V \in v$ there exists an element $W \in w$ so that $\pi^{\#}_{\mathfrak{B}} V = \langle [f^{-1} \pi^{\#}_{\mathfrak{W}} W] \rangle$. Let $x \in V$ and assume that $\tilde{f}x \notin [W]$. Then there exists a neighbourhood $O\tilde{f}x$ which does not intersect W. Then

$$\pi_{\mathfrak{W}} O\tilde{f}x \cap \pi^{\#}_{\mathfrak{W}} W = \phi\,,$$

and consequently

$$\pi_{\mathfrak{B}}^{-1} f^{-1} \pi_{\mathfrak{W}} O\tilde{f}x \cap \pi_{\mathfrak{B}}^{-1} f^{-1} \pi^{\#}_{\mathfrak{W}} W = \phi\,.$$

But the set

$$\pi_{\mathfrak{B}}^{-1} f^{-1} \pi_{\mathfrak{W}} O\tilde{f}x = \tilde{f}^{-1} \pi_{\mathfrak{W}}^{-1} \pi_{\mathfrak{W}} O\tilde{f}x \supset \tilde{f}^{-1} O\tilde{f}x$$

contains a neighbourhood Ox of the point x, because the mapping $\widetilde{f}$ is continuous. Therefore, $Ox \cap \pi_{\mathfrak{V}}^{-1} f^{-1} \pi_{\mathfrak{W}}^{\#} W = \phi$, and thus $\pi_{\mathfrak{V}}^{\#} Ox \cap \cap f^{-1} \pi_{\mathfrak{W}}^{\#} W = \phi$, which means that $\pi_{\mathfrak{V}}^{\#} Ox \cap \langle [f^{-1} \pi_{\mathfrak{W}}^{\#} W] \rangle = \phi$. But $\langle [f^{-1} \pi_{\mathfrak{W}}^{\#} W] \rangle = \pi_{\mathfrak{V}}^{\#} V$, consequently $\pi_{\mathfrak{V}}^{\#} Ox \cap \pi_{\mathfrak{V}}^{\#} V = \phi$. However, this contradicts $x \in V$. Thus we have $\widetilde{f} V \subset [W]$ and $v > f^{-1}[w]$. This implies the uniform continuity of the mapping $\widetilde{f}$, since for every uniform cover of the space $Y_{\mathfrak{W}}$ there is another uniform cover, whose "closure" refines it. The proof of theorem 3 is completed.

Theorem 4. *Let* $f\colon (X, \mathfrak{V}) \to (Y, \mathfrak{W})$ *be a* ϑ*-uniformly continuous mapping and assume that* fX *is dense in* Y. *Furthermore let us suppose that the mapping* $f\colon X \to fX$ *is closed* and irreducible. Then there exists a uniformly continuous mapping*

$$\widetilde{f}\colon (X_{\mathfrak{V}}, \widetilde{\mathfrak{V}}) \to (Y_{\mathfrak{W}}, \widetilde{\mathfrak{W}})$$

such that $\pi_{\mathfrak{W}} \widetilde{f} = f \pi_{\mathfrak{V}}$.

The proof of this repeats that of the previous theorem with the exception of its beginning in which theorem 1 of [10] must be applied.

The mapping $f\colon (X, \mathfrak{V}) \to (Y, \mathfrak{W})$ will be called ϑ-uniformly regular if it is ϑ-uniformly continuous and for every ϑ-cover $v \in \mathfrak{V}$ there exists a ϑ-cover $w \in \mathfrak{W}$, which is a refinement of the system $f^{\#} v$. We shall say that a ϑ-uniformly regular mapping is a uniformly regular ϑ-mapping if it is ϑ-perfect and irreducible.

As a corollary of theorem 2 we obtain

Proposition 5. *Let* $(X, \mathfrak{V})$ *be a* ϑ*-uniform space. Then the mapping* $\pi_{\mathfrak{V}}\colon (X_{\mathfrak{V}}, \widetilde{\mathfrak{V}}) \to (X, \mathfrak{V})$ *is a uniformly regular* ϑ*-mapping.*

We recall that (cf. [8]) a ϑ-perfect, irreducible and ϑ-proximately continuous mapping $f\colon (X, \delta) \to (Y, \delta')$ is called a regular ϑ-mapping if the inverse images of two sets near in Y are themselves near in X.

Now we have the following obvious

Proposition 6. *If the mapping* $f\colon (X, \mathfrak{V}) \to (Y, \mathfrak{W})$ is a uniformly

*Here the closedness of the mapping f does not presuppose its continuity.

regular ϑ*-mapping, then* $f\colon (X, \delta_{\mathfrak{B}}) \to (Y, \delta_{\mathfrak{W}})$ *is a regular* ϑ*-mapping.*

Theorem 5. *The uniform spaces are projective objects with respect to the uniformly regular* ϑ*-mappings.*

Proof. We have to show that if $f_i\colon (X^i, \mathfrak{B}_i) \to (Y, \mathfrak{W})$, $i = 1, 2$, are uniformly regular ϑ-mappings and $(X^1, \mathfrak{B}_1)$ is a uniform space, then there exists a uniformly regular mapping $g\colon (X^1, \mathfrak{B}_1) \to (X^2, \mathfrak{B}_2)$ such that $f_1 = f_2 g$. By virtue of proposition 6, the mapping $f_2\colon (X^2, \delta_{\mathfrak{B}_2}) \to$ $\to (Y, \delta_{\mathfrak{W}})$ is a regular ϑ-mapping. Thus by theorem 7 of [8] the mapping $\widetilde{f}_2$ in the commutative diagram

$$
\begin{array}{ccc}
X^2_{\mathfrak{B}_2} & \xrightarrow{\ \widetilde{f}_2\ } & Y_{\mathfrak{W}} \\
{\scriptstyle \pi_{\mathfrak{B}_2}}\big\downarrow & & \big\downarrow{\scriptstyle \pi_{\mathfrak{W}}} \\
X^2 & \xrightarrow{\ \ f\ \ } & Y
\end{array}
\tag{2}
$$

is a homeomorphism. Moreover, from the definition of ϑ-uniformly regular mappings and lemma 1 it follows that $\pi^{\#}_{\mathfrak{W}} \widetilde{\mathfrak{W}} = \mathfrak{W}$. Therefore, from theorem 2 we obtain $(Y_{\mathfrak{W}}, \widetilde{\mathfrak{W}}) = (X^1, \mathfrak{B}_1)$ and $\pi_{\mathfrak{W}} = f_1$. Furthermore the ϑ-uniformly continuous homeomorphism $\widetilde{f}_2$ is a uniformly regular ϑ-mapping since in diagram (2) all the remaining mappings are uniformly regular ϑ-mappings ($\pi_{\mathfrak{B}_2}$ is so in view of proposition 5.) But then the mapping $\widetilde{f}_2$ is an isomorphism, for the uniform regularity of a one-to-one mapping implies the uniform continuity of its inverse. Now it remains to put $g = \pi_{\mathfrak{B}_2} \widetilde{f}_2^{-1}$, and the proof of theorem 5 is completed.

§3. PRODUCTS OF ϑ-UNIFORM SPACES

Let X_i, $i = 1, \ldots, k$ be finitely many topological spaces and ν_i be a ϑ-cover of locally finite type of the space X_i. Consider the system

$$\nu_1 \times \nu_2 \times \ldots \times \nu_k = \{V_1 \times V_2 \times \ldots \times V_k \mid V_i \in \nu_i\}.$$

Lemma 3. *The system* $\nu = \nu_1 \times \ldots \times \nu_k$ *is a* ϑ*-cover of locally finite type of the space* $X = X_1 \times \ldots \times X_k$.

Proof. Let us remark first that the members of the system ν, as products of canonical open sets, are themselves canonical open. Let $x = \{x_i\}$ be an arbitrary point of the space X. As ν_i is a ϑ-cover of locally finite type, there exist neighbourhoods U_i of the points $x_i \in X_i$ and finite subcollections $\{V_i^j \mid j = 1, \dots, l_i\}$ of the ϑ-covers ν_i so that

$$U_i = \bigcup_{j=1}^{l_i} [V_i^j] .$$

It is easy to see then that the neighbourhood $U = \prod_{i=1}^{k} U_i$ of the point x is contained in the union of the finite collection $\{[V_1^{j_1}] \times \dots \times [V_k^{j_k}] \mid j_i = 1, \dots, l_i\}$, whose members are closures of elements of the system ν. This proves the lemma.

Now let $\{(X_i, \mathfrak{V}_i) \mid i \in I\}$ be an arbitrary family of ϑ-uniform spaces. Let $\{i_1, \dots, i_k\}$ be any finite set of indices from the set I, and ν_{i_j} be some ϑ-covers from the ϑ-uniformities $\mathfrak{V}_{i_j}$, $j = 1, \dots, k$. It follows from lemma 3 that the system

$$(\nu_{i_1} \nu_{i_2} \dots \nu_{i_k}) = \{V_{i_1} \times \dots \times V_{i_k} \times \prod_{i \neq i_j} X_i \mid V_{i_j} \in \nu_{i_j}\}$$

is a ϑ-cover of the space $X = \prod_{i \in I} X_i$ of locally finite type.

It is easy to verify that the system $\mathfrak{V}_0 = \{(\nu_{i_1} \dots \nu_{i_k})\}$, where $\{i_1, \dots, i_k\}$ runs through all finite subsets of the set I and ν_{i_j} is an arbitrary element of the ϑ-uniformity $\mathfrak{V}_{i_j}$, satisfies axioms $\mathrm{II_U}$, $\mathrm{III_U}$ and $\mathrm{IV_U}$ of ϑ-uniformities, hence it forms a basis of some ϑ-uniformity $\mathfrak{V}$. We shall call this ϑ-uniformity $\mathfrak{V}$ the product of the ϑ-uniformities $\mathfrak{V}_i$. and the space $X = \prod_{i \in I} X_i$ provided with the ϑ-uniformity $\mathfrak{V}$ we shall call the product of the ϑ-uniform spaces $(X_i, \mathfrak{V}_i)$. It is obvious that every projection $\pi_i \colon (X, \mathfrak{V}) \to (X_i, \mathfrak{V}_i)$ is a ϑ-uniformly continuous mapping.

Proposition 7. *The product $\mathfrak{V}$ of the ϑ-uniformities $\mathfrak{V}_i$ is the coarsest of all ϑ-uniformities $\mathfrak{V}'$ on the space $X = \prod_i X_i$ for which all the projections $\pi_i \colon (X, \mathfrak{V}') \to (X_i, \mathfrak{V}_i)$ are ϑ-uniformly continuous.*

Proof. For any ϑ-cover $\nu_i \in \mathfrak{B}_i$ there exists a ϑ-cover $\nu' \in \mathfrak{B}'$ such that $[\nu'] > \pi_i^{-1}[\nu_i]^*$. Consequently all ϑ-covers of the form $(\nu_i) = \pi_i^{-1}\nu_i$, where $\nu_i \in \mathfrak{B}_i$, belong to the ϑ-uniformity $\mathfrak{B}'$. Then the ϑ-uniformity $\mathfrak{B}'$ must contain all finite intersections of the ϑ-covers (ν_i), i.e. all ϑ-covers of the form $(\nu_{i_1} \ldots \nu_{i_k})$. Thus the ϑ-uniformity $\mathfrak{B}'$ includes a basis of the ϑ-uniformity $\mathfrak{B}$, i.e. $\mathfrak{B}' \geqslant \mathfrak{B}$. Proposition 7 is proven.

Let again $\{(X_i, \mathfrak{B}_i) \mid i \in I\}$ be a family of ϑ-uniform spaces. According to theorem 2 there exist uniform spaces $(X_{\mathfrak{B}_i}, \widetilde{\mathfrak{B}}_i)$ and ϑ-perfect irreducible mappings $\pi_{\mathfrak{B}_i}\colon X_{\mathfrak{B}_i} \to X_i$ such that $\mathfrak{B}_i = \pi_{\mathfrak{B}}^{\#}\widetilde{\mathfrak{B}}_i$, $i \in I$. It is easy to verify that the product mapping

$$\prod_i \pi_{\mathfrak{B}_i}\colon \prod_i X_{\mathfrak{B}_i} \to \prod_i X_i$$

is also ϑ-perfect and irreducible. In addition, for any finite choice $\nu_{i_1}, \ldots, \nu_{i_k}$ of canonical uniform covers** of the spaces $(X_{\mathfrak{B}_{i_1}}, \widetilde{\mathfrak{B}}_{i_1}), \ldots, (X_{\mathfrak{B}_{i_k}}, \widetilde{\mathfrak{B}}_{i_k})$ we have

$$\pi^{\#}(\nu_{i_1} \ldots \nu_{i_k}) = (\pi^{\#}_{\mathfrak{B}_{i_1}} \nu_{i_1}, \ldots, \pi^{\#}_{\mathfrak{B}_{i_k}} \nu_{i_k}),$$

where $\pi = \prod_i \pi_{\mathfrak{B}_i}$. Consequently the operator $\pi^{\#}$ transforms the basis $(\nu_{i_1} \ldots \nu_{i_k})$ of the uniformity $\prod_i \widetilde{\mathfrak{B}}_i$ into a basis of the ϑ-uniformity $\prod_i \mathfrak{B}_i$ on the space $X = \prod_i X_i$. Thus we have a proof of

Theorem 6. *The mapping*

$$\pi_{\prod_i \mathfrak{B}_i}\colon \left(X_{\prod_i \mathfrak{B}_i}, \prod_i \widetilde{\mathfrak{B}}_i\right) \to \left(X, \prod_i \mathfrak{B}_i\right)$$

generating the ϑ-uniformity $\prod_i \mathfrak{B}_i$ coincides with the mapping

$$\prod_i \pi_{\mathfrak{B}_i}\colon \left(\prod_i X_{\mathfrak{B}_i}, \prod_i \widetilde{\mathfrak{B}}_i\right) \to \left(X, \prod_i \mathfrak{B}_i\right)$$

*If ω is a system of subsets of a space Y and $f\colon X \to Y$ is a mapping then we put $[\omega] = \{[0] \mid 0 \in \omega\}$ and $f^{-1}\omega = \{f^{-1}0 \mid 0 \in \omega\}$.

**i.e. uniform covers, whose members are canonical open sets.

§4. COMPLETENESS OF ϑ-UNIFORM SPACES

A filter $\mathscr{F}$ on a ϑ-uniform space $(X, \mathfrak{B})$ is called a Cauchy filter if every uniform ϑ-cover $\nu \in \mathfrak{B}$ containes a set V which belongs to $\mathscr{F}$. It follows easily from the axioms of ϑ-uniformity that every Cauchy filter has no more than one δ-limit point.* If in a ϑ-uniform space $(X, \mathfrak{B})$ every Cauchy filter has a δ-limit point then we say that the space $(X, \mathfrak{B})$ is complete.

Proposition 8. *Every Cauchy filter* $\mathscr{F}$ *on the space* $(X, \mathfrak{B})$ *contains a unique minimal (by inclusion) Cauchy filter* $\mathscr{F}_0$. *The canonical open members of this filter* $\mathscr{F}_0$ *constitute a basis for it.*

Proof. Let us consider the family

$$\widetilde{\mathscr{F}} = \{\langle[\mathrm{st}_\nu\, M]\rangle \mid \nu \in \mathfrak{B},\ M \in \mathscr{F}\}.$$

This family $\widetilde{\mathscr{F}}$ is obviously centered, hence it forms the basis of a filter, which we shall denote by $\mathscr{F}_0$.

Now we show that if $\mathscr{F}_1$ is any Cauchy filter contained in $\mathscr{F}$, then we must have $\mathscr{F}_0 \subset \mathscr{F}_1$. By the definition of Cauchy filters, at least one member of any ϑ-cover ν, intersecting a given set $M \in \mathscr{F}$, belongs to the filter $\mathscr{F}_1$. Then we have $\langle[\mathrm{st}_\nu\, M]\rangle \supset \mathrm{st}_\nu M \in \mathscr{F}_1$, consequently $\langle[\mathrm{st}_\nu\, M]\rangle \in \mathscr{F}_1$. Thus the filter $\mathscr{F}_0$ is indeed contained in $\mathscr{F}_1$.

Next we show that $\mathscr{F}_0$ is a Cauchy filter. For any ϑ-cover $\nu \in \mathfrak{B}$ there exists a ϑ-cover $w \in \mathfrak{B}$, which is a double star refinement of ν. Since $\mathscr{F}$ is a Cauchy filter, there exists a set $M \in w$ belonging to $\mathscr{F}$. There exists a member V of the ϑ-cover ν such that $\mathrm{st}_w\, M \subset V$. Since the set V is canonical open and $\mathrm{st}_w\, M$ is open, we have $\langle[\mathrm{st}_w\, M]\rangle \subset V$, consequently $V \in \mathscr{F}_0$. Thus in an arbitrary ϑ-cover $\nu \in \mathfrak{B}$ we have found an element V belonging to the filter $\mathscr{F}_0$. This shows that $\mathscr{F}_0$ is indeed a Cauchy filter.

From the above proved inclusion relation $\mathscr{F}_0 \subset \mathscr{F}_1$, where $\mathscr{F}_1$ is an arbitrary Cauchy filter contained in $\mathscr{F}$, the minimality and unique-

*The point x is called a δ-limit point of the filter $\mathscr{F}$ if for each $M \in \mathscr{F}$ the point x is a δ-limit point of the set M (cf. N.V. Veličko [3]), i.e. $Ox \cap M \neq \phi$ for an arbitrary canonical open neighbourhood of the point x.

ness of the filter $\mathscr{F}_0$ follows immediately. The proof of proposition 8 is completed.

Proposition 9. *In order that the point* x *be a* δ-*limit point of the Cauchy filter* $\mathscr{F}$ *it is necessary and sufficient that the point* x *be a limit point of the minimal Cauchy filter* $\mathscr{F}_0$ *contained in* $\mathscr{F}$.

Proof. Necessity. Let Ox be an arbitrary neighbourhood of the point x and let M be an arbitrary element of the filter $\mathscr{F}_0$. There exists a (canonical) open member M' of the filter $\mathscr{F}_0$ which is contained in M. But $\mathscr{F}_0 \subset \mathscr{F}$, hence $M' \in \mathscr{F}$, too. Since by our assumption x is a δ-limit point of the filter $\mathscr{F}$, we have $\langle[Ox]\rangle \cap M' \neq \phi$, and a fortiori $[Ox] \cap M' \neq \phi$. Since the set M' is open, this implies $Ox \cap M' \neq \phi$ and consequently $Ox \cap M \neq \phi$. This means that the point x is indeed a limit point of the filter $\mathscr{F}_0$.

Sufficiency. Let Ox be an arbitrary canonical open neighbourhood of the point x and M be an arbitrary element of the filter $\mathscr{F}$. There exist a neighbourhood $O'x$ of the point x and a uniform ϑ-cover v such that $\text{st}_v O'x \subset Ox$. The filter $\mathscr{F}_0$ is a Cauchy filter, hence there exists an element V of the ϑ-cover v, which belongs to the filter $\mathscr{F}_0$. Therefore, $V \subset \text{st}_v O'x \subset Ox$, hence $Ox \in \mathscr{F}_0 \subset \mathscr{F}$, and consequently the sets Ox and M, as members of the filter $\mathscr{F}$, have a non-empty intersection. This shows that the point x is a δ-limit point of the filter $\mathscr{F}$, hence proposition 9 is proven.

Corollary. *For the completeness of a* ϑ-*uniform space* $(X, \mathfrak{B})$ *it is both necessary and sufficient that every minimal Cauchy filter in it had a* δ-*limit point.*

Theorem 7. *Let* $(X, \mathfrak{B})$ *be a* ϑ-*uniform space. Then it is complete if and only if the uniform space* $(X_{\mathfrak{B}}, \widetilde{\mathfrak{B}})$ *is complete.*

Proof. Let us suppose first that the ϑ-uniform space $(X, \mathfrak{B})$ is complete. We have to show that every minimal Cauchy filter $\mathscr{F}$ in the space $(X_{\mathfrak{B}}, \widetilde{\mathfrak{B}})$ has a δ-limit point. Consider the family

$$\pi_{\mathfrak{B}}^{\#} \mathscr{F} = \left\{ \pi_{\mathfrak{B}}^{\#} M \mid M \in \mathscr{F} \right\}$$

of subsets of X. The family $\pi_{\mathfrak{B}}^{\#} \mathscr{F}$ is a filter*. Indeed, since the mapping $\pi_{\mathfrak{B}}$ is irreducible and the filter $\mathscr{F}$ has a basis consisting of open sets, we have $\pi^{\#}_{\mathfrak{B}} M \neq \phi$ for all $M \in \mathscr{F}$. Moreover, from the equality

$$\pi^{\#}_{\mathfrak{B}}(M_1 \cap M_2) = \pi^{\#}_{\mathfrak{B}} M_1 \cap \pi^{\#}_{\mathfrak{B}} M_2 \,,$$

which is valid for an arbitrary mapping, the multiplicativity of the family $\pi^{\#}_{\mathfrak{B}} \mathscr{F}$ follows. Finally let us have $\pi^{\#}_{\mathfrak{B}} M \subset A$ for a certain set $M \in \mathscr{F}$. Then $M \cup \pi^{-1}_{\mathfrak{B}} A \in \mathscr{F}$ and

$$\pi^{\#}_{\mathfrak{B}}(M \cup \pi^{-1}_{\mathfrak{B}} A) = \pi^{\#}_{\mathfrak{B}} M \cup A = A \in \pi^{\#}_{\mathfrak{B}} \mathscr{F} \,.$$

Now we show that $\pi^{\#}_{\mathfrak{B}} \mathscr{F}$ is a Cauchy filter in the space $(X, \mathfrak{B})$. Let ν be an arbitrary uniform ϑ-cover of the space $(X, \mathfrak{B})$. By theorem 2, there exists a uniform cover w of the space $(X_{\mathfrak{B}}, \widetilde{\mathfrak{B}})$ such that $\nu = \pi^{\#}_{\mathfrak{B}} w$. Some member $W \in w$ must then belong to the Cauchy filter $\mathscr{F}$, and thus the set $\pi^{\#}_{\mathfrak{B}} W$ belongs both to the filter $\pi^{\#}_{\mathfrak{B}} \mathscr{F}$ and the ϑ-cover $\pi^{\#}_{\mathfrak{B}} w = \nu$. This means that $\pi^{\#}_{\mathfrak{B}} \mathscr{F}$ is a Cauchy filter (and even a minimal Cauchy filter).

Let x be a δ-limit point of the filter $\pi^{\#}_{\mathfrak{B}} \mathscr{F}$. Then every canonical open neighbourhood Ox of the point x intersects any set $\pi^{\#}_{\mathfrak{B}} M \in \pi^{\#}_{\mathfrak{B}} \mathscr{F}$. Since the mapping $\pi_{\mathfrak{B}}$ is closed and ϑ-continuous, this implies that an arbitrary neighbourhood of the compact set $\pi^{-1}_{\mathfrak{B}} x$ intersects any set $M \in \mathscr{F}$. This means that we have $\pi^{-1}_{\mathfrak{B}} x \cap [M] \neq \phi$ for all $M \in \mathscr{F}$. As $[M_1 \cap M_2] \subset [M_1] \cap [M_2]$, the system $\{\pi^{-1}_{\mathfrak{B}} x \cap [M] \,|\, M \in \mathscr{F}\}$ is a centered system of closed subsets of the compactum $\pi^{-1}_{\mathfrak{B}} x$. Thus the point belonging to all the sets $\pi^{-1}_{\mathfrak{B}} x \cap [M]$ will be the (unique) limit point of the Cauchy filter $\mathscr{F}$.

Sufficiency. Now let $(X_{\mathfrak{B}}, \widetilde{\mathfrak{B}})$ be a complete space and $\mathscr{F}$ be a minimal Cauchy filter on the space $(X, \mathfrak{B})$. The system $\pi^{-1}_{\mathfrak{B}} \mathscr{F} = \{\pi^{-1}_{\mathfrak{B}} M \,|\, M \in \mathscr{F}\}$ serves as a basis of some filter which we shall denote by $\widetilde{\mathscr{F}}$. Then this filter $\widetilde{\mathscr{F}}$ is a Cauchy filter on the space $(X_{\mathfrak{B}}, \widetilde{\mathfrak{B}})$. Indeed, let w be a uniform cover of the space $(X_{\mathfrak{B}}, \widetilde{\mathfrak{B}})$. Then $\pi^{\#}_{\mathfrak{B}} w$ is a uniform ϑ-cover of the space $(X, \mathfrak{B})$. Some member $\pi^{\#} W$ of this ϑ-

*If $\mathscr{F}$ is a filter on the set X and $f\colon X \to Y$ is a mapping onto the set Y such that $f^{\#} M \neq \phi$ for all $M \in \mathscr{F}$, then $f^{\#} \mathscr{F}$ is a filter.

cover belongs to the Cauchy filter $\mathscr{F}$. Then the set $W \supset \pi_{\mathfrak{V}}^{-1} \pi_{\mathfrak{V}}^{\#} W$ is an element of the filter $\widetilde{\mathscr{F}}$, hence $\widetilde{\mathscr{F}}$ is a Cauchy filter.

Let x be a limit point of the filter $\widetilde{\mathscr{F}}$. Then the point $\pi_{\mathfrak{V}} x$ is a δ-limit point of the filter $\mathscr{F}$. Indeed, let $O\pi_{\mathfrak{V}} x$ be an arbitrary canonical open neighbourhood of the point $\pi_{\mathfrak{V}} x$. As the mapping $\pi_{\mathfrak{V}}$ is closed and ϑ-continuous, there exists a neighbourhood U of the compactum $\pi_{\mathfrak{V}}^{-1} \pi_{\mathfrak{V}} x$ so that $\pi_{\mathfrak{V}}^{\#} U \subset O\pi_{\mathfrak{V}} x$.

Since x is a limit point of the filter $\widetilde{\mathscr{F}}$, the intersection $U \cap \cap \pi_{\mathfrak{V}}^{-1} M$ is non-empty for any $M \in \mathscr{F}$, and so is the set

$$\pi_{\mathfrak{V}}(U \cap \pi_{\mathfrak{V}}^{-1} M) = \pi_{\mathfrak{V}} U \cap M .$$

However the irreducibility of the mapping $\pi_{\mathfrak{V}}$ and the inclusion $\pi_{\mathfrak{V}}^{\#} U \subset \subset O\pi_{\mathfrak{V}} x$ imply $\pi_{\mathfrak{V}} U \subset [O\pi_{\mathfrak{V}} x]$, and thus the closure of any canonical open neighbourhood of the point $\pi_{\mathfrak{V}} x$ intersects every element of the filter $\mathscr{F}$. Since the (minimal) Cauchy filter $\mathscr{F}$ has a basis consisting of open sets, the point $\pi_{\mathfrak{V}} x$ is indeed a δ-limit point of the filter $\mathscr{F}$. The proof of theorem 7 is completed.

If X is a dense subset of the ϑ-uniform space $(Y, \mathfrak{V})$, then the family $\mathfrak{V}_X = \{ \nu_X \mid \nu \in \mathfrak{V} \}$, where $\nu_X = \{ V \cap X \mid V \in \nu \}$, forms a ϑ-uniformity on X. In this case the space $(Y, \mathfrak{V})$ is called an extension of the space $(X, \mathfrak{V}_X)$. Obviously, we have $V = \langle [V \cap X]_Y \rangle_Y$, if V is a canonical open subset of Y and X is dense in Y.

An extension of a ϑ-uniform space, which is complete we shall call a complete extension.

Proposition 10. *Let $(Y, \mathfrak{W})$ be an extension of the ϑ-uniform space $(X, \mathfrak{V})$ and let Z be an extension of the space X. If there exists a ϑ-continuous compact mapping $f\colon Y \to Z$ onto Z, which is the identity on X and which carries canonical closed sets into closed sets, then there is a (unique) ϑ-uniformity $\mathfrak{W}_0$ on the space Z such that the pair $(Z, \mathfrak{W}_0)$ is an extension of the space $(X, \mathfrak{V})$. If in addition the extension $(Y, \mathfrak{W})$ is complete, so is the extension $(Z, \mathfrak{W}_0)$.*

Proof. For an arbitrary canonical open set $W \subset Y$ we put $W^* = \langle [f^{\#} W] \rangle$. Let us define $\mathfrak{W}_0 = \mathfrak{W}^* = \{ w^* \mid w \in \mathfrak{A} \}$, where $w^* =$

$= \{W^* \mid W \in w\}$. The system w^* consists of canonical open sets. With obvious alterations in the proof of the lemma 1, we obtain that w^* is a ϑ-cover of locally finite type.

Now we show that $w^*_X = w_X$, i.e. that $W \cap X = \langle [f^\# W] \rangle \cap X$ for all $W \in w$. Since W is a canonical open set, $f^\# W$ is open, because the mapping f carries canonical closed sets into closed sets. Consequently we have $f^\# W \subset \langle [f^\# W] \rangle$, and thus $X \cap f^\# W \subset X \cap \langle [f^\# W] \rangle$. But $X \cap f^\# W = W \cap X$, hence $W_X \subset W^*_X$. On the other hand, the set W_X is obviously dense in the open set W^*_X, hence, as W_X is canonical open, we obtain $W_X = W^*_X$.

Consequently, the system $\mathfrak{W}^*$ consisting of covers of Z of locally finite type is an expansion of the ϑ-uniformity $\mathfrak{V} = \mathfrak{W}_X$ from the everywhere dense subspace X of Z. Therefore the system $\mathfrak{W}^*$ automatically satisfies axioms $\mathrm{I_U}$ and $\mathrm{II_U}$ of ϑ-uniformities. Axioms $\mathrm{III_U}$ and $\mathrm{IV_U}$ can be checked in the same way as in the proof of theorem 1. The uniqueness of the ϑ-uniformity $\mathfrak{W}_0$, whose trace on X is the ϑ-uniformity $\mathfrak{V}$, is obvious.

Now assume that the extension $(Y, \mathfrak{W})$ is complete, and let $\mathscr{F}$ be any minimal Cauchy filter in the space $(Z, \mathfrak{W}_0)$. Then the trace $\mathscr{F}_X$ of the filter $\mathscr{F}$ on the space X is a Cauchy filter for the space $(X, \mathfrak{V})$ and also the basis of a Cauchy filter $\mathscr{F}_0$ for the space $(Y, \mathfrak{W})$. If y is a δ-limit point of the filter $\mathscr{F}_0$, then fy is a δ-limit point of the filter $\mathscr{F}$. Thus proposition 10 is proven.

It follows from proposition 10 that a given ϑ-uniform space in general possesses a number of different complete extensions.

The following theorem shows us that among these there exists a maximal one.

Theorem 8. *There exists an, up to isomorphism uniquely determined, complete extension* $\hat{X}$ *of the space* X *possessing the following properties* (P) *and* (Q):

(P) For any continuous and ϑ-uniformly continuous mapping f of the space X into a complete space Y there exists a unique ϑ-uniformly continuous mapping $g \colon \hat{X} \to Y$ such that $f = gi$, where $i \colon X \to \hat{X}$

is the inclusion mapping.

(Q) If $\overset{*}{X}$ is an arbitrary complete extension of the space X having the property (P), then there exists a one-to-one continuous mapping of the extension $\hat{X}$ onto the extension $\overset{*}{X}$, which is the identity on X.

Proof. 1) The uniqueness of the extension satisfying condition (Q) follows from the fact that two Hausdorff extensions of X which can be mapped continuously onto each other by mappings identical on X must coincide.

2) Definition of the space $\hat{X}$. Let us denote by $\hat{X}$ the set obtained by adding to the set X the family X^1 of all minimal Cauchy filters on the space X which have no limit points. We shall introduce a topology in $\hat{X}$ as follows. A neighbourhood basis of a point $x \in X$ is formed by all its neighbourhoods in X, hence X will be open in $\hat{X}$. A basis of neighbourhoods for a filter $\mathscr{F} \in X^1$ will be formed by all sets of the form $\{\mathscr{F}\} \cup U$, where U is an open set in X such that $\langle [U] \rangle \in \in \mathscr{F}$. It is obvious that the system of all these subsets of $\hat{X}$ will form the basis of a topology on the set $\hat{X}$, moreover that this space $\hat{X}$ will be an extension of the space X.

Now we shall define a ϑ-uniformity on the space $\hat{X}$. If U is any open subset of the space X, let us denote by $\hat{U}$ the largest open subset of $\hat{X}$ such that $U = X \cap \hat{U}$. In particular, if U is a canonical open set in X, then we have $\hat{U} = \langle [U]_X \rangle_{\hat{X}}$. Now let $\nu \in \mathfrak{B}$, where $\mathfrak{B}$ denotes the ϑ-uniformity on the space X according to which the extension $\bar{\bar{X}}$ has been constructed. Let us put $\hat{\nu} = \{\hat{V} \mid V \in \nu\}$ and $\hat{\mathfrak{B}} = \{\hat{\nu} \mid \nu \in \mathfrak{B}\}$. We shall prove that the family $\hat{\mathfrak{B}}$ yields a ϑ-uniformity on the space $\hat{X}$.

To see this we first prove that if $\nu \in \mathfrak{B}$ then ν is a ϑ-cover of locally finite type of the space $\hat{X}$. We remark that themmembers of $\hat{\nu}$ are canonical open subsets of $\hat{X}$. Now let y be an arbitrary point of the space $\hat{X}$. If $y \in X$, then there are a finite number of elements $V_1, \dots$ $\dots, V_n$ of the ϑ-cover ν such that the set $\left\langle \bigcup_{i=1}^{n} [V_i]_X \right\rangle_X$ is a neighbourhood of the point y in the space X. However

$$\bigcup_{i=1}^{n} [V_i]_X \subset \bigcup_{i=1}^{n} [\hat{V}_i]_{\hat{X}} ,$$

hence a fortiori

$$\Big\langle \bigcup_{i=1}^{n} [V_i]_X \Big\rangle_X \subset \bigcup_{i=1}^{n} [\hat{V}_i]_{\hat{X}} ,$$

and as $\Big\langle \bigcup_{i=1}^{n} [V_i]_X \Big\rangle_X$ is open is $\hat{X}$, we have

$$\Big\langle \bigcup_{i=1}^{n} [V_i]_X \Big\rangle_X \subset \Big\langle \bigcup_{i=1}^{n} [\hat{V}_i]_{\hat{X}} \Big\rangle_{\hat{X}} .$$

Consequently, the set $\Big\langle \bigcup_{i=1}^{n} [\hat{V}_i]_{\hat{X}} \Big\rangle_{\hat{X}}$ is a neighbourhood of the point u in the space $\hat{X}$.

Now assume $y \notin X$, i.e. y is a minimal Cauchy filter on the space X. There exists an element V of the ϑ-cover v, which belongs to the filter y. The set $V \cup \{y\}$ is a neighbourhood of the point y in the space $\hat{X}$, hence we have

$$y \in \Big\langle \big[V \cup \{y\}\big]_{\hat{X}} \Big\rangle_{\hat{X}} .$$

But $y \in [V]_{\hat{X}}$, hence $\Big\langle \big[V \cup \{y\}\big]_{\hat{X}} \Big\rangle_{\hat{X}} = \hat{V}$, consequently the point y is contained in the element $\hat{V}$ of the system $\hat{v}$. This shows that the system $\hat{v}$ is indeed a ϑ-cover of locally finite type of the space $\hat{X}$. Let us note that the remainder $\hat{X} \setminus X$ is completely covered by the system $\hat{v}$. Now we shall verify that the family $\hat{\mathfrak{V}}$ satisfies all the axioms of a ϑ-uniformity.

Axiom $\mathrm{I_U}$. Suppose that the ϑ-cover $\hat{v} \in \hat{\mathfrak{V}}$ is a refinement of the ϑ-cover w which is of locally finite type. It is obvious then that the system v is a refinement of the system $w_X = \{W \cap X \mid W \in w\}$. The elements of the system w_X are canonical open sets in the space X, and the ϑ-cover v is a refinement of w_X. Consequently $w_X \in \mathfrak{V}$, by virtue of axiom $\mathrm{I_U}$ for the family $\mathfrak{V}$. Then we have $\hat{w}_X \in \hat{\mathfrak{V}}$ and, since the obvious equality $\langle [W \cap X]_{\hat{X}} \rangle_{\hat{X}} = W$ implies $\hat{w}_X = w$, axiom $\mathrm{I_U}$ has indeed been established.

Axiom $\mathrm{II_U}$. To see that it is valid it is sufficient to verify the following obvious fact: if the ϑ-cover $w \in \mathfrak{V}$ is a double star refinement of the ϑ-cover $v \in \mathfrak{V}$, then $\hat{w}$ is a star refinement of $\hat{v}$.

Axiom III_U. If x and y are two different points of X, then there exist neighbourhoods G and H of these points, respectively, in the space X and a ϑ-cover $\nu \in \mathfrak{B}$ so that $H \cap \mathrm{st}_\nu G = \phi$. Let us note that if U and V are open subsets of the space X, then $U \cap V = \phi \Leftrightarrow \Leftrightarrow \hat{U} \cap \hat{V} = \phi$, hence $\hat{H} \cap \mathrm{st}_{\hat{\nu}} \hat{G} = \phi$. Now let $x \in X$ and $y \in \hat{X} \setminus X$. Since y is a filter with no limit points, there exists an element $M \in y$ such that $x \notin [M]$. As we have shown above, the minimal Cauchy filter y has a basis consisting of canonical open sets. Therefore, without any loss of generality we might assume that this set M is open, consequently the set $X \setminus [M]$ is a canonical open neighbourhood of the point x. As axiom IV_U is satisfied for the ϑ-uniformity $\mathfrak{B}$, there exists a neighbourhood H of the point x and a ϑ-cover $\nu \in \mathfrak{B}$ such that $\mathrm{st}_\nu H \subset X \setminus [M]$, i.e. $\mathrm{st}_\nu H \cap [M] = \phi$, hence $\mathrm{st}_{\hat{\nu}} \hat{H} \cap \hat{M} = \phi$. Thus we have found neighbourhoods $\hat{H}$ and $\hat{M}$ of the points x and y, respectively, and a ϑ-cover $\hat{\nu} \in \hat{\mathfrak{B}}$ as required.

Finally, let $x, y \in \hat{X} \setminus X$. Since x and y are now different minimal Cauchy filters, there are sets $U \in x$ and $V \in y$ such that $U \cap V = \phi$. Indeed, otherwise the system

$$\{U \cap V \mid U \in x, V \in y\}$$

would form the basis of a Cauchy filter $\mathscr{F}$, which should contain both filters x and y, which contradicts the fact that there is a unique minimal Cauchy filter contained in $\mathscr{F}$. Moreover, without any loss of generality we can assume that the sets U and V are both canonical open sets. A basis of the minimal Cauchy filter x is constituted by sets of the form $\langle[\mathrm{st}_\nu M]\rangle$, where the set $M \in x$ can again be assumed to be open. Thus there exist open sets $M \in x$ and $V \in y$, and a ϑ-cover $\nu \in \mathfrak{B}$ so that $V \cap \mathrm{st}_\nu M = \phi$. Now the sets $\hat{M}$ and $\hat{V}$ are neighbourhoods of the points x and y, respectively, such that $\hat{V} \cap \mathrm{st}_{\hat{\nu}} \hat{M} = \phi$. Thus axiom III_U is completely verified.

Axiom IV_U. If the point x belongs to X and G is its canonical open neighbourhood in $\hat{X}$, then the existence of another neighbourhood H and a ϑ-cover $\hat{\nu} \in \hat{\mathfrak{B}}$ with $\mathrm{st}_{\hat{\nu}} H \subset G$ follows immediately from the fact that axiom IV_U is satisfied by the ϑ-uniformity $\mathfrak{B}$. Now let $x \in \in \hat{X} \setminus X$ and G be a canonical open neighbourhood of the point x in

$\hat{X}$. There exists in X a canonical open set V with $V \in x$ and $G = \hat{V}$. According to the definition of the basis of the minimal Cauchy filter x, there exist an (again open) set $M \in x$ and a ϑ-cover $\nu \in \mathfrak{B}$ such that $\mathrm{st}_\nu M \subset V$. However, then the set $\hat{M}$ is a neighbourhood of the point x, for which $\mathrm{st}_{\hat{\nu}} \hat{M} \subset \hat{V} = G$.

Thus all the axioms are shown to be valid, consequently the family $\hat{\mathfrak{B}}$ is indeed a ϑ-uniformity on the space $\hat{X}$. It follows from the definition of the ϑ-covers $\hat{\nu} \in \hat{\mathfrak{B}}$ that the restriction of the ϑ-uniformity $\hat{\mathfrak{B}}$ to the space X coincides with $\mathfrak{B}$, hence the ϑ-uniform space $(\hat{X}, \hat{\mathfrak{B}})$ forms an extension of the ϑ-uniform space $(X, \mathfrak{B})$.

3) Now we show that the space $(\hat{X}, \hat{\mathfrak{B}})$ is complete. Let $\mathscr{F}$ be a Cauchy filter in the space $(\hat{X}, \hat{\mathfrak{B}})$. Its trace $\mathscr{F}_X$ on the subspace X is also a Cauchy filter. Let us denote by $\mathscr{F}_0$ the minimal Cauchy filter in the space X which is contained in $\mathscr{F}_X$. If the point $x \in X$ is a limit point of the filter $\mathscr{F}_0$ then it is also a δ-limit point of the filter $\mathscr{F}$, and if $\mathscr{F}_0$ has no limit points then $\mathscr{F}_0 \in \hat{X}$, hence $\mathscr{F}_0$ is obviously a δ-limit point of the filter $\mathscr{F}$.

4) Now let us verify property (P). Let $(Y, \mathfrak{W})$ be a complete ϑ-uniform space and $f\colon (X, \mathfrak{B}) \to (Y, \mathfrak{W})$ be a continuous and ϑ-uniformly continuous mapping. We shall construct an extension of f to $\hat{X}$. Let $\mathscr{F}$ be a minimal Cauchy filter of the space $(X, \mathfrak{B})$, i.e. $\mathscr{F} \in \hat{X} \setminus X$. Consider the family $\tilde{f}\mathscr{F}$ of all such canonical open subsets W of the space Y, for which $\langle [f^{-1} w] \rangle \in \mathscr{F}$. The system $\tilde{f}\mathscr{F}$ is non-empty and obviously it contains non-empty sets only. The system $\tilde{f}\mathscr{F}$ is centered. Indeed, if $W_1, W_2 \in \tilde{f}\mathscr{F}$, then $\langle [f^{-1} W_i] \rangle \in \mathscr{F}$, $i = 1, 2$, consequently

$$\langle [f^{-1} W_1] \rangle \cap \langle [f^{-1} W_2] \rangle \in \mathscr{F}\,.$$

But

$$\langle [f^{-1} W_1] \rangle \cap \langle [f^{-1} W_2] \rangle = \langle [f^{-1} W_1] \cap [f^{-1} W_2] \rangle\,.$$

The sets $f^{-1} W_i$, $i = 1, 2$, are open, because f is continuous, hence

$$\langle [f^{-1} W_1] \cap [f^{-1} W_2] \rangle = \langle [f^{-1}(W_1 \cap W_2)] \rangle\,.$$

Thus we obtain $\langle [f^{-1}(W_1 \cap W_2)] \rangle \in \mathscr{F}$, consequently $\langle [f^{-1}(W_1 \cap W_2)] \rangle \neq$ $\neq \phi$, and therefore $f^{-1}(W_1 \cap W_2) \neq \phi$, which implies $W_1 \cap W_2 \neq \phi$. As

$\mathscr{F}$ is a Cauchy filter, condition 2° of the definition of ϑ-uniformly con-tinuous mappings implies that for any ϑ-cover $w \in \mathfrak{W}$ there exists a set $W \in w$ such that $W \in \tilde{f}\mathscr{F}$. Therefore, the family $\tilde{f}\mathscr{F}$ forms the basis of a Cauchy filter $\tilde{g}\mathscr{F}$ of the space $(Y, \mathfrak{W})$. The δ-limit point of the family $\tilde{f}\mathscr{F}$, which exists because of the completeness of the space $(Y, \mathfrak{W})$. we shall denote by $g\mathscr{F}$. Thus the mapping $f\colon X \to Y$ has been extended to a mapping $g\colon X \to Y$.

We shall show next that the mapping $g\colon (\hat{X}, \hat{\mathfrak{V}}) \to (Y, \mathfrak{W})$ is ϑ-uniformly continuous. Let w be an arbitrary uniform ϑ-cover of the space Y. The mapping f is ϑ-uniformly continuous, hence there exists a ϑ-cover $v \in \mathfrak{V}$ for which the cover $\{f[V]_X \mid V \in v\}$ is a refinement of the cover $\{[W] \mid W \in w\}$. Then the cover $\{g[\hat{V}]_{\hat{X}} \mid \hat{V} \in \hat{v}\}$ is also a refinement of the cover $\{[W] \mid W \in w\}$. Indeed, let $f[V]_X \subset [W]$, where $V \in v$ and $W \in w$. Since $\hat{V} = \langle [V]_{\hat{X}} \rangle_{\hat{X}}$, we have $[\hat{V}]_{\hat{X}} = [V]_{\hat{X}}$. Therefore, the set $[\hat{V}]_{\hat{X}} \setminus [V]_X$ consists of those minimal Cauchy filters all of whose members intersect the set V. Let $\mathscr{F}$ be any one of these filters and let U belong to the family $\tilde{f}\mathscr{F}$. Then we have $\langle [f^{-1}U] \rangle \cap V \neq \phi$, hence $f^{-1}U \cap V \neq \phi$ as well, i.e. $U \cap f^{-1}V \neq \phi$. This shows that every member of the family $\tilde{f}\mathscr{F}$ intersects the set fV, hence a fortiori the set $[W]$. Let us assume now that the point $g\mathscr{F}$ (the δ-limit point of the filter $\tilde{g}\mathscr{F}$) does not belong to the set $[W]$. Then, according to axiom IV_{U} there exists a neighbourhood H of the point $g\mathscr{F}$ and a ϑ-cover $w_1 \in \mathfrak{W}$ so that $\mathrm{st}_{w_1} H \subset Y \setminus [W]$. Since $\tilde{g}\mathscr{F}$ is a Cauchy filter there exists an element W_1 of the ϑ-cover w_1 which belongs to the filter $\tilde{g}\mathscr{F}$. By the definition of the point $g\mathscr{F}$ and the choice of its neighbourhood H we must have $W_1 \subset Y \setminus [W]$, i.e. $W_1 \cap [W] = \phi$. But the family $\tilde{f}\mathscr{F}$ forms a basis of the filter $\tilde{g}\mathscr{F}$, hence there exists a set $U \in \tilde{f}\mathscr{F}$ for which $U \subset W_1$. From this we get $U \cap [W] = \phi$, which contradicts the above verified property of the members of the family $\tilde{f}\mathscr{F}$. Therefore, we have $g\mathscr{F} \in [W]$, consequently $g[\hat{V}]_{\hat{X}} \subset [W]$. Condition 1° of the definition of a ϑ-uniformly continuous mapping has thus been verified.

Now we show that

$$\{\langle [g^{-1}W]_{\hat{X}} \rangle_{\hat{X}} \mid W \in w\} \in \hat{\mathfrak{V}}$$

if $w \in \mathfrak{W}$. Since f is ϑ-uniformly continuous, we have

$$\nu = \{\langle [f^{-1} W]_X \rangle_X \mid W \in w\} \in \mathfrak{V} .$$

However, then

$$\{\langle [g^{-1} W]_{\hat{X}} \rangle_{\hat{X}} \mid W \in w\} = \hat{\nu} \in \hat{\mathfrak{V}} .$$

Indeed, we shall show that

$$\langle [\langle [f^{-1} W]_X \rangle_X]_{\hat{X}} \rangle_{\hat{X}} = \langle [g^{-1} W]_{\hat{X}} \rangle_{\hat{X}} .$$

The set $f^{-1} W$ is a dense open subset of the set $g^{-1} W$, since the mapping f is continuous and X is open in $\hat{X}$. Consequently the set $\langle [f^{-1} W]_X \rangle_X$ is dense in $\langle [g^{-1} W]_{\hat{X}} \rangle_{\hat{X}}$, and thus

$$\langle [\langle [f^{-1} W]_X \rangle_X]_{\hat{X}} \rangle_{\hat{X}} = \langle [g^{-1} W]_{\hat{X}} \rangle_{\hat{X}} .$$

Condition 2° of the definition of ϑ-uniformly continuous mappings has thus been verified.

5) Finally we prove property (Q). Let Y_1 and Y_2 be two extensions of the space X, which both satisfy condition (P). As the embeddings $X \to Y_i$, $i = 1, 2$, are continuous and ϑ-uniformly continuous, there exist ϑ-uniformly continuous mappings $f_1 \colon Y_1 \to Y_2$ and $f_2 \colon Y_2 \to Y_1$ which are equal to the identity mapping on X. The mapping $f_2 f_1 \colon Y_1 \to Y_1$ is also identical because it is ϑ-continuous and identical on the everywhere dense subspace X. Consequently the ϑ-continuous mappings f_1 and f_2 are mutually inverse and both one-to-one. Thus the extensions Y_1 and Y_2 are ϑ-homeomorphic. But it is easy to see from the construction of the extension $\hat{X}$ that it can be condensed onto any extension with which it is homeomorphic. Therefore, the extension $\hat{X}$ possesses property (Q), and the proof of theorem 8 has been completed.

This extension $\hat{X}$ will be called the completion of the ϑ-uniform space X.

Remark. It is not possible to extend to the category of ϑ-uniform spaces the classical theorem of uniform topology (cf. Bourbaki [2]) saying that any uniformly continuous mapping f from a dense subset A of a complete space X into a complete space Y is extendable to X. Indeed, we can take as A any non-complete space, whose remainder $\hat{A} \setminus A$ contains at least two points a_1 and a_2. As Y we take the completion $\hat{A}$, as f the embedding of A into $\hat{A}$, and as X the space obtained from $\hat{A}$ by identifying the points a_1 and a_2. Proposition 10 implies that X is a complete extension of the space A. Assume that there is an extension $\tilde{f}\colon X \to \hat{A}$ of the mapping f. Furthermore, the identification map $g\colon \hat{A} \to X$ is identical on A and ϑ-uniformly continuous. Then the ϑ-uniformly continuous mapping $\tilde{f}g\colon \hat{A} \to \hat{A}$ is identical on A, and therefore identical on $\hat{A}$. But this contradicts $\tilde{f}g(a_1) = \tilde{f}g(a_2)$, and thus proves the impossibility to extend the mapping f to X.

This remark shows us that in general a continuous and ϑ-uniformly continuous mapping f of a space A into a complete space Y cannot be extended to an arbitrary complete extension, however it can always be extended, by virtue of theorem 8, to its completion (maximal extension).

Applying proposition 1 of [10] we obtain easily

Proposition 11. *Let us be given on the set X two ϑ-homeomorphic topologies $\mathscr{T}_1$ and $\mathscr{T}_2$. Then every ϑ-uniformity for the space $(X, \mathscr{T}_1)$ will also be a ϑ-uniformity for the space $(X, \mathscr{T}_2)$.*

We shall denote by $\check{X}$ the incondensable extension of the space X, which is ϑ-homeomorphic to the extension $\hat{X}$. (cf. V.K. Belnov [1]). In view of proposition 11, the ϑ-uniformity of the space X can be extended to the extension $\check{X}$, which in this way becomes an extension of the ϑ-uniform space X (if the spaces $\hat{X}$ and $\check{X}$ are assumed to have the same underlying set of points, then the ϑ-uniformities of the spaces $\hat{X}$ and $\check{X}$ coincide). The extension $\check{X}$, which is obviously complete, we shall call the incondensable completion of the ϑ-uniform space X. It follows from theorem 8 that the extension $\check{X}$ can be mapped ϑ-continuously onto any complete extension of the space X.

Proposition 12. *If X is a uniform space then the extension $\check{X}$ coincides with the completion of the space X in the category of uniform spaces.*

Proof. First we show that $\check{X}$ is a uniform space. We recall that when extending a uniform ϑ-cover $\nu \in \mathfrak{B}$ to $\hat{X}$, the remainder $\hat{X} \setminus X = \check{X} \setminus X$ was covered by the family $\hat{\mathfrak{B}} = \check{\mathfrak{B}}$. Therefore, uniform covers of the space X are extended to uniform covers of the spaces $\hat{X}$ and $\check{X}$, hence the ϑ-uniformity $\hat{\mathfrak{B}} = \check{\mathfrak{B}}$ is indeed a uniformity. It remains to prove that the space $\check{X}$ is semiregular. The space X is semiregular being a uniform space. The extension $\check{X}$ is a Hausdorff extension of the space X, incondensable to any extension ϑ-homeomorphic with it. Therefore $\check{X}$ is semiregular (cf. V. K. Belnov [1]). The proof is completed.

Remark. If the mapping $f\colon (X, \mathfrak{B}) \to (Y, \mathfrak{W})$ sets up an isomorphism between the ϑ-uniformities $\mathfrak{B}$ and $\mathfrak{W}$, the mapping $\tilde{f}$ in the commutative diagram

$$\begin{array}{ccc} X_{\mathfrak{B}} & \xrightarrow{\tilde{f}} & Y_{\mathfrak{W}} \\ {\scriptstyle \pi_{\mathfrak{B}}}\big\downarrow & & \big\downarrow{\scriptstyle \pi_{\mathfrak{W}}} \\ X & \xrightarrow{f} & Y \end{array}$$

is an isomorphism between the uniform spaces $X_{\mathfrak{B}}$ and $Y_{\mathfrak{W}}$. Therefore, in particular the uniform spaces $\hat{X}_{\hat{\mathfrak{B}}}$ and $\check{X}_{\check{\mathfrak{B}}}$ are isomorphic.

Theorem 9. *Let* $(X, \mathfrak{B})$ *be a* ϑ*-uniform space and* $\delta = \delta_{\mathfrak{B}}$. *Then the completion* $(\check{X}_{\mathfrak{B}}, \check{\tilde{\mathfrak{B}}})$ *of the uniform space* $(X_{\mathfrak{B}}, \tilde{\mathfrak{B}})$ *coincides with the uniform space* $(\check{X}_{\check{\mathfrak{B}}}, \tilde{\check{\mathfrak{B}}}) = (\hat{X}_{\hat{\mathfrak{B}}}, \tilde{\hat{\mathfrak{B}}})$ *generating the completion* $(\hat{X}, \hat{\mathfrak{B}})$, *and there exist embeddings*

$$j_0 : \check{X} \to h_\delta X, \quad \tilde{j}\colon \check{X}_{\mathfrak{B}} \to b_\delta X_\delta$$

and

$$j\colon \hat{X} \to h^\delta X$$

such that the following diagram is commutative.*

*Here $h^\delta X$ and $h_\delta X$ respectively denote the maximal H-closed and the maximal incondensible H-closed extensions of the ϑ-proximity space (X, δ) (cf. [10]).

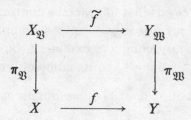

$$
\begin{array}{ccccc}
X_{\mathfrak{B}} & \xrightarrow{\tilde{i}} & \check{X}_{\mathfrak{B}} & \xrightarrow{\tilde{j}} & b_\delta X_\delta \\
\Big\downarrow{\scriptstyle \pi_{\mathfrak{B}}} & & \Big\downarrow{\scriptstyle \pi_{\hat{\mathfrak{B}}}} & & \Big\downarrow{\scriptstyle p} \\
X & \xrightarrow{i} & \hat{X} & \xrightarrow{j} & h^\delta X \\
 & & \Big\downarrow & & \Big\downarrow{\scriptstyle q} \\
 & & \check{X} & \longrightarrow & h^\delta X
\end{array}
$$

Proof. According to theorem 4 the embedding $i: X \to \hat{X}$ can be extended into the commutative diagram

$$
(4) \qquad
\begin{array}{ccc}
X_{\mathfrak{B}} & \xrightarrow{\tilde{i}} & \hat{X}_{\hat{\mathfrak{B}}} \\
\Big\downarrow{\scriptstyle \pi_{\mathfrak{B}}} & & \Big\downarrow{\scriptstyle \pi_{\hat{\mathfrak{B}}}} \\
X & \xrightarrow{i} & \hat{X}
\end{array}
$$

As the mappings $\pi_{\mathfrak{B}}$, i, and $\pi_{\hat{\mathfrak{B}}}$ are irreducible, the set $\tilde{i}X_{\mathfrak{B}}$ is dense in $\hat{X}_{\hat{\mathfrak{B}}}$. The mapping $\tilde{i}$ is an embedding because the subspace $\tilde{i}X_{\mathfrak{B}}$ is isomorphic to $X_{\mathfrak{B}}$, as the space generating the ϑ-uniformity on iX by means of the mapping $\pi_{\hat{\mathfrak{B}}}$ onto $\tilde{i}X_{\mathfrak{B}} = \pi_{\hat{\mathfrak{B}}}^{-1}\, iX$. By theorem 7, the uniform space $(\hat{X}_{\hat{\mathfrak{B}}}, \tilde{\mathfrak{B}})$ is complete. Therefore, the mapping $\tilde{i}$ is a mapping of the uniform space $X_{\mathfrak{B}}$ onto a dense subspace of the complete uniform space $\hat{X}_{\hat{\mathfrak{B}}}$, i.e. the space $(\hat{X}_{\hat{\mathfrak{B}}}, \hat{\tilde{\mathfrak{B}}})$ is a completion of the space $(X_{\mathfrak{B}}, \tilde{\mathfrak{B}})$. From proposition 12, because of the uniqueness of the completion of a uniform space, we obtain that $(\hat{X}_{\hat{\mathfrak{B}}}, \hat{\tilde{\mathfrak{B}}}) = (\check{X}_{\mathfrak{B}}, \check{\tilde{\mathfrak{B}}})$. This establishes the first claim of theorem 9.

The uniform space $\check{X}_{\mathfrak{B}} = \check{X}_{\check{\mathfrak{B}}} = \hat{X}_{\hat{\mathfrak{B}}}$ as a uniform extension of the space $X_{\mathfrak{B}}$ is embeddable into the compact extension $b_\delta X_\delta$ of the proximity space $(X_{\mathfrak{B}}, \delta_{\tilde{\mathfrak{B}}})$. Let us denote this embedding by $\tilde{j}$ and consider the commutative diagram

$$
(5)\qquad
\begin{array}{ccccc}
X_{\mathfrak{B}} & \xrightarrow{\ \tilde{\imath}\ } & \check{X}_{\mathfrak{B}} & \xrightarrow{\ \tilde{\jmath}\ } & b_\delta X_\delta \\
\Big\downarrow{\scriptstyle \pi_{\mathfrak{B}}} & & & & \Big\downarrow{\scriptstyle p} \\
X & & \xrightarrow{\qquad k \qquad} & & h^\delta X
\end{array}
$$

where k denotes the embedding of the ϑ-proximity space (X, δ) into its maximal H-closed extension $h^\delta X$. On the extension $p\tilde{\jmath}\check{X}_{\mathfrak{B}}$ of the space X the uniform space $\check{X}_{\mathfrak{B}}$ generates a ϑ-uniformity $\overset{*}{\mathfrak{B}}$ by means of the ϑ-perfect irreducible mapping $p\tilde{\jmath}$. It is obvious that the trace of the ϑ-uniformity $\overset{*}{\mathfrak{B}}$ on the subspace $kX \subset p\tilde{\jmath}\check{X}_{\mathfrak{B}}$ is the ϑ-uniformity $k\mathfrak{B}$, i.e. the ϑ-uniform space $(p\tilde{\jmath}\check{X}_{\mathfrak{B}}, \overset{*}{\mathfrak{B}})$, is an extension of the space $(X, \mathfrak{B})$. Since the space $(\check{X}_{\mathfrak{B}}, \check{\tilde{\mathfrak{B}}})$ is complete, by theorem 7 so is the space $(p\tilde{\jmath}\check{X}_{\mathfrak{B}}, \overset{*}{\mathfrak{B}})$, hence the space $p\tilde{\jmath}\check{X}_{\mathfrak{B}}$ is a complete extension of the space X. According to theorem 8 the completion $\hat{X}$ can be mapped onto the extension $p\tilde{\jmath}\check{X}_{\mathfrak{B}}$ by means of a continuous mapping j such that $k = ji$. Both ϑ-uniform spaces $\hat{X}$ and $p\tilde{\jmath}\check{X}_{\mathfrak{B}}$ are generated by the same uniform space $\check{X}_{\mathfrak{B}}$. In the commutative diagram

$$
(6)\qquad
\begin{array}{ccccc}
 & & \check{X}_{\mathfrak{B}} & & \\
 & \swarrow{\scriptstyle \pi_{\hat{\mathfrak{B}}}} & & \searrow{\scriptstyle p\tilde{\jmath}} & \\
\hat{X} & & \xrightarrow{\quad j \quad} & & p\tilde{\jmath}\check{X}_{\mathfrak{B}} \\
 & \nwarrow{\scriptstyle i} & & \nearrow{\scriptstyle k} & \\
 & & X & &
\end{array}
$$

the mapping $p\tilde{\jmath}$ maps in a one-to-one manner onto the remainder $\check{X}_{\mathfrak{B}} \setminus \tilde{\imath}X_{\mathfrak{B}}$. Therefore, the mapping j is also one-to-one on the remainder $\hat{X} \setminus iX$, and so, being a homeomorphism on iX it is one-to-one on the whole space $\hat{X}$. Thus the mapping $j^{-1}\colon p\tilde{\jmath}\check{X}_{\mathfrak{B}} \to \hat{X}$ is defined. The mapping $\pi_{\hat{\mathfrak{B}}}$ can be written as the composition of the mappings $p\tilde{\jmath}$ and j^{-1}. Hence, applying lemma 4 below, we get that j^{-1} is ϑ-continuous.

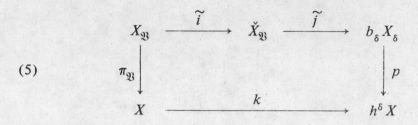

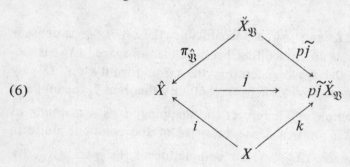

Lemma 4. *If a compact ϑ-continuous mapping $f\colon A \to B$ is equal to the composition $\psi\varphi$ of two mappings such that $\varphi\colon A \to C$ is a closed*

mapping onto C, *then the mapping* $\psi: C \to B$ *is* ϑ*-continuous.*

Proof. Let $x \in C$ and Oy be an arbitrary neighbourhood of the point $y = \psi x \in B$. For every point z of the compact set $f^{-1}y$ there exists a neighbourhood Oz such that $f[Oz] \subset [Oy]$. Consequently there is a neighbourhood U of the compact set $f^{-1}y$, for which $f[U] \subset [Oy]$. Since $x \in \psi^{-1}y$, we have $\varphi^{-1}x \subset \varphi^{-1}\psi^{-1}y = f^{-1}y$. Therefore, U is a neighbourhood of the set $\varphi^{-1}x$ as well. As the mapping φ is closed, the set $\varphi^{\#}U$ is a neighbourhood of the point x and $\varphi[U] \subset [\varphi^{\#}U]$. Then $\psi[\varphi^{\#}U] \subset \psi\varphi U = f[U] \subset [Oy]$, hence the mapping $\cdot\ \psi$ is indeed ϑ-continuous. The lemma is thus proven.

So we have that the extensions $\hat{X}$ and $p\tilde{j}\check{X}$ of the space X are ϑ-homeomorphic. But the extension $p\widetilde{j\check{X}}_{\mathfrak{B}}$ is contained in the Katětov type extension $h^{\delta}X$, that means every ϑ-homeomorphism from such an extension onto any other extension of the space X is a condensation (cf. Mioduszewski and Rudolph [6]).

Lemma 5. *Let* Y *be a Katětov type extension of the space* X *and let* $X \subset Z \subset Y$. *Then* Z *is also a Katětov type extension of the space* X.

Proof. Let $f: Z \to \tilde{Z}$ be a ϑ-homeomorphism of the extension Z onto an extension $\tilde{Z}$. We have to show that the mapping f is continuous. Let us consider the space $\tilde{Y}$ defined on the set $\tilde{Z} \cup (Y \setminus Z)$, whose topology is given by the following stipulations: 1) $\tilde{Z}$ is an open subspace of $\tilde{Y}$; 2) the subspace $X \cup (Y \setminus Z)$ of the extension Y also lies as an open subset in $\tilde{Y}$. Now define a one-to-one mapping $g: Y \to \tilde{Y}$ in the following manner:

$$g(y) = \begin{cases} f(y), & \text{if } y \in Z \\ y, & \text{if } y \in Y \setminus Z. \end{cases}$$

We need to prove on auxiliary lemma.

Lemma 6. *Let* $f: A \to B$ *be a mapping of the topological space* A *onto the topological space* B *with the property that the space* A *is the union of two open subsets* C *and* D, *moreover the mappings* $f|C$ *and* $f|D$ *are continuous and* ϑ*-continuous, respectively. Then the map-*

ping f is ϑ-continuous.

Proof. We remark that the continuity of the mapping $f|C$ alone does not imply the ϑ-continuity of the mapping f in the points of the set C. Let $x \in C$, and Ofx be an arbitrary neighbourhood of the point fx. In view of the continuity of the mapping $f|C$ there exists a neighbourhood Ox of x in C, hence also in A, such that $fOx \subset Ofx$. We shall show that $f[Ox]_A \subset [Ofx]$. Indeed, let $y \in [Ox]_A$ and Ofy be an arbitrary neighbourhood of the point fy. If $y \in C$, then by virtue of the continuity of the mapping $f|C$ there exists a neighbourhood Oy for which $fOy \subset Ofy$. But $Oy \cap Ox \neq \phi$, hence $fOy \cap fOx \neq \phi$, and a fortiori $Ofy \cap Ofx \neq \phi$, consequently $fy \in [Ofx]$. If on the other hand $y \in D$, then there exists a neighbourhood Oy such that $fOy \subset [Ofy]$. Then we obtain $[Ofy] \cap Ofx \neq \phi$, which implies $Ofy \cap Ofx \neq \phi$ and thus $fy \in [Ofx]$. Next we prove the ϑ-continuity of the mapping f at the points of the set D. Let $x \in D$ and Ofx be an arbitrary neighbourhood of the point fx. In view of the ϑ-continuity of the mapping $f|D$ there exists a neighbourhood $Ox \subset D$ such that $f[Ox]_D \subset [Ofx]$. We shall show that then $[fOx]_A \subset [Ofx]$ as well. Let $y \in [Ox]_A \setminus [Ox]_D)$ i.e. $y \in [Ox]_A \cap C$, and Ofy be an arbitrary neighbourhood of the point fy. Since the mapping $f|C$ is continuous, there exists a neighbourhood Oy such that $fOy \subset Ofy$. Since $y \in [Ox]$, we have $Oy \cap Ox \neq \phi$, consequently $fOy \cap fOx \neq \phi$, and a fortiori $Ofy \cap [Ofx] \neq \phi$. From this we get $Ofy \cap Ofx \neq \phi$, i.e. $fy \in [Ofx]$. The proof of the lemma is completed.

It follows from this lemma that the mapping g: $Y \to \widetilde{Y}$ which is the identity on X is a ϑ-homeomorphism. However, Y is a Katětov type extension, hence the mapping g is continuous. Thus the mapping f: $Z \to \widetilde{Z}$ as the restriction of the mapping g, is also continuous. This proves lemma 5.

Lemma 5 implies that the extension $p\widetilde{j}\check{X}_{\mathfrak{B}}$ is also a Katětov type extension. Consequently the ϑ-homeomorphism j^{-1}: $p\widetilde{j}\check{X}_{\mathfrak{B}} \to \check{X}$ is continuous, hence the continuous mapping j: $\hat{X} \to p\widetilde{j}\check{X}_{\mathfrak{B}}$ is a homeomorphism. This proves the existence of the embeddings $\widetilde{j}$ and j and the commutativity of the diagram

$$
(7)\qquad
\begin{array}{ccccc}
X_{\mathfrak{B}} & \xrightarrow{\tilde{i}} & \check{X}_{\mathfrak{B}} & \xrightarrow{\tilde{j}} & b_{\delta} X_{\delta} \\
\Big\downarrow {\scriptstyle \pi_{\mathfrak{B}}} & & \Big\downarrow {\scriptstyle \pi_{\hat{\mathfrak{B}}}} & & \Big\downarrow {\scriptstyle p} \\
X & \xrightarrow{i} & \hat{X} & \xrightarrow{j} & h^{\delta} X
\end{array}
$$

Now we prove another auxiliary proposition.

Lemma 7. *Let Y be a Hausdorff extension of the space X which cannot be condensed onto any extension with which it is ϑ-homeomorphic and let $X \subset Z \subset Y$. Then the extension Z of the space X cannot be condensed onto any extension with which it is ϑ-homeomorphic, either.*

Proof. Assume that there exists an extension $\tilde{Z}$ of the space X and a condensation $f\colon Z \to \tilde{Z}$ which is also a ϑ-homeomorphism. By the construction of the extension Y incondensible within the given class of ϑ-homeomorphic extensions (cf. V. K. Belnov [1]), the canonical open sets of Y form a neighbourhood basis for the points of the remainder $X \setminus X$. Since for any canonical open subset U of Y the set $Z \cap U$ is canonical open in Z (Z being dense in Y), the canonical open sets in Z form a neighbourhood basis for the points of the remainder $Z \setminus X$. It is obvious, in addition, that for any open set V in X we have the equality $V_Z = V_Y \cap Z$, where V_Y (V_Z) denotes the maximal open set in Y (Z), whose intersection with X is the set V.

Let W be open in Z. We shall show that fW is open in $\tilde{Z}$. If $x \in W \setminus X$, then there exists in Z a canonical open set U such that $x \in U$ and $U \subset W$. Since a ϑ-homeomorphism carries a canonical open set into a canonical open set (cf. [10]), the image fU is open in $\tilde{Z}$. Thus the set fW is a neighbourhood of all points of $fW \setminus X$. Now let $x \in W \cap X$. As the extension Y is incondensible, the sets U for which $U = (U \cap X)_Y$ holds, form a neighbourhood basis for the point x in the space Y. Consequently, there exists such a basic neighbourhood U of the point x in the space Y, for which $U \cap Z \subset W$. Since the mapping $f^{-1}\colon \tilde{Z} \to Z$ is ϑ-continuous, there exists a neighbourhood Ofx such that $f^{-1}[Ofx] \subset [U \cap Z]_Z$. Here we can assume that $Ofx \cap X \subset U \cap X$. Then the set $f^{-1}fOfx$, which is open in view of the continuity of the mapping

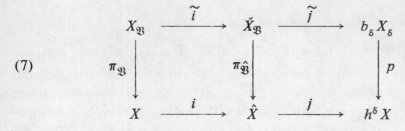

f, is contained in $U \cap Z$. Indeed,

$$(f^{-1}Ofx \cup U) \cap X = (f^{-1}Ofx \cap X) \cup (U \cap X) = U \cap X\,,$$

hence

$$f^{-1}Ofx \cup (U \cap Z) \subset (U \cap X)_Z\,.$$

However, we have

$$(U \cap X)_Z = (U \cap X)_Y \cap Z = U \cap Z\,,$$

consequently

$$f^{-1}Ofx \subset U \cap Z \subset W\,.$$

Then $Ofx \subset fW$, and therefore the set fW is open, which implies that the mapping $f\colon Z \to \widetilde{Z}$ is open and thus a homeomorphism. The proof of the lemma is completed.

It follows from this lemma that the extension $qj\hat{X}$ cannot be condensed onto any extension which it is ϑ-homeomorphic to. The extension $\check{X}$ possesses this same property, and it is obviously ϑ-homeomorphic to $qj\hat{X}$. Thus there exists an embedding $j_0\colon \check{X} \to h_\delta X$, which makes the following diagram commutative

$$\begin{array}{ccc} \hat{X} & \xrightarrow{\;\;j\;\;} & h^\delta X \\ \big\downarrow & & \big\downarrow q \\ \check{X} & \xrightarrow{\;\;j_0\;\;} & h_\delta X \end{array} \tag{8}$$

This ends the proof of Theorem 9.

Corollary. *The remainder* $\check{X} \setminus X$ *is homeomorphic to the remainder* $\check{X}_{\mathfrak{B}} \setminus X_{\mathfrak{B}}$, *and thus is completely regular.*

Theorem 9 can be supplemented with the following statement.

Theorem 10. *Let* $k\colon (X, \mathfrak{B}) \to (Y, \mathfrak{W})$ *be a complete extension of the* ϑ*-uniform space* $(X, \mathfrak{B})$. *Then* $Y_{\mathfrak{W}} = \check{X}_{\mathfrak{B}}$ *and* $\pi_{\mathfrak{W}} = g\pi_{\hat{\mathfrak{B}}}$, *where* g:

$(\hat{X}, \hat{\mathfrak{V}}) \to (Y, \mathfrak{W})$ *is the mapping whose existence is assured by theorem* 8.

Proof. In the commutative diagram

$$(9) \qquad \begin{array}{ccc} X_{\mathfrak{V}} & \xrightarrow{\tilde{k}} & Y_{\mathfrak{W}} \\ {\scriptstyle \pi_{\mathfrak{V}}} \downarrow & & \downarrow {\scriptstyle \pi_{\mathfrak{W}}} \\ X & \xrightarrow{k} & Y \end{array}$$

the space $(Y_{\mathfrak{W}}, \tilde{\mathfrak{W}})$ is complete, $\tilde{k}$ is an embedding and $\tilde{k}X_{\mathfrak{V}}$ is dense in $Y_{\mathfrak{W}}$. These claims can be proven in the same manner as the analogous claims for diagram (4) of theorem 9. Similarly as in the proof of theorem 9, we obtain from proposition 12 that $Y_{\mathfrak{W}} = \check{X}_{\mathfrak{V}}$. Moreover, in the diagram below

$$(10) \qquad \begin{array}{ccccc} X_{\mathfrak{V}} & \xrightarrow{\tilde{i}} & \check{X}_{\mathfrak{V}} & & \\ {\scriptstyle \pi_{\mathfrak{V}}} \downarrow & & \downarrow {\scriptstyle \pi_{\hat{\mathfrak{V}}}} & \searrow {\scriptstyle \pi_{\mathfrak{W}}} & \\ X & \xrightarrow{i} & \hat{X} & & \\ & \searrow {\scriptstyle k} & & \searrow {\scriptstyle g} & \\ & & & & Y \end{array}$$

the following parts are commutative: the inner square (because of theorem 9), the outer quadrilateral (coinciding with diagram (9)), and the lower triangle (because of theorem 8). Thus we have $\pi_{\mathfrak{W}}\tilde{i} = g\pi_{\hat{\mathfrak{V}}}\tilde{i}$, i.e. the ϑ-continuous irreducible mappings $\pi_{\mathfrak{W}}$ and $g\pi_{\hat{\mathfrak{V}}}$ coincide on the dense subspace $\tilde{i}X_{\mathfrak{V}}$. From this we obtain $\pi_{\mathfrak{W}} = g\pi_{\hat{\mathfrak{V}}}$, and theorem 10 is proven.

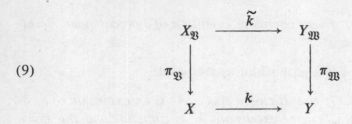

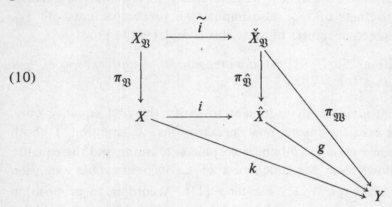

Now assume that the ϑ-uniform space $(X, \mathfrak{V})$ is precompact, i.e. the ϑ-uniformity $\mathfrak{V}$ has a basis consisting of finite ϑ-covers. Then the uniform space $X_{\mathfrak{V}}$ is also precompact, consequently its completion $\check{X}_{\mathfrak{V}}$ is compact (cf. e.g. Bourbaki [2]). In view of theorem 9, the completion $\hat{X}$, as a ϑ-continuous image of the compactum $\check{X}_{\mathfrak{B}}$ is H-closed. By theorem 8 every complete extension of the space $(X, \mathfrak{B})$ is a continuous image of the completion $\hat{X}$ and therefore is H-closed. Thus we obtain

Proposition 13. *Every complete extension of a precompact ϑ-uniform space is H-closed.*

The converse of this proposition is also valid.

Proposition 14. *If the H-closed space X is an extension of a ϑ-uniform space $(Y, \mathfrak{B})$, then the extension X is complete and the space $(Y, \mathfrak{B})$ is precompact.*

Proof. The completion $\widetilde{Y}_{\mathfrak{B}}$ of the space $Y_{\mathfrak{B}}$ has a ϑ-perfect and irreducible mapping onto X, according to theorem 10. In view of the corollary of proposition 2 of [10], the space $\check{Y}_{\mathfrak{B}}$ is compact and hence complete. From this, using theorem 7, we obtain that X is complete. Furthermore the compactness of $\check{Y}_{\mathfrak{B}}$ also implies the precompactness of $Y_{\mathfrak{B}}$, and hence the precompactness of Y. This completes the proof.

Proposition 15. *If $(X, \mathfrak{B})$ is a precompact ϑ-uniform space, then $\hat{X} = h^{\delta_{\mathfrak{B}}}X$, and $\check{X} = h_{\delta_{\mathfrak{B}}}X$.*

Proof. It is obviously sufficient to verify the first equality only, because in this case the incompressible extensions $\check{X}$ and $h_{\delta_{\mathfrak{B}}}X$ both belong to the same class of ϑ-homeomorphic extensions, and the equality $\check{X} = h_{\delta_{\mathfrak{B}}}X$ follows from the uniqueness of an incompressible extension within a given class (cf. V.K. Belnov [1]). According to proposition 13, the extension $\hat{X}$ is H-closed, and by theorem 9 $\hat{X}$ is embeddable in $h^{\delta_{\mathfrak{B}}}X$, which immediately implies $\hat{X} = h^{\delta_{\mathfrak{B}}}X$. Thus the proposition is proven.

The set of all finite ϑ-covers of a given Hausdorff space by canonical open sets forms, as was noticed above, the basis of a ϑ-uniformity $\mathfrak{B}a$. Clearly, this ϑ-uniformity $\mathfrak{B}a$ is a maximal precompact ϑ-uniformity on the space X, and the ϑ-proximity $\delta_{\mathfrak{B}a}$ is maximal as well. Therefore proposition 15 and proposition 21 of [10] imply

Proposition 16. *The completion $\hat{X}$ of the ϑ-uniform space*

*$(X, \mathfrak{B}a)$ coincides with the maximal (Katětov type) H-closed extension τX of the space X.**

§.5. PRODUCTS AND COMPLETENESS OF ϑ-UNIFORM SPACES

Theorem 11. *Any product of complete ϑ-uniform spaces is complete. Conversely, if a product of non-empty ϑ-uniform spaces is complete then so is every factor space.*

Proof. Let $\{X_i \mid i \in I\}$ be a set of complete ϑ-uniform spaces and let $X = \prod_{i \in I} X_i$. Let $\mathscr{F}$ be a Cauchy filter in the product space X and denote by $\mathscr{F}_i$ the projection of the filter $\mathscr{F}$ onto the factor X_i. We shall show that $\mathscr{F}_i$ is a Cauchy filter in the space X_i**. Let ν be a uniform ϑ-cover. Then $\pi_i^{-1}\nu$ is a uniform ϑ-cover of the space X. There exists an element M of the filter $\mathscr{F}$ which is contained in an element $\pi_i^{-1}V$ of the ϑ-cover $\pi_i^{-1}\nu$. Then the set $\pi_i M$ is a member of the filter $\mathscr{F}$ contained in the set $V \in \nu$. Thus $\mathscr{F}_i$ is a Cauchy filter. Now let $x_i \in X_i$ be a δ-limit point of this Cauchy filter. We shall show that the point $x = (x_i)_{i \in I}$, whose coordinates are δ-limit points of the filters $\mathscr{F}_i$, is a δ-limit point of the filter $\mathscr{F}$. Let Ox be an arbitrary canonical open neighbourhood of the point x. There exists a finite collection of indices $\{i_1, \ldots, i_k\} \subset I$, and canonical open neighbourhoods $Ox_{i_1}, \ldots, Ox_{i_k}$ of the points $x_{i_1}, \ldots, x_{i_k}$ such that

$$\bigcap_{j=1}^{k} \pi_{i_j}^{-1} Ox_{i_j} \subset Ox .$$

It is easy to see that every canonical open neighbourhood of a δ-limit point of a Cauchy filter belongs to this same filter. Therefore $Ox_{i_j} \in \mathscr{F}_{i_j}$, $j = 1, \ldots, k$, and this implies $\bigcap_{j=1}^{k} \pi_{i_j}^{-1} Ox_{i_j} \in \mathscr{F}$. Thus every canonical open neighbourhood Ox of the point x belongs to the filter $\mathscr{F}$, hence it is a δ-limit point of the filter $\mathscr{F}$. The completeness of the space X *is thus proven.*

*Concerning the maximal H-closed extension τX, cf. Katětov's works [4] and [5].

**Let us note that the ϑ-uniformly continuous image of a Cauchy filter is in general not a Cauchy filter.

Now, conversely, assume that the complete space X is the product of the non-empty ϑ-uniform spaces X_i, $i \in I$. Let $\mathscr{F}_{i_0}$ be a fixed Cauchy filter in X_{i_0} and for each $i \neq i_0$ take an arbitrary Cauchy filter $\mathscr{F}_i$ in X_i and consider the filter $\mathscr{F} = \prod_{i \in I} \mathscr{F}_i$ in X. $\mathscr{F}$ is obviously a Cauchy filter in X and if $x = (x_i)$ is its δ-limit point, then the point $x_{i_0} \in X_{i_0}$ is a δ-limit point of the filter $\mathscr{F}_{i_0}$. Thus the space X_{i_0} is complete and theorem 11 is proven.

It is not possible to strengthen this theorem so as to obtain the classical theorem of uniform topology saying that a product of completions coincides with the completion of the product. Indeed we have the following

Theorem 12. *Let* $(X, \mathfrak{V})$ *be a non-complete* ϑ*-uniform space, and assume that the* ϑ*-uniformity* $\mathfrak{W}$ *on the space* Y *is not a uniformity. Then the product of the completions* $\hat{X} \times \hat{Y}$ *is not* ϑ*-homeomorphic to the completion of the product* $X \times Y$, *and the product* $\check{X} \times \check{Y}$ *of incompressible completions is not* ϑ*-homeomorphic to the incompressible completion of the product* $X \times Y$.

Proof. First we shall show the second statement. Since the diagrams below are commutative

$$\begin{array}{ccc} X_{\mathfrak{V}} & \xrightarrow{\tilde{i}} & \check{X}_{\mathfrak{V}} \\ \downarrow{\scriptstyle \pi_{\mathfrak{V}}} & & \downarrow{\scriptstyle \pi_{\check{\mathfrak{V}}}} \\ X & \xrightarrow{i_0} & \check{X} \end{array} \qquad \text{and} \qquad \begin{array}{ccc} Y_{\mathfrak{W}} & \xrightarrow{\tilde{k}} & \check{Y}_{\mathfrak{W}} \\ \downarrow{\scriptstyle \pi_{\mathfrak{W}}} & & \downarrow{\scriptstyle \pi_{\check{\mathfrak{W}}}} \\ Y & \xrightarrow{k_0} & \check{Y} \end{array}$$

so is the diagram

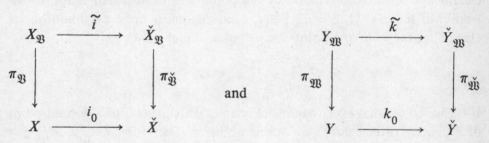

$$(11)\qquad \begin{array}{ccc} X_{\mathfrak{B}} \times Y_{\mathfrak{W}} & \xrightarrow{\tilde{\imath} \times \tilde{k}} & \check{X}_{\mathfrak{B}} \times \check{Y}_{\mathfrak{W}} \\ {\scriptstyle \pi_{\mathfrak{B}} \times \pi_{\mathfrak{W}}} \Big\downarrow & & \Big\downarrow {\scriptstyle \pi_{\check{\mathfrak{B}}} \times \pi_{\check{\mathfrak{W}}}} \\ X \times Y & \xrightarrow{i_0 \times k_0} & \check{X} \times \check{Y} \end{array}$$

Let us denote by Z the product $X \times Y$ and by $\mathfrak{A}$ the product $\mathfrak{B} \times \mathfrak{W}$. According to theorem 6, the uniform space

$$(X_{\mathfrak{B}} \times Y_{\mathfrak{W}},\ \tilde{\mathfrak{B}} \times \tilde{\mathfrak{W}})$$

coincides with the uniform space $(Z_{\mathfrak{A}}, \tilde{\mathfrak{A}})$ generating the ϑ-uniform space $(Z, \mathfrak{A})$, and $\pi_{\mathfrak{B}} \times \pi_{\mathfrak{W}} = \pi_{\mathfrak{A}}$. As the product of completions coincides with the completion of the product within the category of uniform spaces, applying proposition 12 we obtain the equality $\check{X}_{\mathfrak{B}} \times \check{Y}_{\mathfrak{W}} = \check{Z}_{\mathfrak{A}}$. Therefore, diagram (11) can be written in the form

$$(12)\qquad \begin{array}{ccc} Z_{\mathfrak{A}} & \xrightarrow{\tilde{\imath} \times \tilde{k}} & \check{Z}_{\mathfrak{A}} \\ {\scriptstyle \pi_{\mathfrak{A}}} \Big\downarrow & & \Big\downarrow {\scriptstyle \pi_{\check{\mathfrak{B}}} \times \pi_{\check{\mathfrak{W}}}} \\ Z & \xrightarrow{i_0 \times k_0} & \check{X} \times \check{Y} \end{array}$$

Let us denote by $l\colon Z \to \check{Z}$ the embedding of the space Z in the incompressible completion $\check{Z}$. It follows from diagram (12) that the extension $\check{X} \times \check{Y}$ is a complete extension of the ϑ-uniform space Z, for the product of complete spaces is complete. The incompressible completion $\check{Z}$ can be mapped ϑ-uniformly continuously onto the extension $\check{X} \times \check{Y}$. Thus diagram (12) can be extended to the diagram

$$
(13)\qquad
\begin{array}{ccccc}
Z_{\mathfrak{A}} & & \xrightarrow{\ \tilde{i}\times\tilde{k}\ } & & \check{Z}_{\mathfrak{A}} \\
& & & \nearrow^{\pi_{\check{\mathfrak{A}}}} & \\
\Big\downarrow{\scriptstyle \pi_{\mathfrak{A}}} & & \check{Z} & & \Big\downarrow{\scriptstyle \pi_{\check{\mathfrak{B}}}\times\pi_{\check{\mathfrak{W}}}} \\
& {}^{l}\nearrow & & \nwarrow^{f} & \\
Z & & & & \check{X}\times\check{Y}
\end{array}
$$

Now assume that the mapping f is a ϑ-homeomorphism. Since by the corollary of theorem 9 the mapping $\pi_{\check{\mathfrak{A}}}$ is a homeomorphism of the remainder $\check{Z}_{\mathfrak{A}} \setminus Z_{\mathfrak{A}}$ onto the remainder $\check{Z} \setminus Z$, the mapping $\pi_{\check{\mathfrak{B}}} \times \pi_{\check{\mathfrak{W}}}$ is in any case one-to-one on the remainder $\check{Z}_{\mathfrak{A}} \setminus Z_{\mathfrak{A}}$. Now let x be an arbitrary point of the remainder $\check{X} \setminus X$, and y a point of the space Y such that the set $\pi_{\mathfrak{W}}^{-1} y$ consists of more than one point. Such a point y must exist since the ϑ-uniformity $\mathfrak{W}$ is not a uniformity. Then the point (x, y) belongs to the remainder $\check{X} \times \check{Y} \setminus Z$. According to our assumption the inverse image of this point,

$$(\pi_{\check{\mathfrak{B}}} \times \pi_{\check{\mathfrak{W}}})^{-1}(x, y) = \pi_{\check{\mathfrak{B}}}^{-1} x \times \pi_{\check{\mathfrak{W}}}^{-1} y \,,$$

consists of a single point. Consequently the set $\pi_{\check{\mathfrak{W}}}^{-1} y$ also consists of a single point and coincides with the set $\pi_{\mathfrak{W}}^{-1} y$, since the mapping $\pi_{\mathfrak{W}}$ is the restriction of the mapping $\pi_{\check{\mathfrak{W}}}$ onto the complete inverse image. However, this contradicts the fact that the set $\pi_{\mathfrak{W}}^{-1} y$ consists of more than one point, and thus the mapping f cannot be a ϑ-homeomorphism.

The product $\hat{X} \times \hat{Y}$ of the completions being ϑ-homeomorphic to the product $\check{X} \times \check{Y}$ cannot be ϑ-homeomorphic to the completions of the product. This completes the proof of theorem 12.

From the last two theorems we obtain the subsequent

Corollary. *Let* $\{X_i \mid i \in I\}$ *be a family of* ϑ*-uniform spaces and assume that the product* $\prod_{i\in I} \hat{X}_i$ *of their completions coincides with the completion of the product* $\prod_{i\in I} X_i$*. If a* ϑ*-uniformity* $\mathfrak{B}_{i_0}$ *of these is not a uniformity then all spaces* X_i*,* $i \in I \setminus \{i_0\}$*, must be complete.*

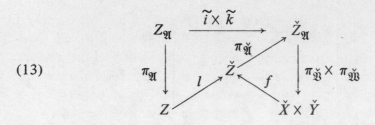

Remark. The necessary condition formulated in this corollary to the effect that the product of completions be identical with the completion of the product is not sufficient even in the case of uniform spaces. Indeed, we can take as X and Y the segment $[0, 1]$ and the half-open interval $[0, 1)$ of the real line, respectively. Consider the uniformities of these spaces induced by their usual metrics. Then the completion of the product $X \times Y$ in the category of ϑ-uniform spaces, although ϑ-homeomorphic to the ordinary square Q^2, does not coincide with the product $X \times \hat{Y}$ of their completions, which is also ϑ-homeomorphic to the square Q^2.

REFERENCES

[1] B.K. Belnov, Classification of Hausdorff extensions, *Vestnik MGU*, 5 (1969), 23-28.

[2] N. Bourbaki, General Topology.

[3] N.V. Veličko, *H*-closed topological spaces, *Math Sbornik*, 70 (1966), 98-112.

[4] M. Katětov, Über *H*-algeschlossene und bikompakte Räume, *Čas. Mat. Fys.*, 69 (1940), 36-49.

[5] M. Katětov, On *H*-closed extensions of topological spaces, *Čas. Mat. Fys.*, 73 (1947), 17-32.

[6] J. Mioduszewski – L. Rudolf, *H*-closed and extremally disconnected Hausdorff spaces, *Dissert. Math.*, 66 (1969), 5-52.

[7] V.V. Fedorčuk, ϑ-proximity and ϑ-absolutes, *DAN SSSR*, 180 (1968), 546-549.

[8] V.V. Fedorčuk, Perfect irreducible mappings and generalized proximities, *Mat. Sbornik*, 76 (1968), 523-536.

[9] V.V. Fedorčuk, Uniform spaces and perfect irreducible mappings of topological spaces. *DAN SSSR*, 192 (1970), 1228-1230.

[10] V.V. Fedorčuk, On *H*-closed extensions of ϑ-proximity spaces, *Mat. Sbornik*, 89 (1972), 400-418.

COLLOQUIA MATHEMATICA SOCIETATIS JÁNOS BOLYAI
8. TOPICS IN TOPOLOGY, KESZTHELY (HUNGARY), 1972.

ON A CERTAIN CLASS OF $C(X)$-MODULES (PRELIMINARY REPORT)

H.R. FISCHER

1. All topological spaces $X, \ldots$ considered here are assumed to be paracompact and T_2, and $C(X)$ denotes the algebra of continuous real-valued functions on X. It is well-known that for a large class of spaces $X, \ldots,$ the "global section functor" Γ establishes a category equivalence between the finite-dimensional vector bundles over X and the finitely generated projective $C(X)$-modules, cf. R. Swan [11]; this is the case e.g. when X is compact and also when X is (paracompact, T_2 and) of finite covering dimension and has a finite number of components, the reason being that in all these cases every vector-bundle over X possesses a *finite* trivialization. This fact then is used to embed a given vector-bundle as a direct summand into some trivial bundle. It is to be noted that the construction of this embedding does not make any essential use of Riemannian structures on the bundles nor of the fact that the fibres are finite-dimensional; therefore, it carries over to Banach bundles without change.

If ξ and η are finite-dimensional vector-bundles, then every $C(X)$-homomorphism $\Gamma(\xi) \to \Gamma(\eta)$ is induced by a bundle-map $\xi \to \eta$. If the fibres are allowed to be infinite-dimensional Banach spaces, this is no longer true and additional conditions of a vaguely "topological" nature have to be imposed. We arrive at these restrictions in the following manner: Let E, F be (real, pour fixer les idées) Banach spaces with admissible norms denoted by $\| \ \|$. We can identify the module of global sections of the trivial bundle $E_X = X \times E$ with the space $C(X, E)$ of all continuous maps $X \to E$ ("principal part" of a section); $C(X, E)$ is a $C(X)$-module in the obvious manner. If E is finite-dimensional, then the submodule $I_x = \{f \in C(X, E) \mid f(x) = 0\}$ is equal to the module $m_x \cdot C(X, E)$ where $m_x \subset C(X)$ is the maximal ideal of the point $x \in X$. This implies immediately that every $C(X)$-homomorphism $C(X, E) \to C(X, F)$ is induced by a bundle map $E_X \to F_X$, cf. [11]. For infinite-dimensional E, the situation is different. One can single out those homomorphisms which arise from bundle-maps by a continuity condition with respect to a suitable notion of convergence in the modules $C(X, E)$: we define, for $f \in C(X, E)$, $\|f\|(x) = \|f(x)\|$, thus obtaining a map $\|$ á: $C(X, E) \to C(X)$ which has the properties of a norm (with respect to the order in $C(X)$). A homomorphism φ from $C(X, E)$ to $C(X, F)$ is called *continuous* if the function $\|\varphi\|(x) = \sup_{\|f\| \leqslant 1} \|\varphi f\|(x)$ is locally bounded. If this is the case, then $f(x) = 0$ implies $(\varphi f)(x) = 0$; this can be used to show that φ is induced by a bundle map $E_X \to F_X$.

The "continuity condition" imposed on φ is somewhat artificial; however, it can be restated in the following manner:

The $C(X)$-valued "norm" $\| \ \|$ defines a convergence structure (Limitierung, cf. e.g. [6], [8]) with respect to which a filter $\mathscr{F}$ on $C(X, E)$ converges to f if and only if $\|\mathscr{F} - f\| \to 0$ *continuously* in $C(X)$; explicitly, this means: given $x \in X$ and $\epsilon > 0$, there exist a member $F \in \mathscr{F}$ and a neighbourhood U of x such that $\|g - f\|(y) < \epsilon$ for $g \in F$ and $y \in U$. It is an easy matter to verify that the structure defined in this manner coincides with standard *continuous convergence* (cf. e.g. [3] and also [1] where continuous convergence is called "convergence uniforme locale", same in [5]). The $C(X)$-homomorphism φ: $C(X, E) \to C(X, F)$ then is continuous if and only if $\|\varphi\|$ is locally bounded. One notes also that continuous convergence is the coarsest *admissible* convergence on $C(X, E)$ for which $\| \ \|$ is continuous, $C(X)$ being equipped with the structure of continuous convergence. Also, the submodule $\{f \mid \|f\|(x) = 0\}$ coincides with I_x and $C(X, E)/I_x = E_x$ is a vector space isomorphic to E, the isomorphism being induced by the evaluation map $f \to f(x)$ whose kernel is I_x. Since $\|f\|(x) = \|f(x)\|$ by definition, $\| \ \|$ induces on E_x a norm for which the isomorphism $E_x \cong E$ becomes an isometry. By topologizing $\cup E_x$ with the product topology of $X \times E$, one obtains a (trivial) bundle whose section module is isomorphic to $C(X, E)$. Note that $m_x \cdot C(X, E)$ is dense in I_x and that this implies that a continuous map $C(X, E) \to$ $\to C(X, F)$ is induced by a bundle map $E_X \to F_X$.

2. The preceding remarks motivate the introduction of a category of $C(X)$-modules with admissible convergence structures as follows:

Definition. Let M be a $C(X)$-module. A *Finsler structure* on M is a map $\| \ \|: M \to C(X)$ with the following properties:

(1) $\|m\| \geqslant 0$ and $\|m\| = 0$ if and only if $m = 0$

(2) $\|m + n\| \leqslant \|m\| + \|n\|$

(3) for $f \in C(X)$ and $m \in M$, $\|fm\| = |f| \cdot \|m\|$

If $(M, \| \ \|)$ is a module with Finsler structure, we define a convergence structure on M- by:

The convergence on M is the coarsest admissible convergence such that $\| \ \|$ is continuous $M \to C(X)$, the latter being given the structure of continuous convergence.

Equivalently and more explicitly, this means:

A filter $\mathscr{F}$ on M converges to $m_0 \in M$ if and only if $\|\mathscr{F} -$ $- m_0\| \to 0$ continuously, that is to say: for $x \in X$ and $\epsilon > 0$, there exist $F \in \mathscr{F}$ and a neighbourhood U of x such that $\|m - m_0\|(y) < \epsilon$ for $y \in U$ and $m \in F$. Modules with Finsler structures will always be equipped with this generalized continuous convergence; this is always a T_2-structure and also is "locally convex" in the sense that if $\mathscr{F} \to 0$, the same holds for the filter $\mathscr{F}^{\#}$ generated by the absolutely convex hulls $F^{\#}$ of the $F \in \mathscr{F}$. Moreover, the convergence is regular in the sense of [2], [6], as is readily seen. Since it is admissible in particular for the additive structure of M, it also is uniformizable ([4]); in fact, a natural Cauchy structure ([7]) is obtained as follows: a filter $\mathscr{F}$ is Cauchy if and only if

$\mathscr{F} - \mathscr{F} \to 0$. cf. also [5]. $(M, \| \ \|)$ is said to be complete if every Cauchy-filter converges. We now give the following

Definition. A *Finsler module* is a $C(X)$-module with a Finsler structure $\| \ \|$ whose natural Cauchy-structure is *complete.*

Examples are, of course, the modules $C(X, E)$ mentioned above (E a Banach space). At this point, there naturally arises the question of *completion* of modules with Finsler structure. As is generally the case with uniform convergence structures, there are some difficulties: the general completion in the sense of Wyler ([12]) is "too fine" for practical purposes. It seems possible and perhaps more reasonable to use completions of Cauchy structures, e.g. in the sense of [7], [9]. In the present situation, one can essentially copy the completion of normed linear spaces, provided that the modules satisfy one additional condition, sufficient for the construction and normally satisfied in the applications: for $m \in M$, we define the support $\mathrm{supp}(m)$ to be $\mathrm{supp}(\| m \|)$. Then we require that

(*) every locally finite family (m_a) in M have a sum $\sum m_a \in M$.

The sum $\sum m_a$ is defined, as usual, as the limit of the net $\left\{\sum_{\Lambda} m_a \mid \Lambda \text{ finite}\right\}$; the notion used thus is "unconditional summability".

Assuming (*), we proceed in the standard manner: if $\widetilde{M}$ denotes the set of Cauchy filters on M, let $\mathscr{F} \sim \mathscr{G}$ mean that $\mathscr{F} - \mathscr{G} \to 0$; clearly, this is an equivalence relation. $\hat{M}$ denotes the quotient $\widetilde{M}/\sim$. Due to the admissibility of the convergence on M, $\hat{M}$ becomes a $C(X)$-module in a natural way. The triangle inequality (2) implies that if $\mathscr{F}$ is Cauchy, so is $\| \mathscr{F} \|$ in $C(X)$, so that $\lim \| \mathscr{F} \|$ exists by the completeness of $C(X)$. Moreover, if $\mathscr{F} \sim \mathscr{G}$, then $\| \mathscr{F} \| \sim \| \mathscr{G} \|$ and therefore $\lim \| \mathscr{F} \|$ only depends on the class $\hat{m}$ of $\mathscr{F}$. With this, one obtains a Finsler structure on $\hat{M}$. If M is embedded by mapping $m \in M$ to the class of the ultrafilter $\dot{m}$, the embedding becomes "isometric". Also, M is dense in $\hat{M}$ since for $\hat{m} \in \hat{M}$ and $\mathscr{F} \in \hat{m}$, the filter generated by $\mathscr{F}$ (on $\hat{M}$) converges to $\hat{m}$. Condition (*) even implies that M is uniformly dense in $\hat{M}$: given $\hat{m} \in \hat{M}$ and $\epsilon > 0$, there is $m \in M$ such that $\| \hat{m} - m \| < \epsilon$ everywhere. Such an m can be obtained as the sum of a suitable locally finite family $\{f_a m_a\}$ where the m_a-s in M, approximate $\hat{m}$ on the sets U_a of some locally finite open covering, and $\{f_a\}$ forms a partition of unity subordinated to $\{U_a\}$. Now, if $A \subset \hat{M}$, let $A_\epsilon = \{\hat{n} \in \hat{M} \mid \exists \hat{m} \in A: \|\hat{m} - \hat{n}\| < \epsilon$ on $X\}$; if $A \neq \phi$, then $A_\epsilon \cap M \neq \phi$. Therefore, if Φ is a Cauchy filter on $\hat{M}$ and if $\tilde{\Phi}$ denotes the filter generated by all sets A_ϵ for $\epsilon > 0$ and $A \in \Phi$, Φ has a trace on $\hat{M}$. Since Φ still is Cauchy, its trace on M is Cauchy and so converges in M. This implies convergence of Φ in the usual way, thus establishing the completeness of $\hat{M}$.

If E is a normed space, $\hat{E}$ its completion, then $C(X, E)$ certainly satisfies (*) and it is easily seen that $C(X, E)^{\wedge} = C(X, \hat{E})$, including the Finsler structures. The interest in completeness and completions lies in the following:

Suppose that M is complete with Finsler structure $\| \ \|$. For each $x \in X$, let $M_x = \{m \in M \mid \|m\|(x) = 0\}$, a closed submodule. Denote M/M_x by E_x. Then the quotient structure makes E_x a Banach space; the triple $(M, \cup E_x, \| \ \|)$ then represents a *continuous field of Banach spaces* in the sense of [5]; in particular, there exists a fibre space $E \xrightarrow{p} X$, $E_x = p^{-1}(x)$, with a Finsler structure $\| \ \|: E \to R$ and "enough" global sections such that $\Gamma(E) \cong M$ including the respective Finsler structures. The construction of $E \to X$ is due to N. Bourbaki and may be found in [5]. The Bourbaki fibre space associated with $(M, \| \ \|)$ is unique up to isomorphism; one thus obtains:

Theorem. *Let X be paracompact, T_2. The functor $E \to \Gamma(E)$ establishes a category equicalence between the "Bourbaki fibre spaces" with Finsler structures and enough sections on the one hand, the Finsler modules $(M, \| \ \|)$ on the other.*

Of course, one still has to argue that the continuous homomorphisms are precisely those which are induced by maps of the corresponding fibrations; this is done in essentially the manner indicated earlier for the special case $M = C(X, E)$. The category equivalence is an extension of Swan's theorem (in those cases where it applies). This follows from the fact that, with the assumptions made here, all continuity considerations become automatic in the case of spaces $C(X, E)$ with finite-dimensional E: Assume that $E \cong R^n$. Such an isomorphism yields an isomorphism $C(X, E) \cong C(X)^n$

which identifies continuous convergence on $C(X, E)$ with the product structure on $C(X)^n$. Moreover, any $C(X)$-homomorphism $C(X)^n \to C(X)^m$ is continuous, being represented by an $(n \times m)$-matrix over $C(X)$. Note here that the product convergence on $C(X)^n$ is characterized by the property that the "coordinate maps" $(f_j)_{1 \leq j \leq n} \to f_i$ are continuous.

The category equivalence is additive: if $E \to X$ and $F \to X$ are two "Bourbaki fibre spaces", their sum $E \oplus F$ is, as usual, given by the fibred product $E \times_X F$ and one has $\Gamma(E \oplus F) = \Gamma(E) \oplus \Gamma(F)$ including the convergences. The construction of tensor products seems to present some difficulties as long as the given fibrations are not necessarily locally trivial. In the trivial case, one can obtain the following: let E, F be Banach spaces, E_X and F_X the trivial bundles with fibre E, F resp. Clearly, $E_X \otimes F_X$ is well-defined: it is the trivial bundle with fibre $E \hat{\otimes} F$, the completed projective tensor product. The map $s, t \to (s \otimes t)(x) = s(x) \otimes t(x)$ defines a natural homomorphism $C(X, E) \otimes_{C(X)} C(X, F) \to C(X, E \hat{\otimes} F)$. Evidently, the map j thus obtained can be described without reference to the bundles. Let us point out here, however, that it is not at all clear how the convergences of $C(X, E)$ and $C(X, F)$ might be "lifted" to the tensor product; the question deserves further investigation. On the other hand, j commutes with direct sum decompositions. Modulo the Swan theorem, this yields as a special case a very simple proof that $\Gamma(\xi) \otimes \Gamma(\eta) = \Gamma(\xi \otimes \eta)$ for finite dimensional bundles ξ, η (over suitable spaces, cf. section 1).

Finally, the bundle maps in the strict sense between, say, E_X and F_X, i.e. those maps φ: $E_X \to F_X$ for which the corresponding map $X \to$ $\to L(E, F)$ is continuous for the norm-topology, correspond to strongly continuous homomorphisms φ: $C(X, E) \to C(X, F)$, i.e. those for which $\| \varphi \|$ is continuous on X rather than just locally bounded. If M is a strongly continuous direct summand of $C(X, E)$, it is the section module of a Banach fibre bundle in the usual sense (cf. e.g. [10]). The classification of arbitrary (continuous) direct summands of $C(X, E)$ is incomplete so far; difficulties arise principally from a lack of precise information about the properties of continuous convergence restricted from $L(E)$ to $GL(E)$.

A more detailed report will be published elsewhere.

REFERENCES

[1] A. Bastiani, Applications différentiables et variétés différentiables de dimension infinite, *Journal d'Analyse Math.*, 13 (1964).

[2] C.H. Cook – H.R. Fischer, Regular convergence spaces, *Math. Ann.*, 174 (1967).

[3] C.H. Cook – H.R. Fischer, On equicontinuity and continuous convergence, *Math. Ann.*, 159 (1965)

[4] C.H. Cook – H.R. Fischer, Uniform convergence structures, *Math. Ann.*, 173 (1967).

[5] J. Dixmier – A. Douady, Champs continus d'espaces hilbertiens . . . , *Bull. Soc. Math. France.*, 91 (1963).

[6] H.R. Fischer, Limesräume, *Math. Ann.*, 137 (1959).

[7] H.H. Keller, Die Limes-Uniformisierbarkeit der Limesräume, *Math.. Ann.*, 176 (1968).

[8] H.-J. Kovalsky, Limesräume und Komplettierung, *Math. Nachr.*, 12 (1954).

[9] F.R. Miller, The approximation of topologies in Functional Analysis, thesis, *Univ. of Massachusetts*, 1968.

[10] R. Palais, Lectures on differential Topology . . . , *Brandeis Univ.*, 1964/65.

[11] R. Swan, Vector bundles and projective modules, *Trans. Am. Math. Soc.*, 105 (1962).

[12] O. Wyler, Ein Komplettierungsfunktor für uniforme Limesräume, *Math. Nach.*, 46 (1970).

COLLOQUIA MATHEMATICA SOCIETATIS JÁNOS BOLYAI
8. TOPICS IN TOPOLOGY, KESZTHELY (HUNGARY), 1972.

ON A PROBLEM OF J. NOVÁK

R. FRIČ

At the *Kanpur Topological Conference* in 1968 J. Novák posed the following problem (cf. [3] Problem 9):

Find a set C_0 of sequentially continuous functionals on $C(R)$ such that $B(R)$ is the C_0 sequential envelope of $C(R)$, where $C(R)$ is the set of all continuous real-valued functions of a real variable, $B(R)$ is the set of all Baire functions and $C(R)$ and $B(R)$ are endowed with the pointwise sequential convergence and the corresponding convergence closure.

The purpose of this note is to give a solution of the problem.

Recall that a convergence space (L, λ) or simply L is a T_1 closure space such that the closure operator λ for L is derived from a sequential (one-valued star) convergence for L. By $C(L)$ it is denoted the set of all continuous (real-valued) functions on L and by $C_0 \subset C$ a subset. The space (L, λ) is said to be C_0 sequentially regular if the defining convergence for L is projectively generated by C_0 (i.e. $\lim x_n = x_0$ iff for each $f \in C_0$ we have $\lim f(x_n) = f(x_0)$).

A convergence space (L_2, λ_2) is said to be C_0 sequential envelope of a C_0 sequentially regular convergence space (L_1, λ_1) where $C_0 \subset \subset C(L_1)$, if

(i) (L_1, λ_1) is a sequentially dense subspace of (L_2, λ_2) (i.e. $L_1 \subset L_2$, $\lambda_1 = \lambda_{2/L_1}$, $\lambda_2^{\omega 1} L_1 = L_2$).

(ii) (L_1, λ_1) is C_0 embedded in (L_2, λ_2) (i.e. each $f \in C_0$ can be extended to a function $\bar{f} \in C(L_2)$) and (L_2, λ_2) is $\bar{C}_0(L_2)$ sequentially regular, where $\bar{C}_0(L_2) = \{ g | g \in C(L_2),\ g/L_1 \in C_0 \}$.

(iii) (L_2, λ_2) is $\bar{C}_0(L_2)$ closed (i.e. there is no convergence space (L_3, λ_3) containing (L_2, λ_2) as a proper subspace and fulfilling (i) and (ii) with regard to (L_1, λ_1) and (L_3, λ_3)).

The detailed exposition of the theory of convergence spaces can be found in [3].

According to Theorem 6 in [1] a convergence space (L_2, λ_2) is a C_0 sequential envelope of a C_0 sequentially regular convergence space (L_1, λ_1) iff

(1) (L_1, λ_1) is a sequentially dense C_0 embedded subspace of (L_2, λ_2).

(2) (L_2, λ_2) has the following property p with respect to $\bar{C}_0(L_2)$:

(p) For each two sequences $< x_n >$, $< y_n >$ of points of L_2 such that

$$(\lambda_2 \cup (x_n)) \cap (\lambda_2 \cup (y_n)) = \phi$$

there is a function $f \in \bar{C}_0(L_2)$ such that

$$\lim f(x_n) = \lim f(y_n)$$

does not hold.

The following Lemma is easy to prove.

Lemma. *Let* (L_2, λ_2) *be a* C_0 *sequential envelope of a* C_0 *sequentially regular convergence space* (L_1, λ_1). *Let* $C_1 \subset C(L_1)$ *be the set of all finite linear combinations of elements of* C_0. *Then* (L_2, λ_2) *is* *a* C_1 *sequential envelope of* (L_1, λ).

Solution. Denote by $E = \{ev_r(f) = f(r) \mid f \in C(R), \ \ r \in R\}$ the set of all evaluations on $C(R)$ at points of R. It is easy to see that $E \subset \subset C(C(R))$, $C(R)$ is E sequentially regular, $C(R)$ is E embedded in $B(R)$ and $B(R)$ has the property p with respect to $\bar{E}$, where $\bar{E} = \{ev_r(f) = f(r) \mid f \in B(R), \ r \in R\}$. Consequently $B(R)$ is an E sequential envelope of $C(R)$.

Finally, denote by $F = \{\sum a_i ev_{r_i}(f) \mid ev_{r_i} \in E, \ a_i, r_i \in R, \ i \in n\}$ the set of all finite linear combinations of the evaluations on $C(R)$. A theorem of S. Mazur – V. Pták (cf. [2]) asserts that every sequentially continuous linear functional on $C(R)$ is u-continuous and F is precisely the set of all u-continuous linear functionals on $C(R)$, where u is the topology of pointwise convergence for $C(R)$. From our Lemma is follows that $B(R)$ is an F sequential envelope of $C(R)$.

REFERENCES

[1] R. Frič, Sequential envelope and subspaces of the Čech-Stone compactification, General Topology and its Relations to Modern Analysis and Algebra, *Proc. Third Prague Topological Sympos.*, 1971. Academia, Prague, 1972, 123-126.

[2] J.R. Isbell, Mazur's theorem, General Topology and its Relations to Modern Analysis and Algebra, (I) *Proc. (First) Prague Topological Sympos.*, 1961. Publishing House of the Czechoslovak Academia of Sciences, Prague, 1962, 221-225.

[3] J. Novák, On some problems concerning the convergence spaces and groups, General Topology and its Relations to Modern Analysis and Algebra, *Proc. Kanpur Topological Conf.*, 1968. Academia, Prague, 1971, 219-229.

COLLOQUIA MATHEMATICA SOCIETATIS JÁNOS BOLYAI
8. TOPICS IN TOPOLOGY, KESZTHELY (HUNGARY), 1972.

COMPLETE PAVINGS

Z. FROLÍK

By a paving of set X we mean a finitely additive and finitely multiplicative cover of X. In what follows let $\mathcal{F}$ be a paving of X. By $\gamma_X \mathcal{M}$ we denote the collection of all $X - M$, $M \in \mathcal{M}$.

1. The following properties of an $\mathcal{F}$-filter Φ are equivalent:

a. If $F_1 \cup F_2 \in \Phi$, $F_i \in \mathcal{F}$ then $F_1 \in \Phi$ or $F_2 \in \Phi$.

b. The collection $\widetilde{\Phi} = X - F | F \in \mathcal{F} - \Phi = \gamma(\mathcal{F} - \Phi)$ is a filter in $\gamma \mathcal{F}$.

c. $\Phi \cup \widetilde{\Phi}$ is a maximal centred collection in $\mathcal{F} \cup \gamma \mathcal{F}$

d. There is an ultrafilter Ψ with $\Psi \cap \mathcal{F} = \Phi$.

The filter with filter with these properties is called *prime.*

If Φ is prime then so is $\widetilde{\Phi}$, and

$$\approx\!\!\!\!\Phi = \Phi .$$

Every maximal filter is prime, and an $\mathcal{F}$-filter Φ is called comaximal if $\widetilde{\Phi}$ is a maximal $\gamma \mathcal{F}$-filter.

2. Separation properties.

A paving $\mathscr{F}$ of X is called *normal* if each prime (comaximal) filter has a unique extension to a maximal filter, or equivalently, if for each each two disjoint sets F_1 and F_2 in $\mathscr{F}$ there exist disjoint U_1 and U_2 in $\gamma\mathscr{F}$ with $F_i \subset U_i$. For example, a topological space X is normal iff closed X is normal, X is extremally disconnected iff open X is normal.

A paving $\mathscr{F}$ is called *separated* if the intersection of each prime (comaximal) filter is empty or a singleton. This is equivalent to usual separation of distinct points by elements of $\gamma\mathscr{F}$.

A paving $\mathscr{F}$ is called *regular* if $\cap \Phi_1 = \cap \Phi_2$ for each prime Φ_1, Φ_2 for which $\Phi_1 \subset \Phi_2$. This is again equivalent to separation of points and the elements of $\mathscr{F}$.

One defines compactness in the usual way. A paving $\mathscr{F}$ is called *comaximal-compact* (another name: *almost compact*) if $\cap \Phi \neq \phi$ for each comaximal Φ. The usual relations hold.

3. A paving $\mathscr{F}$ of X is called

complete if every maximal $\mathscr{F}$-filter with CIP has non-void intersection;

prime-complete if every prime $\mathscr{F}$-filter with CIP has non-void intersection;

comaximal-complete if every comaximal $\mathscr{F}$-filter with CIP has non-void intersection.

Example. If X is a Hausdorff topological space, and if $\mathscr{F}$ is the paving of all closed sets in X then completeness of $\mathscr{F}$ was studied by a number of authors, prime-complete is called x_1-ultracompact by J. van der Slot [1], and comaximal-complete was introduced in Frolík [3] under the name almost realcompact.

It is obvious that compact implies Lindelöf, Lindelöf implies prime complete, and prime-complete implies both complete and comaximal complete.

Theorem. *For regular* $\mathscr{F}$, *prime-complete and comaximal-complete are equivalent, and hence comaximal-complete implies complete. For* $\aleph_0$*-normal (defined obviously) complete implies prime complete, and hence prime-complete and complete are equivalent.*

Corollary. *The following properties of a regular topological space* X *are equivalent:*

a/ X *is* $\aleph_1$*-ultracompact (= closed* X *is prime-complete)*

b/ X *is almost realcompact (= closed* X *is comaximal-complete)*

c/ X *is the perfect image of a realcompact space.*

4. For proofs, references and many results we refer to Frolík [1]. If $\mathscr{F}$ is complemented (i.e. if $\mathscr{F} = \gamma \mathscr{F}$) then prime, maximal and comaximal coincide, and one can prove several completeness theorems for large extensions of $\mathscr{F}$, see Frolík [2].

REFERENCES

[1] Z. Frolík, Prime filters with CIP, *Comment. Math. Univ. Carolinae,* 13 (1972), 553-575.

[2] Z. Frolík, Hyper-extensions of σ-algebras, *Comment. Math. Univ. Carolinae,* 14 (1973), 361-375.

[3] Z. Frolík, A generalization of realcompact spaces, *Czech. Math. J.,* 13 (1963), 127-138.

[4] J. van der Slot, Properties that are closely related to compactness, *Matematisch Centrum Amsterdam,* 1966, No 20.

COLLOQUIA MATHEMATICA SOCIETATIS JÁNOS BOLYAI
8. TOPICS IN TOPOLOGY, KESZTHELY (HUNGARY), 1972.

A CHARACTERIZATION OF BIPERFECT TOPOGENOUS ORDERS

S. GACSÁLYI

A kernelled cover is defined as a triple consisting of a cover of the given set, of a partition of the same set, and of a one-to-one mapping of the partition onto the cover, such that any member of the partition is a subset of its image.

It is shown that a natural one-to-one correspondence exists between the biperfect topogenous orders over a given set and the kernelled covers of that set.

The kernelled covers are partially ordered with the help of the corresponding biperfect topogenous orders, and an "inner" characterization of the partial order introduced is also given.

A detailed exposition has appeared in "*Publicationes Mathematicae, Debrecen*", Vol. 18.

COLLOQUIA MATHEMATICA SOCIETATIS JÁNOS BOLYAI
8. TOPICS IN TOPOLOGY, KESZTHELY (HUNGARY), 1972.

GENERALIZATIONS OF THE UNIFORM CONVERGENCE

W. GÄHLER

We shall introduce several generalizations of the uniform convergence which are of significance in the general differential calculus.

1. LIMIT-SPACES AND LIMIT-UNIFORM SPACES

Let X be a set. Among the filters on X we also admit the improper one, i.e. the power set $\mathfrak{P}X$ of X. A system of subsets of X whose complements form a filter on X will be called a dual filter on X. The filters, as well as the dual filters, on X form a complete distributive lattice with respect to the set-theoretic ordering. A non-empty set $\mathfrak{u}$ of filters (dual filters) on X is called an $\cap$-ideal of filters (dual filters) on X if the following equivalence is satisfied:

$$\mathfrak{F}, \mathfrak{G} \in \mathfrak{u} \Leftrightarrow \mathfrak{F} \cap \mathfrak{G} \in \mathfrak{u}.$$

The $\cap$-ideals of filters, as well as the $\cap$-ideals of dual filters on X also form a complete distributive lattice with respect to the set-theoretic ordering. In what follows the occurring lattice theoretic joins are always meant to be with respect to these or the above lattices.

The principal filter $\{A \in \mathfrak{P}X \mid F \subseteq A\}$ and dual filter $\{A \in \in \mathfrak{P}X \mid A \subseteq F\}$, respectively, generated by a subset F of X will be denoted by $[F]$ and (F), respectively. The $\cap$-ideal generated by a single filter or dual filter $\mathfrak{F}$ will be denoted by $[\mathfrak{F}]$. Instead of $[\{x\}]$ (for $x \in X$) we shall write $[x]$.

Every mapping τ of X into the set of all $\cap$-ideals of filters on X is called a limit-structure of X. If τ is a limit-structure such that for each $x \in X$ the filter $[x]$ belongs to $\tau(x)$, then τ is called a pseudo-topology. Thus we make a distinction between limit-structures and pseudo-topologies. Concerning pseudo-topologies we refer to H. R. Fischer [3]. To every limit-structure τ there corresponds a natural pseudo-topology τ_p for which

$$\tau_p(x) = [[x]] \vee \tau(x) .$$

Obviously, if τ is a pseudo-topology then τ_p coincides with τ.

A pseudo-topology τ is called a multistage topology, if for any $x \in X$ the set $\tau(x)$ is a principal $\cap$-ideal.

With respect to a limit-structure τ of X we can define for every subset A of X its closure $\bar{A}$ as follows:

$$\bar{A} = \{x \in X \mid \exists \mathfrak{F} \in \tau_p(x)\,(F \cap A \neq \phi \quad \text{for all} \quad F \in \mathfrak{F})\} .$$

It is well-known that here the closure axioms $\bar{\phi} = \phi$, $A \subseteq \bar{A}$, and $\bar{A} \cup \bar{B} = \overline{A \cup B}$ are satisfied, moreover it can be shown that the mapping which sends every limit-structure τ to the corresponding closure operation $A \rightsquigarrow \bar{A}$ establishes a one-to-one correspondance between the set of all multistage topologies and the set of all closure operations. If τ is a multistage topology such that $\bar{A} = \bar{\bar{A}}$ is always satisfied, then as is known τ is called a topology.

A limit-structure τ is called separated, if, for any $x, y \in X$, $x \neq y$ implies

$$\tau(x) \cap \tau(y) = \{\mathfrak{P}X\}$$

(note that the improper filter $\mathfrak{P}X$ belongs to every $\cap$-ideal $\tau(x)$).

The set X provided with a limit-structure is called a limit-space, and similarly we can speak about pseudo-topological, multistage topological and topological spaces, respectively.

A limit-uniform structure on X is an $\cap$-ideal $\mathfrak{u}$ of filters on $X \times X$ possessing the following two properties:

1. If $\mathfrak{F}$ belongs to $\mathfrak{u}$, the so does

$$\mathfrak{F}^{-1} = \{ F^{-1} \mid F \in \mathfrak{F} \}$$

as well.

2. If $\mathfrak{F}$ and $\mathfrak{G}$ belong to $\mathfrak{u}$, then so does the filter $\mathfrak{G} \circ \mathfrak{F}$ which has the system

$$\{ G \circ F \mid F \in \mathfrak{F}, G \in \mathfrak{G} \}$$

as a basis.

Here F^{-1} is the converse of the relation F and $G \circ F$ is the usual composition of F and G. A limit-uniform structure $\mathfrak{u}$ is called a pseudo-uniform structure if the principal filter $[\Delta]$ generated by the identical mapping Δ of X onto itself belongs to $\mathfrak{u}$. Concerning pseudo-uniform structures we refer to C.H. Cook and H.R. Fischer [2]. For any limit-uniform structure $\mathfrak{u}$ the family

$$\mathfrak{u}_p = [[\Delta]] \vee \mathfrak{u}$$

is a pseudo-uniform structure. Obviously, every pseudo-uniform structure $\mathfrak{u}$ coincides with the corresponding $\mathfrak{u}_p$. A pseudo-uniform structure is called a uniform structure if it is a principal $\cap$-ideal. The generating filter is then a uniform structure in the usual sense.

To every limit-uniform structure $\mathfrak{u}$ there corresponds in a natural way the uniform structure $\mathfrak{u}_\omega = [\mathfrak{B}_\mathfrak{u}]$, where $\mathfrak{B}_\mathfrak{u}$ is the largest subfilter of $\bigcap_{\mathfrak{F} \in \mathfrak{u}_p} \mathfrak{F}$, with $\mathfrak{B}_\mathfrak{u} \subseteq \mathfrak{B}_\mathfrak{u} \circ \mathfrak{B}_\mathfrak{u}$ and if $\mathfrak{u}$ is a uniform structure then this obviously coincides with $\mathfrak{u}$.

If X is provided with a limit-uniform structure then it is called a limit-uniform space. Similarly, we can talk about pseudo-uniform and uniform spaces.

Limit-uniform structures generate limit-structures, namely to every limit-uniform structure $\mathfrak{u}$ there corresponds the limit-structure $\tau_{\mathfrak{u}}$ for which

$$\tau_{\mathfrak{u}}(x) = \{ \mathfrak{F} \mid \mathfrak{F} \times [x] \in \mathfrak{u} \}$$

holds for all $x \in X$. (Of course, by the product $\mathfrak{F} \times \mathfrak{G}$ of the filters $\mathfrak{F}$ and $\mathfrak{G}$ we understand the filter on $X \times X$ with the basis

$$\{ F \times G \mid F \in \mathfrak{F}, G \in \mathfrak{G} \} .)$$

It is known that if $\mathfrak{u}$ is a pseudo-uniform or uniform structure, respectively, then $\tau_{\mathfrak{u}}$ is a pseudo-topology or a topology, respectively.

A limit-space is called limit-uniformizable or pseudo-uniformizable, respectively, if there exists a limit-uniform structure or a pseudo-uniform structure, respectively, which generates its limit-structure. By H.H. Keller [5] we have that a limit-space is pseudo-uniformizable if and only if the following two conditions are satisfied:

1. $[x] \in \tau(x)$ for any $x \in X$, i.e. τ is a pseudo-topology.

2. $\tau(x) \cap \tau(y) \neq \{ \mathfrak{P}X \}$ always implies $\tau(x) = \tau(y)$.

Furthermore one can show that a limit-space is limit-uniformizable if and only if we have

1. For any $x \in X$ with $\tau(x) \neq \{ \mathfrak{P}X \}$, $[x] \in \tau(x)$ holds.

2. As above.

3. $[y] \in \tau(x)$ always implies $[x] \in \tau(y)$.

2. CONVERGENCE OF SETS

Next we shall introduce a notion of set convergence. Let X be again a set, and $\mathfrak{B}$ be a set of $\cap$-ideals of dual filters on X and $\mathfrak{A}$ be a set of subsets of X. We shall explicate a notion of convergence in $\mathfrak{A}$ with respect to the family $\mathfrak{B}$. For every set $\mathfrak{K}$ of subsets of $\mathfrak{A}$, in particular for every filter $\mathfrak{K}$ on $\mathfrak{A}$, we put

$$\mathfrak{B}_{\mathfrak{K}} = \left\{ \mathfrak{v} \in \mathfrak{B} \mid \exists \mathfrak{F} \in \mathfrak{v} \forall F \in \mathfrak{F} \exists K \in \mathfrak{K} \left(F \subseteq \bigcap_{C \in K} C \right) \right\}.$$

If for an arbitrary $A \in \mathfrak{A}$ the set $\mathfrak{K}$ is the principal filter $[\{A\}]$ generated by $\{A\}$, then instead of $\mathfrak{B}_{\mathfrak{K}}$ we shall also write $\mathfrak{B}_A$, and thus we have

$$\mathfrak{B}_A = \{ \mathfrak{v} \in \mathfrak{B} \mid (A) \in \mathfrak{v} \}.$$

We shall say that a proper filter $\mathfrak{K}$ on $\mathfrak{A}$ has the set $A \in \mathfrak{A}$ as a $\mathfrak{B}$-limit, if we have

$$\mathfrak{B}_A = \mathfrak{B}_{\mathfrak{K}} = \mathfrak{B}_{\mathfrak{K}^*},$$

where $\mathfrak{K}^*$ denotes the set $\{ L \in \mathfrak{P}\mathfrak{A} \mid K \cap L \neq \phi$ for all $K \in \mathfrak{K} \}$.

We define that every set $A \in \mathfrak{A}$ is a $\mathfrak{B}$-limit of the improper filter $\mathfrak{P}\mathfrak{A}$.

For any $A \in \mathfrak{A}$ let $\tau_{\mathfrak{B}}(A)$ be the set of all filters $\mathfrak{K}$ on $\mathfrak{A}$ such that $\mathfrak{K}$ has $\mathfrak{A}$ as a $\mathfrak{B}$-limit.

Theorem 1. *The mapping* $\tau_{\mathfrak{B}}: A \rightsquigarrow \tau_{\mathfrak{B}}(A)$ *is a pseudo-topology. In general* $\tau_{\mathfrak{B}}$ *is not a multistage topology. If for anv two* $A, B \in \mathfrak{A}$ *from* $\mathfrak{B}_A = \mathfrak{B}_B$ *it follows* $A = B$, *then* $\tau_{\mathfrak{B}}$ *is a separated pseudo-topology.*

We are especially interested in the following three particular cases.

Case 1. Suppose that there exists a set $\mathfrak{B}$ of subsets of X such that

$$\mathfrak{B} = \{ [(B)] \mid B \in \mathfrak{B} \}.$$

Then every $\cap$-ideal $\mathfrak{v} \in \mathfrak{B}$ is determined by a set $B \in \mathfrak{B}$. In this case we obtain a set convergence with respect to a system of sets. The notion of convergence will not change in this case if we always replace $\mathfrak{B}_{\mathfrak{K}}$ by

$$\mathfrak{B}_{\mathfrak{K}} = \left\{ B \in \mathfrak{B} \mid \exists K \in \mathfrak{K} \left(B \subseteq \bigcap_{C \in K} C \right) \right\},$$

and in particular $\mathfrak{B}_A$ by

$$\mathfrak{B}_A = \{ B \in \mathfrak{B} \mid B \subseteq A \}.$$

In this case $\tau_{\mathfrak{B}}$ is always a multistage topology. Moreover if for any non-empty subfamily of $\mathfrak{A}$ its intersection also belongs to $\mathfrak{A}$ and $\mathfrak{A} \subseteq \mathfrak{B}$, then in this particular case $\tau_{\mathfrak{B}}$ is even a topology.

Case 2. There exists a set $\mathfrak{L}$ of dual filters on X with

$$\mathfrak{B} = \{[\mathfrak{F}] \mid \mathfrak{F} \in \mathfrak{L}\} .$$

In this case we have a set convergence with respect to a system of dual filters. The notion of convergence will not change in this case if we always replace $\mathfrak{B}_{\mathfrak{K}}$ by

$$\mathfrak{L}_{\mathfrak{K}} = \{\mathfrak{F} \in \mathfrak{L} \mid \forall F \in \mathfrak{F} \exists K \in \mathfrak{K} (F \subseteq \bigcap_{C \in K} C)\} ,$$

and in particular $\mathfrak{B}_A$ by

$$\mathfrak{L}_A = \{\mathfrak{F} \in \mathfrak{L} \mid \mathfrak{F} \subseteq (A)\} .$$

In this special case $\tau_{\mathfrak{B}}$ is a separated pseudo-topology provided that

$$\bigvee_{\mathfrak{F} \in \mathfrak{L}_A} \mathfrak{F} = (A)$$

holds for any $A \in \mathfrak{A}$.

Case 3. There exists a set $\mathfrak{M}$ of filters on X such that $\mathfrak{B}$ consists of exactly those $\cap$-ideals which have the sets of the form

$$\{(F) \mid F \in \mathfrak{F}\} \qquad (\mathfrak{F} \in \mathfrak{M})$$

as a basis. In this case we obtain a set convergence with respect to a system of filters. The notion of convergence will not change in this case if we always replace $\mathfrak{B}_{\mathfrak{K}}$ by

$$\mathfrak{M}_{\mathfrak{K}} = \{\mathfrak{F} \in \mathfrak{M} \mid \exists K \in \mathfrak{K} (\bigcap_{C \in K} C \in \mathfrak{F})\} ,$$

thus in particular $\mathfrak{B}_A$ by

$$\mathfrak{M}_A = \{\mathfrak{F} \in \mathfrak{M} \mid A \in \mathfrak{F}\} .$$

Theorem 2. *If for any* $x \in X$ *we have* $[x] \in \mathfrak{M}$, *then in this case* $\tau_{\mathfrak{B}}$ *is a separated pseudo-topology.*

It is worth mentioning that case 3 encompasses in part the open limits, i.e. the dual of the closed limits. In order to have a more detailed examination of this we shall take X as a limit-space. $\mathfrak{A}$ is, as before a set of subsets of X. As the lower and upper open limits, respectively of a proper filter $\mathfrak{K}$ on X we define

$$o\text{-}\underline{\lim}\,\mathfrak{K} = \Big\{x \in X \mid \forall \mathfrak{F} \in \tau(x) \exists K \in \mathfrak{K} \Big(\bigcap_{C \in K} C \in \mathfrak{F}\Big)\Big\}$$

and

$$o\text{-}\overline{\lim}\,\mathfrak{K} = \Big\{x \in X \mid \forall \mathfrak{F} \in \tau(x) \exists K \in \mathfrak{K}^* \Big(\bigcap_{C \in K} C \in \mathfrak{F}\Big)\Big\},$$

respectively, where $\mathfrak{K}^*$ is defined as above and τ denotes the limit-structure of X. If $o\text{-}\overline{\lim}\,\mathfrak{K} = o\text{-}\underline{\lim}\,\mathfrak{K}$, then we say that this set is the open limit of $\mathfrak{K}$. We define that the improper filter on $\mathfrak{A}$ has every subset of X as an open limit.

For an arbitrary $A \in \mathfrak{A}$ let $\tau_0(A)$ be the set of all filters on $\mathfrak{A}$ which have A as an open limit.

Theorem 3. *The mapping* $\tau_0 \colon A \rightsquigarrow \tau_0(A)$ *is a separated limit-structure. In general it is not a pseudo-topology, although it is if* X *is a pseudo-topological space and* $\mathfrak{A}$ *consists of open sets only.*

(Of course, a subset of a limit-space is called open if its complement is closed, i.e. coincides with its closure.) Concerning open limits we refer to [4].

Now we assume that τ is a pseudo-topology and put

$$\mathfrak{M} = \bigcup_{x \in X} \tau(x),$$

moreover we define $\mathfrak{B}$ as above in case 3. Then we have

Theorem 4. *If a proper filter* $\mathfrak{K}$ *on* $\mathfrak{A}$ *possesses a* $\mathfrak{B}$*-limit, it also possesses an open limit. These two do not have to coincide in general, although they do if the* $\mathfrak{B}$*-limit of* $\mathfrak{K}$ *is open. If the open limit of a proper filter* $\mathfrak{K}$ *on* X *exists and is equal to* X, *then the* $\mathfrak{B}$*-limit of* $\mathfrak{K}$ *also exists and is equal to* X *as well.*

From this theorem it follows, in particular, that $\tau_0(X) = \tau_{\mathfrak{B}}(X)$ for the given $\mathfrak{B}$.

3. CONVERGENCE OF MAPPINGS

In what follows we shall introduce five different types of uniform convergence with respect to a system of $\cap$-ideals of dual filters. Let X be an arbitrary set, $\mathfrak{B}$ be a system of $\cap$-ideals of dual filters on X, Y be a limit-uniform space, $\mathfrak{u}$ be the limit-uniform structure of Y, and $\mathfrak{A}$ be a set of mappings of subsets of X into Y.

For every subset F of X, every subset K of $\mathfrak{A}$ and every mapping $f \in \mathfrak{A}$ we put

$$(K, f)\,[F] = \{(g(x), f(x)) \mid g \in K, x \in F \cap Df \cap Dg\}\,,$$

where Df (Dg) denotes the domain of f (g).

Let $D[\mathfrak{A}]$ be the family of all domains of the mappings $f \in \mathfrak{A}$, and for every filter $\mathfrak{K}$ on $\mathfrak{A}$ let $D\mathfrak{K}$ be the filter on $D[\mathfrak{A}]$ with the basis $\{\{Df \mid f \in K\} \mid K \in \mathfrak{K}\}$.

We shall say that a filter $\mathfrak{K}$ on $\mathfrak{A}$, $\mathfrak{B}$-converges to a mapping $f \in \mathfrak{A}$ in type $k \in \{1, \ldots, 5\}$ if the following conditions are satisfied:

1. Df is the $\mathfrak{B}$-limit of $D\mathfrak{K}$.

2. For $k = 1, \ldots, 5$, respectively, we have

$$\left\{\begin{array}{llllll} k=1: & \underbrace{\exists\mathfrak{G}\in\mathfrak{u}}_{\text{I}} & \underbrace{\forall\mathfrak{v}\in\mathfrak{B}_{Df}}_{1} & \underbrace{\exists\mathfrak{F}\in\mathfrak{v}}_{2} & \underbrace{\forall F\in\mathfrak{F}}_{3} & \underbrace{\forall G\in\mathfrak{G}}_{\text{II}} \\ k=2: & \text{I} & 1 & \text{II} & 2 & 3 \\ k=3: & 1 & \text{I} & 2 & 3 & \text{II} \\ k=4: & 1 & \text{I} & \text{II} & 2 & 3 \\ k=5: & 1 & 2 & 3 & \text{I} & \text{II} \end{array}\right\} \exists K \in \mathfrak{K}\ ((K,f)[F] \subseteq G)\,.$$

It is easy to see that all those combinations of I, II, 1, 2 and 3 which lead to different types of convergence occur here.

The following diagram illustrates how these five different types of convergence depend on each other:

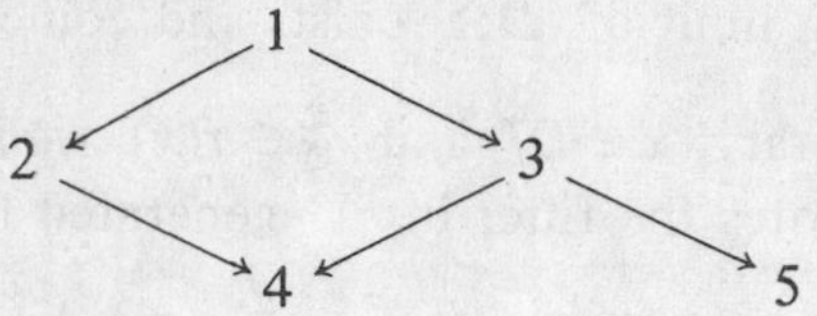

For arbitrary $k = 1, \ldots, 5$ and $f \in \mathfrak{A}$ let $\tau^k_{\mathfrak{B}}(f)$ be the set of all filters $\mathfrak{K}$ on $\mathfrak{A}$ which $\mathfrak{B}$-converge to f in type k.

Theorem 5. *For arbitrary* k *the mapping* $\tau^k_{\mathfrak{B}}: f \rightsquigarrow \tau^k_{\mathfrak{B}}(f)$ *is a limit-structure of* $\mathfrak{A}$. *If* $\mathfrak{u}$ *is a pseudo-uniform structure, then* $\tau^k_{\mathfrak{B}}$ *is a pseudo-topology.*

In the particular case 1 for $\mathfrak{B}$ the types 1, 2 and the types 3, 4, 5, respectively, coincide, hence we have two types of uniform convergence with respect to set systems.

In the particular case 2 for $\mathfrak{B}$ the types 1, 2 and the types 3, 4 coincide, hence there are three types of uniform convergence with respect to systems of dual filters.

In the particular case 3 for $\mathfrak{B}$ the types 3 and 5 coincide, hence we obtain four different types of uniform convergence with respect to systems of filters.

If, in particular, $\mathfrak{u}$ is a uniform structure, then the types 1, 3, 5 and the types 2, 4, respectively, are identical.

Theorem 6. *If* $\tau_{\mathfrak{B}}$ *is a multistage topology,* $\tau^k_{\mathfrak{B}}$ *is a pseudo-topology and* $\mathfrak{u}$ *is a uniform structure, then in the particular case* 2 *for* $\mathfrak{B}$ $\tau^k_{\mathfrak{B}}$ *is always a multistage topology.*

It is interesting that in the particular case 3 the continuous convergence is partly covered, as will be made more clear subsequently. Let X and Y be limit-spaces and τ and τ' be their limit-structures, respectively. Let $\mathfrak{A}$ be a set of mappings of subsets of X into Y. For every $K \subseteq \mathfrak{A}$ and $F \subseteq X$ let

$$[K, F] = \{ f(x) \mid f \in K, x \in F \cap Df \} .$$

We shall say that a filter $\mathfrak{K}$ on $\mathfrak{A}$ continuously converges to a mapping $f \in \mathfrak{A}$ if the following two conditions are satisfied.

1. The open limit of $D\mathfrak{K}$ exists and coincides with Df.

2. For arbitrary $x \in Df$ and $\mathfrak{F} \in \tau(x)$ we have $[\mathfrak{K}, \mathfrak{F}] \in \tau'(f(x))$, where $[\mathfrak{K}, \mathfrak{F}]$ denotes the filter in Y generated by the basis

$$\{[K, F] \mid K \in \mathfrak{K}, F \in \mathfrak{F}\} .$$

For any $f \in \mathfrak{A}$ let $\tau_s(f)$ be the set of all filters on $\mathfrak{A}$ which continuously converge to f.

Theorem 7. *The mapping* $\tau_s : f \rightsquigarrow \tau_s(f)$ *is a limit-structure. If* X *pseudo-topological space and* $\mathfrak{A}$ *consists of continuous mappings with open sets as domains, then* τ_s *is a pseudo-topology.*

Concerning continuous convergence we refer e.g. to [4].

In what follows we assume that Y is provided with an arbitrary pseudo-uniform structure $\mathfrak{u}$, which generates its limit-structure.

From the system

$$\mathfrak{M} = \bigcup_{x \in X} \tau(x)$$

a system $\mathfrak{B}$ of $\cap$-ideals of dual filters may be constructed as above. Then we obtain

Theorem 8. *Let* f *be a continuous mapping belonging to* $\mathfrak{A}$ *with an open set as its domain. We have* $\tau^4_{\mathfrak{B}}(f) \subseteq \tau_s(f)$ *and in case* $Df = X$ *especially* $\tau^4_{\mathfrak{B}}(f) = \tau_s(f)$.

In what follows, let X again be an arbitrary set and $\mathfrak{B}$ be an arbitrary system of $\cap$-ideals of dual filters on X. We shall assume, moreover, that Y is an additive abelian pseudo-topological group and construct a pseudo-uniform structure $\mathfrak{u}$ of Y generating the pseudo-topology τ' of Y. To do this, we shall denote by Γ the set of all those filters $\mathfrak{F}$ on

Y, for which the filter $\mathfrak{F} - \mathfrak{F}$ having the basis $\{G - F \mid F, G \in \mathfrak{F}\}$ belongs to $\tau'(0)$ $(0 \in Y)$. We define $\mathfrak{u}$ as the $\cap$-ideal of filters on $X \times X$ which has the set of all finite non-empty intersections of the filters $[\Delta]$, $(\mathfrak{F} \times \mathfrak{F}) + [\Delta]$ and $\mathfrak{F} \times \mathfrak{F}$ $(\mathfrak{F} \in \Gamma)$ as a basis. (Obviously, $(\mathfrak{F} \times \mathfrak{F}) + [\Delta]$ is the filter on $X \times X$ with the basis $\{(F \times G) + \Delta \mid F, G \in \mathfrak{F}\}$). With respect to the set-theoretic ordering $\mathfrak{u}$ is the smallest pseudo-uniform structure generating the pseudo-topology τ', for which Γ consists exactly of all Cauchy filters (i.e. all filters with $\mathfrak{F} \times \mathfrak{F} = \mathfrak{u}$), and for which together with two filters $\mathfrak{G}$ and $\mathfrak{H}$ the filter $\mathfrak{G} - \mathfrak{H}$ with the basis $\{G - H \mid G \in \mathfrak{G}, H \in \mathfrak{H}\}$ also belongs to $\mathfrak{u}$.

It turns out that with such a choice of $\mathfrak{u}$ the limit-structure $\tau^k_{\mathfrak{B}}$ will not change if in its definition $\mathfrak{G} \in \mathfrak{u}$ is replaced by $\mathfrak{G} \in \tau'(0)$ and $(K, f)[F]$ is replaced by

$$(K - f)[F] = \{(g - f)(x) \mid g \in K, x \in F \cap Df \cap Dg\}\,.$$

If the mappings $f \in \mathfrak{A}$ have the whole set X as their domain, then $\tau^k_{\mathfrak{B}}$ can be easily obtained from a limit-uniform structure $\mathfrak{u}^k_{\mathfrak{B}}$, as we shall show below in detail. To this end, let X, $\mathfrak{B}$, Y and $\mathfrak{A}$ again be chosen as at the beginning of the present section. Of course, we assume that $D[\mathfrak{A}] = \{X\}$. For every $H \subseteq \mathfrak{A} \times \mathfrak{A}$ and $F \subseteq X$ we put

$$[H, F] = \{(g(x), f(x)) \mid (g, f) \in H, x \in F\}.$$

For each $k = 1, \ldots, 5$ a limit-uniform structure $\mathfrak{u}^k_{\mathfrak{B}}$ can be defined on $\mathfrak{A}$ by putting

$$\left[\begin{array}{lccccc}
\mathfrak{H} \in \mathfrak{u}^1_{\mathfrak{B}} \Leftrightarrow & \underbrace{\exists \mathfrak{G} \in \mathfrak{u}}_{\mathrm{I}} & \underbrace{\forall \mathfrak{v} \in \mathfrak{B}}_{1} & \underbrace{\exists \mathfrak{F} \in \mathfrak{v}}_{2} & \underbrace{\forall F \in \mathfrak{F}}_{3} & \underbrace{\forall G \in \mathfrak{G}}_{\mathrm{II}} \\
\mathfrak{H} \in \mathfrak{u}^2_{\mathfrak{B}} \Leftrightarrow & \mathrm{I} & 1 & \mathrm{II} & 2 & 3 \\
\mathfrak{H} \in \mathfrak{u}^3_{\mathfrak{B}} \Leftrightarrow & 1 & \mathrm{I} & 2 & 3 & \mathrm{II} \\
\mathfrak{H} \in \mathfrak{u}^4_{\mathfrak{B}} \Leftrightarrow & 1 & \mathrm{I} & \mathrm{II} & 2 & 3 \\
\mathfrak{H} \in \mathfrak{u}^5_{\mathfrak{B}} \Leftrightarrow & 1 & 2 & 3 & \mathrm{I} & \mathrm{II}
\end{array}\right\} \exists H \in \mathfrak{H}\ ([H, F] \subseteq G).$$

It can be shown that $\mathfrak{u}_{\mathfrak{B}}^k$ generates the limit-structure $\tau_{\mathfrak{B}}^k$, moreover that if $\mathfrak{u}$ is a pseudo-uniform structure so is $\mathfrak{u}_{\mathfrak{B}}^k$.

4. NOTATIONS OF CONVERGENCE IN THE GENERAL DIFFERENTIAL CALCULUS

The numerous notions of differentiability defined for pseudo-topological vector spaces differ, in general, only in the special choice of the notion of convergence they use. It turns out that the above five types of uniform convergence cover those notions of convergence which correspond to the most important notions of differentiability. We shall establish this in detail for the notions of differentiability collected by Averbuh and Smolianov [1] in a table appearing in an appendix to the Russian translation of Frölicher and Bucher's book "Calculus in vector spaces without norm". We shall indicate the names of some authors investigating the corresponding notions of differentiability and add some data concerning the corresponding (or a more general) notion of convergence.

Let X and Y be pseudo-topological vector spaces over the field of real numbers and τ and τ' be the pseudo-topologies of X and Y, respectively. Let $\mathfrak{B}$ be a system of subsets of X and $\mathfrak{C}$ be a system of subsets of Y with the properties

1. $\{0\} \in \mathfrak{C}$,

2. For any $C, C' \in \mathfrak{C}$ there exists a $C'' \in \mathfrak{C}$ with $C - C' \subseteq C''$.

3. $C, C' \in \mathfrak{C}$ implies the existence of a $C'' \in \mathfrak{C}$ with $C \cup C' \subseteq$ $\subseteq C''$.

The properties of $\mathfrak{C}$ guarantee that Y has a pseudo-topology $\tau^{\mathfrak{C}}$ such that $\{[C] \mid C \in \mathfrak{C}\}$ is a basis for $\tau^{\mathfrak{C}}(0)$ and Y is an additive abelian pseudo-topological group with respect to $\tau^{\mathfrak{C}}$ and the vector addition. $\mathfrak{u}$ and $\mathfrak{u}^{\mathfrak{C}}$ stand for the pseudo-uniform structures of Y constructed from τ' and $\tau^{\mathfrak{C}}$ respectively, in the manner shown in the corresponding part of section 3.

	pseudo-uniform structure of Y	particular case for $\mathfrak{B}$	type k	Data about $\mathfrak{B}$
1. Lamadrid, Sebastião e Silva,...	$\mathfrak{u}$	1	3,4,5	$\mathfrak{B}$ arbitrary
2. Michal, Bastiani,...	$\mathfrak{u}$	3	4	$\mathfrak{M} = \bigcup_{x \in X} \tau(x)$
3. Frölicher and Bucher	$\mathfrak{u}$	3	4	$\mathfrak{M} = \{\mathfrak{F} \mid V\mathfrak{F} \in \tau(0)\}$ (V neighbourhood filter of 0 in R)
4. Keller, Eecke	$\mathfrak{u}$	–	1,3 and 5	$\mathfrak{B} = \{\bigvee_{F \in \mathfrak{F}} [\bigvee_{a>0} (aF)] \mid \mathfrak{F} \in \tau(0)\}$
5. Hyers, Fischer,...	$\mathfrak{u}$	–	2 and 4	$\mathfrak{B} = \{\bigvee_{F \in \mathfrak{F}} [\bigvee_{a>0} (aF)] \mid \mathfrak{F} \in \tau(0)\}$
6. Suhinin	$\mathfrak{u}^{\mathfrak{C}}$	–	1,2 and 3,4	$\mathfrak{B} = \{\bigvee_{F \in \mathfrak{F}} [\bigvee_{a>0} (aF)] \mid \mathfrak{F} \in \tau(0)\}$
7. Suhinin	$\mathfrak{u}^{\mathfrak{C}}$	2	1,2	$\mathfrak{L} = \{\bigvee_{a>0} (aB) \mid B \in \mathfrak{B}\}$
8. Sebastião e Silva	$\mathfrak{u}^{\mathfrak{C}}$	2	3,4	$\mathfrak{L} = \{\bigvee_{a>0} (aB) \mid B \in \mathfrak{B}\}$

Concerning Cases 5 to 8 one should observe that for the mapping τ_t appearing in 1 we always have

$$\frac{1}{\epsilon} r_t(h) = r_{\epsilon t}\left(\frac{h}{\epsilon}\right) \qquad (\epsilon > 0).$$

REFERENCES

[1] V.I. Averbuh – O.G. Smolianov, Appendix to the Russian translation of A. Frölicher and W. Bucher's book "Calculus in vector spaces without norm", Moscow, 1970.

[2] C.H. Cook – H.R. Fischer, Uniform Convergence Structures, *Math. Annalen,* 173 (1967), 290-306.

[3] H.R. Fischer, Limesräume, *Math. Annalen,* 137 (1959), 269-303.

[4] W. Gähler, Beiträge zur Theorie der Limesräume, *Theory of sets and topology, Berlin* (1972), 161-197.

[5] H.H. Keller, Die Limes-Uniformisierbarkeit der Limesräume, *Math. Annalen,* 176 (1968), 334-341.

COLLOQUIA MATHEMATICA SOCIETATIS JÁNOS BOLYAI
8. TOPICS IN TOPOLOGY, KESZTHELY (HUNGARY), 1972.

ON G_δ p-SPACES

J. GERLITS

Professor R. W. Heath raised the following problem at the 1971 *Pittsburgh Topology Converence*: is it true that a G_δ p-space is necessarily developpable?

This paper contains a counterexample to this conjecture. We give in addition some equivalent conditions for a Tychonoff space to be a G_δ p-space. One of these conditions (condition C) of Theorem 1) makes it evident that the class of G_δ p-spaces is in a certain sense very close to the class of developpable Tychonoff spaces. For necessary and sufficient conditions assuring a G_δ p-space to be developpable, see D. E. Kullman's paper [1].

We first state some definitions and propositions. A space X is called *developpable* if there is a sequence $\{\mathfrak{U}_n \mid n = 1, 2, \ldots\}$ of open coverings of X such that if $p \in U_n \in \mathfrak{U}_n$ $(n = 1, 2, \ldots)$, then $\{U_n;\ n = 1, 2, \ldots\}$ is a local base at p. Such a sequence of covers is called a development. Regular developpable spaces are the Moore spaces.

A completely regular space X is a *p-space* if in the Stone –

Čech compactification βX *of* X there is a sequence $\{\mathfrak{A}_n \mid n = 1, 2, \ldots\}$ of open families such that each $\mathfrak{A}_n$ covers X and, for each $p \in X$, $\cap \{\mathrm{St}(p, \mathfrak{A}_n) \mid n = 1, 2, \ldots\} \subset X$. The sequence $\{\mathfrak{A}_n\}$ is called a feathering of X in βX.

The definition of p-spaces is due to A.V. Arhangelskiĭ [2].

A space is a G_δ*-space* if $\Delta_x = \{(x, x) \mid x \in X\}$ is a G_δ-subset of $X \times X$. This condition is equivalent to the following one ([3]):

There is a sequence $\{\mathfrak{A}_n \mid n = 1, 2, \ldots\}$ of open coverings of X such that if $x, y \in X$ and $x \neq y$, then, for some n, $y \notin \mathrm{St}(x, \mathfrak{A}_n)$.

The following definition is a slight generalization of the notion of a development.

A sequence $\{\mathfrak{A}_n \mid n = 1, 2, \ldots\}$ of open coverings of a topological space X is called a *subdevelopment* if for each point $p \in X$ and for each sequence $\{U_n \mid n = 1, 2, \ldots\}$ with $p \in U_n \in \mathfrak{A}_n$ $(n = 1, 2, \ldots)$ the family $\{U_n \mid n = 1, 2, \ldots\}$ is a local subbase at p.

We shall need to the following – well-known – proposition.

Lemma. *Let* X *be a compact space,* $p \in X$ *and* $\mathfrak{B}$ *a family of open neighbourhoods of* p *in* X. *If* $\cap\{\mathrm{Cl}\, V \mid V \in \mathfrak{B}\} = \{p\}$ *then* $\mathfrak{B}$ *is a local subbase at* p.

Proof. Let U denote an arbitrary open set in X containing p. The family $\{\mathrm{Cl}\, V - U \mid V \in \mathfrak{B}\}$ is a family of closed sets with empty intersection in the compact space X so it does not have the finite intersection property. This means that for suitable sets $V_1, V_2, \ldots, V_n$ from $\mathfrak{B}$ $\bigcap_1^n (\mathrm{Cl}\, V_i - U) = \phi$, hence $p \in \bigcap_1^n V_i \subset U$.

Theorem 1. *The following conditions are equivalent for a Tychonoff space* X.

a) X *is a* G_δ *p-space.*

b) *The diagonal* Δ *is a* G_δ *set in* $X \times \beta X$.

c) X *has a subdevelopment.*

Proof. a) → b)

If X is a G_δ p-space then there exists a sequence $\{\mathfrak{A}_n \mid n = 1, 2, \dots\}$ of open families in βX with the following properties.

a) Each $\mathfrak{A}_n$ covers X.

b) If $p \in X$ then $\bigcap_1^\infty \mathrm{St}(p, \mathfrak{A}_n) = \{p\}$.

Let us put $V_n = \cup \{(U \cap X) \times U \mid U \in \mathfrak{A}_n\} \subset X \times \beta X$. Evidently V_n is an open *nbd* of Δ in $X \times \beta X$ and $\Delta = \bigcap_1^\infty V_n$.

b) → c). Let $\Delta = \bigcap_1^\infty V_n$ where the V_n's are open in $X \times \beta X$. Fix n and for each point $p \in X$ choose a set U_p^n open in βX such that $p \in U_p^n$ and

$$((\mathrm{Cl}_{\beta X} U_p^n) \cap X) \times (\mathrm{Cl}_{\beta X} U_p^n) \subset V_n\ ,$$

and put $\mathfrak{A}_n = \{U_p^n \cap X \mid p \in X\}$ $(n = 1, 2, \dots)$. We assert that the sequence $\{\mathfrak{A}_n\}$ forms a subdevelopment of X. Indeed, if $p \in X$ and $p \in (U_n \cap X) \in \mathfrak{A}_n$ $(n = 1, 2, \dots)$ then the sets U_n are open *nbd*-s of the point p in βX and $\bigcap_1^\infty \mathrm{Cl}_{\beta X} U_n = \{p\}$. So, by Lemma 1, we can conclude that the U_n's form a *nbd*-subbase of p in βX, hence the sets $U_n \cap X$ form a *nbd*-subbase of p in X.

c) → a) Let $\{\mathfrak{A}_n\}$ be a subdevelopment of X. By our above remark X is surely a G_δ-space.

Select for each $U \in \mathfrak{A}_n$ an open set $\widetilde{U}$ in βX with $\widetilde{U} \cap X = U$ and put $\widetilde{\mathfrak{A}}_n = \{\widetilde{U} \mid U \in \mathfrak{A}_n\}$. We assert that for $p \in X$, $\bigcap_1^\infty \mathrm{St}(p, \widetilde{\mathfrak{A}}_n) \subset X$. Indeed, if $x \in \beta X - X$ and $x \in \bigcap_1^\infty \mathrm{St}(p, \widetilde{\mathfrak{A}}_n)$, then for a suitable sequence $U_n \in \mathfrak{A}_n$ $p, x \in \widetilde{U}_n$ $(n = 1, 2, \dots)$. Select a set U open in βX such that $p \in U$, $x \notin \mathrm{Cl}_{\beta X} U$. Using that $\mathfrak{A}_n$ is a subdevelopment we get that for a suitable integer N $\bigcap_1^N U_n \subset U$. But this is a contradiction because then the non-empty open set $\bigcap_1^N \widetilde{U}_n - \mathrm{Cl}\, U$ in βX does not intersect the dense set X.

Q.e.d.

The following theorem was suggested by a problem in *American Mathematical Monthly* (Problem of R.C. Freiwald 5625, 75 (1968) p. 911).

We begin with a definition:

A topological space X is said to have the G_δ-property if for any space Y and for any function f from Y into X, the set of the points of continuity of f forms a G_δ-set in Y.

Theorem 2. A Tychonoff space X is a G_δ *p-space iff* X *has the* G_δ*-property.*

Proof. The proof of the necessity of the condition is quite simple. For, if X is a G_δ p-space denote by $\{\mathfrak{A}_n\}$ a subdevelopment of X and put

$$V_n = \cup \{V \subset Y \mid V \text{ is open in } Y, \exists U \in \mathfrak{A}_n, f(V) \subset U\}.$$

Now, V_n is open in Y and it is easily seen that the set $\bigcap_1^\infty V_n$ is precisely the set of the points of continuity of f in Y.

Before proving the sufficiency of the condition note the simple fact that both conditions "to have a subdevelopment" and "to have the G_δ-property" are hereditary and $\aleph_0$-productive.

So we can assume without loss of generality that for our Tychonoff-space X with the G_δ-property the relation $X^{\aleph_0} \underset{\text{top}}{=} X$ holds. Disregarding now the trivial case in which X contains only one point we can assume that $X = A \cup B$ where A and B are disjoint dense sets in X.

Now, let us put $Y = X \times \beta X$ and f be the following function from Y onto X:

$$f(x, y) = \begin{cases} y & \text{if } (x, y) \in (A \times A) \cup (B \times B) \\ x & \text{otherwise.} \end{cases}$$

It is very easy to check that the set of the points of continuity of the function f is exactly the diagonal Δ in $X \times \beta X$. Using now the characterization b) of Theorem 1. the proof is completed.

Q.e.d.

We give now an example of a non-developpable G_δ p-space.*

Denote by I the interval $[0, 1]$ of the reals and by $\{K_a;\ a \in A\}$ a family of subsets of I with the following properties.

a) For each $a \in A$ K_a is a convergent sequence in I without its limit point $f(a)$.

b) The intersection of two different K_a's are finite.

c) $\{K_a;\ a \in A\}$ is maximal with respect to the properties a) and b).

Now, put $X = I \cup A$; the points of I are isolated and a local base of a point $a \in A$ is a given by the family $\{U_F(a) = \{a\} \cup (K_a - F) \mid F \subset I$ finite$\}$. It is very easy to see that the space X obtained in this manner is a zero-dimensional locally compact Hausdorff-space.

Put for a positive integer n and for $a \in A$

$$F_n(a) = \{x \mid x \in K_a,\ |x - f(a)| \geqslant \frac{1}{n}\}$$

and

$$V_n(x) = \begin{cases} \{x\} & \text{if } x \in I \\ U_{F_n(x)}(x) & \text{if } x \in A, \end{cases}$$

$$\mathfrak{A}_n = \{V_n(x); x \in X\}.$$

We assert that $\{\mathfrak{A}_n \mid n = 1, 2, \ldots\}$ is a subdevelopment of X. Let $x \in X$, $U_n = U_n(x_n) \in \mathfrak{A}_n$, $x \in U_n$ $(n = 1, 2, \ldots)$. Now, if $x \in A$ then $x_n = x$ $(n = 1, 2, \ldots)$ and $\{U_n \mid n = 1, 2, \ldots\}$ is evidently a local base at X.

*As Prof. Chaber kindly informed me at the *Conference*, Prof. K.D. Burke also constructed a space with the above properties (yet unpublished).

So we can assume that $x \in I$. If for an integer n $x_n \in I$ then necessarily $x_n = x$ and $U_n = \{x\}$. Hence it can be supposed that $x_n \in \in A$ $(n = 1, 2, \ldots)$. We can also suppose that for $n \neq m$ $x_n \neq x_m$. But then $U_1 \cap U_2 = \{y_1, y_2, \ldots, y_N\}$ is finite and if $\frac{1}{n_0} < \frac{1}{2} \inf \{|y_i - y_j|| 1 \leqslant i < j \leqslant N\}$ then

$$U_1 \cap U_2 \cap U_{n_0} = \{x\}$$

and this proves the validity of the assertion.

We must prove finally that X is not developpable. Making use of the fact that an infinite subset of I contains a convergent sequence in the ordinary topology, we can conclude that a subset F of I which is closed in X must be finite. In particular, the open set I is not an F_σ-set in X and this shows that X is not developpable.

Finally, I would mention two problems.

Problem 1. Is it true that a normal G_δ p-space is developpable?

Problem 2. Is it true that a G_δ p-space is quasi-developpable?

(For the notion of quasi-developpable spaces see [4].)

REFERENCES

[1] D.E. Kullman, Developpable Spaces and p-Spaces, *Proc. Amer. Math. Soc.*, 27 (1971), 154-160.

[2] A.V. Arhangelskiĭ, On a Class of Spaces ... *Dokl. Akad. Nauk SSSR*, 151 (1963) 751-754 = *Soviet Math. Dokl.*, 4 (1963) 1051-1055.

[3] C.R. Borges, On Metrizability of Topological Spaces, *Canad. J. Math.*, 20 (1968), 795-804.

[4] H.R. Bennett, On Quasi-Developpable Spaces, *Gen. Top. Appl.*, 1 (1971), 253-262.

COLLOQUIA MATHEMATICA SOCIETATIS JÁNOS BOLYAI
8. TOPICS IN TOPOLOGY, KESZTHELY (HUNGARY), 1972.

EVERY CONTINUUM IS A RETRACT OF SOME IRREDUCIBLE INDECOMPOSABLE CONTINUUM

G.R. GORDH JR.

A continuum is a compact connected Hausdorff space. D.P. Bellamy proved that every metric continuum is a retract of an irreducible metric continuum [1]. Later he used this result to show that every metric continuum is a retract of some indecomposable metric continuum [2].

In this paper we modify Bellamy's proofs to obtain the more general non-metric result indicated by the title. We note that mapping theorems of this type do not always generalize to the non-metric setting. For example, the non-metric version of the familiar Hahn – Mazurkiewicz theorem is false [4].

The reader is referred to [3] for a discussion of continua and for definitions not included in this paper.

Theorem 1. *Every continuum is a retract of some irreducible continuum.*

Proof. Let M denote an arbitrary continuum. We must construct an irreducible continuum I containing a topological copy of M which is a retract of I. The irreducible continuum I will be constructed as a subset of a topological product $M \times A$, where A denotes an appropriately chosen ordered continuum.

We begin by describing the ordered continuum A. Let $X' = x_1, x_2, \ldots, x_\lambda, \ldots$ denote an efficient well-ordering of the points of the continuum M. Let X be the connected ordered space obtained by inserting an open interval of real numbers between each consecutive pair of elements of X' (as in the familiar construction of the "long line" [3]). Let N denote the set of natural numbers $1, 2, \ldots$ and let $A' = N \times X$ with the lexicographic ordering (i.e., (m, x) precedes (n, y) if m precedes n, or if $m = n$ and x precedes y). Now A' is a connected ordered space with no last point. Define A to be the ordered continuum obtained by adding a last point, say ∞, to the space A'. If (m, x) precedes (n, y) in A', then $[(m, x), (n, y)]$ denotes the ordered subcontinuum of A with endpoints (m, x) and (n, y).

For each x_λ in the continuum M, let L_λ denote an arbitrary subcontinuum of M which is irreducible between x_λ and $x_{\lambda+1}$. For each n in N, let I'_n be the subset of the topological product $M \times [(n, x_1), (n+1, x_1)] \subseteq M \times A$ defined as follows.

$$I'_n = \Big[\bigcup_\lambda \big(L_\lambda \times \{(n, x_\lambda)\} \big) \Big] \cup \Big[\bigcup_\lambda \big(\{x_{\lambda+1}\} \times [(n, x_\lambda), (n, x_{\lambda+1})] \big) \Big] .$$

Let I_n be the closure of I'_n in $M \times A$. Clearly

$$I_n \cap \big(M \times \{(n+1, x_1)\} \big) \neq \phi .$$

In particular, $I_1 \cap \big(M \times \{(2, x_1)\} \big) \neq \phi$. We will assume without loss of generality that M was well-ordered in such a way that the point $(x_1, (2, x_1))$ is an element of the set $I_1 \cap (M \times \{(2, x_1)\})$. It follows that $I_1 \cap I_2 \neq \phi$ and, more generally, that $I_n \cap I_{n+1} \neq \phi$ for each n in N.

Now define I to be the closure of $\bigcup_n I_n$ in $M \times A$. By construction, I is a subcontinuum of $M \times A$. Also, $M \times \{\infty\}$ is contained in I. (To see this, observe that for every λ the sequence $\{(x_\lambda, (n, x_\lambda)); n = 1, 2, \ldots\}$ is contained in $\bigcup_n I_n$. The desired conclusion follows from the fact that such a sequence converges to the point (x_λ, ∞) in $M \times A$.) It can now be verified that the continuum I is irreducible between the point $(x_1, (1,x_1))$ and any point of $M \times \{\infty\}$. The irreducible continuum I contains a copy of M, namely $M \times \{\infty\}$, and the restriction to I of the natural projection mapping from $M \times A$ onto $M \times \{\infty\}$ is the desired retraction.

Remark 1. In the proof of Theorem 1, the continuum M is a continuous image of the irreducible continuum I_1. However, M need not be a retract of I_1 since I_1 may not contain a copy of M. The more complicated construction of the irreducible continuum I guarantees that I contains a copy of M.

Lemma 1. *Every irreducible continuum is a retract of some indecomposable continuum.*

Proof. The argument given in [2] for metric continua is easily seen to be valid for non-metric continua as well.

Remark 2. It is not known if every indecomposable continuum is irreducible (or equivalently, if every indecomposable continuum contains more than one composant). However, the construction used by Bellamy in the proof of Lemma 1 (above) yields an irreducible indecomposable continuum.

Combining Theorem 1, Lemma 1, and the second remark, we obtain the main result.

Theorem 2. *Every continuum is a retract of some irreducible indecomposable continuum.*

REFERENCES

[1] D.P. Bellamy, A non-metric indecomposable continuum, *Duke Math. J.*, 38 (1971), 15-20.

[2] D.P. Bellamy, Mappings of indecomposable continua, *Proc. Amer. Math. Soc.*, 30 (1971), 179-180.

[3] J.G. Hocking – G.S. Young, *Topology*, Addison-Wesley, Reading, Mass. 1961.

[4] S. Mardešić, On the Hahn – Mazurkiewicz theorem in non-metric spaces, *Proc. Amer. Math. Soc.*, 11 (1960), 929-937.

COLLOQUIA MATHEMATICA SOCIETATIS JÁNOS BOLYAI
8. TOPICS IN TOPOLOGY, KESZTHELY (HUNGARY), 1972.

THE GENERALIZED SCHOENFLIES THEOREM FOR EUCLIDEAN n-SPACE*

J. de GROOT

Suppose S is a topological mapping of $R^{n-1} \times [0, 1]$ onto a closed subset of R^n $(n \geqslant 2)$. Is $\varphi\left(R^n \times \left\{\frac{1}{2}\right\}\right)$ topologically linear (or flat) in R^n, i.e. does there exist an autohomeomorphism ψ of R^n such that $\psi\varphi\left(R^n \times \left\{\frac{1}{2}\right\}\right)$ is a linear subspace of R^n? The corresponding statement for spheres has been answered in the affirmative by Morton Brown [1] and is called "the generalized Schoenflies theorem".

We shall show that for R^n too the answer is in the affirmative if $n \neq 3$. For R^3 the theorem is false, in general, but it becomes true iff $\varphi\left(R^2 \times \left\{\frac{1}{2}\right\}\right)$ contains a (half) line which is flat in R^3.

In a somewhat different wording we have

Theorem. *If an* R^{n-1} *is closed and (locally) bicollared embedded in an* R^n $(n \geqslant 2,\ n \neq 3)$, *it is flatly embedded. If an* R^2 *is embedded*

*This is the abstract of professor J. de Groot's lecture delivered at the conference. Unfortunately, his untimely death prevented him to produce a complete version of his paper.

in R^3, *it is flat if and only if it is closed, (locally) bicollared embedded and contains a (half) line which is flat in* R^3.

The author is indebted to J. Bruijning for several useful remarks.

In order to see that an R^2 might be closed and bicollared but not flatly embedded in R^3 we refer to a wild arc in an S^3 constructed by Fox – Artin [2]. Around this arc one can shape, close to it in S^3, a wild three-cell, which is "wild" only in one point of S^3. If this point is taken as the point at infinity of R^3 the boundary of the three-cell minus the point at infinity is the R^2 as required.

The proof of the theorem resembles to a certain point M. Brown's original proof in [1]. We outline the steps to be taken.

Definition 1. An R^{n-1} closed embedded in R^n defines two complementary domains. We call the closure of each a *halfspace* H and a domain itself its interior H^0, and R^{n-1} its boundary.

Definition 2. If R^{n-1} is a topologically linear i.e. flat subspace of R^n, we call its halfspaces H *flat halfspaces*

$$\forall i\colon H_{i+1} \subset H_i^0 \quad \text{and} \quad M = \cap_1^\infty H_i .$$

Definition 3. A subset M of R^n is called *cellular* (in R^n) if there exist (flat) halfspaces H_i $(i \in N)$, $H_i \subset R^n$ such that

Lemma 1. *Let* H *be a flat halfspace in* R^n *with boundary* R^{n-1}. *Suppose* M *is cellular in* H^0. *Then there exists a homeomorphism of* $H \setminus M$ *onto* H, *which is the identity on* R^{n-1}.

Lemma 2. *Let* R^{n-1} *be closed embedded in* R^n *and* H *one of its halfspaces. Suppose* M *is cellular in* H^0 *and suppose there exists a homeomorphism of* $H \setminus M$ *onto a flat halfspace. Then* H *itself is a flat halfspace.*

Definition. The suspension σR^m of R^m is defined as usual (double cone) with one slight change to render it metrizable: a neighborhood system of a vertex will be a set of cones "converging" to the vertex, each cone being "similar" to the cone over R^n.

Lemma 3. *Let H be a flat halfspace and M a closed subset of H^0. Suppose f is a continuous map from H into a σR^{n-1} such that fM is a vertex and f is one to one elsewhere. Moreover, fH will contain a neighborhood of one vertex and not meet a neighborhood of the other vertex. Then M is cellular in H.*

Lemma 4. *Let R^{n-1} be closed and bicollared embedded in R^n* $(n > 3)$. *Then the closures of the complementary components of such a double collar are cellular.*

A similar statement holds for R^3 if R^2 moreover, contains a closed line, flat in R^3.

While the proofs of Lemma 1, 2 and 3 do not present great difficulties the crux of the problem is contained in the proof of lemma 4.

This is already clear because of the two cases $n = 3$ and $n > 3$.

Lemma 5. (Probably known, but the author has no reference.) *Every piecewise linear line, closed embedded in R^n* $(n > 3)$, *is flat*. (In R^3 this is not true in general.)

Lemma 6. *If R^{n-1} is bicollared embedded in R^n $(n \geqslant 2)$ and R^{n-1} contains a closed line, flat in R^n (this is certainly the case if R^{n-1} is a linear subspace and in this case the lemma is known), then every closed topological line contained in R^{n-1} is flat in R^n.*

Corollary to the theorem. *If S^{n-1} is locally bicollared embedded in S^n* $(n > 3)$ except *in maybe one point, S^{n-1} is flat (hence bicollared). Hence, there do not exist $(n-1)$-cells in S^n* $(n > 3)$ which are *wild in one point only.*

REFERENCES

[1] Morton Brown, A proof of the generalized Schoenflies theorem, *Bull. A. M. S.*, 66 (196), 74-76.

[2] R.H. Fox – E. Artin, Some wild cells and spheres in 3-dim. space, *Ann. Math.*, 49 (1948), 979-990.

[illegible]

[illegible]

[illegible]

This [illegible] completes the proof of Lemma [illegible]

This is [illegible] of the two cases [illegible]

[illegible] (this [illegible])

[illegible]

[illegible]

REFERENCES

[illegible]

[illegible]

COLLOQUIA MATHEMATICA SOCIETATIS JÁNOS BOLYAI
8. TOPICS IN TOPOLOGY, KESZTHELY (HUNGARY), 1972.

ON INFINITE-DIMENSIONAL CANTOR MANIFOLDS

N. HADZIIVANOV

The notion of "cardinal dimension" was introduced in [1] by saying that a space X has cardinal dimension $\leqslant \tau$ (or shortly that X is at most τ-dimensional), here τ is a cardinal number, if to any system of cardinality $\tau + 1$ of pairs of closed sets F_{+a}, F_{-a} with $F_{+a} \cap F_{-a} = \phi$ we can find dividing sets C_a separating F_{+a} and F_{-a} whose intersection is empty: $\cap C_a = \phi$. The minimal τ such that X is at most τ-dimensional we denote by $k(X)$ and call it the cardinal dimension of the space X.

Remarks.

1) The case in which τ is finite is admitted.

2) If τ is infinite and m is finite then $\tau + m$ is equal to τ.

3) If τ is finite and the space X is normal then, as is well-known, $k(X) = \tau$ if and only if $\dim X = \tau$.

4) The inequality $k(X) \leqslant \aleph_0$ means that the space X is weakly infinite-dimensional in the sense of P.S. Alexandrov.

The fundamental result concerning the cardinal dimension contained in [1] reads as follows:

The Tychonov cube I^τ, where τ is an arbitrary cardinal number, cannot be decomposed into the union of countably many proper closed subsets whose pairwise intersections are at most $(\tau - 2)$-dimensional.

Let us note that this results is interesting and new even in the case in which τ is finite. In this case P. S. Uryson [2] proved only that I^n is not the sum of two proper closed subsets whose intersection is at most $(n - 2)$-dimensional, or in other words that the n-dimensional cube is a Cantor manifold.

In connection with the above cited result we also refer to our earlier publications [3] and [4].

The study of the cardinal dimension was carried on in our paper [5]. There, among other things, we introduced the notion of a "Cantor τ-manifold".

The space X is called a Cantor τ-manifold, if it cannot be decomposed into the union of two proper closed subsets whose intersection is at most $(\tau - 2)$-dimensional.

We shall say that the space X is a strong Cantor τ-manifold if it cannot be decomposed into the sum of finitely or countably many proper closed subsets whose pairwise intersections are at most $(\tau - 2)$-dimensional.

We know it already (by the above result) that the Tychonov cube of weight τ is a strong Cantor τ-manifold. In the present note we are goint to introduce another, wider class of strong Cantor τ-manifolds.

Let us begin with the following definition:

The space X is called a local Cantor τ-manifold if at each point x of the space x there exists a local basis consisting of connected open sets such that their closures are Cantor τ-manifolds.

By the above stated theorem the Tychonov cube I^τ is an example of a local Cantor τ-manifold. However there are trivial examples of Cantor τ-manifolds which are not local Cantor τ-manifolds.

The main aim of this note is to prove the following proposition.

Theorem. *Assume that the locally compact, connected Hausdorff space* X *is a local Cantor* τ*-manifold, where* τ *is an infinite cardinal number. Then* X *is a strong Cantor* τ*-manifold.*

Proof. Suppose that the statement of the theorem is false and let

$$X = \bigcup_{n=1}^{\infty} \Phi_n ,$$

where the Φ_n are proper closed subsets of the space X whose all pairwise intersections are at most τ-dimensional: $k(\Phi_i \cap \Phi_j) \leqslant \tau$ for $i \neq j$.

Under these assumptions the following two lemmas are valid.

Lemma 1. *Let* W *be a connected open subset of the space* X *which is not contained in any of the sets* Φ_n. *The we can find a connected open set* U *such that its clusure* $[U]$ *is contained in* W *but is not contained in any of the sets* Φ_n, *and is a compact Cantor* τ*-manifold.*

Lemma 2. *Let* V *be an open connected subset of the space* X *whose closure* $[V]$ *is a Cantor* τ*-manifold not contained in any of the sets* Φ_n *and* i *be an arbitrary natural number. Then we can find an open connected set* U *whose closure satisfies the following requirements:*

1) $[U] \subset V$;

2) $[U] \cap \Phi_i = \phi$;

3) $[U]$ *is not contained in any of the sets* Φ_n;

4) $[U]$ *is a compact Cantor* τ*-manifold.*

Proof of lemma 1. Since W is an open subset of the locally compact space X it is locally compact too, hence from the equality

$$W = \bigcap_{n=1}^{\infty} (W \cap \Phi_n)$$

we obtain by the Baire category theorem that $\operatorname{Int}(W \cap \Phi_i) \neq \phi$ for at least one index i. The connectedness of the set W implies then $W \cap \operatorname{Fr}(\operatorname{Int}(W \cap \Phi_i)) \neq \phi$. Let x be an arbitrary point of this latter set and

O be a neighbourhood of the point x such that its closure is contained in W. Since X is a local Cantor τ-manifold we can find a connected neighbourhood U of x which is contained in O and whose closure is a Cantor τ-manifold.

We shall prove that U possesses the properties required by the lemma. To do this we only have to show that $[U] \not\subset \Phi_n$ for $n = 1, 2, 3, \ldots$.

Assume, on the contrary, that $[U] \subset \Phi_j$ for certain j. Then $j \neq i$, since $x \notin \operatorname{Int} \Phi_i$. We have

$$U \cap \operatorname{Int} \Phi_i \subset \Phi_j \cap \Phi_i .$$

The set $U \cap \operatorname{Int} \Phi_i$ is open and non-empty as U is a neighbourhood of the point x and $x \in \operatorname{Fr}(\operatorname{Int} \Phi_i)$. This implies that $U \cap \operatorname{Int} \Phi_i$ contains a closed subset of dimension bigger than τ. Indeed, let $y \in U \cap \operatorname{Int} \Phi_i$ and V be a neighbourhood of the point y whose closure is a Cantor τ-manifold contained in $U \cap \operatorname{Int} \Phi_i$.

Obviously, the set $[V]$ cannot be a singleton, because otherwise we would have $\{y\} = V = [V]$, consequently $\{y\}$ would be an open and closed proper subset of the space X, which contradicts its connectedness. Thus $[V]$ is a non-trivial Cantor τ-manifold. But in this case we have $k([V]) > \tau$. In fact, let $z \in [V]$ and $z \neq y$. Since X is Hausdorff we can find a neighbourhood S of the point y such that $z \,\overline{\in}\, [S]$. The equation

$$[V] = ([V] \cap [S]) \cup ([V] \setminus S)$$

yields a decomposition of the set $[V]$ into the sum of two proper closed subsets. If now we assume $k([V]) \leqslant \tau$, then the intersection of these two closed subsets will also have a dimension at least τ, being a closed subset of the set $[V]$. From this we conclude that $[V]$ is not a Cantor τ-manifold, a contradiction, hence we must have $k([V]) > \tau$.

To sum up this we have

$$[V] \subset U \cap \operatorname{Int} \Phi_i \subset \Phi_j \cap \Phi_i \quad \text{and} \quad k([V]) > \tau .$$

On the other hand $k(\Phi_j \cap \Phi_i) \leqslant \tau$, since $j \neq i$. Then $k([V]) \leqslant \tau$, because $[V]$ is a closed subset lying in $\Phi_j \cap \Phi_i$. This contradiction proves our lemma.

Proof of lemma 2. Let us consider the open set $W = V \setminus \Phi_i$. First we prove that it is not contained in any of the sets Φ_n. Assume, on the contrary, that $W \subset \Phi_j$, where j is an arbitrary index. Then $j \neq i$ and we have $V \subset \Phi_i \cup \Phi_j$, consequently

$$[V] = ([V] \cap \Phi_i) \cup ([V] \cap \Phi_j) .$$

By our assumptions $[V]$ is not contained in any of the sets Φ_n, hence the closed sets $[V] \cap \Phi_i$ and $[V] \cap \Phi_i$ are proper subsets of the set $[V]$. Moreover, their intersection is contained in $\Phi_i \cap \Phi_j$, and since it is closed and $i \neq j$ it must be at most τ-dimensional. But this contradicts our assumption that $[V]$ is a Cantor τ-manifold, proving that W is not contained in any of the sets Φ_n.

Now there are the following two possibilities:

1) W is connected,

2) W is disconnected.

In the first case, applying lemma 1, we immediately obtain the desired set U.

It remains to consider the case in which the set W is disconnected. Let us denote then by W' one of the connected components of W.

We shall prove that the set W' satisfies all the requirements of lemma 1.

1) The set W' is connected by its choice.

2) The set W' is open. Indeed, the set W is locally connected since it is an open subset of the locally connected space X,* W' is a connected component of the set W, consequently W' is open in W and therefore is open in the whole space X as well.

3) W' is not contained in any of the sets Φ_n.

*Since X is a local Cantor τ-manifold, it is obviously locally connected.

The proof of this property which we shall now turn to is the most difficult.

Assume, on the contrary, that this is not valid and $W' \subset \Phi_j$ for some j. Clearly, then $i \neq j$ and $\operatorname{Fr} W' \subset \Phi_j$.

We shall prove that $\operatorname{Fr} W' \cap V \subset \Phi_i$. Indeed, $\operatorname{Fr} W' \cap W = \phi$ since W', being a connected component of the set W, is not only an open but also closed subset of this set. Moreover

$$\operatorname{Fr} W' \cap V = (\operatorname{Fr} W' \cap W) \cup (\operatorname{Fr} W' \cap \Phi_i) = \operatorname{Fr} W' \cap \Phi_i \subset \Phi_i .$$

Thus $\operatorname{Fr} W' \cap V \subset \Phi_i$, and consequently

$$\operatorname{Fr} W' \cap V \subset \Phi_i \cap \Phi_j , \qquad i \neq j .$$

The set $\operatorname{Fr} W' \cap V$ is non-empty. Indeed, in the opposite case W' would be an open and closed proper* subset of the set V, contradicting the connectedness of the latter.

Let $x \in \operatorname{Fr} W' \cap V$ and O be a neighbourhood of the point x, whose closure is a Cantor τ-manifold contained in W.** We have the following decomposition of the set $[O]$ into the union of two closed subsets

$$[O] = ([O] \cap [W']) \cup ([O] \setminus W') .$$

Since $x \in \operatorname{Fr} W'$ and O is a neighbourhood of the point x, the first of these sets is non-empty. The second is also non-empty. Indeed, if $[O] \setminus W' = \phi$, then $O \subset W'$ and consequently $x \in W'$, which is impossible since $x \in \operatorname{Fr} W'$ and W' is open.

Thus both sets in the above decomposition of $[O]$ are non-empty. Moreover, neither of them coincides with the set $[O]$. If the first one coincided with $[O]$, we would have $O \subset [W'] \cap W = W'$ which, as we know, is impossible. The second one cannot coincide with $[O]$ because, as we know, $O \cap W' \neq \phi$.

*$W' \subsetneqq W \subset V$, because W is not connected.

**Such a neighbourhood exists because X is a local Cantor τ-manifold.

For the intersection of these two sets we obviously have

$$([O] \cap [W']) \cap ([O] \setminus W') \subset O \cap W' \cap (X \setminus W') =$$
$$= O \cap \operatorname{Fr} W' \subset V \cap \operatorname{Fr} W' \subset \Phi_i \cap \Phi_j .^*$$

Thus, since this intersection is a closed subset of the set $\Phi_i \cap \Phi_j$, $i \neq j$, its dimension cannot exceed τ.

Now we have decomposed the set $[O]$ into the union of two proper closed subsets such that the dimension of their intersection is at most τ, which is a contradiction as $[O]$ is a Cantor τ-manifold.

This contradiction proves that W' is indeed not contained in any of the sets Φ_n.

Thus we see that the set W' satisfies all the conditions of lemma 1, hence applying this lemma we conclude that there exists an open connected set U whose closure is contained in W' and not contained in any of the sets Φ_n, and which is a compact Cantor τ-manifold. Obviously, this set U satisfies all the four requirements of lemma 2.

Thus the proof of lemma 2 is completed.

Now we turn to the proof of the theorem. We shall construct a sequence of non-empty compact sets, $K_1, K_2, K_3, \ldots$ satisfying the following two requirements:

1) $K_n \supset K_{n+1}$ for each n;

2) $K_n \cap \Phi_n = \phi$ for each n.

If we can succeed in doing this, the proof of our theorem is obtained easily. In fact, the intersection $\bigcap_{n=1}^{\infty} K_n$ is non-empty as the system of compact sets $\{K_n\}_{n<1}^{\infty}$ is centered. On the other hand

$$\bigcap_{n=1}^{\infty} K_n \cap \Big(\bigcap_{n=1}^{\infty} \Phi_n \Big) = \phi$$

* All these inclusions are trivial, except the last one, which we have proved above.

i.e.

$$\bigcap_{n=1}^{\infty} K_n \cap X = \phi , \quad \text{or shortly} \quad \bigcap_{n=1}^{\infty} K_n = \phi .$$

The obtained contradiction proves the theorem.

Thus let us construct the desired sequence.

Applying lemma 1 to the set $W = X$ we obtain a connected open set U, whose closure is not contained in any of the sets Φ_n and is a Cantor τ-manifold. Then we apply lemma 2 to the set $V = U$ and the index $i = 1$ and find a connected open set U_1 whose closure satisfies the following three conditions:

1) $[U_1] \subset U \setminus \Phi_1$;
2) $[U_1] \not\subset \Phi_n$, $n = 1, 2, 3, \ldots$;
3) $[U_1]$ is a compact Cantor τ-manifold.

The clause 2) implies that U_1 is non-empty.

Now assume that the connected open sets $U_1, \ldots, U_k$ have already been constructed in such a way that their closures are Cantor τ-manifolds not contained in any of the sets Φ_n, and moreover that $[U_i] \subset \subset U_{i-1} \setminus \Phi_i$, $i = 1, 2, \ldots, k$.

Applying lemma 2 to the set $V = U_k$ and the index $k + 1$, we can find a connected open set U_{k+1} such that

1) $[U_{k+1}] \subset U_k \setminus \Phi_{k+1}$;
2) $[U_{k+1}]$ is not contained in any of the sets Φ_n;
3) $[U_{k+1}]$ is a Cantor τ-manifold.

Obviously, 2) implies $U_{k+1} \neq \phi$.

Thus by induction we can contruct a sequence of open sets U_1, $U_2, U_3, \ldots$ so that $[U_n] \neq \phi$, $[U_n] \cap \Phi_n = \phi$, $[U_{n+1}] \subset [U_n]$, and $[U_n]$ is compact for each n. Now it remains to put $K_n = [U_n]$ and the desired sequence has been constructed.

The proof of the theorem is completed.

REFERENCES

[1] N. Hadziivanov, On extending mappings into spheres and countable decompositions of Tychonov cubes, *Math. Sbornik,* 84 (1971), 119-140 (In Russian).

[2] P.S. Urysohn, Mémoire sur les multiplicés Cantoriennes, *Fund. Math.*, 7 (1925), 30-137.

[3] N. Hadziivanov, The n-dimensional cube is not decomposable into a countable union of proper closed subsets, whose pairwise intersections are at most $(n-2)$ dimensional, *Dokladi An SSSR,* 195 (1970), 43-45 (In Russian).

[4] N. Hadziivanov, The Hilbert cube is not decomposable into a countable union of proper closed subsets, whose pairwise intersections are weakly infinite-dimensional, *Dokladi AN SSSR,* 195 (1970), 1282-1285 (In Russian).

[5] N. Hadziivanov, On infinite-dimensional spaces, *Bull. Acad. Polon. Sci.,* 19 (1971), 491-500 (In Russian).

COLLOQUIA MATHEMATICA SOCIETATIS JÁNOS BOLYAI
8. TOPICS IN TOPOLOGY, KESZTHELY (HUNGARY), 1972.

SOME INDEPENDENCE RESULTS IN SET THEORETICAL TOPOLOGY

A. HAJNAL

The main aim of the talk is to present the proof of two theorems obtained by Hajnal and Juhász.

Theorem 1. Con $(ZF) \Rightarrow$ Con $(ZFC + 2^{\omega} = \omega_1 + 2^{\omega_1} =$ anything *reasonable + there is* $R \subset 2^{\omega_1}$, $|R| = 2^{\omega_1}$, R *is hereditarily separable*).

Theorem 2. Con $(ZF) \Rightarrow$ Con $(ZFC + GCH +$ *there is a* 0-*dimensional* T_2 *space* R, $|R| = \omega_1$ *such that,* R *is hereditarily Lindelöf, every countable subspace of* R *is closed discrete and,* $\omega(R') = \omega_2$ *for every* $R' \subset R$, $|R'| = \omega_1$).

As a corollary of Theorem 1 it follows that the following is consistent with $ZFC + 2^{\omega} = \omega_1$.

For every a, $2^{\omega} \leqslant a < 2^{\omega_1}$, $\mathrm{cf}(a) \neq \omega$ there is $R \subset 2^{\omega_1}$ with $|R| = a =$ the number of all open subsets of R.

A number of related problems and results will be discussed.

A preliminary report containing these results has already appeared in "Two consistency results in topology", *Bull. of the AMS,* 78 (1972), 711. and the defailed proofs will be published elsewhere.

COLLOQUIA MATHEMATICA SOCIETATIS JÁNOS BOLYAI
8. TOPICS IN TOPOLOGY, KESZTHELY (HUNGARY), 1972.

SUPEREXTENSIONS AND PREPROXIMITIES

P. HAMBURGER

INTRODUCTION

In [5], using the notion of linked system, from each normal subbase of a Tychonoff space X, J. de Groot constructed a supercompact superextension $\hat{X}$ of X. In this manner he gave an internal characterization of the Tychonoff spaces. The author generalized this characterization in [8] and [9]. The aim of this paper is to generalize the method of J. de Groot and construct supercompact superextensions of a Tychonoff space from each strong preproximity of the space. We shall show that the theorems proved in [2] and [12] by Á. Császár and A. Verbeek are also valid for these superextensions. In this manner we give a method to get all Hausdorff compactifications of a Tychonoff space. It is no wonder that all Hausdorff compactifications of a Tychonoff space can be got in this manner, because all Hausdorff compactifications of a Tychonoff space can be got using the proximities of the space. But this construction gives a new characterization of the Smirnov's compactification [10], and in the same time, it is a natural generalization of the Wallman – Shanin and Aarts – de Groot type compactification ([11], [4] and [1]). It is unknown whether all Hausdorff compactifications of a Tychonoff space can be got using the method of Aarts – de Groot or Wallman – Shanin.

1. CONSTRUCT SUPEREXTENSIONS

For the sake of simplicity all our spaces will be T_1-spaces.

Definition 1. The space X is called *supercompact relative to* $\mathfrak{B}$ if $\mathfrak{B}$ is a subbase for the closed sets of X such that every subfamily $\mathfrak{T}$ of $\mathfrak{B}$ for which $\cap \mathfrak{T} = \phi$ contains a pair $A, B \in \mathfrak{T}$ such that $A \cap B = \phi$.

A space X *is supercompact* if there exists a subbase $\mathfrak{B}$ for the closed sets of X such that X is supercompact relative to $\mathfrak{B}$. The well-known Alexander's theorem says that a supercompact space is compact.

Definition 2. We shall say that $(\mathfrak{B}, r)$ is a *strong preproximity* in X iff $\mathfrak{B}$ is a subbase for the closed sets of X and r is a symmetrical binary relation on $\mathfrak{B}$ such that

(i) $\phi, X \in \mathfrak{B}, \phi \, r \, X$,

(ii) if $A, B \in \mathfrak{B}$, $A \, r \, B$ implies $A \cap B = \phi$,

(iii) r can be weakly screened in $\mathfrak{B}$, which means that if $A, B \in \in \mathfrak{B}$, $A \, r \, B$ then there exist two finite subfamilies $\{A_i;\ i = 1, \dots, n\}$ and $\{B_j;\ j = 1, \dots, m\}$ of $\mathfrak{B}$ such that $\{A_i;\ i = 1, \dots, n\}$ covers A, $\{B_j;\ j = 1, \dots, m\}$ covers $B, A_i \, r \, B_j$ for every $i = 1, \dots, n$, $j = 1, \dots \dots, m$ and every A_i, B_j can be screened in $\mathfrak{B}$ i.e. for every i, j there exists a finite subfamily $\mathfrak{T}_{i,j}$ of $\mathfrak{B}$ which covers X and for every $C \in \in \mathfrak{T}_{i,j}$ either $C \, r \, A_i$ or $C \, r \, B_j$,

(iv) if $x \in X - A$, $A \in \mathfrak{B}$, then there is a set $B \in \mathfrak{B}$ such that $x \in B$, $B \, r \, A$.

Note. Compare the above definition of strong preproximity with the definition of preproximity given in [8]. We have also to note that there is great analogy between the notion of topology of a space and a subbase for the closed sets of this space and the notion of proximity and preproximity of a space.

First we are going to show that if $(\mathfrak{B}, r)$ is a strong preproximity in X, then a supercompact superextension $\hat{X}$ of X can be constructed.

We need some definitions.

Definition 3. Let $(\mathfrak{B}, r)$ be a strong preproximity in X. A subfamily $\mathfrak{R}$ of $\mathfrak{B}$ will be called a *linked system* in $(\mathfrak{B}, r)$ if $A, B \in \mathfrak{R}$ implies that $A \not r B$, where $\not r$ denotes the negation of the relation r.

It is easy to see that if $\mathfrak{R}$ is a linked system in $(\mathfrak{B}, r)$ then there exists a maximal linked system $\mathfrak{R}'$ in $(\mathfrak{B}, r)$ which contains $\mathfrak{R}$.

Definition 4. The maximal linked systems will be called *ultralinked systems.*

Let $(\mathfrak{B}, r)$ be a strong preproximity in X and $\hat{X}$ the family of all ultralinked systems in $(\mathfrak{B}, r)$. For every $S \in \mathfrak{B}$ put $\hat{S} = \{\mathfrak{R};\ \mathfrak{R}$ is an ultralinked system in $(\mathfrak{B}, r),\ S \in \mathfrak{R}\}$. Let $\hat{\mathfrak{B}} = \{\hat{S};\ S \in \mathfrak{B}\}$ be a subbase for the closed sets of $\hat{X}$, and h the following map: for every $x \in X$ let $h(x) = \mathfrak{R}_x$ where $\mathfrak{R}_x = \{A;\ A \in \mathfrak{B},\ x \in A\}$.

Theorem 1. *Let $(\mathfrak{B}, r)$ be a strong preproximity in X, then $\hat{X}$ is a T_1-superextension of X, and $\hat{X}$ is supercompact relative to $\hat{\mathfrak{B}}$.*

Proof. Let $\hat{\mathfrak{T}}$ be a subfamily of $\hat{\mathfrak{B}}$ such that $\cap \hat{\mathfrak{T}} = \phi$. Suppose that $\hat{T}_1, \hat{T}_2 \in \hat{\mathfrak{T}}$ implies that $\hat{T}_1 \cap \hat{T}_2 \neq \phi$, then $\mathfrak{T}$ is a linked system in $(\mathfrak{B}, r)$. In fact, if $\hat{T}_1 \cap \hat{T}_2 \neq \phi$ then there is a linked system $\mathfrak{R}$ in $(\mathfrak{B}, r)$ such that $T_1, T_2 \in \mathfrak{R}$, and so $T_1 \not r T_2$, i.e. $\mathfrak{T}$ is a linked system. Using the Kuratowski – Zorn lemma there exists an ultralinked system $\mathfrak{R}_{\mathfrak{T}}$ in $(\mathfrak{B}, r)$ which contains $\mathfrak{T}$. Then $\mathfrak{R}_{\mathfrak{T}} \in \hat{X}$. Further, for every $\hat{T} \in \hat{\mathfrak{T}}$, $T \in \mathfrak{R}_{\mathfrak{T}}$, thus $\mathfrak{R}_{\mathfrak{T}} \in \hat{T}$ for every $\hat{T} \in \hat{\mathfrak{T}}$. But this contradicts that $\cap \hat{\mathfrak{T}} = \phi$: We got that $\hat{X}$ is supercompact relative to $\hat{\mathfrak{B}}$.

We show that $\hat{X}$ is a superextension of X, i.e. the map h is a homeomorphism of X onto $h[X] \subset \hat{X}$.

First we have to show that for every $x \in X$ $\mathfrak{R}_x$ is an ultralinked system. It is trivial that $\mathfrak{R}_x$ is a linked system. If $x \notin B$ then, by the condition (iv) of Definition 2, there is a set $A \in \mathfrak{B}$ such that $x \in A$, $A\, r\, B$. This means that $B \notin \mathfrak{R}_x$ i.e. $\mathfrak{R}_x$ is an ultralinked system.

Let $x, y \in X$, $x \neq y$. Since $\mathfrak{B}$ is a subbase for the closed sets of X, and X is a T_1-space, there is $A \in \mathfrak{B}$, $x \in A$, $y \notin A$. This means that $h(x) \neq h(y)$ and we get that h is a one-to-one map.

We shall show that h is a continuous map. Let $h(y) \notin \cup \{ \hat{S}_i;\ i = 1, \dots, n,\ \hat{S}_i \in \hat{\mathfrak{B}} \}$. This means that $S_i \notin h(y)$, $i = 1, \dots, n$ i.e. $y \notin S_i$, $i = 1, \dots, n$. Since $(\mathfrak{B}, r)$ is a strong preproximity in X there exist sets $A_i \in \mathfrak{B}$, $i = 1, \dots, n$, such that $y \in A_i$, $A_i\, r\, S_i$, $i = 1, \dots, n$. By the condition (ii) of Definition 2, r can be weakly screened in $\mathfrak{B}$, so there are two finite subfamilies $\{ A_{i,s};\ s = 1, \dots, l_i \}$, $\{ S_{i,j};\ j = 1, \dots, m_i \}$ of $\mathfrak{B}$ such that $\{ A_{i,s};\ s = 1, \dots, l_i \}$ covers A, $\{ S_{i,j};\ j = 1, \dots, m_i \}$ covers S and $A_{i,s}\, r\, S_{i,j}$ for every $j = 1, \dots, m_i$, $s = 1, \dots, l_i$ and for every s, j there exists a finite subfamily $\mathfrak{T}_{i,j,s}$ of $\mathfrak{B}$ which screens the pair $A_{i,s}$, $S_{i,j}$. Let A_{i,s_0} $(1 \leqslant s_0 \leqslant l_i)$, be a set such that $y \in A_{i,s_0}$ and put

$$D_{i,s_0} = \cup \{ D;\ D \in \mathfrak{T}_{i,j,s_0},\ D\, r\, A_{i,s_0},\ j = 1, \dots, m_i \}.$$

Put $B = \cup \{ D_{i,s_0};\ i = 1, \dots, n \}$. If we show that

$$h[X - B] \subset \left(\hat{X} - \cup \{ \hat{S}_i;\ i = 1, \dots, n \} \right) \cap h[X]$$

then we proved that h is a continuous map. For this we have to show that if $x \in X - B$ then $h(x) \notin \hat{S}_i$, $i = 1, \dots, n$, i.e. $S_i \notin h(x)$. But this means that $x \notin S_i$, $i = 1, \dots, n$. Let $x \in X - B$. $x \notin B$ implies that $x \notin D_{i,s_0}$, $i = 1, \dots, n$. But $S_i \subset D_{i,s_0}$, for if $u \in S_i$ then $u \in S_{i,j_0}$, $1 \leqslant j_0 \leqslant m_i$. Since $\mathfrak{T}_{i,j_0,s_0}$ covers X, so there is a set $D \in \mathfrak{T}_{i,j_0,s_0}$ such that $u \in D \cap S_{i,j_0} \neq \phi$. We got that there is a set $D \in \mathfrak{T}_{i,j_0,s_0}$ such that $u \in D$ and $D\, r\, A_{i,s_0}$ is valid, i.e. we have shown that h is a continuous map.

Now, we are going to show that h^{-1} is also continuous. Let $\mathfrak{R} \in h[X]$. Then there is a unique point $x \in X$ such that $h^{-1}(\mathfrak{R}) = x$ i.e. $\mathfrak{R} = \mathfrak{R}_x$. Let $x \in X - \cup \{ S_i;\ i = 1, \dots, n \}$, $S_i \in \mathfrak{B}$. We have to show that there is a neighbourhood U of $\mathfrak{R}_x$ in $h[X]$ such that $h^{-1}[U] \subset X - \cup \{ S_i;\ i = 1, \dots, n \}$. If $x \notin S_i$, $i = 1, \dots, n$ then $S_i \notin \mathfrak{R}_x$, i.e. $\mathfrak{R}_x \notin \hat{S}_i$, $i = 1, \dots, n$. Then

$$\mathfrak{R}_x \in \left(\hat{X} - \cup \{\hat{S}_i;\ i = 1, \dots, n\}\right) \cap h[X].$$

Let U be the above neighbourhood of $\mathfrak{R}_x$ in $h[X]$. Let $\mathfrak{A}$ be an arbitrary point from U. h is a one-to-one map, so there is a unique point $y \in X$ such that $\mathfrak{A} = \mathfrak{R}$. If $y \notin X - \cup \{S_i;\ i = 1, \dots, n\}$ was valid then $y \in S_i$ would hold for some i. But then $S_i \in \mathfrak{R}$ would be valid which contradicts to $\mathfrak{A} \in U$.

To show that $\hat{X}$ is a T_1-space we need the following. If $\mathfrak{R}_1 \neq \mathfrak{R}_2$, $\mathfrak{R}_1, \mathfrak{R}_2 \in \hat{X}$ then there are sets $A \in \mathfrak{R}_1$, $B \in \mathfrak{R}_2$ such that $A\, r\, B$. In fact, if $\mathfrak{R}_1 \neq \mathfrak{R}_2$ then there exists a set $A \in \mathfrak{B}$ such that A is contained only in one of the families $\mathfrak{R}_1$ and $\mathfrak{R}_2$. Suppose that $A \in \mathfrak{R}_1$, $A \notin \mathfrak{R}_2$. If for every $B \in \mathfrak{R}_2$, $A \not r B$ would hold, then $\mathfrak{R}_2$ would not be a maximal linked system in $(\mathfrak{B}, r)$. So there is $B \in \mathfrak{R}_2$ such that $A\, r\, B$.

Finally we shall show that $\hat{X}$ is a T_1-space. Let $\mathfrak{R}_1 \neq \mathfrak{R}_2$, $\mathfrak{R}_1, \mathfrak{R}_2 \in \hat{X}$. We have seen that there are sets $S_1 \in \mathfrak{R}_1$, $S_2 \in \mathfrak{R}_2$ such that $S_1\, r\, S_2$. Thus $\mathfrak{R}_1 \in \hat{S}_1$, $\mathfrak{R}_2 \in \hat{S}_2$ and $\hat{S}_1 \cap \hat{S}_2 = \phi$. This shows that $\hat{X}$ is a T_1-space.

Q.e.d.

In general X is not dense in $\hat{X}$ and $\hat{X}$ is not Hausdorff. However, in the next theorem we give a sufficient condition for $\hat{X}$ to be Hausdorff. We need the following definitions.

Definition 5. Let $\mathfrak{B}$ be a family and r a symmetrical binary relation on $\mathfrak{B}$. We shall say that r is screened in $\mathfrak{B}$, *by the pairs of* $\mathfrak{B}$, if for every $A, B \in \mathfrak{B}$, $A\, r\, B$, there are $C, D \in \mathfrak{B}$ such that $C \cup D = X$ and $C\, r\, A$, $D\, r\, B$.

r is screened in $\mathfrak{B}$, if for every $A, B \in \mathfrak{B}$, $A\, r\, B$ there is a cover $\{A_i;\ i = 1, \dots, n\} \subset \mathfrak{B}$ of X such that for every $i = 1, \dots, n$ either $A_i\, r\, B$ or $A_i\, r\, A$.

We shall say that r is a descending relation on $\mathfrak{B}$ if $A, B, C, D \in \mathfrak{B}$, $C \subset A$, $D \subset B$, $A\, r\, B$ imply $C\, r\, D$.

Theorem 2. *Let* $(\mathfrak{B}, r)$ *be a strong preproximity in* X. *If* r *is screened in* $\mathfrak{B}$ *by the pairs of* $\mathfrak{B}$, *and* r *is a descending relation on* $\mathfrak{B}$, *then* $\hat{X}$ *is a Hausdorff space.*

Proof. We have seen that if $\mathfrak{R}_1 \neq \mathfrak{R}_2$ then there are sets $S_1 \in \in \mathfrak{R}_1$, $S_2 \in \mathfrak{R}_2$ such that $S_1\, r\, S_2$. Since r can be screened in $\mathfrak{B}$, by the pairs of $\mathfrak{B}$, so there are sets $A_1, A_2 \in \mathfrak{B}$ such that $A_1 \cup A_2 = X$, $S_1\, r\, A_2$, $S_2\, r\, A_1$. This means that $A_2 \notin \mathfrak{R}_1$, $A_1 \notin \mathfrak{R}_2$, i.e. $\mathfrak{R}_1 \notin \hat{A}_2$, $\mathfrak{R}_2 \notin \hat{A}_1$. We have to show that $\hat{A}_1 \cup \hat{A}_2 = \hat{X}$. Suppose that $\mathfrak{A} \in \hat{X} - - (\hat{A}_1 \cup \hat{A}_2)$. This means that $A_1, A_2 \notin \mathfrak{A}$ which implies that there are sets $B_1 \in \mathfrak{A}$, $B \in \mathfrak{A}$ such that $B_1\, r\, A_1$, $B\, r\, A_2$. Obviously $B_1 \subset A_2$. Since r is a descending relation, $B\, r\, B_1$ would be also valid. But this is impossible.

Q.e.d.

2. CONSTRUCT ALL COMPACTIFICATIONS OF A TYCHONOFF SPACE

Theorem 3. *Let* $(\mathfrak{B}, r)$ *be a strong preproximity in* X. *If* r *is screened in* $\mathfrak{B}$ *then* $\overline{h[X]}^{\hat{X}}$ *is a Hausdorff compactification of* X.

Proof. We have only to show that $\overline{h[X]}^{\hat{X}}$ is a Hausdorff space. If $\mathfrak{R}_1, \mathfrak{R}_2 \in \overline{h[X]}^{\hat{X}}$, $\mathfrak{R}_1 \neq \mathfrak{R}_2$ then there are sets $S_1 \in \mathfrak{R}_1$, $S_2 \in \mathfrak{R}_2$ such that $S_1\, r\, S_2$. Since r is screened in $\mathfrak{B}$ so there is a finite subfamily $\mathfrak{T}$ of $\mathfrak{B}$ which covers X and for every $A \in \mathfrak{T}$ either $A\, r\, S_1$ or $A\, r\, S_2$. Obviously $\hat{\mathfrak{T}}$ covers $h[X]$ thus $\hat{\mathfrak{T}}$ also covers $\overline{h[X]}^{\hat{X}}$. Put

$$U = \left(\hat{X} - \cup \{\hat{A};\ A\, r\, S_1,\ A \in \mathfrak{T}\}\right) \cap \overline{h[X]}^{\hat{X}}$$

and

$$V = \left(\hat{X} - \cup \{\hat{A};\ A\, r\, S_2,\ A \in \mathfrak{T}\}\right) \cap \overline{h[X]}^{\hat{X}}.$$

Then U and V are disjoint neighbourhoods of $\mathfrak{R}_1$ and $\mathfrak{R}_2$.

Q.e.d.

In the next theorem we characterize the elements of $\overline{h[X]}^{\hat{X}}$.

Definition 6. We shall say that a subfamily $\mathfrak{R}$ of $\mathfrak{B}$ is a $\mathfrak{B}$-*prime* system if $\mathfrak{R}$ contains at least one element of all finite covers of X by the elements of $\mathfrak{B}$.

Lemma 1. *Let* $(\mathfrak{B}, r)$ *be a strong preproximity in* X *and* r *be screened in* $\mathfrak{B}$. *Then* $\mathfrak{R} \in \overline{h[X]}^{\hat{X}}$ *if and only if* $\mathfrak{R} \in \hat{X}$ *is* $\mathfrak{B}$*-prime.*

Proof. Suppose that $\mathfrak{R} \in \overline{h[X]}^{\hat{X}}$. Let $\{S_i;\ i = 1, \ldots, n\}$, $S_i \in \mathfrak{B}$, $i = 1, \ldots, n$ be a cover of X. Then $\cup\{\hat{S}_i;\ i = 1, \ldots, n\} \supset \overline{h[X]}^{\hat{X}}$. Thus $\mathfrak{R} \in \cup\{\hat{S}_i;\ i = 1, \ldots, n\}$ i.e. there is i_0, $1 \leqslant i_0 \leqslant n$ such that $\mathfrak{R} \in \hat{S}_{i_0}$ and so $S_{i_0} \in \mathfrak{R}$.

Conversely, let $\mathfrak{R}$ be a $\mathfrak{B}$-prime ultralinked system. If $\mathfrak{R} \in \hat{X} - \overline{h[X]}^{\hat{X}}$ was valid, then there would be a neighbourhood U in $\hat{X}$ of $\mathfrak{R}$ such that $U \cap \overline{h[X]}^{\hat{X}} = \phi$. One can suppose that $U = \cap\{\hat{X} - \hat{S}_i;\ i = 1, \ldots, n\}$, $S_i \in \mathfrak{B}$. Then $h[X] \cap [\hat{X} - \cup\{\hat{S}_i;\ i = 1, \ldots, n\}] = \phi$. Thus if $x \in X$ then there is i_0, $1 \leqslant i_0 \leqslant n$ such that $\mathfrak{R}_x \in \hat{S}_{i_0}$ i.e. $S_{i_0} \in \mathfrak{R}_x$ and so $x \in S_{i_0}$. We got that $\{S_i;\ i = 1, \ldots, n\}$ is a finite cover of X by the elements of $\mathfrak{B}$. Since $\mathfrak{R}$ is $\mathfrak{B}$-prime thus there is a set S_{j_0} $(1 \leqslant j_0 \leqslant n)$ such that $S_{j_0} \in \mathfrak{R}$ i.e. $\mathfrak{R} \in \hat{S}_{j_0}$. This contradicts the fact that $\mathfrak{R} \in U$.

Q.e.d.

Lemma 2. *Let* $(\mathfrak{B}, r)$ *be a strong preproximity in* X, *and* r *be screened in* $\mathfrak{B}$. *Then* $\mathfrak{R} \in \overline{h[X]}^{\hat{X}}$ *if and only if* $\mathfrak{R}$ *contains a maximal centered system* $\mathfrak{A}$ *from* $\mathfrak{B}$ *i.e. if* $\mathfrak{R}$ *has the following form:* $\mathfrak{A}$ *is a maximal centered system in* $\mathfrak{B}$ *and* $\mathfrak{R} = \mathfrak{R}_{\mathfrak{A}} = \{S;\ S \in \mathfrak{B},\ T \not r S$ *for every* $T \in \mathfrak{A}\}$.

Proof. We will show that if $\mathfrak{A}$ is a maximal centered system in $\mathfrak{B}$ then $\mathfrak{R}_{\mathfrak{A}}$ is an ultralinked system in $\mathfrak{B}$. Suppose $A, B \in \mathfrak{R}_{\mathfrak{A}}$, $A\, r\, B$. $(\mathfrak{B}, r)$ is a strong preproximity on X and r can be screened in $\mathfrak{B}$ thus there are sets $T_j \in \mathfrak{B}$, $j = 1, \ldots, m$ such that $\cup\{T_j;\ j = 1, \ldots, m\} = X$ and either $T_j\, r\, A$ or $T_j\, r\, B$. Every maximal centered system in $\mathfrak{B}$ is $\mathfrak{B}$-prime therefore there is an index j_0, $1 \leqslant j_0 \leqslant n$ such that $T_{j_0} \in \mathfrak{A}$. Then either $A\, r\, T_{j_0}$ or $B\, r\, T_{j_0}$ would be valid which contradicts the definition of $\mathfrak{R}_{\mathfrak{A}}$ We got that if $\mathfrak{A}$ is a maximal centered system in $\mathfrak{B}$ then $\mathfrak{R}_{\mathfrak{A}} \in \hat{X}$. ($\mathfrak{R}_{\mathfrak{A}}$ is an ultralinked system because if $A \notin \mathfrak{R}_{\mathfrak{A}}$ then there is $B \in \mathfrak{A} \subset \mathfrak{R}_{\mathfrak{A}}$ such that $A\, r\, B$.)

Since $\mathfrak{A} \subset \mathfrak{R}_{\mathfrak{A}}$ and since every maximal centered system in $\mathfrak{B}$ is $\mathfrak{B}$-prime, thus Lemma 1 gives us that $\mathfrak{R}_{\mathfrak{A}} \in \overline{h[X]}^{\hat{X}}$.

If $\mathfrak{R} \in \overline{h[X]}^{\hat{X}}$ then Lemma 1 said that $\mathfrak{R}$ is $\mathfrak{B}$-prime. We shall show that there is a maximal centered system $\mathfrak{A}$ in $\mathfrak{B}$ such that $\mathfrak{A} \subset \mathfrak{R}$.

First we are going to show that $\mathfrak{R}$ contains a minimal $\mathfrak{B}$-prime system. We use the Kuratowski – Zorn lemma. Since $\mathfrak{R}$ is $\mathfrak{B}$-prime we have to show that if $\{\mathfrak{G}_a;\ a \in A\}$ is a chain of $\mathfrak{B}$-prime systems in $\mathfrak{R}$, then $\cap\{\mathfrak{G}_a;\ a \in A\}$ is also a $\mathfrak{B}$-prime system. Let $\{S_i;\ i = 1, \ldots, n\}$ be a finite cover of X by the elements of $\mathfrak{B}$. Each $\mathfrak{G}_a$ is $\mathfrak{B}$-prime thus every $\mathfrak{G}_a$ contains an element of the cover. But then obviously there is a set S_{i_0} such that $S_{i_0} \in \mathfrak{G}_a$ for every $a \in A$, thus $S_{i_0} \in \cap\{\mathfrak{G}_a\colon a \in A\}$.

Let $\mathfrak{G}$ be a minimal $\mathfrak{B}$-prime system in $\mathfrak{R}$, then $\mathfrak{G}$ is a centered system. In fact, if we show that $\mathfrak{P} = \{X - S;\ S \in \mathfrak{B} - \mathfrak{G}\}$ is a maximal centered system in $\mathfrak{T} = \{X - S;\ S \in \mathfrak{B}\}$ then we get that $\mathfrak{G}$ is a centered system because if $A_i \in \mathfrak{G}$, $i = 1, \ldots, n$ and $\cap\{A_i;\ i = 1, \ldots$ $\ldots, n\} = \phi$ would be valid then their complements would cover X. $\mathfrak{P}$ being a maximal centered system in $\mathfrak{T}$ therefore there would be i_0, $1 \leqslant i_0 \leqslant n$ such that the complement of A_{i_0} is an element of $\mathfrak{P}$. But this is a contradiction. We got that $\mathfrak{G}$ is a centered system.

Hence we must show that $\mathfrak{P} = \{X - S;\ S \in \mathfrak{B} - \mathfrak{G}\}$ is a maximal centered system in $\mathfrak{T}$. $\mathfrak{P}$ is centered, because if $X - A_i \in \mathfrak{P}$, $i = 1, \ldots$ $\ldots, n$ and $\cap\{X - A_i;\ i = 1, \ldots, n\} = \phi$, then $\cup\{A_i;\ i = 1, \ldots, n\} =$ $= X$ would hold; $\mathfrak{G}$ is $\mathfrak{B}$-prime, thus there is $A_i \in \mathfrak{G}$, which contradicts the definition of $\mathfrak{P}$. On the other hand, $\mathfrak{P}$ is maximal too, because if $\mathfrak{P}$ is not a maximal centered system then there is a maximal centered system $\mathfrak{P}'$ in $\mathfrak{T}$ which contains $\mathfrak{P}$. Then $\mathfrak{G}' = A;\ X - A \in \mathfrak{B} - \mathfrak{P}' \notin \mathfrak{P} \subset$ $\subset \mathfrak{G} \subset \mathfrak{B}$ would be a $\mathfrak{B}$-prime system. In fact, if $\{S_i;\ i = 1, \ldots, n\}$ is a finite cover of X by elements of $\mathfrak{B}$ and $S_i \notin \mathfrak{G}'$, $i = 1, \ldots, n$ then $X - S_i \in \mathfrak{P}'$, $i = 1, \ldots, n$, $\cap\{X - S_i;\ i = 1, \ldots, n\} = \phi$ but this contradicts the fact that $\mathfrak{P}'$ is a centered system. Therefore $\mathfrak{G}'$ is a $\mathfrak{B}$-prime system and $\mathfrak{G}' \subset \mathfrak{G}$ is valid, but according to the hypothesis this may only be valid in the case $\mathfrak{G} = \mathfrak{G}'$. Thus we got that $\mathfrak{P}$ is a maximal centered system in $\mathfrak{T}$.

Hence we have seen that if $\mathfrak{R} \in \overline{h[X]}^{\hat{X}}$ then there is a $\mathfrak{B}$-prime centered system $\mathfrak{G}$ in $\mathfrak{R}$.

We are going to show that if $\mathfrak{G}$ is a $\mathfrak{B}$-prime system in $\mathfrak{B}$ then there is at most a unique ultralinked system $\mathfrak{R} \in \overline{h[X]}^{\hat{X}}$ which contains $\mathfrak{G}$.

In contradiction, suppose that there are two different ultralinked systems $\mathfrak{R}_1, \mathfrak{R}_2 \in \overline{h[X]}^{\hat{X}}$ which contain $\mathfrak{G}$. If $\mathfrak{R}_1, \mathfrak{R}_2$ are two different ultralinked systems in $\mathfrak{B}$ then there are sets $S \in \mathfrak{R}_1$, $T \in \mathfrak{R}_2$ such that $T\, r\, S$. Since r is screenable in $\mathfrak{B}$ thus there are sets $A_i \in \mathfrak{B}$, $i = 1, \ldots, n$ such that $\cup \{A_i;\ i = 1, \ldots, n\} = X$ and either $A_i\, r\, S$ or $A_i\, r\, T$ is fulfilled. $\mathfrak{G}$ is a $\mathfrak{B}$-prime system so there is an index i_0 $(1 \leqslant i_0 \leqslant n)$ such that $A_{i_0} \in \mathfrak{G}$. By the indirect condition $\mathfrak{G} \subset \mathfrak{R}_1$, $\mathfrak{G} \subset \mathfrak{R}_2$ thus $A_{i_0} \in \mathfrak{R}_1$, $A_{i_0} \in \mathfrak{R}_2$. But this contradicts the fact that $\mathfrak{R}_1, \mathfrak{R}_2$ are linked systems, because either $A_{i_0}\, r\, S$ or $A_{i_0}\, r\, T$ would be fulfilled.

Let $\mathfrak{R} \in \overline{h[X]}^{\hat{X}}$. We have seen that there is a $\mathfrak{B}$-prime centered system $\mathfrak{G}$ contained in $\mathfrak{R}$. Using the Kuratowski – Zorn lemma we get that there is a maximal centered system $\mathfrak{G}' \subset \mathfrak{B}$ which contains $\mathfrak{G}$. We show that $\mathfrak{G}' \subset \mathfrak{R}$. We have shown that $\mathfrak{R}_{\mathfrak{G}'} = \{S;\ S \in \mathfrak{B},\ S \not r\, T$ for every $T \in \mathfrak{G}'\}$ belongs to $\overline{h[X]}^{\hat{X}}$. But $\mathfrak{G} \subset \mathfrak{R}_{\mathfrak{G}'}$ and $\mathfrak{G} \subset \mathfrak{R}$ thus using the above results $\mathfrak{R} = \mathfrak{R}_{\mathfrak{G}'}$.

Q.e.d.

Theorem 4. *Let* $(\mathfrak{B}, r)$ *be a strong preproximity in* X *and* r *be screened in* $\mathfrak{B}$. *Then* $\overline{h[X]}^{\hat{X}}$ *is a Hausdorff compactification of* X *which belongs to* $\delta_{(\mathfrak{B},r)}$ *(see the definition of* $\delta_{(\mathfrak{B},r)}$ *in* [8]*). For every* $S \in \mathfrak{B}$ *let* $\check{S}$ *be the following set:* $\check{S} = \{\mathfrak{R}_{\mathfrak{A}};\ \mathfrak{A}$ *is a maximal centered system in* $\mathfrak{B}$, $S \in \mathfrak{R}_{\mathfrak{A}}\}$. *Let* $\check{\mathfrak{B}} = \{\check{S};\ S \in \mathfrak{B}\}$ *and for* $\check{A}, \check{B} \in \check{\mathfrak{B}}$, $\check{A}\, \check{r}\, \check{B}$ *iff* $A\, r\, B$. *Then* $(\check{\mathfrak{B}}, \check{r})$ *is a strong preproximity in* $\overline{h[X]}^{\hat{X}}$, *and* $\check{r}$ *can be screened in* $\check{\mathfrak{B}}$.

Proof. By Theorem 3 we got that $\overline{h[X]}^{\hat{X}}$ is a Hausdorff compactification of X. For the sake of simplicity X will be identified with $h[X]$, and we put $Y = \overline{h[X]}^{\hat{X}}$.

We show that $(\check{\mathfrak{B}}, \check{r})$ is a strong preproximity in Y. Let $\mathfrak{R} \in \in Y - \check{S}$, $S \in \mathfrak{B}$. Then $\mathfrak{R} \notin \check{S}$, i.e. $S \notin \mathfrak{R}$. Thus there is $T \in \mathfrak{R}$ such that $S\, r\, T$, and $\mathfrak{R} \in \check{T}$. Since $\check{S}\, \check{r}\, \check{T}$ is fulfilled, we got that the condition (iv) of Definition 2 is valid. The condition (i) is trivial and we have seen that the condition (ii) of Definition 2 is also fulfilled, because if $A, B \in \mathfrak{B}$, $A\, r\, B$ then $\hat{A} \cap \hat{B} = \phi$ thus $\check{A} \cap \check{B} = \phi$. Finally if we show that $\check{r}$ is screenable in $\check{\mathfrak{B}}$ then we get that $(\check{\mathfrak{B}}, \check{r})$ is a strong preproximity in Y. Let $\check{A}, \check{B} \in \check{\mathfrak{B}}$, $\check{A}\, \check{r}\, \check{B}$. Then $A, B \in \mathfrak{B}$ and $A\, r\, B$. r is screened in $\mathfrak{B}$ thus there is a screen $\{A_i;\ i = 1, \dots, n\} \subset \mathfrak{B}$ of the pair A, B. Then $Y = \cup \{\check{A}_i;\ i = 1, \dots, n\}$ and $A_i\, r\, A$ or $A_i\, r\, B$ implies that $\check{A}_i\, \check{r}\, \check{A}$ or $\check{A}_i\, \check{r}\, \check{B}$, and we have seen that $\{\check{A}_i;\ i = 1, \dots, n\}$ is a screen of $\check{A}, \check{B}$.

Using the well known theorem of Smirnov (see [10]), we are going to show that Y is equivalent to $\delta_{(\mathfrak{B}, r)}(X)$ (see the definition of $\delta_{(\mathfrak{B}, r)}(X)$ in [8]).

Let δ be the following relation in X: $A \not\delta B$ iff $\bar{A}^Y \cap \bar{B}^Y = \phi$. We show that if $A, B \in \mathfrak{B}$, $A\, r\, B$ then $A \not\delta B$. But $\bar{A}^Y \subset \hat{A}$, $\bar{B}^Y \subset \hat{B}$, thus it is enough to show that if $A\, r\, B$ then $\hat{A} \cap \hat{B} = \phi$. Assume $\mathfrak{R} \in \in \hat{A} \cap \hat{B}$. This means that $A, B \in \mathfrak{R}$ which is impossible. Hence by Theorem 1 of [8] we get that $\delta_{(\mathfrak{B}, r)} < \delta$. Finally, we show that if $A \not\delta B$ then $A \not\delta_{(\mathfrak{B}, r)} B$. $A \not\delta B$ means that $\bar{A}^Y \cap \bar{B}^Y = \phi$. Since $(\check{\mathfrak{B}}, \check{r})$ is a strong preproximity in Y, Y is a compact Hausdorff space, and the family of all closed sets in Y is a normal base (see [4]), using Theorem 2.4 of [8] and the fact that in a compact Hausdorff space there is only one compatible proximity, we get that $\bar{A}^Y \cap \bar{B}^Y = \phi$ implies $\bar{A}^Y \not\delta_{(\check{\mathfrak{B}}, \check{r})} \bar{B}^Y$. But this implies that $A \not\delta_{(\mathfrak{B}, r)} B$. Thus we got $\delta < \delta_{(\mathfrak{B}, r)}$; and in summary $\delta = = \delta_{(\mathfrak{B}, r)}$ is fulfilled. Since Y is the Hausdorff compactification of X belonging to δ, using the well-known theorem of Smirnov (see [10]), Y is equivalent to $\delta_{(\mathfrak{B}, r)}(X)$.

Q.e.d.

It is easy to see that this method is a generalization of the Wallman – Shanin method and the method given by Aarts and de Groot, see [4], [11], [1]. The question is whether this method gives us all Hausdorff compactifications of a Tychonoff space.

Theorem 5. *Let X be a compact Hausdorff space and Y be a dense subset of X. Then there exists a strong preproximity $(\mathfrak{B}_Y, r_Y)$ in Y, in which r_Y can be screened, such that $\bar{Y}^{\hat{Y}}$ is equivalent to X.*

Proof. For $A, B \subset Y$ let $A \not\delta_Y B$ if and only if $\bar{A}^X \cap \bar{B}^X = \phi$. Let $\mathfrak{B}_Y = \{A: A$ is closed in $Y\}$, and $r_Y = \not\delta_Y$. Then $(\mathfrak{B}_Y, r_Y)$ is a strong preproximity in Y, in which r_Y is screened. By Theorem 4, $\bar{Y}^{\hat{Y}}$ is equivalent to $\delta_{(\mathfrak{B}_Y, r_Y)}(Y)$. But since $\delta_Y = \delta_{(\mathfrak{B}_Y, r_Y)}$ is fulfilled, thus X is equivalent to $\bar{Y}^{\hat{Y}}$, too.

Q.e.d.

Finally we should like to note that Theorem 3 gives an internal characterization of T_2-compactifiable spaces, with the help of the notion of strong preproximity, and the proof of the theorem does not use the notion of real numbers or the real functions and also does not use the well known theorem of Smirnov [10]. This was a problem raised by Ervin Deák, see [3].

REFERENCES

[1] J.M. Aarts and J. de Groot, Complete regularity as a separation axiom, *Communication of the International Congress of Math.*, Moscow, 1966.

[2] Á. Császár, Wallman-type compactification and superextension, *Periodica Math. Hung.*, 1 (1) (1967).

[3] E. Deák and P. Hamburger, Interne Charakterisation der kompaktifizierterbaren Räume, *Periodica Math. Hung.*, (to appear).

[4] O. Frink, Compactification and seminormal spaces, *Amer. Math. J.*, 86 (9164), 602-607.

[5] J. de Groot and J.M. Aarts, Complete regularity as a separation axiom, *Canad. J. Math.*, 21 (1969), 96-105.

[6] J. de Groot, G.A. Jensen and A. Verbeek, Superextension, *Report ZW.* 1968. Mathematical Centre Amsterdam.

[7] J. de Groot, Supercompactness and superextension, Berlin, *Topology Symposium.* 1967. 89-90.

[8] P. Hamburger, A general method to give internal characterization of completely regular spaces, *Acta Math. Acad. Sci.,* 23 (1972), 3-4.

[9] P. Hamburger, Completely regularity as a separation axiom, *Periodica Math. Hung.,* 4 (2) (to appear).

[10] Ju.M. Smirnov, On the theory of proximity spaces, *Math. Sb.,* 31 (1952), (73) (in Russian).

[11] N.A. Shanin, On the theory of bicompact extension of topological spaces, *Dokl. C.C.C.P.* 38 (1943).

[12] A. Verbeek, Superextension, (to appear).

[13] W.H. Wallman, Lattices and topological spaces, *Math. Ann.,* (2) 39 (1938).

COLLOQUIA MATHEMATICA SOCIETATIS JÁNOS BOLYAI
8. TOPICS IN TOPOLOGY, KESZTHELY (HUNGARY), 1972.

UNIFORM DIMENSION AND RINGS OF BOUNDED UNIFORMLY CONTINUOUS FUNCTIONS

J. HEJCMAN

M. Katětov examined in [4] (see also [1] or [5]) connections between the dimension of a space X and properties of the ring $C^*(X)$ of all bounded continuous real-valued functions on X. The most interesting case is for the compact metric spaces X when $\dim X$ completely coincides with the so called analytical dimension of $C^*(X)$. We are going to derive a similar characterization of the uniform dimension Δd of metric spaces; this will be also a generalization of the mentioned Katětov's result. The basic idea of the proof will be similar to those used in [5]. First, recall some definitions and notations.

Let C be a topological commutative ring with a unit e and with continuous real scalar multiplication. Then C is called an *analytical ring*. A subring C_1 of C is called *analytically closed* if

1) C_1 is a submodule of C, $e \in C_1$,
2) C_1 is a closed subset of C,
3) $y \in C, y^2 \in C_1$ imply $y \in C_1$.

A subset B of C is called an *analytical base* of C if there is no analytically closed subring C_1 such that $B \subset C_1$ and $C_1 \neq C$. The least cardinal number of an analytical base of C is called the *analytical dimension* of C.

For any uniform space X denote by $C^*(X)$, $U^*(X)$ the set of all continuous, uniformly continuous respectively bounded real-valued functions on X both endowed with the usual topology of uniform convergence. Clearly, $C^*(X)$ and $U^*(X)$ are analytical rings. If X is compact they both coincide with $C(X)$: this indicates the ring of all continuous real-valued functions on X. The original Katětov's definition of an analytical subring required, instead of 3), the following stronger condition

3') $y \in C$, $a_i \in C_1$, $i = 1, \ldots, n$, $y^n + a_1 y^{n-1} + \ldots + a_n = 0$ imply $y \in C_1$.

But it follows just from some theorems in [4] that for any analytical ring $C(X)$ with X compact, the both collections of conditions are equivalent. On the other hand the condition 3) implies (for $C = C^*(X)$ or $C = U^*(X)$) the weaker condition

3'') $y \in C, |y| \in C_1$ imply $y \in C_1$,

which will be quite sufficient for our purposes. We shall often use the fact that any closed subring of $C^*(X)$ or $U^*(X)$ containing the constant functions is a sublattice (see e.g.[1], 16.2).

Concerning uniform dimension Δd, basic definitions and theorems can be found in [3] but we shall need nothing more that the following assertion which is a consequnece of Theorem 4 in [2]. The symbol I^n will always denote an n-dimensional cube.

A) *Let X be a metric space. Then $\Delta dX \leqslant n$ if and only if there exists a mapping $f\colon X \to I^n$ such that $\Delta df = 0$.*

Recall that if X, Y are metric spaces, $f\colon X \to Y$ is a uniformly continuous mapping, then $\Delta df = 0$ means: for each $\epsilon > 0$ there exist $\delta > 0$ and $\eta > 0$ such that whenever $M \subset Y$, $\operatorname{diam} M \leqslant \delta$, then there exists a collection $\mathscr{K}$ such that $\bigcup \mathscr{K} = f^{-1}[M]$, mesh $\mathscr{K} \leqslant \epsilon$ and $\mathscr{K}$ is η-discrete, i.e. $\operatorname{dist}(K, L) \geqslant \eta$ for any distinct K, L from $\mathscr{K}$. A collection is called uniformly discrete if it is η-discrete for some $\eta > 0$.

If X is a uniform space then sX denotes the Samuel compactification of X (see [3]). We shall need this property only: $f \in U^*(X)$ if and only if f is the restriction of a function $g \in C(sX)$.

The proofs of the main theorems will be preceded by several lemmas.

Lemma 1. *Let X be a Hausdorff uniform space, sX its Samuel compactification. For each $f \in U^*(X)$ let $\sigma(f) \in C(sX)$ be the unique extension of f. Then the mapping σ is an isometry onto and also a linear, ring and lattice isomorphism. If $A \subset U^*(X)$, A separates far subsets of X then $\sigma[A]$ separates distinct points of sX.*

Proof. The isometry and the isomorphisms are obvious. Let $a, b \in \in sX$, $a \neq b$. Let W be an open symmetric member of the uniformity of sX such that $b \notin W \circ W \circ W[a]$. Now $W[a], W[b]$ are open in sX, $W[a] \cap X$, $W[b] \cap X$ are open and far in X. Choose $g \in A$ such that, e.g., for some $\alpha < \beta$, $g(x) \leqslant \alpha$ for $x \in W[a] \cap X$, $g(x) \geqslant \beta$ for $x \in \in W[b] \cap X$. Let $h = \sigma(g)$. Since h is continuous and X is dense in sX, $h(a) \leqslant \alpha$, and $h(b) \geqslant \beta$.

Lemma 2. *Let X be a topological space, $f\colon X \to I^n$ be continuous, $f = (f_1, \dots, f_n)$. Let L be a sublattice and submodule of $C^*(X)$ which contains $f_1, \dots, f_n$ and all constant functions. Suppose that J, K are n-dimensional intervals, $K \subset J \subset I^n$, K is closed, J is open in I^n. Then there exists a non-negative $h \in L$ such that $h(x) \geqslant 1$ for $x \in \in f^{-1}[K]$, $h(x) = 0$ for $x \in f^{-1}[I^n \setminus J]$.*

Proof. Let $J = \prod_{k=1}^{n} J_k$. If $J_k = \,]\alpha, \beta[\,$, we put

$$g_k(x) = \max(\min(f_k(x) - \alpha, -f_k(x) + \beta), 0)\,.$$

If e.g. $J_k = [0, \beta[$ we put $g_k(x) = \max(-f_k(x) + \beta, 0)$ etc. Finally, it suffices to take a suitable positive λ and to define

$$h(x) = \lambda \min(g_1(x), \dots, g_n(x))\,.$$

Lemma 3. *Let $\mathscr{A}$ be an η-discrete collection of subsets of a* *metric space* X, $\eta > 0$. *Let* $h \in U^*(X)$, $h(x) \geqslant 0$ *for each* $x \in X$, $h(x) = 0$ *for* $x \in X \setminus \cup \mathscr{A}$. *Let* $L \subset U^*(X)$ *fulfil the condition* 3'') *above*, $h \in L$. *Given any* $\mathscr{A}' \subset \mathscr{A}$, *put*

$$\begin{aligned} h'(x) &= \quad h(x) \quad \text{for} \quad x \in \bigcup \mathscr{A}' \\ &= -h(x) \quad \text{for} \quad x \in X \setminus \bigcup \mathscr{A}'. \end{aligned}$$

Then $h' \in L$.

Proof is obvious.

Theorem 1. *Let* X *be a metric space*, $f\colon X \to I^n$, $\Delta df = 0$, $f = (f_1, \ldots, f_n)$. Then $\{f_1, \ldots, f_n\}$ *is an analytical base of* $U^*(X)$.

Proof. Let L be an analytically closed subring of $U^*(X)$ containing $f_1, \ldots, f_n$. First, we shall prove that L separates far subsets of X. Let C, D be non-void subsets of X, $\operatorname{dist}(C, D) > \epsilon > 0$. Since $\Delta df = 0$ choose, for this ϵ, positive δ and η such that, for any $M \subset I^n$ with $\operatorname{diam} M \leqslant \delta$, $f^{-1}[M]$ can be expressed as a union of an η-discrete collection with mesh less then or equal to ϵ. Let $\{J_i\}_{i=1}^r$ be a finite sequence of subintervals open in I^n such that $\operatorname{diam} J_i \leqslant \delta$, $\bigcup_{i=1}^r J_i = I^n$. Choose, for each i, a closed interval $K_i \subset J_i$ such that $\bigcup_{i=1}^r K_i = I^n$. For each J_i and K_i, choose a function h_i by Lemma 2. Now let $\mathscr{A}_i$, $i = 1, \ldots, r$, be η-discrete collections with $\operatorname{mesh} \mathscr{A}_i \leqslant \epsilon$ such that $f^{-1}[J_i] = \bigcup \mathscr{A}_i$. Then for each i, let $\mathscr{A}'_i = \{A \in \mathscr{A}_i, A \cap C \neq \phi\}$. For each h_i and $\mathscr{A}'_i$, take h'_i by Lemma 3. Finally put $g(x) = \sum_{i=1}^r h'_i(x)$. Since the collection of all $f^{-1}[K_i] \cap A$ where $A \in \mathscr{A}_i$ and $i = 1, \ldots, r$ is a cover of the space X, $g(x) \geqslant 1$ for each $x \in C$. Since $A \cap D \neq \phi$ implies $A \notin \mathscr{A}'_i$, $g(x) \leqslant 0$ for each $x \in D$.

Now apply Lemma 1, use the mapping σ. We know $\sigma[L]$ separates points of sX, hence by the Stone – Weierstrass Theorem, $\sigma[L] = C(sX)$. Therefore $L = U^*(X)$ and the theorem is proved.

Lemma 4. *Let* X, Y *be metric spaces*, Y *compact*, $f\colon X \to Y$ *uniformly continuous. Then* $\Delta df = 0$ *if and only if for each* $\epsilon > 0$ *and*

$q \in Y$ *there exist a neighbourhood* V *of* q *and a uniformly discrete collection* $\mathcal{K}_V$ *with* mesh $\mathcal{K}_V \leqslant \epsilon$ *such that* $f^{-1}[V] = \bigcup \mathcal{K}_V$.

Proof. To prove the sufficiency, using the compactness of Y we take a finite open cover by the neighbourhoods V and use the Lebesgue number of this cover.

Lemma 5. *Let* $\mathcal{M}_n$, $n = 1, 2, \ldots$, *be collections of subsets of a metric space* X, $\mathcal{M}_{n+1}$ *refine* $\mathcal{M}_n$. *Let* $\epsilon > 0$ *and* mesh $\mathcal{M}_n > \epsilon$ *for each* n. *Then there exists* $g \in U^*(X)$ *such that* mesh $\{g[M];\ M \in \in \mathcal{M}_n\} \geqslant 1$ *for each* n.

Proof follows immediately from the following lemma, we may suppose $\epsilon = 1$.

Lemma 6. *Let* $\mathcal{D}$ *be an infinite family of two-point subsets of a metric space* X, $\operatorname{diam} D \geqslant 1$ *for each* $D \in \mathcal{D}$. *Then there exist* $g \in \in U^*(X)$ *and an infinite* $\mathcal{D}_1 \subset \mathcal{D}$ *such that* $\operatorname{diam} g[D] \geqslant 1$ *for each* $D \in \in \mathcal{D}_1$.

Proof. Choose $D_1 \in \mathcal{D}$. If it is possible choose $D_2 \in \mathcal{D}$ such that $\operatorname{dist}(D_1, D_2) \geqslant 1/3$. If $D_1, \ldots, D_k$ are defined choose, if possible, $D_{k+1} \in \mathcal{D}$ such that $\operatorname{dist}(D_i, D_{k+1}) \geqslant 1/3$ for $i = 1, \ldots, k$. Two cases are to be distinguished:

a) We get an infinite sequence $D_1, D_2, \ldots$. Let $D_j = \{a_j, b_j\}$, put $A = \{a_j\}_{j=1}^{\infty}$, $g(x) = \min(3 \operatorname{dist}(x, A), 1)$. Then g has the required properties for $\mathcal{D}_1 = \{D_j\}_{j=1}^{\infty}$.

b) It is possible to construct a finite sequence $D_1, \ldots, D_k$ only. Then there exists h such that $\operatorname{dist}(D, D_h) < 1/3$ for infinitely many D in $\mathcal{D}$. Keep this h fixed, let $D_h = \{a, b\}$. Then either $\operatorname{dist}(a, D) < < 1/3$ for each D from some infinite $\mathcal{D}_1 \subset \mathcal{D}$ or the same with b instead of a. Consider the first case. Put $g(x) = \min(3 \operatorname{dist}(x, a), 3)$. Let $D = \{u, v\} \in \mathcal{D}_1$. Suppose $\operatorname{dist}(u, a) < 1/3$. Then $\operatorname{dist}(v, a) > 2/3$ since $\operatorname{dist}(u, v) \geqslant 1$. Thus $g(v) - g(u) > 1$, $g \in U^*(X)$.

Theorem 2. *Let* X *be a metric space,* $f\colon X \to I^n$, $f = (f_1, \ldots, f_n)$. *Let* $\{f_1, \ldots, f_n\}$ *be an analytical base of* $U^*(X)$. *Then* $\Delta df = 0$.

Proof. Suppose the assertion is false, use Lemma 4. Thus there exist $\epsilon > 0$, $q \in I^n$ such that for no neighbourhood V of q can $f^{-1}[V]$ be expressed as in Lemma 4. Keep these ϵ and q fixed. Let S be the set of all $g \in U^*(X)$ such that for each $\delta > 0$ there exists a neighbourhood V of q and a uniformlv discrete collection $\mathscr{H}$ such that $\bigcup \mathscr{H} = f^{-1}[V]$, mesh $\{g[H];\ H \in \mathscr{H}\} \leqslant \delta$. Each f_i belongs to S (if diam $V \leqslant \delta$ the choice of $\mathscr{H}$ is trivial). Let us prove that S is an analytically closed subring of $U^*(X)$.

a) Let $g_i \in S$, $i = 1, 2$. Given $\delta > 0$ choose a common neighbourhood V and the collections $\mathscr{H}_i$. Then the collection $\{H_1 \cap H_2;\ H_i \in \mathscr{H}_i\}$ is uniformly discrete and always $\operatorname{diam}(g_1 + g_2)[H_1 \cap H_2] \leqslant \leqslant 2\delta$. Thus $g_1 + g_2 \in S$. The proof for product is quite similar, however the boundedness of g_i's must be used too. Real multiples and all constant functions evidently belong to S.

b) Let $g \in \bar{S}$, $\delta > 0$. Choose $g_1 \in S$ such that $|g_1(x) - g(x)| \leqslant \leqslant \delta$ for each $x \in X$. If $\mathscr{H}_1$ is the mentioned collection for g_1, δ and some neighbourhood V then it suffices for g too and mesh $\{g[H]; H \in \in \mathscr{H}_1\} \leqslant 3\delta$.

c) Let $g \in U^*(X)$, $g^2 \in S$. Then $|g|$ alsc belongs to S as it is a compostition of g^2 and a uniformly continuous function. Given $\delta > 0$ let V be a neighbourhood of q and $\mathscr{H}$ an η-discrete collection such that $\bigcup \mathscr{H} = f^{-1}[V]$, mesh $\{|g|[H]; H \in \mathscr{H}\} \leqslant \delta$. For each $H \in \mathscr{H}$ put $H^+ = \{x \in H; g(x) \geqslant 0\}$, $H^- = \{x \in H; g(x) \leqslant 0\}$. Choose $\xi > 0$, $\zeta \leqslant \eta$ such that $\operatorname{dist}(x, y) < \zeta$ implies $|g(x) - g(y)| < \delta$. Let $\mathscr{K}$ be the collection of the following sets: if $H \in \mathscr{H}$ and $|g(x)| \geqslant \delta/2$ for each $x \in \in H$ then let both H^+ and H^- belong to $\mathscr{K}$; if not then let H itself belong to $\mathscr{K}$. Then $\mathscr{K}$ is ζ-discrete, mesh $\{g[K]; K \in \mathscr{K}\} \leqslant 3\delta$. Hence $g \in S$.

Let us use essentially the properties of ϵ and q. Let $V_1 \supset \supset V_2 \supset \ldots$ be a base of neighbourhoods of q. For each j define a relation on the set $f^{-1}[V_j]$: $x \sim y$ means there exist $x = x_0, x_1, \ldots, x_k = y$ such that $x_i \in f^{-1}[V_j]$, $\operatorname{dist}(x_i, x_{i+1}) < 1/j$. Let $\mathscr{M}_j$ be the collection of all classes defined by this equivalence. By the assumptions on ϵ and q, mesh $\mathscr{M}_j > \epsilon$. The sequence $\mathscr{M}_1, \mathscr{M}_2, \ldots$ fulfils the conditions of

Lemma 5. Take the function g from Lemma 5. Then clearly g does not belong to S. Since $f_1, \ldots, f_n$ is an analytical base, we have a contradiction.

Theorem 3. *Let X be a non-void metric space. Then either ΔdX and the analytical dimension of $U^*(X)$ are finite and equal or they are both infinite.*

Proof. It is almost sufficient to prove: $\Delta dX \leqslant n$ if and only if there are $f_i \in U^*(X)$, $i = 1, \ldots, n$ such that $\{f_1, \ldots, f_n\}$ is an analytical base of $U^*(X)$. This is a direct consequence of Theorems 1 and 2 and the above assertion A) but it helps for non-zero dimensions only. For the remaining case, let f_0 be a constant real-valued function on the space X. Then $\Delta dX = 0$ if and only if $\Delta df_0 = 0$. By Theorems 1 and 2, this holds if and only if $\{f_0\}$ is an analytical base of $U^*(X)$. But since any constant function can be excluded from an analytical base, this means that the empty set is an analytical base. Thus the proof is complete.

All the rings of real-valued functions might be considered, from the beginning, as vector lattices only. A sublattice can be said to be analytically closed if it is a submodule containing all constant functions which satisfies conditions 2) and 3″). Then the analytical dimension would be defined quite analogously. The reader can go through all proofs and easily verify that the theorems hold again.

Added in proof. Some related results were obtained in [6].

REFERENCES

[1] L. Gillman – M. Jerison, *Rings of continuous functions,* Princeton 1960.

[2] J. Hejcman, Uniform dimension of mappings, *Proc. of the Second Prague Top. Symp.*, 1966, Praha 1967, 182-183.

[3] J.R. Isbell, *Uniform spaces,* Providence 1964.

[4] M. Katětov, On rings of continuous functions and dimensions of bicompact spaces, *Časopis Pěst. Mat. Fys.*, 75 (1950), 1-16 (in Russian).

[5] J. Nagata, *Modern dimension theory,* Amsterdam 1965.

[6] S.A. Bogatyĭ, Characteristics of topological and uniform dimension in terms of rings of continuous functions, *Sibirskiĭ Mat. Žurnal,* 14 (1973, 289-299 (in Russian).

COLLOQUIA MATHEMATICA SOCIETATIS JÁNOS BOLYAI
8. TOPICS IN TOPOLOGY, KESZTHELY (HUNGARY), 1972.

PERFECT SUBCATEGORIES AND FACTORIZATIONS

H. HERRLICH

In modern topology properties of topological spaces and relations between certain classes of topological spaces are usually studied by means of certain classes of maps. Among those classes which in this respect have proved to be most useful the class of perfect maps plays an outstanding role. Cf. e.g. the beautiful results of Arhangelskiĭ [1,2], Banaschewski [3], Borges [5], Coban [7], Flachsmeyer [11], Franklin [12], Frolík and Liu [14], Frolík and Mrówka [15]. Henriksen and Isbell [18], Hušek [23], Mancuso [28], Michael [29, 31, 33], Morita [35], Mrówka [36, 37], Nagata [38], Okuyama [40], Rainwater [44], van der Slot [45, 46], Strauss [47], and Whyburn [52]. No wonder that recently several authors have tried to generalize the concept of perfect maps. The characterization given by Henriksen and Isbell [18] has been used by Błaszczyk and Mioduszewski [4], Franklin [13], Hager [16], Mioduszewski and Rudolf [34], and Tsai [50]. Nakagawa [39] and the author [20] independently have suggested another generalization.

In the first part of this paper we study the relations among these concepts. Next we study for nice categories $\mathfrak{C}$ the relations between full subcategories $\mathfrak{K}$ of $\mathfrak{C}$ and the corresponding class $P\mathfrak{K}$ of all $\mathfrak{K}$-perfect morphisms in $\mathfrak{C}$. It turns out that this relation can best be described by means of two Galois-correspondences π and δ, each being of independent interest. Whereas π describes the relation between socalled "universal" and "special" problems (cf. Pumplün [42] and Sonner [49]), δ helps to characterize those morphism classes $\mathfrak{E}$ and $\mathfrak{M}$ of $\mathfrak{C}$ for which each $\mathfrak{C}$-morphism has a unique $(\mathfrak{E}, \mathfrak{M})$-factorization (Cf. Collins [8], Ehrbar and Wyler [10], Herrlich [19], Herrlich and Strecker [22], Isbell [24], Jurchescu and Lascu [25], Kelly [26], Kennison [27], Michael [29], Raasch [43], and Whyburn [51, 53]). Finally we characterize the "perfect" subcategories of $\mathfrak{C}$, i.e. those subcategories $\mathfrak{P}$ of $\mathfrak{C}$ for which there exists a full subcategory $\mathfrak{K}$ of $\mathfrak{C}$ with $\mathfrak{P} = P\mathfrak{K}$.

I. PERFECT MAPS

I.1. Theorem. *Let $\mathfrak{C}$ be the category of completely regular spaces and continuous maps, let $\mathfrak{K}$ be the full subcategory of $\mathfrak{C}$ whose objects are the compact spaces, let $\beta_A : A \to \beta A$ denote the $\mathfrak{K}$-reflection (= Čech–Stone-compactification) of A, and let $f\colon X \to Y$ be a $\mathfrak{C}$-morphism. Then the following conditions are equivalent:*

(P1)
$$\begin{array}{ccc} X & \xrightarrow{\beta_X} & \beta X \\ {\scriptstyle f}\downarrow & & \downarrow{\scriptstyle \beta f} \\ Y & \xrightarrow[\beta_Y]{} & \beta Y \end{array}$$
is a pullback in $\mathfrak{C}$, i.e. $\beta f(\beta X \setminus X) \subset (\beta Y \setminus Y)$

(P2) There exist $\mathfrak{C}$-morphisms g and h and a $\mathfrak{K}$-morphism k such that

$$\begin{array}{ccc} \bullet & \xrightarrow{g} & \bullet \\ {\scriptstyle f}\downarrow & & \downarrow{\scriptstyle k} \\ \bullet & \xrightarrow[h]{} & \bullet \end{array}$$
is a pullback in $\mathfrak{C}$

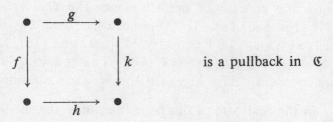

(P3) *There exist* $\mathfrak{C}$*-morphisms* g *and* h *and an extremal monomorphism (= closed embedding)* m *such that the diagram*

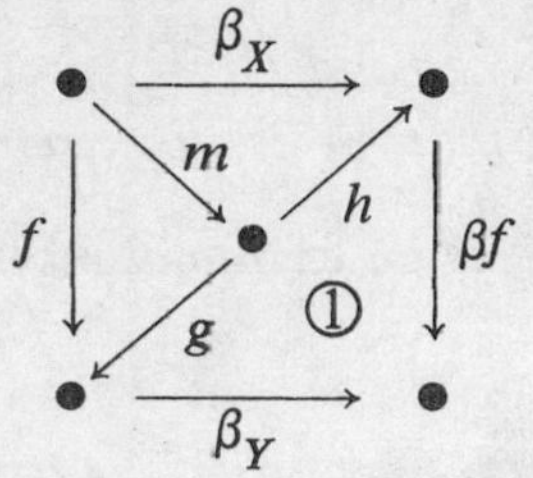

commutes and ① *is a pullback in* $\mathfrak{C}$

(P4) f *belongs to the smallest subcategory of* $\mathfrak{C}$ *which contains* $\mathfrak{K}$ *and all extremal monomorphisms of* $\mathfrak{C}$ *and is closed under the formation of pullbacks in* $\mathfrak{C}$

(P5) f *belongs to the smallest subcategory of* $\mathfrak{C}$ *which contains* $\mathfrak{K}$ *and is closed under the formation of both-sided restrictions* and of pullbacks in* $\mathfrak{C}$

(P6) *There exist* $\mathfrak{C}$*-morphisms* g, h, *and* l, $\mathfrak{C}$*-extremal monomorphisms* m *and* n *and a* $\mathfrak{K}$*-morphism* k *such that the diagram*

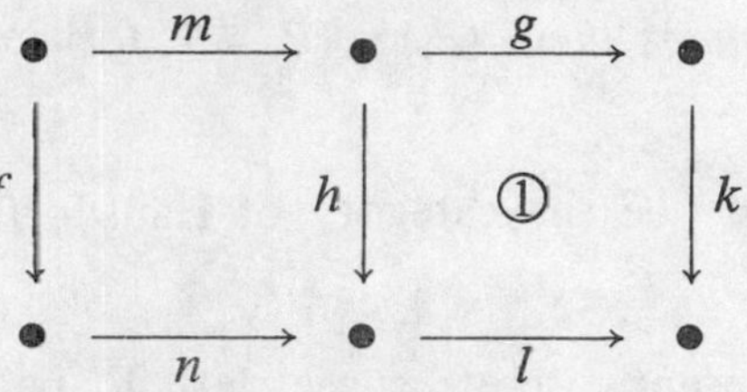

commutes and ① *is a pullback in* $\mathfrak{C}$

(P7) *For any* $\mathfrak{K}$*-extendable epimorphism* e *and any* $\mathfrak{C}$*-morphisms* g *and* h *with* $f \cdot g = h \cdot e$ *there exists a* $\mathfrak{C}$*-morphism* l *such that the diagram*

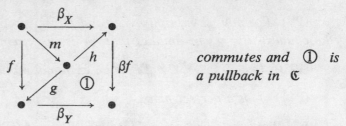

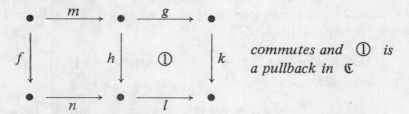

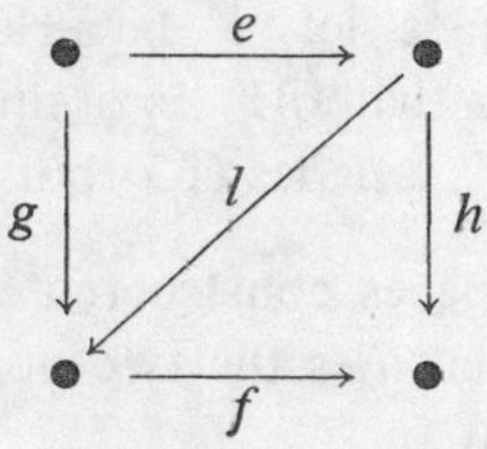

commutes

*$f\colon A \to B$ is called a both-sided restriction of $g\colon C \to D$ iff there exist extremal monomorphisms $m\colon A \to C$ and $n\colon B \to D$ with $n \circ f = g \circ m$.

(P8) *For any* $\mathfrak{C}$*-morphism* g *and any* $\mathfrak{K}$*-extendable epimorphism* e *the equality* $f = g \cdot e$ *implies that* e *is an isomorphism*

(P9) *For any* $\mathfrak{C}$*-morphism* g *and any* $\mathfrak{K}$*-extendable morphism* m *the equality* $f = g \cdot m$ *implies that* m *is an extremal monomorphism*

(P10) $(f, \beta_X)\colon X \to Y \times \beta X$ *is an extremal monomorphism*

(P11) f *is a perfect map.*

I.2. Remark. The equivalence of (P1) and (P11) has been observed first by Henriksen and Isbell [18]. Błaszczyk and Mioduszewski [4], Franklin [13], Hager [16], Mioduszewski and Rudolf [34], and Tsai [50] have used this fact to generalize the concept of perfect map. The equivalence of (P5), (P6), and (P11) was observed by Nakagawa [39], who generalized perfect maps accordingly and proved the equivalence of (P5) and (P6) in a general categorical setting. The equivalence of (P7), (P8), and (P11) was used by the author [20] to generalize perfect maps.

I.3. Proposition. *If we replace in* I.1. $\mathfrak{C}$ *by the category of Hausdorff spaces and continuous maps then* $(P_i) \Rightarrow (P_{i+1})$ *for* $i = 1, \ldots, 10$. *Moreover* (P10) ⇒ (P3).

I.4. Examples. Let $\mathfrak{C}$ be the category of Hausdorff spaces and continuous maps.

a) Let X be a two-point discrete space, let Y be a Hausdorff space containing X such that the Čech–Stone-compactification of Y is obtained set-theoretically by identifying the two points of X, and let $f\colon X \to Y$ be the embedding. Then f satisfies (P2) but not (P1).

b) Let X be a one-point space, let Y be a Hausdorff space containing at least two points and such that βY is a singleton, and let $f\colon X \to Y$ be an arbitrary map. Then f satisfies (P3) but not (P2).

c) Let X and Y be the spaces constructed in example a) and let f be the quotient map which identifies the two points of X in Y. Then f satisfies (P11) but not (P10).

I.5. Theorem. *If we replace in* I.1 $\mathfrak{C}$ *by any complete, cocomplete, wellpowered, and cowellpowered category,* $\mathfrak{K}$ *by any epireflective full subcategory of* $\mathfrak{C}$, *and the Čech–Stone-compactification by the* $\mathfrak{K}$-*reflection then* $(P_i) \Rightarrow (P_{i+1})$ *for* $i = 1, \ldots, 9$ and $(P10) \Rightarrow (P3)$.

II. GALOIS-CORRESPONDENCES

In this section we will provide some useful terminology. Let A and B be arbitrary classes and let ρ be a subclass of $A \times B$. Then ρ induces a function $L_\rho : PB \to PA$ between the power classes of B and A, defined by

$$L_\rho(X) = \{a \in A \mid a\rho x \text{ for all } x \in X\}$$

for all $X \subset B$. Analogously δ induces a function $R_\delta : PA \to PB$. A subclass X of A is called L_ρ-closed provided that $X = L_\rho Y$ for some subclass Y of B (equivalently: iff $X = L_\rho R_\rho X$). Analogously R_ρ-closedness is defined for subclasses of B.

III. EPIREFLECTIVE SUBCATEGORIES VIA STANDART CLASSES OF EPIMORPHISMS

From now on let $\mathfrak{C}$ be a complete, cocomplete, wellpowered, and cowellpowered category.

Let $\mathfrak{A} = \text{Ob}\,\mathfrak{C}$ be the class of $\mathfrak{C}$-objects, let $\mathfrak{B} = \text{Epi}\,\mathfrak{C}$ be the class of epimorphisms in $\mathfrak{C}$, and let π be the following subclass of $\mathfrak{A} \times \mathfrak{B}$:

$$\pi = \{(K, e) \mid K \in \mathfrak{A} \text{ and } e: X \to Y \in \mathfrak{B} \text{ and for all } f: X \to K \text{ there exists } g: Y \to K \text{ with } f = g \cdot e\}.$$

Then for any $\mathfrak{K} \subset \mathfrak{A}$ the class $R_\pi \mathfrak{A}$ is precisely the class of $\mathfrak{K}$-extendable epimorphisms in $\mathfrak{C}$, and for any $\mathfrak{E} \subset \mathfrak{B}$ the class $L_\pi \mathfrak{E}$ is precisely the class of $\mathfrak{E}$-injective $\mathfrak{C}$-objects.

We will not distinguish between a subclass of $\mathfrak{A} = \text{Ob}\,\mathfrak{C}$ and the corresponding full subcategory of $\mathfrak{C}$. Also we will not distinguish between a subclass of $\mathfrak{B} = \text{Epi}\,\mathfrak{C}$ which is closed under composition and contains with any $e: X \to Y$ the identities 1_X and 1_Y and the corresponding subcategory of $\mathfrak{C}$.

The following theorems provide satisfactory characterizations of L_π-closed and R_π-closed classes:

III.1. Theorem. *For any full, isomorphism-closed subcategory* $\mathfrak{K}$ *of* $\mathfrak{C}$ *the following conditions are equivalent:*

a) $\mathfrak{K}$ *is* L_π*-closed*

b) $\mathfrak{K}$ *is epireflective in* $\mathfrak{C}$

III.2. Theorem. *For any subclass* $\mathfrak{E}$ *of* Epi $\mathfrak{C}$ *the following conditions are equivalent:*

a) $\mathfrak{E}$ *is* R_π*-closed*

b) $\mathfrak{E}$ *satisfies the following conditions:*

(E1) $\mathfrak{E}$ *is a subcategory of* $\mathfrak{C}$ *which contains all* $\mathfrak{C}$*-isomorphisms*

(E2) $\mathfrak{E}$ *is closed under pushouts in* $\mathfrak{C}$

(E3) $\mathfrak{E}$ *has "special" solutions, i.e. for each* $\mathfrak{C}$*-object* X *there exists* $e_X: X \to \widetilde{X}$ *in* $\mathfrak{E}$ *such that for any* $e: X \to Y \in \mathfrak{E}$ *there exists a unique* $\mathfrak{C}$*-morphism* $f: Y \to \widetilde{X}$ *with* $e_X = f \cdot e$

(E4) $\mathfrak{E}$ *is left-cancellative with respect to epis, i.e.* $f \cdot g \in \mathfrak{E}$ *and* $g \in$ Epi $\mathfrak{C}$ *implies* $g \in \mathfrak{E}$

c) $\mathfrak{E}$ *satisfies the conditions* (E1), (E2), (E4) *and* (E3*) $\mathfrak{E}$ is *closed under multiple pushouts (= cointersections).*

III.3. Remark. The relations between epireflective subcategories (= subcategories with "universal" solutions) and subcategories with "special" solutions have been studied systematically first by Pumplün [42]. The equivalence of conditions b) and c) was first observed by Strecker [48] who called classes of epimorphisms satisfying these conditions "standart" classes.

IV. FACTORIZATIONS AND THE DIAGONALRELATION δ

If $\mathfrak{M}$ is a class of $\mathfrak{C}$-morphisms which contains all $\mathfrak{C}$-isomorphisms and if $\mathfrak{E}$ is a class of epimorphisms in $\mathfrak{C}$ which contains all $\mathfrak{C}$-isomorphisms then $\mathfrak{C}$ is called an $(\mathfrak{E}, \mathfrak{M})$-*category* provided that $\mathfrak{E}$ and $\mathfrak{M}$ are subcategories of $\mathfrak{C}$ and every $\mathfrak{C}$-morphism has a unique $(\mathfrak{E}, \mathfrak{M})$-factorization. In this case $\mathfrak{E}$ is called a *"left factor"* of $\mathfrak{C}$ and $\mathfrak{M}$ is called a *"right factor"* of $\mathfrak{C}$. Let $\mathfrak{A} = \mathrm{Epi}\,\mathfrak{C}$ be the class of all epimorphisms in $\mathfrak{C}$, let $\mathfrak{B} = \mathrm{Mor}\,\mathfrak{C}$ be the class of all $\mathfrak{C}$-morphisms and let δ be the following subset of $\mathfrak{A} \times \mathfrak{B}$:

$\delta = \{(e, f) \mid e \in \mathfrak{A}$ and $f \in \mathfrak{B}$ and whenever r and s are $\mathfrak{C}$-morphisms with $f \cdot r = s \cdot e$ then there exists a $\mathfrak{C}$-morphism g such that the diagram

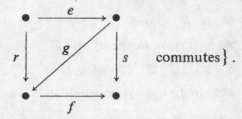

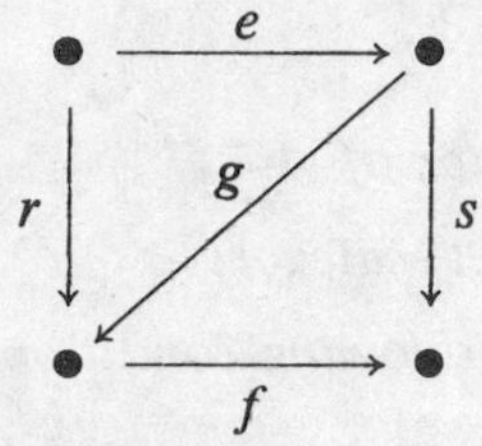

commutes$\}$.

IV.1. Remark. $(\mathfrak{E}, \mathfrak{M})$-categories and the corresponding δ-relation have been investigated by various authors. Cf. Ehrbar and Wyler [10], Herrlich and Strecker [22], Isbell [24], Jurchescu and Lascu [25], Kelly [26]. Kennison [27] and Raasch [43].

IV.2. Proposition. *Let* $\mathfrak{C}$ *be a* $(\mathfrak{E}, \mathfrak{M})$*-category and let* f *be a* $\mathfrak{C}$*-morphism. Then the following conditions are equivalent:*

a) $f \in \mathfrak{M}$

b) $f = g \cdot e$ *and* $e \in \mathfrak{E}$ *imply that* e *is an isomorphism.*

IV.3. Proposition: *Let* $\mathfrak{C}$ *be a* $(\mathfrak{E}, \mathfrak{M})$*-category and let* e *be an epimorphism in* $\mathfrak{C}$*. Then the following codnitions are equivalent:*

a) $e \in \mathfrak{E}$

b) $e = m \cdot g$ *and* $m \in \mathfrak{M}$ *imply that there exists* h *with* $g = h \cdot e$.

IV.4. Theorem. *For any subclass* $\mathfrak{E}$ *of* $\mathrm{Epi}\,\mathfrak{C}$ *the following con-*

ditions are equivalent:

a) $\mathfrak{E}$ *is* L_δ*-closed*

b) $\mathfrak{E}$ *is a left factor of* $\mathfrak{C}$

c) $\mathfrak{C}$ *is a* $(\mathfrak{E}, \mathfrak{R}_\delta\mathfrak{E})$*-category*

d) $\mathfrak{E}$ *satisfies the conditions* (E1), (E2), *and* (E3*) of III.2

e) *the full subcategory of* $[2, \mathfrak{C}]$ *whose objects are the elements of* $\mathfrak{E}$ *is isomorphism-closed and coreflective in* $[2, \mathfrak{C}]$ such *that the first component of each coreflection map is a monomorphism in* $\mathfrak{C}$.

IV.5. Theorem. *For any class of* $\mathfrak{C}$*-morphisms the following conditions are equivalent:*

a) $\mathfrak{M}$ *is* R_δ*-closed*

b) $\mathfrak{M}$ *is a right factor of* $\mathfrak{C}$

c) $\mathfrak{C}$ *is a* $(L_\delta\mathfrak{M}, \mathfrak{M})$*-category*

d) $\mathfrak{M}$ *satisfies the following conditions:*

(M1) $\mathfrak{M}$ *is a subcategory of* $\mathfrak{C}$ *which contains all extremal monomorphisms of* $\mathfrak{C}$

(M2) $\mathfrak{M}$ *is closed under pullbacks in* $\mathfrak{C}$

(M3) $\mathfrak{M}$ *is closed under multiple pullbacks in* $\mathfrak{C}$

e) $\mathfrak{M}$ *satisfies the conditions* (M1), (M2) and (M4) $\mathfrak{M}$ *is closed under products in* $\mathfrak{C}$

f) $\mathfrak{M}$ *satisfies the conditions* (M1), (M2), (M3), (M4) *and the following conditions:*

(M5) $f\colon X \to Y \in \mathrm{Mor}\,\mathfrak{C}$ *and* $m\colon X \to Z \in \mathfrak{M}$ *imply* $(f, m)\colon X \to Y \times Z \in \mathfrak{M}$

(M6) $f \cdot g \in \mathfrak{M}$ *implies* $g \in \mathfrak{M}$

g) *the full subcategory of* $[2, \mathfrak{C}]$ *whose objects are the elements of* $\mathfrak{M}$ *is isomorphism-closed and epireflective in* $[2, C]$

IV.6. Remark. In the category $\mathfrak{C} = Haus$ the class $\mathfrak{M}$ of perfect maps satisfies the conditions of the above theorem. Michael [31] has pointed out that condition (M5) enables us to give a trivial proof of the following theorem of Arhangelskiĭ [2]: if f: $X \to Y$ is perfect and g: $X \to Z$ is one-to-one then (f, g): $X \to Y \times Z$ is (up to a homeomorphism) a closed embedding.

V. $\mathfrak{K}$-PERFECT MORPHISMS AND PERFECT SUBCATEGORIES

Let $\mathfrak{K}$ be a full subcategory of $\mathfrak{C}$.

$E\mathfrak{K} = R_\pi \mathfrak{K}$ denotes the subcategory of all $\mathfrak{K}$-extendable epimorphisms in $\mathfrak{C}$.

$P\mathfrak{K} = R_\delta(E\mathfrak{K})$ denotes the subcategory of all $\mathfrak{K}$-*perfect* morphisms in $\mathfrak{C}$.

V.1. Theorem. *Let* $\mathfrak{K}$ *be the epireflective hull of a subcategory* $\mathfrak{L}$ *of* $\mathfrak{C}$. *Then:*

a) $E\mathfrak{K} = E\mathfrak{L}$

b) $P\mathfrak{K} = P\mathfrak{L}$

c) *any of the conditions* (P3)–(P10) *of* I.1 *is equivalent to the condition* $f \in P\mathfrak{K}$

d) $\mathfrak{C}$ *is a* $(E\mathfrak{K}, P\mathfrak{K})$*-category*

V.2. Remark. The $(E\mathfrak{K}, P\mathfrak{K})$-factorization of a $\mathfrak{C}$-morphism f: $X \to Y$ can be obtained in several ways:

1) Let $(X \xrightarrow{f} Y = X \xrightarrow{e_i} X_i \xrightarrow{f_i} Y)_{i \in I}$ be a representative family of all factorizations of f with first factor e_i belonging to $E\mathfrak{K}$, let $X \xrightarrow{e} Z$ be the multiple pushout (= cointersection) of the family $(X \xrightarrow{e_i} X_i)_{i \in I}$, and let p: $Z \to Y$ be the unique morphism with $f = p \cdot e$. Then $f = p \cdot e$ is the desired factorization.

2) Let $(X \xrightarrow{f} Y = X \xrightarrow{e_i} Y_i \xrightarrow{p_i} Y)_{i \in I}$ be a representative family of all (epi, $P\mathfrak{K}$)-factorizations of f, let p: $Z \to Y$ be the multiple pullback of the family $(Y_i \xrightarrow{p_i} Y)_{i \in I}$, let g be a morphism with $f = p \cdot g$, and let $g = m \cdot e$ be the (epi, extremal mono)-factorization of g. Then $f = (p \cdot m) \cdot e$ is the desired factorization.

3) Let $k_X: X \to kX$ be the $\mathfrak{K}$-reflection of X, let (f, k_X) be the unique morphism for which the diagram

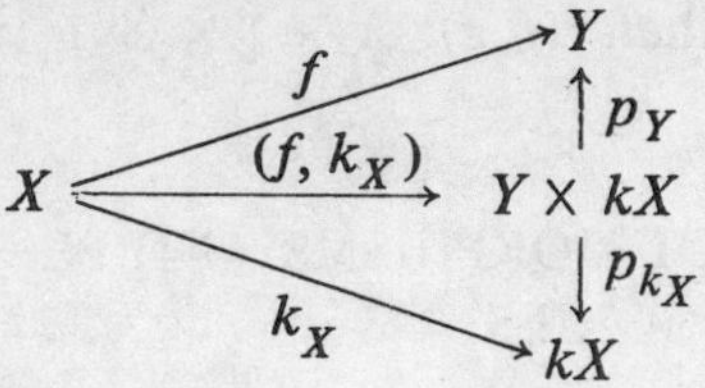

commutes, and let $(f, k_X) = m \cdot e$ be the (epi, extremal mono)-factorization of (f, k_X). Then $f = (p_Y \cdot m) \cdot e$ is the desired factorization.

4) Let $k_X: X \to kX$ resp. $k_Y: Y \to kY$ be the $\mathfrak{K}$-reflections of X resp. Y, let g be the unique $\mathfrak{C}$-morphism such that the following diagram (where ① is supposed to be a pullback) commutes

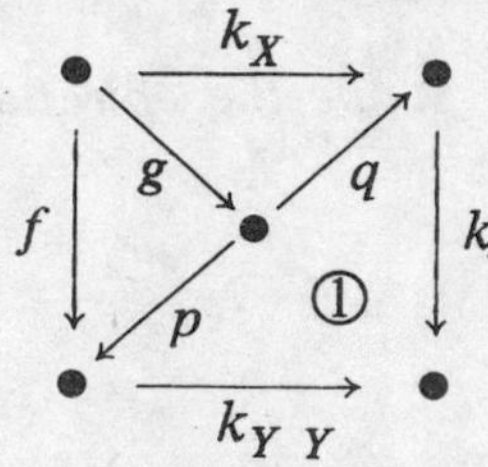

and let $g = m \cdot e$ be the (epi, extremal mono)-factorization of g.

Then $f = (p \cdot m) \cdot e$ is the desired factorization. Quite often it turns out that g is already an epimorphism, so that $f = p \cdot g$ is already the desired factorization.

Examples. a) $\mathfrak{C} = CReg$ and $\mathfrak{K} = Comp$. In this case p is exactly what Whyburn [53] calls the "compactification" of f b) More general: $\mathfrak{C} = \mathfrak{E}\text{-}Reg$ and $\mathfrak{K} = \mathfrak{E}\text{-}Comp$ for any subcategory $\mathfrak{E}$ of the category *Haus* of Hausdorff spaces and continuous maps. c) $\mathfrak{C} = Unif.$ and $\mathfrak{K} = Compl.\ Unif.$ d) $\mathfrak{C}$ is the category of ordered sets and sup-preserving maps, $\mathfrak{K}$ is the full subcategory of sup-complete ordered sets.

V.3. Definition. A subcategory $\mathfrak{P}$ of $\mathfrak{C}$ is called *perfect* iff there exists a full subcategory $\mathfrak{K}$ of $\mathfrak{C}$ with $\mathfrak{P} = P\mathfrak{K}$.

V.4. Theorem. *For any class* $\mathfrak{P}$ *of* $\mathfrak{C}$*-morphisms the following conditions are equivalent:*

a) $\mathfrak{P}$ *is a perfect subcategory of* $\mathfrak{C}$

b) $\mathfrak{P}$ *satisfies the conditions* (M1), (M2), (M3) *of* IV.5 *and the following condition (P)* $\mathfrak{P} \cap L_\delta\{f \in \mathfrak{P} \mid$ *codomain* f *is terminal*$\} = \mathrm{Iso}\ \mathfrak{C}$.

V.5. Remark. The last theorem answers the problem posed in [20]. A similar characterization has been obtained independently by Strecker [48].

VI. PERFECT PREIMAGES AND $\mathfrak{K}$-FITTING PROPERTIES

Let $\mathfrak{A}$ and $\mathfrak{B}$ be full subcategories of $\mathfrak{C}$ and let $\mathfrak{K}$ be the epireflective hull of $\mathfrak{B}$ in $\mathfrak{C}$

VI.1. Proposition. *If* X *and* A *are* $\mathfrak{C}$*-objects then the following conditions are equivalent:*

a) *there exists a* $\mathfrak{B}$*-perfect morphism* $X \to A$

b) X *is an extremal subobject of a product of* A *with a* $\mathfrak{K}$*-object.*

VI.2. Definition. $\mathfrak{A}$ is called $\mathfrak{B}$*-fitting* provided it contains with any object X all $\mathfrak{B}$-perfect preimages of X.

VI.3. Proposition. *The following conditions are equivalent:*

a) $\mathfrak{A}$ *is* $\mathfrak{B}$*-fitting*

b) $\mathfrak{A}$ *is closed under formation of extremal subobjects and under formation of products with* $\mathfrak{K}$*-objects.*

VI.4. Proposition. *If there exists a morphism with terminal domain and codomain in* $\mathfrak{A}$ *the following conditions are equivalent:*

a) $\mathfrak{A}$ *is* $\mathfrak{A}$*-fitting*

b) $\mathfrak{A}$ *is epireflective in* $\mathfrak{C}$.

Final Remark. For simplicity we have assumed throughout this article that $\mathfrak{C}$ is complete, cocomplete, wellpowered, and cowellpowered. But many of the above results hold without any or under much weaker restrictions on $\mathfrak{C}$.

VII. EXAMPLES

VII.1. If $\mathfrak{C}$ is the category of completely regular spaces and $\mathfrak{K}$ is the full subcategory of compact Hausdorff spaces then $P\mathfrak{K}$ is the class of perfect maps. The perfect preimages of metrizable spaces are the paracompact M-spaces.

VII.2. If $\mathfrak{C}$ is as above and $\mathfrak{K}$ is the full subcategory of k-compact spaces then a $\mathfrak{C}$-morphism $f: X \to Y$ is $\mathfrak{K}$-perfect (= k-perfect) iff each maximal zerofilter in X with k-intersection-property whose f-image converges in Y must converge in X.

VII.3. If $\mathfrak{C}$ is the category of uniform Hausdorff spaces and $\mathfrak{K}$ is the full subcategory of complete uniform spaces then a $\mathfrak{C}$-morphism $f: X \to Y$ is $\mathfrak{K}$-perfect iff each Cauchy-filter in X whose f-image converges in Y must converge in X.

VII.4. If $\mathfrak{C}$ is the category of compact Hausdorff spaces and $\mathfrak{K}$ is the full subcategory of totally disconnected compact Hausdorff spaces then a $\mathfrak{C}$-morphism $f: X \to Y$ is $\mathfrak{K}$-perfect iff the preimage of each $y \in$ $\in Y$ meets each component of X in at most one point.

VII.5. If $\mathfrak{C}$ is the category of all topological spaces and $\mathfrak{K}$ is the full subcategory of T_0-spaces then a $\mathfrak{C}$-morphism $f: X \to Y$ is $\mathfrak{K}$-perfect iff the preimage of each $y \in Y$ is T_0.

VII.6. If $\mathfrak{C}$ is the category of Hausdorff spaces and p-maps in the sense of Harris [17] (= τ-proper maps in the sense of Błaszczyk and Mioduszewski [4]) and $\mathfrak{K}$ is the full subcategory of H-closed spaces then a $\mathfrak{C}$-morphism $f: X \to Y$ is $\mathfrak{K}$-perfect iff it is τ-perfect in the sense of Błaszczyk and Mioduszewski [4], i.e. iff each maximal open filter in X whose f-image converges in Y must converge in X; equivalently: iff $f^{-1}(y)$ is far from the remainder for each $y \in Y$ and $f[A]$ is closed for each regularly closed subset A of X (see [4]).

VII.7. If $\mathfrak{C}$ is the category of groups and $\mathfrak{K}$ is the full subcategory of Abelian groups then a $\mathfrak{C}$-morphism $f: X \to Y$ is $\mathfrak{K}$-perfect iff $\operatorname{Ker} f \cap X' = e$.

VII.8. If $\mathfrak{C}$ is the category of Hausdorff groups and $\mathfrak{K}$ is the full subcategory of compact Hausdorff groups then a $\mathfrak{C}$-morphism f: $X \to Y$ is $\mathfrak{K}$-perfect iff it satisfies the following three conditions:

1) $\operatorname{Ker} f \cap \operatorname{Ker} r_X = \{e\}$, where $r_X: X \to \widetilde{X}$ denotes the Bohr-compactification of X, 2) f is closed, 3) $\operatorname{Ker} f$ is compact. A locally compact Hausdorff group X is a $\mathfrak{K}$-perfect preimage of an Abelian Hausdorff group iff 1) the topological commutator subgroup X' of X is compact and 2) X is maximal almost periodic, i.e. r_X is one-to-one. These topological groups are known as Takahashi groups (D. Poguntke [41]).

VII.9. If $\mathfrak{C}$ is the category of ordered sets and sup-preserving maps and $\mathfrak{K}$ is the full subcategory of sup-complete ordered sets then a $\mathfrak{C}$-morphism f: $X \to Y$ is $\mathfrak{K}$-perfect iff any subset of X whose f-image has a supremum in Y must have a supremum in X.

VII.10. If $\mathfrak{C}$ is the category of topological spaces and $\mathfrak{K}$ is the full subcategory of discrete spaces then a $\mathfrak{C}$-morphism f: $X \to Y$ is $\mathfrak{K}$-coperfect iff a subset of Y is open in Y whenever its f-preimage is open in X.

REFERENCES

[1] A.V. Arhangelskiĭ, On a class of spaces containing all metric and all locally bicompact spaces, *Dokl. Akad. Nauk SSSR*, 151 (1963), 751-754 = *Soviet Math. Dokl.*, 4 (1963), 1051-1055.

[2] A.V. Arhangelskiĭ, Perfect mappings and injections, *Dokl. Akad. Nauk SSSR*, 176 (1967), 983-986 = *Soviet Math. Dokl.*, 8 (1967), 1217-1220.

[3] B. Banaschewski, Projective covers in categories of topological spaces and topological algebras, *Proc. Conf. Kanpur*, 1968 (1971), 63-91.

[4] A. Błaszczyk – J. Mioduszewski, On factorization of maps through τX, *Colloq. Math.*, 23 (1971), 45-52.

[5] C.J.R. Borges, On stratifiable spaces, *Pacific J. Math.*, 17 (1966), 1-16.

[6] N. Bourbaki, *Topologie générale,* Ch. I, II, 3rd ed Hermann, Paris, 1961.

[7] M.M. Čoban, Perfect mappings and spaces of countable type, *Vestnik Moskov Univ.,* 22 (1967), 87-93. (Russian).

[8] P.J. Collins, Concordant mappings and the concordant-dissonant factorization of an arbitrary continuous function, *Proc. Amer. Math. Soc.,* 27 (1971), 587-591.

[9] B.J. Day – G.M. Kelly, On topological quotient maps preserved by pullbacks or products, *Proc. Camb. Phil. Soc.,* 67 (1970), 553-558.

[10] H. Ehrbar – O. Wyler, On subobjects and images in categories; to appear.

[11] J. Flachsmeyer, Topologische Projektivräume, *Math. Nachr.,* 26 (1963), 57-66.

[12] S.P. Franklin, On epireflective hulls, *Gen. Top. Appl.,* 1 (1971), 29-31.

[13] S.P. Franklin, On epireflective hulls, II, *Notes Meerut Univ. Summar Inst. Topol.,* 1971.

[14] Z. Frolík – C.T. Liu, An embedding characterization of almost realcompact spaces, to appear.

[15] Z. Frolík – S. Mrówka, Perfect images of *R*- and *N*-compact spaces, *Bull. Acad. Polon. Sci. Sér. Sci. Math. Astron. Phys.,* 19 (1971), 369-371.

[16] A.W. Hager, Perfect maps and epireflective hulls, *Math. Zeitschr.,* to appear.

[17] D. Harris, Katětov extension as a functor, *Math. Ann.,* 193 (1971), 171-175.

[18] M. Henriksen – J.R. Isbell, Some properties of compactifications, *Duke Math. J.,* 25 (1958), 83-106.

[19] H. Herrlich, Factorizations of morphisms $f: B \to FA$. *Math. Zeitschr.*, 114 (1970), 180-186.

[20] H. Herrlich, A generalization of perfect maps, *Proc. Third Prague Top. Symp.*, 1971 (1972), 187-191.

[21] H. Herrlich – J. van der Slot, Properties which are closely related to compactness, *Indag. Math.*, 29 (1967), 524-529.

[22] H. Herrlich – G.E. Strecker, Coreflective subcategories, *Trans. Amer. Math. Soc.*, 157 (1971), 205-226.

[23] M. Hušek, Perfect images of *E*-compact spaces, *Bull. Acad. Polon. Sci. Sér. Sci. Math. Astr. Phys.*, 20 (1972), 41-45.

[24] J.R. Isbell, Subobjects, adequacy, completeness and categories of algebras, *Rozprawy Math.*, 36 (1964).

[25] M. Jurchescu – A. Lascu, Morfisme stricte, categorii cantoriene, functori de completare, *Studii Cerc. Mat.*, 18 (1966), 219-234.

[26] G.M. Kelly, Monomorphisms, epimorphisms and pullbacks, *J, Austral. Math. Soc.*, 9 (1969), 124-142.

[27] J.F. Kennison, Full reflective subcategories and generalized covering spaces, *Illinois J. Math.*, 12 (1968), 353-365.

[28] V.J. Mancuso, Inverse images and first countability, *Gen. Top. Appl.*, 2 (1972), 29-44.

[29] E. Michael, Cuts, *Acta Math.*, 111 (1964), 1-36.

[30] E. Michael, Biquotient maps and cartesian products of quotient maps, *Ann. Inst. Fourier/Grenoble*, 18 (1968), 287-302.

[31] E. Michael, A theorem on perfect maps, *Proc. Amer. Math. Soc.*, 28 (1971), 633-634.

[32] E. Michael, A quintuple quotient test, *Gen. Topol. Appl.*, 2 (1972), 91-138.

[33] E. Michael, On representing spaces as images of metrizable and related spaces, *Gen. Topol. Appl.*, 1 (1971), 329-343.

[34] J. Mioduszewski – L. Rudolf, *H*-closed and extremally disconnected Hausdorff Spaces, *Dissertationes Math.*, 66 (1969).

[35] K. Morita, Products of normal spaces with metric spaces, *Math. Ann.*, 154 (1964), 365-382.

[36] S. Mrówka, Some properties of *Q*-spaces, *Bull. Acad. Polon. Sci. Sér. Sci. Math. Astr. Phys.*, 5 (1957), 947-950.

[37] S. Mrówka, Some comments on the author's example of a non-*R*-compact space, *Bull. Acad. Polon. Sci. Sér. Sci. Math. Astr. Phys.*, 18 (1970), 443-448.

[38] J. Nagata, A note on *M*-space and topologically complete space, *Proc. Japan. Acad.*, 5 (1969), 541-543.

[39] R. Nakagawa, Classes of morphisms and reflections, *Gen. Topol. Appl.*, to appear.

[40] A. Okuyama, On metrizability of *M*-spaces, *Proc. Japan Acad.*, 40 (1964), 176-179.

[41] D. Poguntke, A universal property of the Takahashi quasi-dual, *Canad. J. Math.*, 24 (1972), 530-536.

[42] D. Pumplün, Universelle und spezielle Probleme, *Math. Ann.*, 198 (1972), 131-146.

[43] J. Raasch, Faktorisierbarkeit und Galois-Paare Diplomarbeit, *Freie Universität*, Berlin, 1971.

[44] J. Rainwater, A note on projective resolutions, *Proc. Amer. Math. Soc.*, 10 (1959), 734-735.

[45] J. van der Slot, Some properties related to compactness, *Thesis, Math. Centrum*, Amsterdam, 1968.

[46] J. van der Slot, A note on perfect irreducible mappings, *Bull. Acad. Polon. Sci. Sér. Sci. Math. Astr. Phys.*, 19 (1971), 377-382.

[47] D.P. Strauss, Extremally disconnected spaces, *Proc. Amer. Math. Soc.*, 18 (1967), 305-309.

[48] G.E. Strecker, Epireflection operators vs. perfect morphisms and closed classes of epimorphisms, *Bull. Austral Math. Soc.*, 7 (1972), 359-366.

[49] J. Sonner, Universal and special problems, *Math. Zeitschr.*, 82 (1963), 200-211.

[50] J.H. Tsai, On a generalization of perfect maps, to appear.

[51] G.T. Whyburn, Open and closed mappings, *Duke Math. J.*, 17 (1950), 69-74.

[52] G.T. Whyburn, On compactness of mappings, *Proc. Nat. Acad. Sci. USA*, 52 (1964), 1426-1431.

[53] G.T. Whyburn, Compactification of mappings, *Math. Ann.*, 166 (1966), 168-174.

COLLOQUIA MATHEMATICA SOCIETATIS JÁNOS BOLYAI
8. TOPICS IN TOPOLOGY, KESZTHELY (HUNGARY), 1972.

NON-CANCELLATION PHENOMENA IN TOPOLOGY

P. HILTON

Suppose given any category C in which we may form products of objects. The question then naturally arises as to whether $X \times A \cong Y \times$ $\times A$ implies $X \cong Y$. The classical case is that of finite, non-empty sets, where the implication holds and expresses the fact that the positive integers constitute a *cancellation* semigroup under multiplication. An interesting example where cancellation is valid is that of abelian groups of finite type. In this category the product is the direct sum and the cancellation property is an easy consequence of the structure theorem for abelian groups of finite type.

Plainly some sort of 'finiteness' condition would need to be imposed to give any prospect of the cancellation property holding. For if A is the product of countably many copies of X and countably many copies of Y, then certainly $X \times A \cong Y \times A$ (in any category with countable products). On the other hand, there are unexpected examples of failure of cancellation in categories whose objects satisfy a finiteness condition which looks rather strong. For example, there exist finitely presentable groups G_1, G_2 such that

(1.1) $$G_1 \not\cong G_2\,, \quad G_1 \times Z \cong G_2 \times Z\,.$$

Charlap [1] found examples of flat Riemannian manifolds M_1, M_2 such that

(1.2) $$M_1 \not\simeq M_2\,, \quad M_1 \times S^1 \overset{d}{\cong} M_2 \times S^1\;;$$

that is, M_1, M_2 have different *homotopy types* but $M_1 \times S^1$ and $M_2 \times$ $\times\, S^1$ are *diffeomorphic.* Indeed, Charlap's example reflects the phenomenon (1.1) since M_i is the Eilenberg-MacLane space $K(G_i, 1)$. We will give below examples of the phenomenon

(1.3) $$M_1 \not\simeq M_2\,, \quad M_1 \times N \overset{d}{\cong} M_2 \times N\,,$$

where M_1, M_2, N are 1-connected closed manifolds. It should be noted that if M_1, M_2, N are 1-connected compact polyhedra such that $M_1 \times$ $\times\, N \simeq M_2 \times N$, then, in view of the cancellation property in the category of abelian groups of finite type, it follows that M_1 and M_2 have isomorphic homology and homotopy groups. Thus the phenomenon (1.3) cannot be detected by applying the conventional functors of algebraic topology.

Examples of failure of cancellation in topology have long existed. By pinching to a point wild arcs in Euclidean space, one constructs examples of the phenomenon

$$X \not\cong Y\,, \quad X \times A \cong Y \times A\,,$$

where $\cong$ is the relation of homeomorphism; but, in these classical examples, X and Y always have the same homotopy type. Recent work using the *h*-cobordism theorem and the *s*-cobordism theorem yield examples of manifolds M_1, M_2, N such that

$$M_1 \overset{d}{\not\cong} M_2\,, \quad M_1 \times N \overset{d}{\cong} M_2 \times N\,;$$

that is, M_1 and M_2 have different topological types but $M_1 \times N$ and $M_2 \times N$ are diffeomorphic. Indeed if M_1 and M_2 are lens spaces of the same homotopy type, then $M_1 \times N \overset{d}{\cong} M_2 \times N$ if $N = S^3 \times S^k$, for k large. For 1-connected examples of this phenomenon, see Levine [8].

Consideration of the cancellation problem permits one also to formulate the dual question. Writing $X_1 + X_2$ for the coproduct (sum) of

X_1 and X_2, one may ask whether $X + A \cong Y + A$ implies $X \cong Y$. Freyd [2] gave examples of the failure of sum-cancellation in the stable homotopy category. We will discuss this question, related to the (unstable) homotopy category, in Section 2, since the construction there motivates the approach to the construction of examples of (1.3).

Failure of cancellation is in no wise to be regarded as a defect of the category. Rather it suggests new equivalence relations to be obtained from old ones. Thus, given the equivalence relation $\simeq$ in the category C (for example, the homotopy relation in the category of based compact polyhedra), we from the equivalence relation $\overset{P}{\simeq}$ by declaring $X \overset{P}{\simeq} Y$ if $X \times A \simeq Y \times A$ for some A; and the equivalence relation $\overset{S}{\simeq}$ by declaring $X \overset{S}{\simeq} Y$ if $X + A \simeq Y + A$ for some A. We may then seek invariants of '*P*-type' and '*S*-type'; and this problem is of new interest precisely in the absence of cancellation. Of course, we may interpret '*P*-type' and '*S*-type' in terms of the appropriate Grothendieck groups.

One should mention that it is also possible and sensible to consider 'local' versions of the cancellation question. One may ask 'For which objects A may we infer $X \cong Y$ from $X \times A \cong Y \times A$?' For example, it is known that an abelian group of finite type may be cancelled in the category of all abelian groups. One may also ask, 'Given A, when may we infer $X \cong Y$ from $X \times A \cong Y \times A$?' Another interesting type of question is suggested by Freyd's theorem that, in the *stable* homotopy category (of finite polyhedra), if $X + A \simeq Y + A$, then $X + B \simeq Y + B$, where B is a sum of spheres.

A final general remark refers to a different but related problem of cancellation. We may ask whether $X^n \cong Y^n$ implies $X \cong Y$, or, dually, whether $nX \cong nY$ implies $X \cong Y$. That is, we may ask whether we can cancel the operations of the natural numbers, or whether there is torsion in the semigroup $(C, \times)$ or $(C, +)$. This, of course, is not the same question as that of the existence of torsion in the associated Grothendieck groups. If C is the category of finite, non-empty sets, there is no torsion in $(C, \times)$ or $(C, +)$. If C is the category of abelian groups of finite type there is no torsion. But there is torsion in the homotopy categories referred to.

Much of the work reported on in this survey is joint work with J. Roitberg or G. Mislin.

2. THE DUAL PROBLEM IN HOMOTOPY THEORY

Let C_h be the category of based compact polyhedra (or spaces of the same homotopy type as such) and based homotopy classes of (continuous) maps. Then we form $P + Q$ by taking disjoint copies of P and Q and identifying their base points.

Let C_f be the (based) mapping cone of the map $f\colon X \to Y$; thus $C_f = Y \cup_f CX$ Then the (based) homotopy type of C_f depends only on the homotopy class of f, so we may write

$$C_\alpha = Y \cup_\alpha CX$$

for the homotopy type containing C_f for any $f \in \alpha$.

Let $\alpha \in \pi_{n-1}(S^m)$ be an element of order $k < \infty$, let l be prime to k, let $\beta = l\alpha$. Then the following results are proved in [3].

Theorem 2.1.

$$C_\alpha + S^n \simeq C_\beta + S^n .$$

Theorem 2.2. $C_\alpha \simeq C_\beta$ *if and only if there is a commutative diagram*

$$\begin{array}{ccc} S^{n-1} & \xrightarrow{\alpha} & S^m \\ \Big\downarrow{\scriptstyle \pm 1} & & \Big\downarrow{\scriptstyle \pm 1} \\ S^{n-1} & \xrightarrow{\beta} & S^m \end{array}$$

Corollary 2.3. *Suppose in addition that* α *is a suspension element. Then* $C_\alpha \simeq C_\beta$ *if and only if* $l \equiv \pm 1 \pmod{k}$.

These results give us a ready source of examples of failure of cancellation in C_h. The construction may be readily generalized [4]. There is also proved in [3] a result which relates to the question of torsion in C_h. We revert to C_α, C_β above; the result we quote generalizes Corollary 2.3.

Theorem 2.4. *Let* a *be a suspension element. Then* $qC_a \simeq qC_\beta$ *if and only if* $l^q \equiv \pm 1 \pmod{k}$. *Thus, in particular,*

$$\varphi(k)C_a \simeq \varphi(k)C_\beta ,$$

where φ *is Euler's function.*

On the other hand, these examples do not give any indication about the presence of torsion in the Grothendieck group, in view of Theorem 2.1. Freyd has shown that, in the stable case, the Grothendieck group is free.

It is possible, in principle, to dualize the construction given to yield examples of failure of cancellation in C_h with respect to the product. However, this procedure leads to examples which are 'of finite type', in the sense that the polyhedra involved have compact skeleta, but they are nevertheless infinite-dimensional. For dualization leads to the consideration of spaces with two non-vanishing homotopy groups, and such spaces are infinite-dimensional. Thus, while these examples are significant, we need to modify the dual construction in order to get finite-dimensional examples. We give this modification in the next section.

3. PRINCIPAL BUNDLES

We consider a principal G-bundle $\xi_a\colon G \to E_a \xrightarrow{f_a} S^n$, over the sphere S^n, where G is a Lie group. Then ξ_a is classified by $a \in \pi_{n-1}(G)$. If B_G is the classifying space for G then there is a natural isomorphism

(3.1) $$\pi_{n-1}(G) \cong \pi_n(BG) .$$

Let $a_0 \in \pi_n(BG)$ correspond to a under (3.1). Our objective is to find examples of principal G-bundles E_a, E_β such that

$$E_a \times G \overset{d}{\cong} E_\beta \times G\, , \quad E_a \not\simeq E_\beta ,$$

using arguments somewhat dual to those employed in Section 2. However, a substantial new difficulty arises, due to the fact that B_G is not a group-like space; in the strictly dual construction its place would be taken by an Eilenberg-MacLane space, thus a group-like space.

In [6] only the case $G = S^3$ is considered. Then the analogue of Corollary 2.3 is the following; we take $a \in \pi_{n-1}(S^3)$ of order $k < \infty$, l prime to k, $\beta = la$.

Theorem 3.1. $E_a \simeq E_\beta$ *if and only if* $l \equiv \pm 1 \bmod k$. *Indeed, then,* $E_a \overset{d}{\cong} E_\beta$.

However, we cannot proceed immediately to prove the analogue of Theorem 2.1. Indeed, it turns out to be false, in general, that

$$\beta_0 \circ f_a = 0 \colon E_a \to S^n \to B_{S^3} ;$$

and this is a key step in the argument. Indeed, it is proved [6] that

Theorem 3.2. $\beta_0 \circ f_a = 0$ *if and only if* l *may be chosen to satisfy*

$$\frac{l(l-1)}{2}\,\omega \circ \Sigma^3 a = 0 , \tag{3.2}$$

where $\omega \in \pi_6(S^3)$ *measures the non-commutativity of* S^3.

Corollary 3.3. *If* (3.2) *is satisfied, then* $E_a \times S^3 \overset{d}{\cong} E_\beta \times S^3$.

We obtain immediate examples by taking $n = 2p + 1$, $a \in \pi_{2p}(S^3)$ of order p, $p \neq 2, 3$, and choosing l so that $2 \leqslant l \leqslant p - 2$. Condition (3.2) is automatically satisfied since ω is of order 12, so that $\omega \circ \Sigma^3 a = 0$. A further, particularly interesting example is furnished by taking $n = 7$, $a = \omega \in \pi_6(S^3)$, $l = 7$. Then (3.2) is satisfied because $\omega \circ \Sigma^3 \omega$ is of order 3. But $E_\omega = \mathrm{Sp}(2)$, the symplectic group. Thus, if $M = E_{7\omega}$,

$$M \times S^3 \overset{d}{\cong} \mathrm{Sp}(2) \times S^3 , \quad M \not\simeq \mathrm{Sp}(2) . \tag{3.3}$$

It follows that M is a Hopf manifold, but not a Lie group. It follows from methods of Stasheff and Zabrodsky that, in fact, M has the homotopy type of a topological group. Since M does not have the homotopy type of a Lie group, it follows that any topological group of the homotopy type of M must be infinite-dimensional.

Scheerer has generalized much of the technique of [6] to provide examples of principal G-bundles for any Lie group G such that

(3.4) $$E_a \times G \overset{d}{\cong} E_\beta \times G\,,\quad E_a \not\simeq E_\beta\,.$$

His main result is the following [10].

Theorem 3.4. *Consider the principal G-bundle* $\xi_a : G \to E_a \xrightarrow{f_a} S^n$, *characterized by* $a \in \pi_{n-1}(G)$. *Then* $\beta_0 \circ f_a = 0$ *provided the maps* $S^{n-1} \times G \to G$ *given by* $(x, y) \mapsto a(x)^l y^l$, $(x, y) \mapsto (a(x)y)^l$ *are homotopic, where* $a\colon S^{n-1} \to G$ *represents* a, $x \in S^{n-1}$, $y \in G$.

(Here, of course, it is supposed, as above, that β is expressible as la.) Remark that the condition is surely satisfied if the map $y \mapsto y^l$ is an H-map $G \to G$. It is not difficult to see that, given any Lie group, there exists a positive integer t such that $y \mapsto y^l$ is an H-map provided $t \mid l$.

Scheerer's generalization still leaves open the question of whether there exist examples of (1.3) of arbitrarily high connectivity. Such examplex are not to be had by considering principal boundles over S^n. For we must take $n \geqslant 7$ and then E_a, if it is 1-connected, has connectivity 2, since $\pi_3(G)$ contains a cyclic infinite summand. Thus we are constrained to go outside the domain of principal bundles to achieve examples of (1.3) of arbitrary connectivity. This step is described in the next section.

4. SPHERE–BUNDLES

There is no prospect of extending the arguments used to obtain the results of the previous section to general fibre bundles. For while, for principal bundles $G \to E \xrightarrow{f} X$, it is always the case that $gf \simeq 0$, where $g\colon X \to B_G$ is the classifying map, it is false, in general, that, if $F \to$ $\to Y \xrightarrow{f} X$ is a G-bundle, then $gf \simeq 0$. This property, $gf \simeq 0$, of principal bundles is crucial to the arguments used in [6,10], and thus we are led to define a G-bundle $F \to Y \xrightarrow{f} X$ to be *quasi-principal* [5] if $gf \simeq 0$, where $g\colon X \to B_G$ classifies the principal bundle to which the given bundle is associated. It turns out that we can construct, for a given *odd* number $q \geqslant 3$, an interesting collection of quasi-principal bundles,

(4.1) $$S^q \to Y_a \xrightarrow{f_a} S^n\,.$$

Thus we choose a prime $p > q$ and take $n = q + 2p - 2$. Then there are maps

$$S^q \overset{s}{\to} SO(q+1) \overset{r}{\to} S^q .$$

where $SO(q+1)$ is the rotation group and r is the canonical projection of $SO(q+1)$ onto S^q; and elements $a \in \pi_{n-1}(SO(q+1))$, $\bar{a} \in \pi_{n-1}(S^q)$, of order p, such that

$$s_*(\bar{a}) = a\,, \quad r_*(a) = \bar{a}\,. \tag{4.2}$$

We then take (4.1) to be the sphere-bundle associated with the principal $SO(q+1)$-bundle classified by a. Let us change notation and write

$$\xi_i\colon S^q \to Y_i \xrightarrow{f_i} S^n \tag{4.3}$$

to be the sphere-bundle associated with the principal bundle classified by ia, $i = 1, 2, \ldots, p-1$. Then the bundles (4.3) turn out to be quasi-principal; moreover, we are able to prove [5].

Theorem 4.1. *With reference to the bundles* (4.3),

$$Y_i \times S^q \overset{d}{\cong} Y_j \times S^q\,, \quad Y_i \simeq Y_j \quad \textit{if and only if} \quad i \equiv \pm j \bmod p\,.$$

Plainly Y_i is $(q-1)$-connected so that we have indeed obtained non-cancellation examples of arbitrarily high connectivity. Notice also that, given any positive integer N, we may impose the supplementary condition $p > 2N$, and then the manifolds $Y_1, Y_2, \ldots, Y_N$ have the property

$$Y_i \times S^q \overset{d}{\cong} Y_2 \times S^q \overset{d}{\cong} \ldots \overset{d}{\cong} Y_N \times S^q\,, \quad Y_i \simeq Y_j \quad \text{if} \quad i \neq j\,. \tag{4.4}$$

Thus we may find, for any N and any odd $q \geqslant 3$, N manifolds of connectivity $(q-1)$, all of different homotopy type, and such that they all become diffeomorphic on multiplication with S^q.

The attempt to extend this argument to the case of q even founders on the obstruction that we cannot, in similar fashion, construct quasi-principal fibrations. In the case of q even, we construct sphere-bundles as follows. We choose a prime $p > q + 1$ and take $n = 2q + 2p - 3$.

Then we have elements $a \in \pi_{n-1}(SO(q+1))$, $\bar{a} \in \pi_{n-1}(S^q)$, of order p, such that $r_*(a) = \bar{a}$, where $r: SO(q+1) \to S^q$ is the projection. However, no 'quasisection' $s: S^q \to SO(q+1)$ (see (4.2)) exists in this case. We take

$$\xi_i : S^q \to Y_i \xrightarrow{f_i} S^n \tag{4.5}$$

to be the sphere-bundle associated with the principal bundle classified by ia, $i = 1, 2, \ldots, p-1$. The bundles (4.5) are not quasi-principal, so we cannot infer (by our methods) the analogue of Theorem 4.1. Nevertheless the bundles (4.5) do have an interesting, and related property, which we proceed to describe.

Given a space X and a prime p, we may *localize* X *at* p [11]; that is, we associate with X, in functorial manner, a space X_p and a homotopy class of maps $e_p : X \to X_p$ such that (i) the homology and homotopy groups of X_p are p-local, and (ii) the homology and homotopy homomorphisms induced by e_p localize the homology and homotopy groups of X. Some restriction on X is necessary for the validity of this construction, but it is certainly applicable to 1-connected polyhedra, and we will be concerned with this case.

Localization behaves well with respect to standard homotopy constructions; in particular, fibrations and cofibrations are preserved. Thus we may localize a fibration $F \to E \to B$ to obtain a fibration $F_p \to E_p \to B_p$, provided the spaces involved are 1-connected. We will say that two fibrations $F \to E_i \to B$, $i = 1, 2$, have the same *genus* if, for each prime p, their localizations at p are fibre-homotopy-equivalent. We then have [5].

Theorem 4.2.

(i) *q being odd, the fibrations* (4.3) *have the same genus;*

(ii) *q being even, the fibrations ξ_i, ξ_j* (4.5) *have the same genus, provided j/i is a quadratic residue* $\bmod p$.

On the other hand, we know that, in both cases, $Y_i \simeq Y_j$ only if $i \equiv \pm j \bmod p$. Thus certainly, if $i \not\equiv \pm j \bmod p$, the fibration ξ_i and ξ_j are not themselves fibre-homotopy-equivalent. Thus we see that distinct fibrations may have equivalent localizations at all primes. We also see, of course, that a genus may contain arbitrarily many homotopy types. In case q is odd, as already explained, we supplement the assumptions of Theorem 4.1 with the condition $p > 2N$ and consider the fibrations ξ_i (4.3) with $i = 1, 2, \ldots, N$; in case q is even, we impose $p > 2N^2$ and consider the fibrations ξ_i (4.5) with $i = 1, 4, \ldots, N^2$. We have no example where the genus is infinite.

We may pass from Theorem 4.2 to Theorem 4.1 by means of the following result [5].

Theorem 4.3. *Let* ξ_i: $F \to Y_i \to B$, $i = 1, 2$, *be quasi-principal bundles for the Lie group* G, *all spaces being* 1*-connected smooth manifolds. If* ξ_1 *and* ξ_2 *have the same genus, then* $Y_1 \times F \overset{d}{\cong} Y_2 \times F$.

5. APPENDIX

Of course, the notion of genus may be applied to spaces as well as to fibrations (or cofibrations); indeed, M i s l i n [9] originally defined the genus for spaces. Then all the cases discussed in this article, where two spaces X, Y stand in the relation $\overset{S}{\simeq}$ or $\overset{P}{\simeq}$, in fact involve spaces X, Y of the same genus. It is thus natural to investigate the 'relation' between the relations

(i) $X \overset{S}{\simeq} Y$,

(ii) $X \overset{P}{\simeq} Y$,

(iii) $X \overset{G}{\simeq} Y$, (meaning X, Y have the same genus).

One might conjecture that (i) or (ii) implies (iii). Such a conjecture is practically equivalent to the conjecture that one may cancel summands and factors in the (homotopy) category of p-local spaces. K o z m a [7] has established an interesting connection between (i) and (ii), working in the category of spaces of finite type (it is essential for his method that polyhedra of infinite dimension be admitted; in this article we have excluded such objects). K o z m a proves

Theorem 5.1.

(i) *If* X, Y, A *are* 1*-connected polyhedra of finite type, such that* $X + A \simeq Y + A$, *then there exists a connected polyhedron* B *of finite type such that* $\Omega X \times B \simeq \Omega Y \times B$ *(where* Ω *is the loop space functor);*

(ii) *If* X, Y, A *are connected polyhedra of finite type, such that* $X \times A \simeq Y \times A$, *then there exists a* 1*-connected polyhedron* B *of finite type such that* $\Sigma X + B \simeq \Sigma Y + B$ *(where* Σ *is the suspension functor).*

BIBLIOGRAPHY

[1] L. Charlap, Compact flat Riemannian manifolds I, *Ann. of Math.*, 81 (1965), 15-30.

[2] P.J. Freyd, Stable homotopy, *Proc. Conf. Cat. Alg.*, La Jolla, Springer (1966), 121-172.

[3] P.J. Hilton, On the Grothendieck group of compact polyhedra, *Fund. Math*, 61 (1967), 199-214.

[4] P.J. Hilton, Note on the homotopy type of mapping cones, *Comm. Pure App. Math.*, 11 (1968), 516-619.

[5] P.J. Hilton – G. Mislin – J. Roitberg, Sphere-bundles over spheres and non-cancellation phenomena, Proc. Conf. Alg. Top., Seattle (1971), *Springer Lecture Notes* 249 (1971), 34-46.

[6] P.J. Hilton – J. Roitberg, On principal S^3-bundles over spheres, *Ann. of Math.*, 90 (1969), 91-107.

[7] I. Kozma, Some relations between semigroups of polyhedra, Ph.D. Thesis, Cornell University (1971).

[8] J. Levine, Self-equivalences of $S^n \times S^k$, *Trans. Amer. Math. Soc.*, 143 (1969), 523-543.

[9] G. Mislin, The genus of an *H*-space, Proc. Conf. Alg. Top., Seattle (1971), *Springer Lecture Notes*, 249 (1971), 75-83.

[10] H. Scheerer, On principal bundles over spheres, *Indag. Math.*, 32 (4) (1970), 353-355.

[11] D. Sullivan, Geometric topology, Part 1: Localization, peridoicity and Galois symmetry, M.I.T. (1971).

COLLOQUIA MATHEMATICA SOCIETATIS JÁNOS BOLYAI
8. TOPICS IN TOPOLOGY, KESZTHELY (HUNGARY), 1972.

FIXED POINT AND CENTRALIZING THEOREMS IN COMPACT SEMIGROUP THEORY

K.H. HOFMANN – M. MISLOVE

In the theory of topological transformation groups one encounters the following problem: Let G be a compact group operating continuously on a compact connected space X which is acyclic, i.e. has the Cech cohomology of a singleton, relative to a fixed coefficient module. Under which additional hypthesis is the fixed point space $\mathrm{Fix}(G, X) = \{x \in X : Gx = \{x\}\}$ also acyclic? Very little is known is general; in the case that G is abelian, due to results by Borel, Conner, Smith one has some pretty good insights into the problem: If G is connected and abelian and the coefficient group is the rationals, the answer is affirmative; if G is abelian and has a dense p-group and the coefficient group is the cyclic group with p elements for some prime p, the answer is likewise affirmative. Examples show that one cannot expect much beyond this situation, unless additional information is made available. In the theory of compact semigroups one encounters such additional information under very natural circumstances; namely, X is a compact connected monoid with zero 0 (different from the identity 1) and the elements of G act as automorphisms. For numerous special classes of groups G and monoids X and for all explicitly known examples (of which there are a great number), the following has been verified:

Fixed Point Conjecture. The fixed point set $\mathrm{Fix}(G, X)$ of a compact group of automorphisms of a compact connected monoid with zero (different from the identity) is connected.

Since $\mathrm{Fix}(G, X)$ is a compact submonoid containing the zero and the identity, its connectivity implies its acyclicity for every choice of coefficients. If S is a compact monoid and H its group of units (invertible elements), then H acts as a compact automorphism group on S under inner automorphisms $(h, s) \to hsh^{-1}$. In this situation, the fixed point set becomes the centralizer $Z(H, S) = \{s \in S \mid hs = sh \text{ for all } h \in H\}$ of H in S. Therefore, if the Fixed Point Conjecture holds in general, then so does the following.

Centralizing Conjecture. If S is a compact connected monoid with zero (different from the identity) and H its group of units, then $Z(H, S)$ is connected.

By a rather elementary semigroup-theoretical split extension construction, it is not hard to see that the universal validity of the Centralizing Conjecture implies that of the Fixed Point Conjecture. Thus the two conjectures are equivalent in this sense. However, it should be noted that it is quite conceivable, and indeed seems to be the fact that more partial progress can be made with the Centralizing Conjecture than with the Fixed Point Conjecture, since special cases may be settled for the former which do not have reasonable counterparts in the latter. As an example, we mention the following theorem [5].

Left Normality Theorem. *If S is a compact connected monoid with zero $(\neq 1)$ and if $G \subseteq H$ is a group of units satisfying $sG \subseteq Gs$ (i.e. being left normal), then $Z(G, S)$ is connected.*

As an illustration of a partial result which equally contributes to both versions of the main conjecture, we record one which emanates from the transformation group results of Borel, Conner, Smith mentioned before:

Commutativity Theorem. *If X is a compact connected monoid with zero ($\neq 1$), if G is a compact abelian group acting on X as a group of automorphisms, and if G has no p-adic direct factor for any prime p, then* $\mathrm{Fix}(G, X)$ *is connected.*

The p-adic case is still open, although partial results are known in this case, provided that X satisfies special hypothesis.

The proof of the commutativity theorem uses sophisticated machinery from algebraic topology such as the cohomology ring of a universal space of a compact group and its fine structure. On the other hand, the left normality theorem uses results from the structure theory of compact groups and semigroup-theoretical methods common to the area.

In the present contribution we establish a partial result towards the centralizing conjecture, known as the $H \times H$-theorem. An alleged proof given in [2] was found to be invalid by J.H. Carruth in 1971. We offer an entirely different approach, which we will outline; the details and some additional results will appear elsewhere. Thus, the remainder of this discussion is devoted to

The $H \times H$-Theorem. *Let S be a compact connected monoid with precisely the two idempotents 0 and 1. Let H be its group of units and let D be the equivalence relation whose cosets are precisely the sets HsH. Assume that the quotient space S/D has a total order such that the quotient topology and the order topology agree. Then $Z(H, G)$ is connected.*

In fact, we show in the process that $Z(H, G)$ contains a subsemigroup I which is isomorphic to the ordinary unit interval $[0, 1]$ under multiplication or its homomorphic image obtained by collapsing the ideal $[0, 1/2]$ to a point, and that $S = H \cdot I$. We note that the relation D, in general is defined by the stipulation that $s \in D(t)$ iff $SsS = StS$; however, it can be shown that under our other assumptions the cosets of D have the form described in the theorem (see [2], p. 39).

SECTION 1. A TRANSFORMATION GROUP THEOREM

Theorem 1. *Let G be a compact group acting on an arcwise connected topological space X. If an orbit Gx of a point x is a retract of X, and if there exists some fixed point, then x is a fixed point (i.e. Gx is singleton).*

In a slightly improved version of the theorem, the hypothesis of the arc connectivity of X may be replaced by the weaker hypothesis of the arc connectivity of the orbit space X/G; the necessary reduction is effected by an application of the arc lifting theorem by Montgomery and Yang [6], respectively, a slight generalization thereof to the case of arbitrary compact groups in place of Lie groups. However, we need only the version of the theorem given above, and we now outline its proof.

i) We show that the function $f\colon G \to X$ given by $f(g) = gx$ is homotopic to a constant function: If $a\colon [0,1] \to X$ is an arc with $a(0) = x$ and $a(1) \in \mathrm{Fix}(G, X)$ (such an arc exists by our hypothesis), then $h_t(g) = ga(t)$ defines a homotopy with $h_0 = f$ and $h_1 =$ constant function with value $a(1)$.

ii) We observe that if Y is a retract of a space X and $f\colon G \to Y$ is a function which is X is homotopic to a constant, then it is homotopic to a constant in Y (just apply the retraction to the homotopy).

iii) Thus since Gx is a retract, the function $f\colon G \to Gx$ is homotopic to a constant, $f(g) = gx$. But if K is the isotropy group $\{g \in G \mid gx = x\}$, then f is equivalent to the coset projection $p\colon G \to G/K$, $p(g) = gk$. Hence we have the following situation:

iv) Let G be a compact group and K a closed subgroup. Assume that the coset projection $p\colon G \to G/K$ is null-homotopic. We claim that then $K = G$. We indicate how this is proved in steps (the details of this proof have been published in [2], p. 300, p. 342, in slightly different form): a) First reduction: One observes easily that w.l.g. one may assume that G is connected. b) Second reduction: We show that it is sufficient to treat the case that G is semisimple; this is done by showing that the identity component Z of the center of G has to be contained in $G'K$; in the course of this proof one uses the fact that no non-zero character of a compact connected abelian group is null-homotopic. c) Final step: Proof of the assertion for compact connected semisimple G. By utilizing the existence of a semisimple Lie group local direct factor S of G and the homotopy lifting lemma we produce a homotopy $k_t: S \to S$ with $k_0 = id_S$ and k_1 factoring through the closed subgroup $\overline{K}$ of S.

We find a compact semisimple non-degenerate Lie subgroup $S \subseteq G$ with its finite center C and have a covering map $S \to G \to S/C$; we derive a null homotopy of the quotient map $S \to S/K_1$ where $K_1 \subseteq S$ is that subgroup which under the covering map goes onto the image in S/C of K. With the aid of the covering homotopy lemma for fiber bundles we arrive at a homotopy $k_t: S \to S$ with $k_0 = id_S$ and k_1 factoring through the closed submanifold K_1. By considering the induced map $1 = H^n(k_0) = H^n(k_1)$ in cohomology (or homology) we deduce $K_1 = S$. An approximation argument will then finish the proof by showing $K = G$.

We remark that the Brouwer fixed point theorem is an easy consequence of Theorem 1: Let X be the unit ball C^n in R^n, $G = S_0(n)$ acting on C^n by its natural action. Then 0 is a fixed point, and the orbit S^{n-1}, being not singleton, cannot be a retract of C^n. By an elementary argument, this implies the Brouwer fixed point theorem.

SECTION 2. A TRANSFORMATION SEMIGROUP THEOREM

An I-semigroup is a topological semigroup on a totally ordered compact connected space with identity and zero as endpoints. Examples are the ordinary unit interval $[0, 1]$ under the multiplication of real numbers or under the multiplication $(x, y) \to \min\{x, y\}$. We say that a monoid S acts on a space X if there is a continuous function $S \times X \to X$ such that $1x = x$ and $(st)x = s(tx)$ for all $s, t \in S$, $x \in X$. We say that an open set $A \subseteq X$ is transversal for an I-semigroup T acting on X if i) $0X \subseteq A$, ii) $TA \subseteq A$, iii) for each b in the boundary B of A the relation $tb \in B$ implies $tb = b$ (i.e. each T-orbit hits the boundary in precisely one point).

Theorem 2. *If an I-semigroup acts on a compact space X and A is a transversal open set, then the boundary B of A is a retract of the complement C of A. More accurately, B is a deformation retract* *of* C *in the sense that there is a continuous function* $F\colon T \times C \to C$ *with* $F(1, c) = c$, $F(t, b) = b$, $F(\{0\} \times C) \subseteq B$ *for all* $c \in C$, $b \in B$, $t \in T$.

In our applications we bring Theorem 1 and Theorem 2 together to yield the following

Corollary 3. *Let* X *be a compact space,* G *a compact group, and* T *an I-semigroup both acting on* X. *Suppose that* A *is an open set in* X *with boundary* B *and complement* C *and assume the following hypotheses:*

i) A *is transversal for the action of* T *and invariant under the action of* G.

ii) C *contains a fixed point for* G.

iii) B *is a* G*-orbit.*

Then B *is singleton, i.e. consists of precisely one* G*-fixed point.*

SECTION 3. THE PROOF OF THE $H \times H$-THEOREM

Let S be as in the $H \times H$-Theorem. There exists an I-semigroup $T \subseteq S$ with $1, 0 \in T$ and T isomorphic to $[0, 1]$ under ordinary multiplication or to the homomorphic image of $[0, 1]$ obtained by collapsing $[0, 1/2]$ to a single point. Let $s \in S$, $s \neq 0$. Define A' to be the largest ideal not containing s; such an ideal exists and is open. One has $HA'H = A'$. Now let X be the orbit space of the action of H on the left on S under translation; the elements of X are the cosets Hs. Let A be the image of A' in X. The group H operates on X on the right under translation $(Hs)g = Hsg$ and so does the I-semigroup T. We show that the boundary B' of A' in S is precisely HsH; thus the boundary B of A in X (which is precisely the image of B') is an H-orbit of X. The complement C of A in X contains the fixed point $H = H1$; and with a few elementary arguments involving techniques from compact semigroup theory one shows that A' is transversal for the right action of T on S, whence A is transversal for the action of T on X. By Corollary 3 we then conclude that B is singleton, i.e. that $HsH = Hs$. This equivalent to $sH = Hs$. Thus we have shown that for all $s \in S$ with $s \neq 0$ we have $sH \subseteq Hs$, and for $s = 0$ this is trivially correct. Thus H is left normal in S and by the Left Normality Theorem, $Z(H, S)$ is connected. One finds a one parameter semigroup $I \subseteq Z(H, S)$ containing 0 and 1, and since S/D is totally ordered and $D(s) = HsH$, then $S = HIH = H^2 I = HI$.

SECTION 4. FURTHER RESULTS

We describe, without going into details, a generalization of the $H \times H$-theorem.

Generalized $H \times H$-Theorem. *Let S be a compact connected monoid with 0. Then the following conditions are equivalent:*

1) *The space $S/(H \times H)$ of orbits HsH has a total order such that the order topology and the quotient topology agree and $\{0\}$ is an end point of $S/(H \times H)$; moreover, for each idempotent e the subset HeH is a subsemigroup.*

2) *The space S/H of all orbits Hs has a total order so that the order topology and the quotient topology agree, and $\{0\}$ is an endpoint of S/H.*

3) *There is an I-semigroup I containing 1 and 0 all of whose elements commute with all elements of H, and $S = HI$.*

Problem. If condition 1) in the Generalized $H \times H$ theorem is modified by dropping the postulate about the cosets HsH of idempotents, is the modified condition still equivalent to 3)? We do not know the answer at this time, if the Centralizing Conjecture is correct in general, then the answer is affirmative.

REFERENCES

[1] K.H. Hofmann, Automorphic actions of compact groups on compact monoids, *Proc. 2nd Florida Symposium Automorphisms and Semigroups,* 53 pp.

[2] K.H. Hofmann and P.S. Mostert, *Elements of Compact Semigroups,* Ch. E. Merrill, Columbus, 1966.

[3] K.H. Hofmann and P.S. Mostert, *Problems about compact semigroups,* Semigroups, ed. by K.W. Folley, Academic Press, 1969. 85-100.

[4] K.H. Hofmann and P.S. Mostert, Applications of transformation groups to problems in topological semigroups, *Proc. Conf. Transf. Gr.*, ed. by P.S. Mostert, Springer-Verl. 1968, 370-380.

[5] K.H. Hofmann and M. Mislove, The centralizing theorem for left normal groups of units in compact monoids, *Semigroup Forum,* 3 (1971), 31-42.

[6] D. Montgomery and C.T. Yang, The existence of a slice, *Ann. of Math.*, 65 (1957), 108-116.

COLLOQUIA MATHEMATICA SOCIETATIS JÁNOS BOLYAI
8. TOPICS IN TOPOLOGY, KESZTHELY (HUNGARY), 1972.

SOME SUFFICIENT CONDITIONS FOR EMBEDDING POLYHEDRA IN EUCLIDEAN SPACE

K. HORVATIĆ

It is well-known that every n-dimensional polyhedron can be piecewise linearly embedded in the Euclidean space E^{2n+1} and that this result is in general the best possible. We prove the following theorem: Let X be an n-polyhedron, $n \neq 2$, such that the integral cohomology group $H^n(X - \operatorname{Int} A) = 0$ for some n-simplex A in a triangulation of X. Then X embeds piecewise linearly in E^{2n}. From the theorem we derive that a large class of polyhedra actually embeds in the ambient space of a double dimension. Among these are various generalizations of piecewise linear manifolds as trinagulated manifolds, polyhedral homology manifolds, pseudomanifolds and manifolds with singular boundary.

COLLOQUIA MATHEMATICA SOCIETATIS JÁNOS BOLYAI
8. TOPICS IN TOPOLOGY, KESZTHELY (HUNGARY), 1972.

HEWITT REALCOMPACTIFICATION OF PRODUCTS

M. HUŠEK

This contribution is devoted to the "topological equation" $v(P \times Q) = vP \times vQ$. One way how to get the solution of such an equation is to approximate it from both sides, i.e., to find classes of spaces as large as possible which satisfy or do not satisfy the equation. I want to describe several such classes here but I must admit that I was not able to deduce a general rule from the classes. For details and other results I refer to papers [5], [3], [4], [14], [15], [16] and [20].

All the spaces under consideration are supposed to be uniformizable Hausdorff. By $\mathfrak{m}$ we denote the first measurable cardinal if it exists, otherwise $\mathfrak{m}$ is a symbol following all the cardinals. By a pseudo-a-compact space [18], [6, under a different name] we mean a space any locally finite open collection of which is of cardinality less than a. A mapping is said to be z-closed [7] if it carries zero-sets (exact closed sets in the terminology of [2]) to closed sets. If each open cover of a space has a subfamily of power less than a with dense union, then the space is called weakly a-compact [6, under a different name]. A space is $< a$-discrete [25], [18], [11], [23] if intersections of less than a open sets are open. A k'-space is the uniformizable modification of a k-space (i.e., k'-spaces are just the spaces the continuous real-valued functions of which coincide with functions continuous on compact subsets) [21], [1], [22], [17]. Other concepts used in the sequel can be found in [2], [9] and [19].

We shall prove now that the property for $P \times Q$ to satisfy $\upsilon(P \times \times Q) = \upsilon P \times \upsilon Q$ is not topological. This is a big difference in comparison with the compact case, where, for infinite spaces, $\beta(P \times Q) = \beta P \times \beta Q$ if and only if $P \times Q$ is pseudocompact [10], [8]. As a consequence we can answer in the negative the conjecture by S. Negrepontis and W.W. Comfort (oral communication) that $\upsilon(P \times Q) = \upsilon P \times \upsilon Q$ if and only if $\upsilon(P \times Q) = r(P \times Q)$ where r means the ω_2-compactification [12] and also the conjecture of the author [15] that for spaces of measurable cardinalities $\upsilon(P \times Q) = \upsilon P \times \upsilon Q$ if and only if P and Q have a property and $P \times Q$ is pseudo-$\mathfrak{m}$-compact.

First of all we recall two results from [15] (the second one is a strengthening of a result from [5], [3]):

If $\upsilon(P \times Q) = \upsilon P \times \upsilon Q$ *then either* $\min(\operatorname{card} P, \operatorname{card} Q) < \mathfrak{m}$ *or* $P \times Q$ *is pseudo-*$\mathfrak{m}$*-compact.*

If P *is locally compact realcompact then* $\upsilon(P \times Q) = P \times \upsilon Q$ *if and only if either* $\operatorname{card} P < \mathfrak{m}$ *or* Q *is pseudo-*$\mathfrak{m}$*-compact.*

Now the negative results:

Proposition 1. *If* P, Q *are spaces such that* $\upsilon(P \times Q) \neq \upsilon P \times \upsilon Q$ *and* X *is a locally compact realcompact space such that either* $\operatorname{card} X < \mathfrak{m}$ *or* $P \times Q$ *is pseudo-*$\mathfrak{m}$*-compact, then* $\upsilon(X \times (P \times Q)) = X \times \upsilon(P \times Q)$, $\upsilon((X \times P) \times Q) \neq \upsilon(X \times P) \times \upsilon Q$.

Proof. The first equality follows from the second result stated above. If $\upsilon((X \times P) \times Q) = \upsilon(X \times P) \times \upsilon Q$, then $\upsilon(P \times Q) = \upsilon P \times \upsilon Q$ because $\upsilon(X \times P) = X \times \upsilon P$ and $\upsilon(P \times Q)$, is, roughly speaking, the closure of $P \times Q$ in $\upsilon(X \times P \times Q)$.

Corollary. *The property for products* $M \times N$: $\upsilon(M \times N) = \upsilon M \times \upsilon N$, *is not topological.*

Theorem 1. *Let* $\mathcal{K}$ *be a class of spaces containing spaces* X, P, Q *such that* X *is locally compact realcompact,* $\upsilon(P \times Q) \neq \upsilon P \times \upsilon Q$, $P \times Q \in \in \mathcal{K}$, $X \times P \in \mathcal{K}$ *and either* $\operatorname{card} X < \mathfrak{m}$ *or* $P \times Q$ *is pseudo-*$\mathfrak{m}$*-compact. Then there are no topological properties* V, W *such that for* $M, N \in \mathcal{K}$:

$$\upsilon(M \times N) = \upsilon M \times \upsilon N \text{ if and only if } M \in V,\ N \in V \text{ and } M \times N \in W .$$

Proof. Suppose such properties exist. Since $\upsilon(X \times (P \times Q)) = X \times \times \upsilon(P \times Q)$ we get $X \times P \times Q \in W$; since either $\operatorname{card} X < \mathfrak{m}$ or P is pseudo-$\mathfrak{m}$-compact $\upsilon(X \times (X \times P)) = X \times \upsilon(X \times P)$ and so $X \times P \in V$; finally $\upsilon(X \times Q) = X \times \upsilon Q$ entails $Q \in V$. Thus by our assumption, $\upsilon((X \times P) \times Q) = \upsilon(X \times P) \times \upsilon Q$, which contradicts Proposition 1.

If we drop e.g. the condition "$P \times Q \in \mathcal{K}$" then the assertion is not true (take $\mathcal{K}$ to be all pseudocompact infinite spaces). Corollary 3 to Theorem 2 shows that we cannot miss the condition "either $\operatorname{card} X < < \mathfrak{m}$ or $P \times Q$ is pseudo-$\mathfrak{m}$-compact".

Corollary 1. *Let* $\mathcal{K}$ *be a finitely productive class of spaces containing all the compact spaces and a pair* P, Q *such that* $\upsilon(P \times Q) \neq \upsilon P \times \times \upsilon Q$. *Then there are no topological properties* V, W *such that for* $M, N \in \mathcal{K}$:

$$\upsilon(M \times N) = \upsilon M \times \upsilon N \text{ if and only if } M, N \in V \text{ and } M \times N \in W .$$

Corollary 2. *Let* $\mathcal{K}$ *be as in the preceding corollary. Then there are no reflective subcategories* R, S *of Top (with reflections* r, s*) such that for* $M, N \in \mathcal{K}$:

$$\upsilon(M \times N) = \upsilon M \times \upsilon N \text{ if and only if } r(M \times N) = s(M \times N) .$$

Corollary 3. *Let* $\mathcal{K}$ *contain all spaces* M *with* $\alpha \leqslant \operatorname{card} M < \beta$ *where* $\alpha < \beta$ *are uncountable cardinals. Then there are no topological properties* V, W *such that for* $M, N \in \mathcal{K}$:

$$\upsilon(M \times N) = \upsilon M \times \upsilon N \text{ if and only if } M, N \in V \text{ and } M \times N \in W .$$

For the proof one must realize only that for any uncountable cardinal α there exists a pair P, Q of spaces of cardinalities α such that $\upsilon(P \times Q) \neq \upsilon P \times \upsilon Q$ and $P \times Q$ is pseudo-$\mathfrak{m}$-compact (see Example in [15] where X may be chosen with an arbitrary cardinality).

Now we turn our attention to positive results: The only way, I know, for obtaining sufficient conditions for P, Q to satisfy $\upsilon(P \times Q) = \upsilon P \times \upsilon Q$ is the method of function spaces that was used by I. Glicksberg [10] and Z. Frolík [8] for $\beta(P \times Q)$, i.e. the following procedure: to find a topology on $C(Q)$ such that $C(P \times Q) \subset C(P, C(Q))$, $C(P, C(Q)) = C(\upsilon P, C(Q))$ and $C(\upsilon P, C(Q)) \subset C(\upsilon P \times Q)$. The best way would be to find such a topology which has a complete uniformity; in that case the equality $C(P, C(Q)) = C(\upsilon P, C(Q))$ holds if either P is pseudo-$\mathfrak{m}$-compact or $C(Q)$ is pseudo-$\mathfrak{m}$-compact (since by a Shirota's theorem [24], [9] a space is realcompact if and only if it is pseudo-$\mathfrak{m}$-compact and has a complete uniformity) – the converse is true whenever $C(P \times Q) \subset C(P, C(Q))$ and $C(\upsilon P, C(Q)) \subset C(\upsilon P \times Q)$. Hence *if we can find a topology on* $C(Q)$ *having a complete uniformity and such that* $C(P \times Q) = C(P, C(Q))$, $C(\upsilon P \times Q) = C(\upsilon P, C(Q))$, *then* $P \times Q$ *is C-embedded in* $\upsilon P \times Q$ *if and only if either* P *or* $C(Q)$ *is pseudo-*$\mathfrak{m}$*-compact.* I do not know whether the existence of such a topology is a necessary condition for the equality $\upsilon(P \times Q) = \upsilon P \times \upsilon Q$ and either a general method how to find such a topology. So I will try to make use of a converse way: to find spaces P for a given complete uniformity on $C(Q)$ such that an inclusion or an equality between the above function spaces is fulfilled. In this case, the preceding conclusions will provide sufficient conditions for $\upsilon(P \times Q) = \upsilon P \times \upsilon Q$; necessary conditions will follow from the first result stated before Proposition 1.

First take the topology of uniform convergence on $C(Q)$. It is known [23] that in this case $C(P \times Q) \subset C(P, C(Q))$ if and only if the projection $pr_P: P \times Q \to P$ is z-closed; since the inclusion $C(\upsilon P, C(Q)) \subset C(\upsilon P \times Q)$ holds automatically and $C(Q)$ is pseudo-$\mathfrak{m}$-compact if and only if $\operatorname{card} Q < \mathfrak{m}$ [16, Theorem 1] we have (for $\operatorname{card} P$, $\operatorname{card} Q < \mathfrak{m}$ see [5]):

Theorem 2. *If the projection* $pr_P: P \times Q \to P$ *is z-closed, then* $\upsilon(P \times Q) = \upsilon P \times \upsilon Q$ *if and only if either* P *is pseudo-*$\mathfrak{m}$*-compact or* $\operatorname{card} Q < \mathfrak{m}$.

Thus by Theorems 3.1 and 4.2 from [11]:

Corollary 1. *Let* P *be* $< \mathfrak{a}$*-discrete and* Q *be weakly* $\mathfrak{a}$*-compact.*

Then $\upsilon(P \times Q) = \upsilon P \times \upsilon Q$ *if and only if either* P *is pseudo-*$\mathfrak{m}$*-compact or* $\operatorname{card} Q < \mathfrak{m}$.

Corollary 2. *Let* $P \times Q$ *be pseudo-*$\mathfrak{a}$*-compact and* P *be* $< \mathfrak{a}$*-discrete. Then* $\upsilon(P \times Q) = \upsilon P \times \upsilon Q$ *if and only if either* P *is pseudo-*$\mathfrak{m}$*-compact or* $\operatorname{card} Q < \mathfrak{m}$.

The last, corollary implies the following weak analogy to the Glicksberg-Frolík theorem on $\beta(P \times Q)$:

Corollary 3. *Let* $\operatorname{card} P, \operatorname{card} Q \geqslant \mathfrak{m}$ *and* P *or* Q *be* $< \mathfrak{m}$*-discrete. Then* $\upsilon(P \times Q) = \upsilon P \times \upsilon Q$ *if and only if* $P \times Q$ *is pseudo-*$\mathfrak{m}$*-compact.*

It is easy to prove that $pr_P \colon P \times Q \to P$ is z-closed provided $\upsilon(P \times Q) = \upsilon P \times \upsilon Q$ and $pr_{\upsilon P} \colon \upsilon P \times \upsilon Q \to \upsilon P$ is z-closed [16, Theorem 5] and hence

Corollary 4. *Let* Q *be pseudocompact. Then* $\upsilon(P \times Q) = \upsilon P \times \upsilon Q$ *if and only if* $pr_P \colon P \times Q \to P$ *is* z*-closed and either* P *is pseudo-*$\mathfrak{m}$*-compact or* $\operatorname{card} Q < \mathfrak{m}$.

Conversely, if $pr_P \colon P \times Q \to P$ is z-closed and either P is pseudo-$\mathfrak{m}$-compact or $\operatorname{card} Q < \mathfrak{m}$, then $pr_{\upsilon P} \colon \upsilon P \times \upsilon Q \to \upsilon P$ is z-closed.

Corollary 5. *If* P *is* $< \mathfrak{a}$*-discrete for an* $\mathfrak{a} \leqslant \mathfrak{m}$, *then* υP *has the same property.*

Proof follows from the fact that a space X is β-discrete if and only if $pr_X \colon X \times D(\beta) \to X$ is z-closed [23], where $D(\beta)$ is discrete, $\operatorname{card} D(\beta) = \beta$.

Clearly, Corollary 5 is false for $\mathfrak{a} > \mathfrak{m}$.

If P, Q are pseudo-$\mathfrak{a}$-compact, P is $< \mathfrak{a}$-discrete, non-discrete, then $P \times Q$ is pseudo-$\mathfrak{a}$-compact provided pr_P is z-closed [11, Theorem 4.4], [23, Theorem 3.4]. For $\mathfrak{a} = \mathfrak{m}$ we can now state more:

Corollary 6. *Let* P, Q *be pseudo-*$\mathfrak{m}$*-compact spaces. If* pr_P *is* z*-closed then* $P \times Q$ *is pseudo-*$\mathfrak{m}$*-compact.*

The next very nice topology on $C(Q)$ is the compact-open topology. In this case always $C(P \times Q) \subset C(P, C(Q))$. The remaining inclusion is rather difficult to decide (it holds e.g. if Q is locally compact): when υP and Q are k'-spaces then $C(\upsilon P, C(Q)) \subset C(\upsilon P \times Q)$ if and only if $\upsilon P \times Q$ is a k'-space [17] – this is not the condition we expect because it contains a property on υP but I do not know a better one (for a discussion when υP is locally compact see [3], [4]).

Theorem 3. *Let P be a k'-space and either υQ be locally compact or $\upsilon P \times \upsilon Q$ be a k'-space. Then $\upsilon(P \times Q) = \upsilon P \times \upsilon Q$ if and only if either* $\min(\operatorname{card} P, \operatorname{card} Q) < \mathfrak{m}$ *or P and Q are pseudo-$\mathfrak{m}$-compact.*

Proof. By the first result before Proposition 1, our conditions are necessary. The assumptions of the Theorem entail $C(\upsilon P, C(\upsilon Q)) = C(\upsilon P \times \upsilon Q)$ and if moreover the conditions are fulfilled, then

$$C(P \times Q) \subset C(Q, C(P)) = C(\upsilon Q, C(P)) = C(P, C(\upsilon Q)) =$$

$$= C(\upsilon P, C(\upsilon Q)) = C(\upsilon P \times \upsilon Q)$$

and, thus, $\upsilon(P \times Q) = \upsilon P \times \upsilon Q$. We have used a Brown's result [1] in the third relation.

Corollary 1. *If P is a k'-space and Q is pseudocompact, then $\upsilon(P \times Q) = \upsilon P \times \upsilon Q$ if and only if either P is pseudo-$\mathfrak{m}$-compact or* $\operatorname{card} Q < \mathfrak{m}$.

Getting together the foregoing corollary and Corollary 4 to Theorem 2 we obtain:

If P is a k'-space and Q is pseudocompact and either P is pseudo-$\mathfrak{m}$-compact or $\operatorname{card} Q < \mathfrak{m}$, then $pr_P \colon P \times Q \to P$ is z-closed.

If P is a pseudocompact k'-space and Q is pseudocompact, then $P \times Q$ is pseudocompact [22].

Corollary 2. *If P is a pseudocompact k'-space and υQ a k'-space, then $\upsilon(P \times Q) = \upsilon P \times \upsilon Q$ if and only if either P is of nonmeasurable cardinality or Q is pseudo-$\mathfrak{m}$-compact.*

Corollary 3. *Let P be a realcompact k'-space and υQ locally compact or $P \times \upsilon Q$ a k'-space. Then $\upsilon(P \times Q) = P \times \upsilon Q$ if and only if* *either* P *is of nonmeasurable cardinality or* Q *is pseudo-*$\mathfrak{m}$*-compact.*

Corollary 4. *Let* P, Q *be* k'*-spaces and either* υQ *be locally compact or* $\upsilon P \times \upsilon Q$ *be a* k'*-space. If* $\upsilon(P \times Q) = \upsilon P \times \upsilon Q$ *then* $P \times Q$ *is a* k'*-space.*

If Q is locally weakly α-compact then the uniformity on $C(Q)$ of uniform convergence on weakly α-compact subsets is complete, $C(P \times Q) \supset C(P, C(Q))$ for any P and $C(P \times Q) = C(P, C(Q))$ provided P is $< \alpha$-discrete; in this case $C(Q)$ is pseudo-$\mathfrak{m}$-compact if and only if card $Q < \mathfrak{m}$ [16, Theorem 1].

Theorem 4. Let Q *be a realcompact locally weakly* α*-compact space and* P *be* $< \alpha$*-discrete. Then* $\upsilon(P \times Q) = \upsilon P \times Q$ *if and only if either* P *is pseudo-*$\mathfrak{m}$*-compact of* card $Q < \mathfrak{m}$.

Corollary 1. *Let* Q *be locally Lindelöf realcompact and* P *be a P-space. Then* $\upsilon(P \times Q) = \upsilon P \times Q$ *if and only if either* P *is pseudo-*$\mathfrak{m}$*-compact or* card $Q < \mathfrak{m}$.

Corollary 2. *Let* Q *be a metrizable locally separable space of non-measurable cardinality. Then* $\upsilon(P \times Q) = \upsilon P \times Q$ *for any P-space* P.

Corollary 3. *Let* Q *be a realcompact locally weakly* α*-compact and* P *be a pseudo-*$\mathfrak{m}$*-compact* $< \alpha$*-discrete space. Then* $P \times Q$ *is pseudo-*$\mathfrak{m}$*-compact.*

REFERENCES

[1] R. Brown, Function spaces and product topologies, *Quart. J. Math., Oxford* (2), 15 (1964), 238-250.

[2] E. Čech, *Topological spaces,* revised ed. by M. Katětov and Z. Frolík (Academia, Prague, 1966).

[3] W.W. Comfort, On the Hewitt realcompactification of a product space, *Trans. Amer. Math. Soc,* 131 (1968), 107-118.

[4] W.W. Comfort, Locally compact realcompactifications, *Proc. 2nd Prague Top. Symp.* 1966 (Academia, Prague, 1967), 95-100.

[5] W.W. Comfort – S. Negrepontis, Extending continuous functions on $X \times Y$ to subsets of $\beta X \times \beta Y$, *Fund. Math.*, 59 (1966), 1-12.

[6] Z. Frolík, Generalizations of compact and Lindelöf spaces, *Czech. Math. J.*, 9 (1959), 172-217.

[7] Z. Frolík, Applications of complete families of continuous functions to the theory of Q-spaces, *Czech. Math. J.*, 11 (1961), 115-133.

[8] Z. Frolík, The topological product of two pseudocompact spaces, *Czech. Math. J.*, 10 (1960), 339-349.

[9] L. Gillman – M. Jerison, Rings of continuous functions, (Van Nostrand, Princeton, 1960).

[10] I. Glicksberg, Stone – Čech compactifications of products, *Trans. Amer. Math. Soc.*, 90 (1959), 369-382.

[11] A.W. Hager, Projections of zero-sets (and the fine uniformity on a product), *Trans. Amer. Math. Soc.*, 140 (1969), 87-94.

[12] H. Herrlich, Fortsetzbarkeit stetiger Abbildungen und Kompaktheitsgrad topologischer Raume, *Math. Z.*, 96 (1967), 64-72.

[13] E. Hewitt, Rings of real-valued continuous functions. I, *Trans. Amer. Math. Soc.*, 64 (1948), 45-99.

[14] M. Hušek, The Hewitt realcompactification of a product, *Comment. Math. Univ. Carolinae,* 11 (1970), 393-395.

[15] M. Hušek, Pseudo-$\mathfrak{m}$-compactness and $v(P \times Q)$, *Indag. Math.*, 33 (1971), 320-326.

[16] M. Hušek, Realcompactness of function spaces and $v(P \times Q)$, *J. Gen. Top.*, 2 (1972), 165-179.

[17] M. Hušek, Products of quotients and of k'-spaces, *Comment. Math. Univ. Carolinae,* 12 (1971), 61-68.

[18] J.R. Isbell, Uniform spaces, *Amer. Math. Soc.*, Providence, 1964.

[19] J.L. Kelley, General topology, Van Nostrand, Princeton, 1955.

[20] W.G. McArthur, Hewitt realcompactifications of products, *Canad. J. Math.*, 22 (1970), 645-656.

[21] E. Michael, A note on k-spaces and k_R-spaces, Topology Conf., Arizona State Univ., (1967), 247-249.

[22] N. Noble, Countably compact and pseudocompact products, *Czech. Math. J.*, 19 (1969), 390-397.

[23] N. Noble, Products with closed projections, *Trans. Amer. Math. Soc.*, 140 (1969), 381-391.

[24] T. Shirota, A class of topological spaces, *Osaka Math. J.*, 4 (1952), 23-40.

[25] R. Sikorski, Remarks on some topological spaces of high power, *Fund. Math.*, 37 (1950), 125-136.

COLLOQUIA MATHEMATICA SOCIETATIS JÁNOS BOLYAI
8. TOPICS IN TOPOLOGY, KESZTHELY (HUNGARY), 1972.

APOSYNDETIC CONTINUA

F.B. JONES

In 1938 I attended my first meeting of the *American Mathematical Society* and presented a paper [18] about the boundaries of complementary domains of locally connected continua in spaces that satisfied R. L. Moore's Axioms 0 – 4. Ahead of me on the program was G. T. Whyburn who talked about semi-locally-connected continua in the plane [36]. During the short time between our talks I was unable to adapt his generalization of local connectivity to my situation. And in fact it was six months before I discovered [19] the required notion which was truly a generalization of local connectivity as a point-wise property – a sort of complementary notion to semi-local-connectivity. Whyburn's terminology was not very good because a continuum could be locally connected at a point without being semi-locally-connected $(s - 1 - c)$ there. Also Whyburn had previously called this notion $(s - 1 - c$ at the point $p)$ *locally divisible in* p. Either he didn't like the original terminology or it was so non-suggestive that he had forgotten it. I was determined to do better! So, after having almost worn out a large Webster's Dictionary with no success, I consulted Dr. Leon in the *Classics Department* at Texas. It wasn't easy to explain to someone who knew almost nothing about mathematics what it meant to have a continuum contain p in its interior but miss the point q entirely. After about thirty minutes though he grasped the idea, assembled the word *aposyndetic* and I have suffered ever since from sly digs and sarcastic remarks like: When was I going to write another paper about "apo-what-ever-it-was"?

Actually the terminology is really pretty good. "Deo" means "to bind," "syn" means "together," and "apo" means "away from." The continuum M is aposyndetic at p with respect to q means that M is bound together at p away from q. To a topologist this happens when some subcontinuum H of M contains p in its interior and misses q, i.e., $\{p\} \subset H° \subset H \subset M - \{q\}$. Obviously the notion can be generalized to disjoint sets P and Q. In fact, M is connected im kleinen at p [see 24] when $\{p\} \subset H° \subset H \subset M - Q$ for arbitrary closed sets Q not containing p. It follows from Whyburn's definition that if M is $s - 1 - c$ at p, then M is aposyndetic at each point x of $M - p$ with respect to p. (The converse requires some sort of compactness, e.g., local peripheral compactness at p.) We say that M is aposyndetic if it is aposyndetic at each point with respect to each other point.

After the war (World War II) I *did* write another paper on aposyndesis and in 1949 I gave a talk on the subject to the Society. In this talk I outlined a crude sort of classification of continua arranged in a spectrum of increasing aposyndetic strength from left to right:

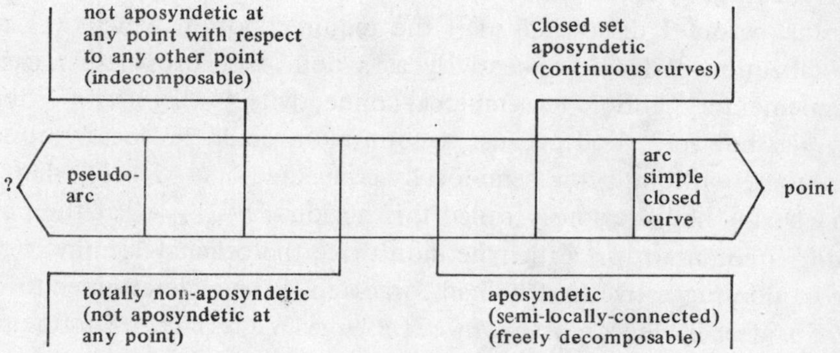

Spectrum

Since then a considerable number of new lines have been added to that spectrum. Also the notion has been rediscovered from a different point of view and several of the old theorems duplicated [5]. So it is the purpose of this talk to summarize insofar as possible the results of the past twentytwo or three years. Some of the new lines are indicated below:

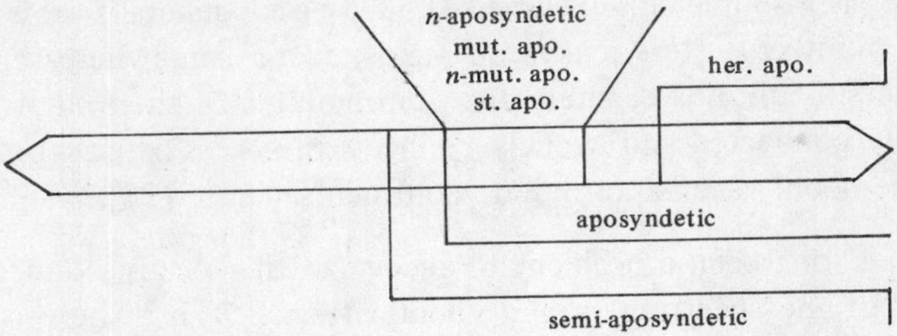

Definitions. A continuum M is *semi-aposyndetic* provided that given any two (distinct) points of M, M is at least aposyndetic at one of them with respect to the other. A continuum M is *n-aposyndetic* provided that given a point p and an n-point set Q not containing p there exists a subcontinuum H such that $\{p\} \subset H^\circ \subset H \subset M - Q$. If Q were a countable closed set, M would be called $\aleph_0$*-aposyndetic.* If every two points of a continuum belong to the interiors of disjoint subcontinua, the continuum is said to be *mutually aposyndetic.* An analogous notion would apply to a set of n points (instead of 2) and such a continuum would be said to be *n-mutually aposyndetic.*

Vought, who was interested in the classification of curves [see 37], discovered that hereditarily aposyndetic continua are all hereditarily locally connected [33]. But this is not true of hereditarily semi-aposyndetic continua [e.g., the Cantor fan]. Lee Rogers quickly observed that the proposition does hold for aposyndetic continua: Suppose that M is aposyndetic. Them M is hereditarily locally connected if and only if M is hereditarily semi-aposyndetic [27]. (The suspension of the Cantor set is *not* a counterexample to this theorem.) Don Bennett defined a continuum to be strongly aposyndetic if it is decomposable and each continuum of any such decomposition is aposyndetic [2]. He proved that every strongly aposyndetic continuum is $\aleph_0$-aposyndetic. In settling a question raised by Bennett, Vought showed that every strongly aposyndetic continuum is, in fact, hereditarily locally connected [34]. Actually, Vought has a somewhat stronger theorem.

There is another pair of theorems that belong in this group. Bing showed that if no point cuts and no subcontinuum separates a continuum M then M is a simple closed curve [4]. Vought generalizes by proving that if no point cuts (this makes M aposyndetic somewhere) and only aposyndetic subcontinua separate the continuum M, then M is hereditarily locally connected and cyclicly connected [32]. Rogers proves the same is true if only semi-aposyndetic continua separate [27].

It has long been a problem to recognize the locally connected division line of the spectrum in aposyndetic terms. Bing showed that if the continuum M is aposyndetic at p then M would be locally connected at p provided that no subcontinuum of M separates M [4]. Vought proved the same is true provided that only aposyndetic subcontinua separate [32]. Rogers tried to enlarge the collection of separating continua to include semi-aposyndetic continua but this didn't work. However, he did get the following: If the continuum M is both aposyndetic and $s-1-c$ at p and only semi-aposyndetic continua separate, then M is connected im kleinen at p (that is, M possesses arbitrarily small closed connected neighborhoods of p) [27]. I had stated that if a continuum were aposyndetic at each point with respect to subcontinua not containing it then the continuum would have to be locally connected Davis published a proof, showing that if the continuum M is aposyndetic at p with respect to continua and $s-1-c$ at p then M is locally connected at p. Davis' theorem is for compact Hausdorff continua and Hagopian showed that only regularity is required [13].

Nearly all of what has been said so far to this point is for compact metric spaces. However, in what follows many of the theorems apply to quite general situations. Only in a few cases will I attempt to point this out. We shall now consider three groups of theorems. The first of these will indicate something of the role played by aposyndesis in Cartesian products.

Product theorems. Each factor in a product should always be understood to be non-degenerate. Whyburn showed that in a compact metric space the product of an aposyndetic continuum by an arbitrary continuum is aposyndetic [36]. For the $s-1-c$ notion a certain amout of compactness is useful and sometimes necessary. For aposyndesis the situation is neater: the product of any two continua is aposyndetic whether compact or not, no matter how abstract the space [20]. In fact, this is not really topology but just words – linguistics. Once in class when we had nothing better to do, Vought showed that the product of an m-aposyndetic continuum with an n-aposyndetic continuum was $(m+n+1)$-aposyndetic. This is a sort of number theory theorem. Fitzgerald showed (for the compact Hausdorff case) that the product of two arbitrary continua was n-aposyndetic (in fact, $\aleph_0$-aposyndetic) [6]. Rogers did the same for Hausdorff continua (not necessarily compact) [25]. Hagopian proved that the product of aposyndetic continua is mutually aposyndetic and that the product of *three* arbitrary continua is mutually aposyndetic [10]. Rogers extended the first of these to semi-aposyndetic continua and the last to conclude that the product is n-mutually aposyndetic (for every n) [25]. I believe these theorems are for Hausdorff spaces except the last, which requires regularity.

Rogers proved some rather special sort of theorems: if the product of two chainable continua is mutually aposyndetic then both factors are arcs [22]: the product of a simple closed curve with any continuum is always n-mutually aposyndetic: and []^2 is not mutually aposyndetic [25]. Hagopian showed that the product of two chainable indecomposable continua is strictly non-mutually aposyndetic and posed the still open question: Is the product of two indecomposable continua (whether chainable or not) always strictly non-mutually aposyndetic, e.g., every two subcontinua with non-void interior intersect [10].

Cut point theorems. I discovered that every continuum that was aposyndetic at no point had to contain at least one cut point and raised the question: what aposyndetic properties or non-aposyndetic properties cause almost all points to be cut points. [A point *cuts* if its complement is not *continuumwise* connected.] Grace has done a lot of work on cut points generalizing into the most abstract situations. His best answer to this question is the following: If a continuum is aposyndetic but non-$s-1-c$ at each point of a dense G_δ-set then it contains a dense G_δ-set of cut points [9]. This is a curious combination of good and bad properties. Hagopian dropped the good condition (aposyndesis) but could only prove that there were c cut points in every open subset of the continuum [13]. Recently Shirley under the same hypothesis proved that there is a sub-continuum of cut points and goes on to get a sufficient condition for almost all points to be cut points [28]. But his condition is pretty far out and the problem is *really* still unresolved.

For the plane there are a number of theorems related to some of the above propositions. Hagopian shows [15] that each point of K_x – (the x-component of K_x) is a cut point of the continuum M. [If x is a point of M then K_x is the set of all points y such that M is not aposyndetic at x with respect to y.] This is not a vacuous theorem because K_x need not be connected. But I really don't know what it means relative to the structure of continua. It should have some consequences.

Also in the plane there are some theorems related to those dealing with local connectivity. I think Whyburn was hoping that plane $s-1-c$ continua would have to be arcwise connected [36]. He did point out that the boundary of a complementary domain would be locally connected and hence arcwise connected. I showed that if an aposyndetic continuum had only a finite number of complementary domains it was locally connected and hence arcwise connected [19]. Hagopian has proven the same is true of semi-aposyndetic continua [14]. Of course, these need *not* be locally connected. He also showed that every semi-aposyndetic continuum whose complementary domains form a null sequence (diameters go to zero) is also arcwise connected [16]. I believe that I had shown that if a plane continuum were 2-aposyndetic it would have to be locally connected and hence arcwise connected. Hagopian puts together this rather odd combination of properties: if a continuum is aposyndetic and 2-aposyndesis fails with respect to a finite set (for each point) then the continuum is arcwise connected; or if a continuum is either not $s-1-c$ or not aposyndetic at each point but $s-1-c$ at all except finitely many points than it is arcwise connected [12].

I would like to close by discussing a third group of theorems:

namely, *decomposition theorems.* In particular, how can one improve the quality of a continuum (move it to the right in the spectrum) with monotone maps? McAuley was the first to attack this problem. When he first started out to move continua into the aposyndetic end of the spectrum he had a good definition for the decomposition but couldn't prove that the elements were connected. I thought this was just stupidity and insisted that surely they were connected. About two-thirty one moring he constructend an example (actually in the plane) which he showed me the next afternoon in which the elements were *not* connected. [A fantastic example which should have told us more than it did.] However, the components of these elements did the job and McAuley had his aposyndetic decomposition, i.e., the decomposition space was aposyndetic [22]. It possessed a flaw: certain continua which were already aposyndetic were changed. In a later paper he successfully modified his procedure so as to leave aposyndetic continua unchanged [23]. Roughly his theorem is as follows: In the continuum M we say that $x \sim y$ if there exists no collection G of separators with the properties that (1) some element of G separates x from y in M and (2) if $g \in G$ separates the subcontinuum N from the point b then N is interior to a continuum C which is separated from b by three elements of G in an ordered fashion. Then $M/\sim$ is aposyndetic.

Fairly recently Fitzgerald and Swingle attacked this problem more directly really defining the elements of the decomposition using aposyndesis. However, things are still not simple. To simplify, their theorem specialized to an irreducible continuum M is as follows: If $x \sim y$ when $T^2(x) \supset \{y\}$ and $T^2(y) \supset x$, then $M/\sim$ is an arc. Whyburn pointed out every irreducible aposyndetic continuum is an arc; so this is the form an aposyndetic decomposition would naturally take [7].

For this special case – irreducible continua – several rather nice theorems exist. Thomas [30] gets an aposyndetic decomposition when no indecomposable subcontinuum has nonvoid interior; Fugate [8], when when there are uncountably many minimal separating subcontinua and Vought [35], when there is a saturated collection of separators, all with void interior.

If one specialized to continua which are separated by no subcontinuum then an aposyndetic decomposition would yield a simple closed curve.

The first to get an aposyndetic decomposition for continua of this sort was Stratton [29]. Vought [35] defined "bisaturated" as follows: a collection G is bisaturated provided that if p is a point not in g_1 of G there exist elements g_2 and g_3 of G such that $g_2 \cup g_3$ separates p from g_1. His theorem is: if the continuum M is separated by no subcontinuum and there exists a bisaturated collection of subsets all with void interior then M has an aposyndetic decomposition.

I feel that for these specialized continua we have about all we need but that for continua in general other good theorems are still to be found.

Finally, I wish to mention two rather startling and recent results, Schlais has shown that whenever K_x has nonvoid interior for some x in M, then M contains an indecomposable continuum. (More about K_x would be welcome.) And Bellamy has shown [1] that $\beta E^n - E^n$ for $n \geqslant 2$ is aposyndetic. Here is an example of aposyndetic continua existing in nature, so to speak. Of course, Cartesian products exist in nature but for the most part they are locally connected also. I have often thought that somewhere in analytic function theory others might be found. So if you run across any others there or elsewhere please send me a postcard.

REFERENCES

[1] D.P. Bellamy, Aposyndesis in the remainder of Stone – Čech compatifications, *Bulletin de'l'Academia Polanaise des Sciences,* 19 (1971), 941-944.

[2] D.E. Bennett, A sufficient condition for countable-set aposyndesis, *Proc. Amer. Math. Soc.*, 32 (1972), 578-584.

[3] D.E. Bennett, Aposyndetic properties of unicoherent continua, *Pacific J. Math.,* 37 (1971), 585-589.

[4] R.H. Bing, Some characterizations of arcs and simple closed curves, *Amer. J. Math.,* 70 (1948), 497-506.

[5] J. de Groot, Connectedly generated spaces, *Proceedings of the Topological Symposium of Herceg Novi,* Yugoslavia, (1968), 171-175.

[6] R.W. Fitzgerald, The Cartesian product of non-degenerate compact continua is *n*-point aposyndetic, *Proc. A.S.U. Topology Conference,* Tempe, 1967, 324-326.

[7] R.W. Fitzgerald – P.M. Swingle, Core decompositions of continua, *Fund. Math.,* 61 (1967), 33-50.

[8] J.B. Fugate, Irreducible continua, *Proc. A.S.U. Topology Conference,* Tempe, (1967), 100-103.

[9] E.E. Grace, Cut points in totally non-semi-locally-connected continua, *Pacific J. Math.,* 14 (1964), 1241-1244.

[10] C.L. Hagopian, Mutual aposyndesis, *Proc. Amer. Math. Soc.,* 23 (1969), 615-622.

[11] C.L. Hagopian, Arcwise connected plane continua, *Proc. Topology Conference,* Emory University, 1970, 41-44.

[12] C.L. Hagopian, Concerning arcwise connectedness and the existence of simple closed curves in plane continua, *Trans. Amer. Math. Soc.,* 147 (1970), 389-402. [See also ibid., 157 (1971), 507-509.]

[13] C.L. Hagopian, On generalized forms of aposyndesis, *Pacific J. Math.,* 34 (1970), 97-108.

[14] C.L. Hagopian, A class of arcwise connected continua, *Proc. Amer. Math. Soc.,* 30 (1971), 164-168.

[15] C.L. Hagopian, A cut point theorem for plane continua, *Duke Math. J.,* 38 (1971), 509-512.

[16] C.L. Hagopian, Arcwise connectivity of semi-aposyndetic plane continua, *Trans. Amer. Math. Soc.,* 158 (1971), 161-165.

[17] C.L. Hagopian, Semiaposyndetic nonseparating plane continua are arcwise connected, *Bull. Amer. Math. Soc.,* 77 (1971), 593-595.

[18] F.B. Jones, Concerning the boundary of a complementary domain of a continuous curve, *Bull. Amer. Math. Soc.,* 45 (1939), 428-435.

[19] F.B. Jones, Aposyndetic continua and certain boundary problems, *Amer. J. Math.,* 53 (1941), 545-553.

[20] F.B. Jones, Concerning non-aposyndetic continua, *ibid.*, 70 (1948), 403-413.

[21] F.B. Jones, Concerning aposyndetic and non-aposyndetic continua, *Bull. Amer. Math. Soc.*, 58 (1952), 137-151.

[22] L.F. McAuley, On decomposition of continua into aposyndetic continua, *Trans. Amer. Math. Soc.*, 81 (1956), 74-91.

[23] L.F. McAuley, An atomic decomposition of continua into aposyndetic continua, *ibid.*, 88 (1958), 1-11.

[24] R.L. Moore, Foundations of point-set theory, *Amer. Math. Soc. Colloquium Publications,* 13 (1932); revised 1962.

[25] L.E. Rogers, Concerning *n*-mutual aposyndesis in products of continua, *Trans. Amer. Math. Soc.*, 162 (1971), 239-251.

[26] L.E. Rogers, Mutually aposyndetic products of chainable continua, *Pacific J. Math.*, 37 (1971), 805-812.

[27] L.E. Rogers, Continua in which only semi-aposyndetic subcontinua separate, *Pacific J. Math.*, 43 (1972), 493-502.

[28] E.D. Shirley, Semi-local-connectedness and cut points in metric continua, *Proc. Amer. Math. Soc.*, 31 (1972), 291-296.

[29] H.H. Stratton, On continua which resemble simple closed curves, *Fund. Math.*, 68 (1970), 121-128.

[30] E.S. Thomas, Jr., Monotone decompositions of irreducible continua, *Rozprawy Math.*, 50 (1966).

[31] E.J. Vought, *n*-Aposyndetic continua and cutting theorems, *Trans. Amer. Math. Soc.*, 140 (1969), 127-135.

[32] E.J. Vought, Concerning continua not separated by any non-aposyndetic subcontinuum, *Pacific J. Math.*, 31 (1969), 257-262.

[33] E.J. Vought, A classification scheme and characterization of certain curves, *Colloq. Math.*, 20 (1969), 91-98.

[34] E.J. Vought, Strongly semi-aposyndetic continua are hereditarily locally connected, *Proc. Amer. Math. Soc.*, 33 (1972), 619-622.

[35] E.J. Vought, Monotone decompositions of continua into arcs and simple closed curves, sumbitted to *Fund. Math.*

[36] G.T. Whyburn, Semi-locally-connected sets, *Amer. J. Math.*, 61 (1939), 733-749.

[37] G.T. Whyburn, Analytic Topology, *Amer. Math. Soc. Colloquium Publications,* 28 (1942).

COLLOQUIA MATHEMATICA SOCIETATIS JÁNOS BOLYAI
8. TOPICS IN TOPOLOGY, KESZTHELY (HUNGARY), 1972.

REMARKS ON CARDINAL FUNCTIONS

I. JUHÁSZ

The interrelations between different cardinal functions are interesting both in themselves and from the point of view of several set-theoretic problems in topology. An exposition of these interrelations has been given in [1], chapter 2, and the aim of this talk is to present some further developments in this area, including solutions of a few problems raised in [1]. (Notation is the same as in [1]).

Problem 2.4 of [1] asked about the interrelation between the weight and density of Hausdorff spaces. This problem was solved by K. Kunen and the present author (cf. [2]), who proved

Theorem 1. *For every cardinal* α *there is a Hausdorff space* X *such that* $d(X) = \alpha$ *and*

$$w(x) = \exp_3(\alpha)$$

Problem 2.17 of [1] concerned the interrelation between the pseudocharacter and the width of spaces. I could settle this by proving

Theorem 2. *If* $2^\alpha = \alpha^+$, *then there exists a 0-dimensional* T_2

space X *such that* $\psi(X) = \mathfrak{a}^+ > z(X) = \mathfrak{a}$.

I do not know whether X can also be chosen compact. The following result can be proved easily, but it is new and has interesting consequences. ($\sigma(X)$ denotes the number of all open subsets of).

Theorem 3. *For any space* X *we have* $\sigma(X) \leqslant |X|^{z(X)}$.

Corollary. *If* X *is Hausdorff, then*

$$\sigma(X) \leqslant \exp\exp s(X) .$$

This improves a result of A. Hajnal and the present author [3].

Finally, we have some results concerning the following problem of K. Prikry (oral communication): Let X be Hausdorff, $|X| = \omega_2$ and assume that every $Y \subset X$ with $|Y| = \omega_2$ is separable, is then X hereditarily separable? I proved

Theorem 4. *If* X *is a Uryson space,* $|X| = \mathfrak{a}$ *and* $\mathrm{cf}(\mathfrak{a}) \neq \omega_1$, *then if every subspace* $Y \subset X$ with $|Y| = \mathfrak{a}$ *is separable then* X *is hereditarily separable.*

It would be very interesting to know whether the condition $\mathrm{cf}(\mathfrak{a}) \neq \omega_1$ is essential here or not. Analogous question can be raised as above if we replace "separable" by "Lindelöf". For this I proved

Theorem 5. *If* X *is strongly Hausdorff,* $|X| = \mathfrak{a}$ *and every subspace* $Y \subset X$ *with* $|Y| = |X| = \mathfrak{a}$ *is Lindelöf, then* X *is hereditarily Lindelöf.*

REFERENCES

[1] I. Juhász, Cardinal Functions in Topology, *Math. Centre Tract*, 34, Amsterdam 1971.

[2] I. Juhász – K. Kunen, On the weight of Hausdorff spaces, *Gen. Top. Appl*, (to appear)

[3] A. Hajnal – I. Juhász, Discrete subspaces of topological spaces, *Indig. Math.*, 29 (1967), 343-356.

COLLOQUIA MATHEMATICA SOCIETATIS JÁNOS BOLYAI
8. TOPICS IN TOPOLOGY, KESZTHELY (HUNGARY), 1972.

ON TREE-LIKE SPACES AND THE INTERSECTION OF CONNECTED SUBSETS OF A CONNECTED T_1-SPACE

H. KOK

In this paper a topological space X will always be assumed to be connected and T_1.

We use the following abbreviations for properties of a topological space X:

S – $\forall x, y \in X$, $(x \neq y)$: $\exists z \in X$: z separates x and y. (Such a space is called a "tree-like" space. Tree-like spaces have been studied by G.T. Whyburn in [7], G.L. Gurin in [3] and V.V. Proizvolov in [6] under the additional assumption that X is locally connected, resp. peripherally bicompact.)

Int – The intersection of an arbitrary collection of connected subsets of X is connected. (This "connected intersection property" occurs in Whyburn [7] for locally connected spaces.)

Int* – The *closure* of the intersection of an arbitrary collection of connected subsets of X is connected.

Int′ – The intersection of an arbitrary collection of *closed* connected subsets of X is connected.

W – (A, B connected subsets of X and $A \cap B = \phi) \Rightarrow |\bar{A} \cap \bar{B}| \leqslant 1$. (An equivalent form of this property is discussed by A.E. Brouwer [1].)

Theorem 1. Int ⇒ Int* ⇒ Int′.

Proof. Immediate from the definitions.

Theorem 2. Int* ⇒ W; $S \Rightarrow W$.

Proof. Let A and B be disjoint connected subsets of X. Let $p, q \in \bar{A}' \cap \bar{B}$. $(p \neq q)$. Then $A_1 = A \cup \{p, q\}$ and $B_1 = B \cup \{p, q\}$ are connected.

It is easy to see that p and q cannot be separated by a third point. Moreover, $A_1 \cap B_1$ is not connected.

Theorem 3. Int ⇒ S.

Proof. see [2].

Theorem 4. Int* ⇔ Int.

Proof. This proof will appear in our doctoral dissertation in 1973.

The following counterexamples make clear that there are no other implications between any pair of these properties:

1. $X_1 = \{(x, y) \in I^2 \mid \frac{1}{x} \in N \vee (y = 1 \wedge x \neq 0) \vee (x = y = 0)\}$ with the subspace topology. X_1 satisfies S but does not satisfy Int′.

2. Let X_2 be the subset of the plane constructed by E.W. Miller in [5]. Then X_2 is a biconnected space without dispersion point which is also widely connected (i.e.: every non-trivial connected subset is dense in the space.) Since a biconnected set without dispersion point cannot contain any cut point it is easily seen that X_2 satisfies Int′ and W, but does not satisfy S.

3. $X_3 = N$ with the following topology τ: Let $\{F_a\}_{a \in A}$ be a

free ultrafilter on N. Let $\tau = \{F_a\}_{a \in A} \cup \{\phi\}$. X_3 satisfies Int' and W, but does not satisfy S.

4. $X_4 = N$ with the cofinite topology. X_4 satisfies Int', but does not satisfy Int* or W.

Theorem 5. Int' + $S \Leftrightarrow$ Int.

Proof. This proof will also appear in our thesis.

Remark. *It is not possible to replace* S *by* W *in Theorem* 5. X_2 *is a space satisfying Int' and* W *but not satisfying* Int.

If we assume the space X to be locally connected, property S is equivalent to Int, as was proved by Whyburn [7]. That this is not the case for the properties Int' and W, follows from the examples X_3 and X_4. However, if we replace T_1 by T_2 we have the following

Theorem 6. *If* X *is a connected, locally connected* T_2*-space, then all five properties* S, Int, Int*, Int' *and* W *coincide.*

Proof.

(i) $W \Rightarrow S$: (see also [1].) Let p and q be two distinct points in X. Let U and V be two disjoint open connected neighbourhoods of p resp. q. Let A be that component of $X \setminus \bar{U}$ that contains q (and hence V). Then A is open in X.

Since $X \setminus A$ is connected there exists, by property W, exactly one point $r \in \bar{A} \setminus A$. Hence A is clopen (= closed-and-open) in $X \setminus r$, so r separates p and q.

(ii) Int' $\Rightarrow$ Int: Let p and q be two distinct points in X. Let $K(p, q)$ be the intersection of all closed connected subsets of X containing p and q. Then $K(p, q)$ is closed and connected. Let $C(p, q)$ denote the intersection of all connected subsets of X containing p and q. Then $C(p, q) \subset K(p, q)$. We have to prove the connectedness of $C(p, q)$. In fact we will show that $C(p, q) = K(p, q)$.

Suppose, on the contrary, there exists a point $r \in K(p, q)$ such that $r \notin C(p, q)$. Then there exists a connected subset $S \subset X$ such that $p, q \in S$, but $r \notin S$.

For every $x \in S$ let U_x be an open connected neighbourhood of x such that $r \notin \bar{U}_x$.

Then $\{U_x\}_{x \in S}$ is an open covering of the connected set S, hence there exists a simple chain $U_{x_1}, \ldots, U_{x_n}$ from p to q.

The union of the members of that chain is connected, contains p and q, but the closure does not contain r. Hence $r \notin K(p, q)$.

Remark. Note that all five properties studied above are satisfied in orderable connected spaces. For an investigation of other properties of orderable connected spaces we refer to [2] and [4].

REFERENCES

[1] A.E. Brouwer, On a property of tree-like spaces, rapport nr. 19 (1970), *Wiskundig Seminarium der Vrije Universiteit,* Amsterdam.

[2] A.E. Brouwer – H. Kok, On some properties of orderable connected spaces, rapport nr. 21, (1971), *Wiskundig Seminarium der Vrije Universiteit,* Amsterdam.

[3] G.L. Gurin, On tree-like spaces, *Vestnik Moskov. Univ. Ser. I Mat. Meh.,* (1969), no. 1, 9-12, (Russian).

[4] H. Kok, On conditions equivalent to the orderability of a connected space, *Nieuw Archief voor Wiskunde,* (3), XVIII, 250-270. (1970).

[5] E.W. Miller, Concerning biconnected sets, *Fund. Math.,* 29 (1937), 123-133.

[6] V.V. Proizvolov, On peripherally bicompact tree-like spaces, *Dokl. Akad. Nauk SSSR,* 189 (1969), no. 4 = *Soviet Math. Dokl.* 10 (1969), no. 6, 1491-1493.

[7] G.T. Whyburn, Cut points in general topological spaces, *Proc. Nat. Acad. of Sciences,* 61 (1968), 380-387.

COLLOQUIA MATHEMATICA SOCIETATIS JÁNOS BOLYAI
8. TOPICS IN TOPOLOGY, KESZTHELY (HUNGARY), 1972.

REMARK ON PRODUCT OF PROXIMITIES

W. KULPA

1. It is known that $\beta(X \times Y) \neq \beta X \times \beta Y$, in general. In this note we shall shown that if X and Y are not pseudocompact spaces then there exist at least continuum compactifications of $X \times Y$ finer than $\beta X \times \beta Y$, where β means the Čech – Stone compactification. This result follows from an analogous theorem for proximities.

The idea of this note proceeds from a paper of E.E. Reed and W.J. Thron [2].

2. Let $\mathcal{U}$ be a uniformity on a set X. The uniformity $\mathcal{U}$ induces a proximity relation $\delta(\mathcal{U})$; for every $A, B \subset X$, $A\delta(\mathcal{U})B$ iff for every $P \in \mathcal{U}$ $\operatorname{st}(A, P) \cap B \neq \phi$. Let $P\mathcal{U} = \{P \in \mathcal{U} : P$ has a finite refinement belonging to $\mathcal{U}\}$. A family $p\mathcal{U}$ is the greatest totally bounded uniformity contained in $\mathcal{U}$ and for every $A, B \subset X$.

$$A\delta(\mathcal{U})B \quad \text{iff} \quad A\delta(p\mathcal{U})B .$$

If δ is a proximity relation, than there exists a unique totally bounded uniformity $\mathcal{U}$ such that $\delta(\mathcal{U}) = \delta$. Thus there exists a $1-1$ correspondence between the proximity relations and totally bounded uniformities.

Let us put $[\mathscr{U}] = \{\mathscr{V} : p\mathscr{V} = p\mathscr{U}\}$ and let us write $b[\mathscr{U}] \geq \mathfrak{m}$ iff there exists a uniformity $\mathscr{U}' \in [\mathscr{U}]$ which has no base consisting of coverings of cardinality less than $\mathfrak{m}$. If $b[\mathscr{U}] \geq \mathfrak{m}$ then there exists a uniformly discrete set $D \subset X$ in sense of $\mathscr{U}'$ with card $D = \mathfrak{m}$, i.e. such that there exists $P \in \mathscr{U}'$ such that for every $d, d' \in D$ if $d \neq d'$ then $\mathrm{st}(d, P) \cap \mathrm{st}(d', P) = \phi$ (see [1], pp. 24-26).

Theorem. *Let $\mathscr{U}$ and $\mathscr{V}$ be uniformities on sets X and Y, respectively. If $b[\mathscr{U}] \geq \mathfrak{m}$ and $b[\mathscr{V}] \geq \mathfrak{m}$, $\mathfrak{m} \geq \chi_0$, then there exist $2^{\mathfrak{m}}$ uniformities $\mathscr{U}_a$ belonging to $[\mathscr{U}]$ and there exist $2^{\mathfrak{m}}$ uniformities $\mathscr{V}_a$ belonging to $[\mathscr{V}]$ such that if $a \neq a'$ then*

$$p(\mathscr{U}_a \times \mathscr{V}_a) \neq p(\mathscr{U}_{a'} \times \mathscr{V}_{a'}) .$$

Proof. Let $\mathscr{U}' \in [\mathscr{U}]$ be a uniformity such that there exists a uniformly discrete set D with card $D = \mathfrak{m}$. Let $\{D_a : a \in M\}$ with card $M = 2^{\mathfrak{m}}$ be a family of subsets of D, each of cardinality $\mathfrak{m}$ and such that the symmetric difference of distinct elements of the family has cardinality $\mathfrak{m}$. The existence of such family follows from the fact that there exist $2^{\mathfrak{m}}$ subsets of D hence there exist $2^{\mathfrak{m}}$ subsets of $D \times D$ such that the symmetric difference of distinct sets has cardinality $\mathfrak{m}$, but card $D \times D = \mathrm{card}\, D$.

Let C be a uniformly discrete set in sense of $\mathscr{V}' \in [\mathscr{V}]$ and let card $C = \mathfrak{m}$. Let $\varphi: D \to C$ be $1-1$ map. Put $C_a = \varphi(D_a)$.

For every $P \in \mathscr{U}'$, $Q \in \mathscr{V}'$ let $P^a = \{\mathrm{st}(D_a, P)\} \cup \{u \in P : u \cap \cap D_a = \phi\}$, and $Q^a = \{\mathrm{st}(C_a, Q)\} \cup \{v \in Q : v \cap C_a = \phi\}$. Notice, that if $P_1 <^* P$ then $P_1^a <^* P^a$ (symbols $<$ and $<^*$ stand for refinement and star-refinement).

Indeed, let $u' \in P_1^a$. Then $u' = \mathrm{st}(D_a, P_1)$ or $u' \in P_1$. If $u' = \mathrm{st}(D_a, P_1)$, then $\mathrm{st}(u', P_1^a) = \mathrm{st}[\mathrm{st}(D_a, P_1), P_1^a] \subset \bigcup \{\mathrm{st}(u, P_1) : \mathrm{st}(u, P_1) \cap D_a \neq \phi, u \in P_1\} \subset \mathrm{st}(D_a, P) \in P^a$. If $u' \in P_1^a$ and $u' \in P_1$, then $u' \cap \mathrm{st}(D_a, P_1) \neq \phi$ or $u' \cap \mathrm{st}(D_a, P_1) = \phi$; if $u' \cap \mathrm{st}(D_a, P_1) \neq \phi$, then $\mathrm{st}(u', P_1) \cap D_a \neq \phi$ and $\mathrm{st}(u', P_1^a) = \mathrm{st}(D_a, P_1) \cup \mathrm{st}(u', P_1) \subset \subset \mathrm{st}(D_a, P)$; if $u' \cap \mathrm{st}(D_a, P_1) = \phi$, then $\mathrm{st}(u', P_1^a) = \mathrm{st}(u', P_1) \subset v \in \in P < P^a$.

There exist coverings $P_0 \in \mathscr{U}'$, $Q_0 \in \mathscr{V}'$ such that if $d, d \in D$,

$c, c' \in C$, $d \neq d'$ and $c \neq c'$ then $\mathrm{st}(d, P_0) \cap \mathrm{st}(d', P_0) = \phi$ and $\mathrm{st}(c, Q_0) \cap \mathrm{st}(c', Q_0) = \phi$. Choose sequences $P_0^* > P_1^* > P_2^* > \ldots, P_i \in \in \mathscr{U}'$, $Q_0^* > Q_1^* > Q_2^* > \ldots, Q_i \in \mathscr{V}'$.

Let us consider uniformities $\mathscr{U}_a$, $\mathscr{V}_a$, $a \in M$, generated by the subbases

$$p\mathscr{U} \cup \{P_i^a \colon i = 0, 1, 2, \ldots\}$$

$$p\mathscr{V} \cup \{Q_i^a \colon i = 0, 1, 2, \ldots\}.$$

Let $a \neq a'$. Without loss of generality we may assume that card $(D_a - D_{a'}) = \mathfrak{m}$. Let us consider the sets

$$F_1 = \{(d, \varphi(d)) \colon d \in D_a - D_{a'}\}$$

$$F_2 = \{(d', \varphi(d)) \colon d' \neq d,\ d, d' \in D_a - D_{a'}\}.$$

Let δ_a and $\delta_{a'}$ mean the proximity relations induced by $\mathscr{U}_a \times \mathscr{V}_a$ and $\mathscr{U}_{a'} \times \mathscr{V}_{a'}$, respectively.

Notice that $F_1 \not\delta_{a'} F_2$, because $\mathrm{st}(F_1, P_0^{a'} \times Q_0^{a'}) \cap F_2 = \phi$. But $F_1\ \delta_a\ F_2$ holds. In fact, since every basic covering $P' \in \mathscr{U}_a$ is of the form $P' = P_i^a \wedge Q_1 \wedge \ldots \wedge Q_n$, $Q_i \in p\mathscr{U}$ and card $Q_i < \aleph_0$, hence for every $P \in \mathscr{U}_a$ there exists u_0 which contains $\mathfrak{m}$ elements of $D_a - D_{a'}$, and for every $Q \in \mathscr{V}_a$ there exists v_0 which contains $\mathfrak{m}$ elements of $\varphi(u_0 \cap D_a - D_{a'}) \subset C_a$. Let $d, d' \in \varphi^{-1}[v_0 \cap \varphi(u_0 \cap D_a - D_{a'})]$. Then $(d, \varphi(d))$ and $(d', \varphi(d))$ belong to $u_0 \times v_0$. Thus $\mathrm{st}(F_1, P \times Q) \cap F_2 \neq \neq \phi$, for each $P \in \mathscr{U}_a$, $Q \in \mathscr{V}_a$. Hence $F_1\ \delta_a\ F_2$.

To show that the uniformities $\mathscr{U}_a$ are distinct, let us put $\mathscr{U}' = \mathscr{V}'$, $D = C$, and $\varphi(x) = x$, and $P_i = Q_i$. Then we have $p(\mathscr{U}_a \times \mathscr{V}_a) \neq \neq p(\mathscr{U}_{a'} \times \mathscr{U}_{a'})$, for $a \neq a'$. For the same reasons the uniformities $\mathscr{V}_a$ are distinct.

Corollary. *If the completely regular spaces X and Y are not pseudocompact then there exist at least continuum compactifications of the product $X \times Y$ finer then $\beta X \times \beta Y$.*

Proof. If X and Y are not pseudocompact then the finest uniformities $\mathscr{U}$ on X and $\mathscr{V}$ on Y compatible with the topologies are not totally bounded. This means that $b[\mathscr{U}] \geqslant \aleph_0$ and $b[\mathscr{V}] \geqslant \aleph_0$. The completion of $p\mathscr{U}$ leads to βX, the completion of $p\mathscr{V}$ to βY and the completion of $p\mathscr{U} \times p\mathscr{V}$ to $\beta X \times \beta Y$. Applying the Theorem, there exists at least $2^{\aleph_0}$ uniformities $\mathscr{U}_a \times \mathscr{V}_a$ on $X \times Y$ such that all the uniformities $p(\mathscr{U}_a \times \mathscr{V}_a)$ are different and finer that $p\mathscr{U} \times p\mathscr{V}$. The compactifications of $X \times Y$ corresponding to $p(\mathscr{U}_a \times \mathscr{V}_a)$ are different and each majorizes $\beta X \times \beta Y$.

REFERENCES

[1] J.R. Isbell, *Uniform spaces,* Providence 1964.

[2] E.E. Reed – W.J. Thron, *m*-bounded uniformites between two given uniformities, *Transactions of the American Mathematical Society,* 141 (1969), 71-77.

COLLOQUIA MATHEMATICA SOCIETATIS JÁNOS BOLYAI
8. TOPICS IN TOPOLOGY, KESZTHELY (HUNGARY), 1972.

DEVELOPMENT OF THE RESEARCH ON INDECOMPOSABLE CONTINUA

K. KURATOWSKI

Definition. A (metric) continuum is called indecomposable if it cannot be represented as the union of two proper subcontinua.

Historical development. Theorems by Schönflies on the common boundary of three or more domains in the plane (1908). Counterexamples given by L.E.J. Brouwer using indecomposable continua.

Further development of the theory of indecomposable continua due to Janiszewski, Mazurkiewicz, Knaster (who defined the simplest indecomposable continuum) and others. Final solution of the problem of the common boundary of three domains in the plane (Knaster – Kuratowski). Other definitions of indecomposable continua (Wada, van Dantzig, Nadler).

The composants of indecomposable continua. Problems of their accessibility. Classical theorems by Mazurkiewicz and their generalizations. Recent results by Krasinkiewicz, H. Cook and others. The topological Strong Transitivity for sets of composants.

Chainable continua, pseudo-arcs (Bing, Moise).

Problems involving indecomposable continua; in particular, the fixed-point property of plane continua (Bell, Sieklucki, Hagopian, Krasinkiewicz), non-trivial homogeneous plane continua (Bing, Moise, F.B. Jones, Fugate), non-trivial continuum homeomorphic to each of its subcontinua (G.W. Henderson).

Hereditarily indecomposable continua.

COLLOQUIA MATHEMATICA SOCIETATIS JÁNOS BOLYAI
8. TOPICS IN TOPOLOGY, KESZTHELY (HUNGARY), 1972.

A UNIVERSAL PROPERTY OF THE LATTICE OF ALL TOPOLOGIES (BOOLEAN-VALUED TOPOLOGIES)

R. LIEDL – K. KUHNERT

INTRODUCTION

In some recent papers a generalization of set theory and by virtue of this a generalization of point set topology has been carried out (cf. D. Klaua [1], C.L. Chang [1], L.A. Zadeh [1]). Here we want to proceed in the following way:

It is a well-known fact that the power set $\mathfrak{P}(X)$ of an arbitrary set X is isomorphic as a boolean algebra to the set $\{0, 1\}^X$ of all characteristic functions from X into the simple boolean algebra $\{0, 1\}$, where boolean operations are defined pointwise. In order to generalize this notion we replace the complete, simple boolean algebra $\{0, 1\}$ by the complete boolean algebra formed by the power set $\mathfrak{P}(M)$ of an arbitrary set M. We define a "$\mathfrak{P}(M)$-valued" subset of X as a function from X into $\mathfrak{P}(M)$. If $M = \{\phi\}$ and therefore $\mathfrak{P}(M)$ is isomorphic to $\{0, 1\}$, we obtain the usual case, where the subsets of X are identified with the characteristic functions. The general case is made clearer by the identification of $\mathfrak{P}(M)^X$ with $\mathfrak{P}(X \times M)$ via the boolean isomorphism

$$\phi: f \to \{(x, m) \mid x \in X, m \in f(x)\} \quad \text{for each} \quad f \in \mathfrak{P}(M)^X$$

(Exponential law!).

Now we shall construct a complete boolean homomorphism χ from our generalized power set $\mathfrak{P}(M)^X$ into the ordinary power set $\{0, 1\}^X$. For this purpose we make use of a complete boolean homomorphism χ' from $\mathfrak{P}(M)$ into $\{0, 1\}$. Then $\chi: \mathfrak{P}(M)^X \to \{0, 1\}^X$ is defined by $\chi(f) = \chi' \circ f$ for every $f \in \mathfrak{P}(M)^X$. The complete boolean homomorphism $\chi': \mathfrak{P}(M) \to \{0, 1\}$ is determined by the ultrafilter $\mathfrak{U} = = \chi'^{-1}(1) \subset \mathfrak{P}(M)$. From the completeness of χ' we conclude that

$$\chi'\Big(\bigcap_{U \in \mathfrak{U}} U\Big) = \bigcap_{U \in \mathfrak{U}} \chi'(U) = \bigcap_{U \in \mathfrak{U}} 1 = 1$$

and consequently $\bigcap_{U \in \mathfrak{U}} U \in \mathfrak{U}$. Thus $\bigcap_{U \in \mathfrak{U}} U = \{m_0\}$, where $m_0 \in M$. In this manner each element $m_0 \in M$ determines a complete boolean homomorphism χ from $\mathfrak{P}(M)^X$ into $\{0, 1\}^X$.

Replacing $\mathfrak{P}(M)^X$ by $\mathfrak{P}(X \times M)$ the relative homomorphism $\chi \circ \phi^{-1}: \mathfrak{P}(X \times M) \to \{0, 1\}^X$ transforms every generalized subset $A \in \in \mathfrak{P}(X \times M)$ of X into an ordinary subset $\chi \circ \phi^{-1}(A) \in \mathfrak{P}(X)$ of X by drawing the horizontal section of A at $m_0 \in M$. That means $\chi \circ \phi^{-1}(A) = \{x \mid (x, m_0) \in A\}$.

We call $\mathfrak{P}(X \times M)$ the $\mathfrak{P}(M)$-power set of X and call the boolean homomorphism $\chi: \mathfrak{P}(X \times M) \to \{0, 1\}^X$ the m_0-realization of $\mathfrak{P}(X \times M)$. If $B \in \mathfrak{P}(X \times M)$, then $\{x \mid (x, m) \in B\}$ is termed the m_0-realization of B.

For the following a certain duality is of importance. There is an obvious boolean isomorphism from $\mathfrak{P}(X \times M)$ onto $\mathfrak{P}(M \times X)$. Thus, by changing the roles of X and M, the $\mathfrak{P}(M)$-power set of X with m_0-realizations for every $m_0 \in M$ is associated with the $\mathfrak{P}(X)$-power set of M with x_0-realizations for every $x_0 \in X$.

§1. THE TOPOLOGY $\mathfrak{T}_0^+(x_0, X)$ ON THE LATTICE OF ALL TOPOLOGIES ON X

Let X and M be sets as before.

Definition 1. *A* $\mathfrak{P}(M)$*-topology on* X *is a subset* $\mathfrak{T} \subset \mathfrak{P}(M)^X$ *which is closed under arbitrary suprema and finite infima.*

Via the isomorphism ϕ an ordinary topology $\mathfrak{T}^\#$ on $\mathfrak{P}(X \times M)$ belongs to each $\mathfrak{P}(M)$-topology $\mathfrak{T}$ on X and vice versa. Therefore, by duality, to a $\mathfrak{P}(M)$-topology $\mathfrak{T}$ on X there belongs a $\mathfrak{P}(X)$-topology $\tilde{\mathfrak{T}}$ on M and vice versa.

If $\mathfrak{T}$ is a $\mathfrak{P}(M)$-topology on X and $m_0 \in M$, then

$$\mathfrak{T}_{m_0} = \{T_{m_0} \mid T_{m_0} = \{x \mid (x, m_0) \in T\},\ T \in \mathfrak{T}\}.$$

is an ordinary topology on X, which we call the m_0-realization of $\mathfrak{T}$.

If $\{\mathfrak{T}^m\}_{m \in M}$ is an arbitrary family of topologies on X, we can construct a $\mathfrak{P}(M)$-topology $\mathfrak{T}$ on X such that the m_0-realization $\mathfrak{T}_{m_0}$ of $\mathfrak{T}$ is equal to the given topology $\mathfrak{T}^{m_0}$ for every $m_0 \in M$.

One way of such a construction is the following: Let be

$$\mathfrak{B} = \{T_A \mid T_A = \{(x, m) \mid A \text{ is a } \mathfrak{T}^m\text{-neighbourhood of } x\},\ A \subset X\};$$

then $T_{A \cap B} = T_A \cap T_B$ for every $A, B \subset X$ and therefore $\mathfrak{B}$ is a basis for a topology $\mathfrak{T}^\#$ on $X \times M$, the m_0-realization $\mathfrak{T}_{m_0}$ of which is equal to $\mathfrak{T}^{m_0}$ for each $m_0 \in M$.

In this manner we introduce the topology $\mathfrak{T}^\#$ on $X \times \tau(X)$, where $\tau(X)$ denotes the lattice of all topologies on X, such that the $\mathfrak{T}^*$-realization of $\mathfrak{T}^\#$ is equal to $\mathfrak{T}^*$ for each $\mathfrak{T}^* \in \tau(X)$.

But we are interested in the x_0-realizations $\mathfrak{T}_{x_0}$ of $\mathfrak{T}^\#$ for each $x_0 \in X$, which are ordinary topologies on $\tau(X)$. Obviously the topology $\mathfrak{T}_{x_0}$ on $\tau(X)$ is independent of the choice of $x_0 \in X$ up to a homeomorphism.

If $x_0 \in X$ and $A \subset X$, then with

$$T_A = \{(x, \mathfrak{T}^*) : A \text{ is a } \mathfrak{T}^*\text{-neighbourhood of } x\}$$

the set

$$A^* = \{\mathfrak{T}^* \mid A \text{ is a } \mathfrak{T}^*\text{-neighbourhood of } x_0\}$$

is a $\mathfrak{T}_{x_0}$-open subset of $\tau(X)$, because A^* is the x_0-realization of T_A. The set $\{A^* \mid A \subset X\}$ forms a basis $\mathfrak{B}^*$ of the topology $\mathfrak{T}_{x_0}$ on $\tau(X)$. With respect to a later aim we denote the topology $\mathfrak{T}_{x_0}$ on $\tau(X)$ by $\mathfrak{T}_0^+(x_0, X)$.

We can characterize the topology $\mathfrak{T}_0^+(x_0, X)$ on $\tau(X)$ in terms of convergence classes in the sense of Kelley.

Theorem 1. *A net $\mathfrak{N}: D \to \tau(X)$ of topologies on X converges to a topology $\mathfrak{T}^*$ on X with respect to $\mathfrak{T}_0^+(x_0, X)$ iff for every $\mathfrak{T}^*$-neighbourhood A of x_0 there exists a $d_0 \in D$ such that for every $d \geqslant d_0$ the subset A of X is a $\mathfrak{N}(d)$-neighbourhood of x_0, too.*

This immediately follows from the characterization of $\mathfrak{T}_0^+(x_0, X)$ by the basis $\mathfrak{B}$.

The next question about this topology $\mathfrak{T}_0^+(x_0, X)$ on $\tau(X)$ is: How does it look? The following representation theorem gives a certain insight.

Representation theorem for $\mathfrak{T}_0^+(x_0, X)$:

If $(Y, \mathfrak{T})$ is any topological space, then there exists a set X and a point $x_0 \in X$ such that $(Y, \mathfrak{T})$ is homeomorphic to some subspace of $(\tau(X), \mathfrak{T}_0^+(x_0, X))$.

Proof. Set $X = Y \cup \{\omega\}$, where $\omega \notin Y$.

Furthermore choose $x_0 = \omega$. We have to construct an embedding $\psi: Y \to \tau(X)$ and for this purpose define $\psi(y) = \mathfrak{T}_y = \{A \cup \{\omega\} \mid y \in A \in \mathfrak{T}\} \cup \{\{y\}\} \cup \{\phi\}$.

Obviously ψ is injective. Next we show that ψ is continuous. Suppose $A \subset X$ and thus $A^* \in \mathfrak{B}^*$, then $\psi^{-1}(A^*) = \{y \mid \mathfrak{T}_y \in A^*\} = \{y \mid A$ is a $\mathfrak{T}_y$-neighbourhood of $x_0 = \omega\} = \{y \mid \exists B, y \in B \in \mathfrak{T}, B \cup \{\omega\} \subset A\} = \mathfrak{T}$-interior of $A \cap Y$, if $\omega \in A$, and is empty otherwise. Hence ψ is continuous. If $B \in \mathfrak{T}$ we choose $A = B \cup \{\omega\}$ and get $\psi^{-1}(A^*) = B$. That completes the proof.

§2. TOPOLOGIES ON THE LATTICE OF ALL TOPOLOGIES ON X THAT ARE RELATED TO $\mathfrak{T}_0^+(x_0, X)$

We consider quasiuniform structures (cf. Á. Császár [1] and M.G. Murdeshwar – S.A. Naimpally [1]) inducing the topology $\mathfrak{T}_0^+(x_0, X)$ on $\tau(X)$ and related topologies. For this purpose we define a reflective and transitive relation $\leqslant_{(x,A)}$ on $\tau(X)$ for each $x \in X$ and each $A \subset X$:

$$\mathfrak{T}_1 \leqslant_{(x,A)} \mathfrak{T}_2 \Leftrightarrow (A \text{ is a } \mathfrak{T}_1\text{-neighbourhood of } x \Rightarrow A \text{ is a } \mathfrak{T}_2\text{-neighbourhood of } x).$$

Furthermore we define

$$\leqslant_x = \bigcap_{A \subset X} \leqslant_{(x,A)} \quad \text{for each } x \in X,$$

$$\leqslant_A = \bigcap_{x \in X} \leqslant_{(x,A)} \quad \text{for each } A \subset X \text{ and}$$

$$\leqslant = \bigcap_{\substack{x \in X \\ A \subset X}} \leqslant_{(x,A)}.$$

These relations are reflective and transitive on $\tau(X)$, too. We form subbases of quasiuniform structures on $\tau(X)$ that are families of relations of the preceeding types.

Definition. *Suppose that* $x_0 \in X$ *and* $A_0 \subset X$. *Then*

$$\mathfrak{B}_0^+(x_0, A_0, X) = \{\leqslant_{(x_0, A_0)}\}$$

$$\mathfrak{B}_0^+(x_0, X) = \{\leqslant_{(x_0, A)}\}_{A \subset X}$$

$$\mathfrak{B}_0^+(A_0, X) = \{\leqslant_{(x, A_0)}\}_{x \in X}$$

$$\mathfrak{B}_0^+(X) = \{\leqslant_{(x, A)}\}_{x \in X, A \subset X}$$

$$\mathfrak{B}_1^+(x_0, X) = \{\leqslant_{x_0}\}$$

$$\mathfrak{B}_1^+(A_0, X) = \{\leqslant_{A_0}\}$$

$$
\begin{aligned}
\mathfrak{B}^{+}_{1\pi}(X) &= \{\leqslant_x\}_{x \in X} \\
\mathfrak{B}^{+}_{1u}(X) &= \{\leqslant_A\}_{A \subset X} \\
\mathfrak{B}^{+}_{2}(X) &= \{\leqslant\}
\end{aligned}
$$

These are subbases for 9 different quasiuniform structures on $\tau(X)$, whose induced topologies are denoted by replacing the letter $\mathfrak{B}$ by $\mathfrak{T}$ in the preceeding definition. However, the definition of the relations $\leqslant_{(x,A)}$, $\leqslant_x$, $\leqslant_A$, $\leqslant$ immediately yields the following

Diagram 1.

$$
\begin{array}{ccccc}
\mathfrak{T}^{+}_{0}(x_0, A_0, X) & \hookrightarrow & \mathfrak{T}^{+}_{0}(A_0, X) & \hookrightarrow & \mathfrak{T}^{+}_{1}(A_0, X) \\
\downarrow & & \downarrow & & \downarrow \\
\mathfrak{T}^{+}_{0}(x_0, X) & \hookrightarrow & \mathfrak{T}^{+}_{0}(X) & \hookrightarrow & \mathfrak{T}^{+}_{1u}(X) \\
\downarrow & & \downarrow & & \downarrow \\
\mathfrak{T}^{+}_{1}(x_0, X) & \hookrightarrow & \mathfrak{T}^{+}_{1\pi}(X) & \hookrightarrow & \mathfrak{T}^{+}_{2}(X)
\end{array}
$$

The following derived topologies were introduced by Chr. Kollreider [1]. Replacing the considered relations on $\tau(X)$ by their conjugated (= inverse) ones, in the analogous way we get 9 new topologies on $\tau(X)$. These topologies are denoted by replacing the superscript $+$ by $-$. (Cf. conjugated quasiuniform structures in M.G. Murdeshwar – S.A. Naimpally.)

Finally, replacing the considered relations on $\tau(X)$ by the meet of them with their conjugated relations, we get another set of 9 topologies on $\tau(X)$, which are denoted by replacing the superscript $+$ by 0. Since the topologies with superscript 0 are finer than their related topologies with superscripts $+$ and $-$, respectively, the following diagrams hold

Diagram 2. Replace the superscript $+$ by $-$ in diagram 1.

Diagram 3. Replace the superscript $+$ by 0 in diagram 1.

Diagram 4.

$$\begin{array}{ccc} & & \text{diagram 1} \\ & & \downarrow \\ \text{diagram 2} & \hookrightarrow & \text{diagram 3} \end{array}$$

There each inclusion sign represents a scheme of 9 inclusions between the related topologies belonging to one another.

Using the following lemmas, the investigation of these 27 topologies is not difficult.

Lemma 1. *All relations we consider are reflective and transitive.*

The proof is trivial.

Lemma 2. *If a subbasis of a quasiuniform structure consists of reflexive and transitive relations only, then the vertical sections of these relations are open sets with respect to the induced topology.*

The proof follows classical lines.

Lemma 2 enables us to see that the two topologies $\mathfrak{T}_0^+(x_0, X)$ defined in § 1 and § 2 are really the same.

It is a simple but labourious work to provide bases for the other 26 topologies on $\tau(X)$ by lemma 2. However, we think it is attractive to characterize the 27 topologies in terms of convergence classes. For all these characterizations can be obtained from one by varying the quantifyers and implications as shown below. (Varying the implications is due to Chr. Kollreider [1].)

By $\mathfrak{A}(\mathfrak{T})$ we abbreviate: "*A net* $\mathfrak{N}: D \to \tau(X)$ *converges to a topology* $\mathfrak{T}^*$ *on* X *with respect to the topology* $\mathfrak{T}$ *on* $\tau(X)$."

By $\mathfrak{J}^+$ we abbreviate: "*A is a* $\mathfrak{T}$*-neighbourhood of* $x \Rightarrow A$ *is a* $\mathfrak{N}(d)$*-neighbourhood of* x."

By $\mathfrak{J}^-$ we abbreviate: "*A is a* $\mathfrak{T}$*-neighbourhood of* $x \Leftarrow A$ *is a* $\mathfrak{N}(d)$*-neighbourhood of* x."

By $\mathfrak{J}^0$ we abbreviate: *"A is a $\mathfrak{T}$-neighbourhood of $x \Leftrightarrow A$ is a $\mathfrak{N}(d)$-neighbourhood of x."*

Theorem 2. *If $A_0 \subset X$ and $x_0 \in X$, then the 27 topologies on $\tau(X)$ are characterized by*

$\mathfrak{A}(\mathfrak{T}_0^{+-0}(x_0, A_0, X))$ iff $\exists d_0 \in D,\ \forall d \geqslant d_0,\ x = x_0,\ A = A_0\ \ \mathfrak{J}^{+-0}$

$\mathfrak{A}(\mathfrak{T}_0^{+-0}(x_0, X))$ iff $\forall A \subset X\ \exists d_0 \in D,\ \forall d \geqslant d_0,\ x = x_0\ \ \mathfrak{J}^{+-0}$

$\mathfrak{A}(\mathfrak{T}_0^{+-0}(A_0, X))$ iff $\forall x \in X\ \exists d_0 \in D,\ \forall d \geqslant d_0,\ A = A_0\ \ \mathfrak{J}^{+-0}$

$\mathfrak{A}(\mathfrak{T}_0^{+-0}(X))$ iff $\forall x \in X,\ \forall A \subset X,\ \exists d_0 \in D,\ \forall d \geqslant d_0\ \ \mathfrak{J}^{+-0}$

$\mathfrak{A}(\mathfrak{T}_1^{+-0}(x_0, X))$ iff $\exists d_0 \in D,\ \forall d \geqslant d_0\ \forall A \subset X,\ x = x_0\ \ \mathfrak{J}^{+-0}$

$\mathfrak{A}(\mathfrak{T}_1^{+-0}(A_0, X))$ iff $\exists d_0 \in D,\ \forall d \geqslant d_0\ \forall x \in X,\ A = A_0\ \ \mathfrak{J}^{+-0}$

$\mathfrak{A}(\mathfrak{T}_{1\pi}^{+-0}(X))$ iff $\forall x \in X\ \exists d_0 \in D,\ \forall d \geqslant d_0\ \forall A \subset X\ \ \mathfrak{J}^{+-0}$

$\mathfrak{A}(\mathfrak{T}_{1u}^{+-0}(X))$ iff $\forall A \subset X\ \exists d_0 \in D,\ \forall d \geqslant d_0\ \forall x \in X\ \ \mathfrak{J}^{+-0}$

$\mathfrak{A}(\mathfrak{T}_2^{+-0}(X))$ iff $\exists d_0 \in D,\ \forall d \geqslant d_0\ \forall x \in X\ \forall A \subset X\ \ \mathfrak{J}^{+-0}$

§3. REPRESENTATION THEOREMS*

Observing the logical relations in the characterization of the 27 topologies on $\tau(X)$ and the fact that $\mathfrak{T}_0^+(x_0, X)$ is one of them, we may ask if it is possible to obtain representation theorems for the other 26 topologies, too.

The second named author's contribution to this paper consists in the proofs of the representation theorems for the topologies $\mathfrak{T}_0^+(x_0, A_0, X)$; $\mathfrak{T}_0^-(x_0, A_0, X)$, $\mathfrak{T}_0^-(x_0, X)$, $\mathfrak{T}_0^-(A_0, X)$, $\mathfrak{T}_1^-(x_0, X)$, $\mathfrak{T}_1^-(A_0, X)$, $\mathfrak{T}_0^-(X)$, $\mathfrak{T}_{1\pi}^-(X)$, $\mathfrak{T}_{1u}^-(X)$, $\mathfrak{T}_2^-(X)$; $\mathfrak{T}_0^0(x_0, A_0, X)$, $\mathfrak{T}_0^0(A_0, X)$, $\mathfrak{T}_1^0(x_0, X)$, $\mathfrak{T}_1^0(A_0, X)$.

B. Roider [1], [2], has proved the representation theorems for the topologies $\mathfrak{T}_0^+(A_0, X)$ and $\mathfrak{T}_{1\pi}^+(X)$.

*By the "universal property of the lattice of all topologies" we denote, in an intuitive sense, the fact that the representation theorems hold.

The representation theorem for $\mathfrak{T}_2^0(X)$ is trivial because $\mathfrak{T}_2^0(X)$ is discrete.

Open problems are the representation theorems for the topologies $\mathfrak{T}_0^0(x_0, X)$, $\mathfrak{T}_{1\pi}^0(X)$, $\mathfrak{T}_{1u}^0(X)$, $\mathfrak{T}_0^0(X)$.

It is straightforward to prove that some of the 27 topologies have certain hereditary properties. For instance all topologies with superscript 0 are underlying topological spaces of K-structures in the sense of J.-P. Olivier [1]. Therefore only topologies with the same hereditary properties can be represented as subspaces of such topologies on $\tau(X)$.

The representation theorems for the topologies without special hereditary properties show that every topological space is quasiuniformizable. For instance the consequences of the representation theorem for the topology $\mathfrak{T}_0^+(x_0, X)$ lead to the Pervin quasiuniformity (cf. W.J. Pervin [1]).

What is rather difficult in proving the representation theorems is to find the embedding in question.

However, verifying that the function we have constructed really is an embedding is a straightforward matter.

Representation theorem for $\mathfrak{T}_0^+(x_0, A_0, X)$: $\mathfrak{T}_0^+(x_0, A_0, X)$ *has the following hereditary property: There is not more than one non-trivial open set in it. If* $(Y, \mathfrak{T})$ *is a topological space with the same hereditary property, then there exists a set* X *with a subset* $A_0 \subset X$ *and a point* $x_0 \in X$ *such that* $(Y, \mathfrak{T})$ *is homeomorphic to some subspace of* $(\tau(X), \mathfrak{T}_0^+(x_0, A_0, X))$.

Proof. We have to distinguish between two cases according as $(Y, \mathfrak{T})$ has or has not a non-trivial open set.

Case 1. $\mathfrak{T} = \{\phi, B, Y\}$ with $B \neq \phi$, $B \neq Y$. Set $X = Y$ and $A_0 \doteq B$ and let x_0 be any element of B. The wished embedding $\psi: Y \to \tau(X)$ is defined by $\psi(y) = \{\phi, \{y, x_0\}, X\}$.

Case 2. $\mathfrak{T} = \{\phi, Y\}$. Set $X = Y$ and $A_0 = Y$ and choose an arbitrary $x_0 \in X$. The embedding ψ is defined by $\psi(y) = \{\phi, \{y\}, X\}$.

Representation theorem for $\mathfrak{T}_0^-(x_0, A_0, X)$: *$\mathfrak{T}_0^-(x_0, A_0, X)$ has the same hereditary property as $\mathfrak{T}_0^+(x_0, A_0, X)$. Thus the analogous theorem holds.*

Proof.

Case 1. $\mathfrak{T} = \{\phi, B, Y\}$ with $B \neq \phi$, $B \neq Y$. Set $X = Y$, $A_0 = B^c$ and let x_0 be any element of B^c. Then the embedding ψ is defined by $\psi(y) = \{\phi, \{y\}, B^c \cup \{y\}, X\}$.

Case 2. $\mathfrak{T} = \{\phi, Y\}$. Set $X = Y$, $A_0 = \phi$ and choose an arbitrary $x_0 \in Y$. Then the embedding ψ is defined by $\psi(y) = \{\phi, \{y\}, X\}$.

Representation theorem for $\mathfrak{T}_0^0(x_0, A_0, X)$: *$\mathfrak{T}_0^0(x_0, A_0, X)$ has the following hereditary property: There are not more than two non-trivial open sets in it and all open sets are closed.*

If $(Y, \mathfrak{T})$ is a topological space with the same hereditary property, then there exists a set X with a subset $A_0 \subset X$ and a point $x_0 \in X$ such that $(Y, \mathfrak{T})$ is homeomorphic to some subspace of $(\tau(X), \mathfrak{T}_0^0(x_0, A_0, X))$.

Proof.

Case 1. $\mathfrak{T} = \{\phi, B, B^c, Y\}$ with $B \neq \phi$, $B \neq Y$. Set $X = Y$ and $A_0 = B$ and let x_0 be any element of B. Then the embedding ψ is defined by $\psi(y) = \{\phi, \{y, x_0\}, X\}$.

Case 2. $\mathfrak{T} = \{\phi, Y\}$. Set $X = Y$, $A_0 = Y$ and choose an arbitrary $x_0 \in X$. Then the embedding ψ is defined by $\psi(y) = \{\phi, \{y\}, X\}$.

Representation theorem for $\mathfrak{T}_0^-(x_0, X)$: *Let $(Y, \mathfrak{T})$ be an arbitrary topological space. Then there exists a set X with a point $x_0 \in X$ such that $(Y, \mathfrak{T})$ is homeomorphic to some subspace of $(\tau(X), \mathfrak{T}_0^-(x_0, X))$.*

Proof. Without restriction of generality we suppose that $a \notin b$ for every $a, b \in Y$ and set $X = \mathfrak{T} \cup Y \cup \{\omega\}$, where $\omega \notin \mathfrak{T} \cup Y$, and $a \notin \omega$ for every $a \in Y$. Furthermore we set $x_0 = \omega$ and define the embedding ψ by $\psi(y) = \{B^c \cup \{x_0\} \mid B \subset X, y \notin \bigcup_{V \in B} V\} \cup \{\phi, \{y\}, X\} \cup \cup \{B^c \cup \{x_0, y\} \mid B \subset X,\ Y \notin \bigcup_{V \in B} V\}$.

Representation theorem for $\mathfrak{T}_0^+(A_0, X)$ (cf. B. Roider [1]):

Let $(Y, \mathfrak{T})$ be an arbitrary topological space. Then there exists a set X with a subset A_0 of X such that $(Y, \mathfrak{T})$ is homeomorphic to some subspace of $(\tau(X), \mathfrak{T}_0^+(A_0, X))$.

Proof. We set $X = \mathfrak{P}(Y) \cup \{\{Y\}\}$ and $A_0 = \mathfrak{T}$. Then the embedding ψ is defined by $\psi(y) = \{\phi, \{A \mid A \in \mathfrak{T}, y \in A\}, \{\{y\}\}, \{\{y\}\} \cup \cup \{A \mid A \in \mathfrak{T}, y \in A\}, X\}$.

Representation theorem for $\mathfrak{T}_0^-(A_0, X)$: *Let $(Y, \mathfrak{T})$ be an arbitrary topological space. Then there exists a set X with a subset A_0 of X such that $(Y, \mathfrak{T})$ is homeomorphic to some subspace of $(\tau(X), \mathfrak{T}_0^-(A_0, X))$.*

Proof. Without loss of generality we suppose that $a \notin b$ for every $a, b \in Y$ and set $X = \mathfrak{T} \cup Y$ and $A_0 = \{B \mid B \in \mathfrak{T}, B \neq Y\} \cup Y$. Then the embedding ψ is defined by $\psi(y) = \{\phi, \{A \mid A \in X, y \notin A\}, \{y\}, X\}$.

Representation theorem for $\mathfrak{T}_0^0(A_0, X)$: *$(\tau(X), \mathfrak{T}_0^0(A_0, X))$ is a zero-dimensional space.*

*If $(Y, \mathfrak{T})$ is any zero dimensional space, then there exists a set X with a subset A_0 of X such that $(Y, \mathfrak{T})$ is homeomorphic to some subspace of $(\tau(X), \mathfrak{T}_0^0(A_0, X))$.**

Proof. Without loss of generality we suppose that $a \notin b$ for every $a, b \in Y$ and assume that a clopen base $\mathfrak{B}$ of $\mathfrak{T}$ contains Y as an element. We set $X = \mathfrak{B} \cup Y$, $A_0 = \{B \mid B \in \mathfrak{B}, B \neq Y\} \cup Y$ and define the embedding ψ by $\psi(y) = \{\phi, \{A \mid A \in X, y \notin A\}, \{y\}, X\}$.

Representation theorem for $\mathfrak{T}_0^+(X)$: *$(\tau(X), \mathfrak{T}_0^+(X))$ is a T_0 space.*

If $(Y, \mathfrak{T})$ is any T_0 space, then there exists a set X such that $(Y, \mathfrak{T})$ is homeomorphic to some subspace of $(\tau(X), \mathfrak{T}_0^+(X))$.

Proof. We set $X = Y$ and define the embedding ψ by $\psi(y) = \{B \mid y \in B \in \mathfrak{T}\} \cup \{\phi\}$.

*From this theorem it results that every zero dimensional space is uniformizable by the aid of the uniform subbasis $\{B \times B \cup B^c \times B^c \mid B \in \mathfrak{B}\}$, $\mathfrak{B}$ being any clopen basis of $\mathfrak{T}$. (Cf. J.-P. Olivier [1].)

Representation theorem for $\mathfrak{T}_0^-(X)$: *$(\tau(X), \mathfrak{T}_0^-(X))$ is a T_0 space.*

If $(Y, \mathfrak{T})$ is any T_0 space, then there exists a set X such that $(Y, \mathfrak{T})$ is homeomorphic to some subspace of $(\tau(X), \mathfrak{T}_0^-(X))$.

Proof. We set $X = \mathfrak{T}$ and define the embedding ψ by $\psi(y) = = \{\phi, X\} \cup \{B^c \mid y \notin \bigcup_{V \in B} V, B \subset X\}$.

Representation theorem for $\mathfrak{T}_1^+(x_0, X)$: *The topology $\mathfrak{T}_1^+(x_0, X)$ is saturated, that means that each point y of the space has a minimal neighbourhood $U_{\min}(y)$. The saturated spaces form a monocoreflexive subcategory of the category of all topological spaces* (cf. H. Herrlich [1], F. Lorrain [1]).

Obviously this property is hereditary.

If $(Y, \mathfrak{T})$ is any saturated topological space, then there exists a set X with a point $x_0 \in X$ such that $(Y, \mathfrak{T})$ is homeomorphic to some subspace of $(\tau(X), \mathfrak{T}_1^+(x_0, X))$.

Proof. We set $X = Y \cup \{\omega\}$, where $\omega \notin Y$. Furthermore we choose $x_0 = \omega$ and define the embedding ψ by $\psi(y) = \{\phi, \{y\}, U_{\min}(Y) \cup \{\omega\}, X\}$.

Representation theorem for $\mathfrak{T}_1^-(x_0, X)$: *$\mathfrak{T}_1^-(x_0, X)$ is a saturated topology such as $\mathfrak{T}_1^+(x_0, X)$. If $(Y, \mathfrak{T})$ is any saturated topological space, then there exists a set X with a point $x_0 \in X$ such that $(Y, \mathfrak{T})$ is homeomorphic to some subspace of $(\tau(X), \mathfrak{T}_1^-(x_0, X))$.*

Proof. We set $X = Y \cup \{\omega\}$, where $\omega \notin Y$, and set $x_0 = \omega$. Then the embedding ψ is defined by $\psi(y) = \{\phi, X - U_{\min}(y), X - \{y\}, X\}$.

Representation theorem for $\mathfrak{T}_1^0(x_0, X)$: *$(\tau(X), \mathfrak{T}_1^0(x_0, X))$ has the following hereditary property: It is a zero dimensional, saturated space.*

If $(Y, \mathfrak{T})$ is any zero dimensional, saturated topological space, then there exists a set X with a point $x_0 \in X$ such that $(Y, \mathfrak{T})$ is homeomorphic to some subspace of $(\tau(X), \mathfrak{T}_1^0(x_0, X))$.

Proof. We set $X = Y \cup \{\omega\}$, where $\omega \notin Y$, and $x_0 = \omega$ and define the embedding ψ by $\psi(y) = \{\phi, \{y\}, U_{\min}(y) \cup \{x_0\}, X\}$.

Representation theorem for $\mathfrak{T}_1^+(A_0, X)$: *$\mathfrak{T}_1^+(A_0, X)$ is a saturated topology such as $\mathfrak{T}_1^+(x_0, X)$.*

If $(Y, \mathfrak{T})$ is any saturated space, then there exists a set X with a subset A_0 such that $(Y, \mathfrak{T})$ is homeomorphic to some subspace of $(\tau(X), \mathfrak{T}_1^+(A_0, X))$.

Proof. We set $X = Y \cup \{\omega\}$, where $\omega \notin Y$, and choose $A_0 = Y$. Then the embedding ψ is defined by $\psi(y) = \{\phi, Y - U_{\min}(y), X - \{y\}, X\}$, where $U_{\min}(y)$ is the minimal neighbourhood of y with respect to the topology $\mathfrak{T}$.

Representation theorem for $\mathfrak{T}_1^-(A_0, X)$: *$\mathfrak{T}_1^-(A_0, X)$ is a saturated topology such as $\mathfrak{T}_1^+(A_0, X)$.*

If $(Y, \mathfrak{T})$ is any saturated topological space, then there exists a set X and a subset A_0 of X such that $(Y, \mathfrak{T})$ is homeomorphic to some subspace of $(\tau(X), \mathfrak{T}_1^-(A_0, X))$.

Proof. We set $X = Y \cup \{\omega\}$ with $\omega \notin Y$ and $A_0 = Y$, and define the embedding ψ by $\psi(y) = \{\phi, \{y\}, U_{\min}(y), X\}$.

Representation theorem for $\mathfrak{T}_1^0(A_0, X)$: *$(\tau(X), \mathfrak{T}_1^0(A_0, X))$ is a zero dimensional, saturated space such as $(\tau(X), \mathfrak{T}_1^0(x_0, X))$.*

If $(Y, \mathfrak{T})$ is any zero dimensional, saturated topological space, then there exists a set X and a subset A_0 of X such that $(Y, \mathfrak{T})$ is homeomorphic to some subspace of $(\tau(X), \mathfrak{T}_1^0(A_0, X))$.

Proof. We set $X = Y \cup \{\omega\}$ with $\omega \notin Y$ and $A_0 = Y$. Then the embedding ψ is defined by $\psi(y) = \{\phi, \{y\}, U_{\min}(y), X\}$.

Representation theorem for $\mathfrak{T}_{1\pi}^+(X)$ (cf. B. Roider [2]): *$(\tau(X), \mathfrak{T}_{1\pi}^+(X))$ is a T_0 space.*

If $(Y, \mathfrak{T})$ is any T_0 space, then there exists a set X such that $(Y, \mathfrak{T})$ is homeomorphic to some subspace of $(\tau(X), \mathfrak{T}_{1\pi}^+(X))$.

Proof. We set $X = \mathfrak{T}$ and define the embedding ψ by $\psi(y) = \{W \mid W \subset X, \forall A \in W\ y \in A, \forall A \in W\ \forall B \in X\ (B \supset A \Rightarrow B \in W)\} \cup \{\phi, X\}$.

Representation theorem for $\mathfrak{T}^-_{1\pi}(X)$: *$(\tau(X), \mathfrak{T}^-_{1\pi}(X))$ is a T_0 space.*

If $(Y, \mathfrak{T})$ is any T_0 space, then there exists a set X such that $(Y, \mathfrak{T})$ is homeomorphic to some subspace of $(\tau(X), \mathfrak{T}^-_{1\pi}(X))$.

Proof. We set $X = \mathfrak{T}$ and define the embedding ψ by $\psi(y) = \{W \mid W \subset \{B \mid B \in X, y \notin B\}\} \cup \{X\}$.

Representation theorem for $\mathfrak{T}^+_{1u}(X)$: *$(\tau(X), \mathfrak{T}^+_{1u}(X))$ is a T_0 space.*

If $(Y, \mathfrak{T})$ is any T_0 space, then there exists a set X such that $(Y, \mathfrak{T})$ is homeomorphic to some subspace of $(\tau(X), \mathfrak{T}^+_{1u}(X))$.

Proof. We set $X = Y$ and define the embedding ψ by $\psi(y) = \{B \mid y \in B \in \mathfrak{T}\} \cup \{\phi\}$.

Representation theorem for $\mathfrak{T}^-_{1u}(X)$: *$(\tau(X), \mathfrak{T}^-_{1u}(X))$ is a T_0 space.*

If $(Y, \mathfrak{T})$ is any T_0 space, then there exists a set X such that $(Y, \mathfrak{T})$ is homeomorphic to some subspace of $(\tau(X), \mathfrak{T}^-_{1u}(X))$.

Proof. We set $X = \mathfrak{T}$ and define the embedding ψ by $\psi(y) = \{B^c \mid B \subset X, y \notin \bigcup_{V \in B} V\} \cup \{\phi\}$.

Representation theorem for $\mathfrak{T}^+_2(X)$: *$(\tau(X), \mathfrak{T}^+_2(X))$ has the following hereditary property: It is a saturated T_0 space.*

If $(Y, \mathfrak{T})$ is any saturated T_0 space, then there exists a set X such that $(Y, \mathfrak{T})$ is homeomorphic to some subspace of $(\tau(X), \mathfrak{T}^+_2(X))$:

Proof. We set $X = Y$ and define the embedding ψ by $\psi(y) = \{B \mid y \in B \in \mathfrak{T}\} \cup \{\phi\}$.

Representation theorem for $\mathfrak{T}^-_2(X)$: *$(\tau(X), \mathfrak{T}^-_2(X))$ is a saturated T_0 space such as $(\tau(X), \mathfrak{T}^+_2(X))$.*

If $(Y, \mathfrak{T})$ is any saturated T_0 space, then there exists a set X

such that $(Y, \mathfrak{T})$ *is homeomorphic to some subspace of* $(\tau(X), \mathfrak{T}_2^-(X))$.

Proof. We set $X = Y \cup \{\omega\}$, where $\omega \notin Y$, and define the embedding ψ by $\psi(y) = \{B \mid B \in \mathfrak{T}, U_{\min}(y) \supset B\} \cup \{X\}$.

REFERENCES

C.L. Chang [1], Fuzzy topological spaces, *J. Math. Anal. Appl.*, 24 (1968), 182-190.

Á. Császár [1], *Fondements de la topologie générale,* Akadémiai Kiadó, Budapest, 1960.

H. Herrlich [1], *Topologische Reflexionen und Coreflexionen,* Lecture Notes in Mathematics 78, Springer, Berlin, 1968.

D. Klaua [1], Stetige Gleichmächtigkeiten kontinuierlichwertiger Mengen, *Monatsber. Deutsch. Akad. Wiss. Berlin,* 12 (1970), 749-758.

Chr. Kollreider [1], Topologische und uniforme Strukturen auf Mengensystemen, Thesis, Innsbruck, 1971.

F. Lorrain [1], Notes on topological spaces with minimum neighbourhoods, *Amer. Math. Monthly,* 76 (1969), 616-727.

M.G. Murdeshwar – S.A. Naimpally [1], *Quasi-Uniform Topological Spaces,* Nordhoff, 1966.

J.-P. Olivier [1], *K*-structures, *Bull. Sci. Math.*, (2) 91 (1967), 25-31.

W.J. Pervin [1], Quasi-uniformization of topological spaces, *Math. Annalen,* 147 (1962), 316-317.

B. Roider [1], Jeder topologische Raum ist $\mathfrak{T}_6(A_0, X)$-darstellbar, *Anzeiger der math.-naturw. Klasse der Österr. Akad. Wiss.*, Jahrg. 1970, Nr. 12, 237-239.

B. Roider [2], T_0-Räume und $\mathfrak{T}_{1\pi}^+(X)$, *Anzeiger der math.-naturw. Klasse der Österr. Akad. Wiss.*, Jahrg. 1972, Nr. 10, 235-236.

L.A. Zadeh [1], Fuzzy sets, *Information and Control,* 8 (1965), 338-353.

COLLOQUIA MATHEMATICA SOCIETATIS JÁNOS BOLYAI
8. TOPICS IN TOPOLOGY, KESZTHELY (HUNGARY), 1972.

ON P.S. ALEXANDROFF'S AXIOMATIC DEFINITION OF THE DIMENSION

O.V. LOKUCIEVSKIĬ – E.V. ŠČEPIN

1. In his works P.S. Alexandroff [1], [2] has proposed the axiomatic definition of the dimension* for the class $\mathscr{X}_c^m$ of finite dimensional metrisable compacta.

This report is dedicated to those alterations, which must be brought in his axioms in order to make them suitable for the class $\mathscr{X}^m$ of arbitrary finite dimensional metrisable spaces, or for the class $\mathscr{X}_c$ of arbitrary finite dimensional bicompacta.

At first we shall describe Alexandroff's axioms, and then do the alterations, which were mentioned above.

Let $\mathscr{X} = \{X\}$ be any class of normal spaces, which contains together with each space X all of its closed subspaces, and all spaces homeomorfic to them. Let us suppose also that $\mathscr{X} \supseteq \mathscr{X}_c^m$. Let, further, $Q = \{-1, 0, 1, \dots, n, \dots\}$.

*The dimension of a normal space X means everywhere its covering dimension (dim X).

P. S. Alexandroff considers the function $d\colon \mathscr{X} \to Q$, which satisfies the following axioms:

A_1 (axiom of initial conditions). If T^n is a closed n-simplex $(T^{-1} = \phi)$, then $dT^n = n$; if X' is homeomorfic to X, then $dX' = dX$.

A_2 (finite union axiom). If X is the union of its closed subspaces X_1 and X_2, then $dX = \max\{dX_1, dX_2\}$.

A_3 (Brouwer's axiom). Let $dX = n$. Then there exists a finite open covering ω of X such that for an arbitrary ω-mapping $f\colon X \to Y (Y \in \mathscr{X})$ we have $d(fX) \geqslant n$.

A_4 (Poincaré's axiom). If X contains more that one point, and $dX = n$, then there exists a closed $B \subset X$ such that $dB < dX$, and $X \setminus B$ is disconnected.

P.S. Alexandroff has proved that for the class $\mathscr{X}_c^m$ there exists only one function dX satisfying the axioms $A_1 - A_4$, and that this function is $\dim X$*.

The system of axioms, which was described, is incompatible for the class $\mathscr{X}_c$ (P.S. Alexandroff), and it is incomplete for the class $\mathscr{X}^{m\,0}$ of all finite dimensional metrisable spaces of countable weight (E.V. Ščepin). In connection with this P.S. Alexandroff has formulated two problems about the axiomatic definition of the dimension: on the one hand for the class $\mathscr{X}^{m\,0}$, and on the other hand for the class $\mathscr{X}_c$. Below we shall give solutions for each of these problems. The solution of the first of them (both for the class $\mathscr{X}^{m\,0}$ and for the class $\mathscr{X}^m$) was given by E.V. Ščepin, the solution of the second of them was given by O.V. Lokucievskiĭ.

2. For the classes $\mathscr{X}^{m\,0}$ and $\mathscr{X}^m$ instead of axiom A_2 the following one may be used:

A_2' (countable union axiom). If X is the finite or countable union

*The independence of axioms $A_1 - A_4$ was established by E.V. Ščepin.

of its closed subspaces X_k $(k = 1, 2, \dots)$, then $dX = \sup_k dX_k$.

The following statement is true:

Theorem 1 (special). *for the class* $\mathscr{X}^{m0}$ *there exists only one function* dX *satisfying the axioms* A_1, A'_2, A_3, A_4, *and this function is* $\dim X$.

The proof of this theorem is based on the following statement, which has, in our opinion, independent interest:

There exists for each $X \in \mathscr{X}^{m0}$ a space $X^\sigma \in \mathscr{X}^{m0}$ such that:

(a) X^σ is a countable union of its disjointed closed subspaces, which are homeomorphic to X;

(b) If $\dim X = n$, F is closed in X^σ and $X^\sigma \setminus F$ is disconnected, then $\dim F \geqslant (n - 1)$.

Theorem 2 (general). *For the class* $\mathscr{X}^m$ *there exists only one function* dX *satifying the axioms* A_1, A'_2, A_3, A_4 *and this function is* $\dim X$.

Remark. The special theorem is not a direct consequence of the general one.

3. Let us consider now the class $\mathscr{X}_c$.

Preliminary definition. Let $X \in \mathscr{X}$ and let ω be an open finite covering of X. The closed set $B \subset X$ is called ω-thin in X, if there exists an ω mapping $f: B \to Y (Y \in \mathscr{X})$ such that $d(fB) < dX$.

For the class $\mathscr{X}_c$ instead of axiom A_4 the following one may be used:

A_4^* (weakened form of Poincaré's axiom). If X contains more than one point, and $dX = n$, then there exist two non-empty disjoint canonical* closed subsets $F_1, F_2 \subset X$ such that for every finite open covering ω of X one can find a closed set $B \subset X$, which is ω-thin in X and separates them.

*The closed set $F \subset X$ is called canonical, if it is equal to the closure of its interior.

The following statement is true:

Theorem 3. *For the class* $\mathscr{X}_c$ *there exists only one function* dX *satisfying the axioms* A_1, A_2, A_3, A_4^*, *and this function is* $\dim X$.

The proof of this statement requires the use of the continua (V^p), constructed by P.S. Alexandroff in [3] (see also V.I. Kuzminoff [4]), and of the dimensional invariant $\mathrm{Ind}^* X$, which was considered in O.V. Lokucievskiĭ's work [5].

Both alterations (A_2' instead of A_2, and A_4^* instead of A_4) keep the system of axioms applicable for the class $\mathscr{X}_c^m$.

REFERENCES

[1] P.S. Alexandroff, Dimensionstheorie, *Math. Ann.*, 106 (1932).

[2] P.S. Alexandroff, On some old problems of the homological theory of dimension, *Proceedings of the International Symposium on topology and its applications* held at Herceg-Novi (Yugoslavia) 25-31 August, 1968, Beograd (1969) (Russian).

[3] P.S. Alexandroff, Die Kontinua (V^p) – eine Verschärfung der Kantorschen Mannigfaltigkeiten, *Monatshefte für Math.*, 61 (1957), 1.

[4] V.I. Kuzminoff, On V^n continua, *Dokl. Akad. Nauk SSSR*, 139 (1961) (Russian).

[5] O.V. Lokucievskiĭ, On dimension theory, *Dokl. Akad. Nauk SSSR*, 179 (1968) (Russian).

COLLOQUIA MATHEMATICA SOCIETATIS JÁNOS BOLYAI
8. TOPICS IN TOPOLOGY, KESZTHELY (HUNGARY), 1972.

SHAPES FOR TOPOLOGICAL SPACES*

S. MARDEŠIĆ

The notion of shape of compact metric spaces as a modification of homotopy type, has been introduced in 1968 by K. Borsuk [1], [2]. An alternate approach and generalization to compact Hausdorff spaces was developed by S. Mardešić and J. Segal [8], [9]. A rather complete survey of results on shapes of compact spaces is given in [7].

K. Borsuk has also studied several possible definitions of the notion of shape for arbitrary metric spaces [3], [4], [5].

The purpose of this paper is to exhibit a notion of shape for arbitrary topological spaces.

One forms a category $\mathfrak{S}$, the *shape category*, whose objects are topological spaces and whose morphisms are the *shape maps* $f: X \to Y$. Moreover, one defines a covariant functor $S: \mathfrak{H} \to \mathfrak{S}$ from the homotopy category $\mathfrak{S}$ of topological spaces to $\mathfrak{S}$, having the following properties:

(i) $S(X) = X$, for every topological space X.

*A detailed version will appear in General Topology and its Applications.

(ii) If Y is a CW-complex (or equivalently an ANR for metric spaces), then for every shape map $f: X \to Y$ there is a unique homotopy class of maps $\varphi: X \to Y$, such that $S(\varphi) = f$. Isomorphic objects in $\mathfrak{H}$ are said to be of the same shape.

It is shown that the pair $(\mathfrak{S}, S)$ has the following universal property, which characterizes it up to an isomorphism: If $\mathfrak{S}'$ is a category and $S': \mathfrak{H} \to \mathfrak{S}'$ a covariant functor such that (i) and (ii) hold, then there is a unique covariant functor $\Phi: \mathfrak{S}' \to \mathfrak{S}$, such that $S = \Phi S'$.

It is also shown that for compact Hausdorff spaces this notion of shape coincides with the one introduced in previous papers and that various results generalize from the compact case.

Essential for the above development was an idea of A. Dold to the effect that the shape functor $\mathfrak{S}(X, .)$ should be the Kan extension of the homotopy functor $[X, .]$. The actual construction of $\mathfrak{S}$ is in fact a construction which appears already in a paper by W. Holsztyński ([6], §4). Considerable contributions are due to H. Breger who systematically studied this notion of shape.

REFERENCES

[1] K. Borsuk, Concerning the homotopy properties of compacta, *Fund. Math.*, 62 (1968), 223-254.

[2] K. Borsuk, Concenring the notion of the shape of compacta, *Proc. Internat. Sympos. Topology and its Applications, Herceg-Novi* 1968, Beograd, 1969, 98-104.

[3] K. Borsuk, On the concept of shape for metrizable spaces, *Bull. Acad. Polon. Sci. Ser. Sci. Math. Astr. Phys.*, 18 (1970), 127-132.

[4] K. Borsuk, Theory of shape, Lecture Notes Series No. 28 Mat. Inst., *Aarhus Univ.*, 1971, 1-145.

[5] K. Borsuk, Some remarks concerning the theory of shape in arbitrary metrizable spaces, Proc. 3, *Prague Symp. on General Topology*, 77-81.

[6] W. Holsztyński, An extension and axiomatic characterization of Borsuk's theory of shape, *Fund. Math.*, 70 (1971) 157-168.

[7] S. Mardešić, A survey of the shape theory of compacta, Proc. 3. *Prague Symp. on General Topology*, 291-300.

[8] S. Mardešić – J. Segal, Shapes of compacta and *ANR*-systems, *Fund. Math.*, 72 (1971), 41-59.

[9] S. Mardešić – J. Segal, Equivalence of the Borsuk and the *ANR*-system approach to shapes, *Fund. Math.*, 72 (1971), 61-68.

COLLOQUIA MATHEMATICA SOCIETATIS JÁNOS BOLYAI
8. TOPICS IN TOPOLOGY, KESZTHELY (HUNGARY), 1972.

IRREDUCIBLE COVERS: AN APPLICATION OF A FACTORIZATION LEMMA

J. MIODUSZEWSKI

A factorization lemma for skeletal maps [3] due recently by Błaszczyk [1] and a procedure with using maximum principle such as in authors note [2] leeds after some categorial considerations to extremally disconnected covers for arbitrary Hausdorff spaces.

BIBLIOGRAPHY

[1] A. Błaszczyk, A factorization lemma and extremally disconnected, covers, *Colloquium Mathematicum,* (in print),

[2] J. Mioduszewski, A method which leads to extremally disconnected covers, *Proceedings of the Third Prague Symposium,* (1971), 309-311.

[3] J. Mioduszewski – L. Rudolf, *H*-closed and extremally disconnected Hausdorff spaces, *Dissertations Mathematicae,* 66 (1969).

COLLOQUIA MATHEMATICA SOCIETATIS JÁNOS BOLYAI
8. TOPICS IN TOPOLOGY, KESZTHELY (HUNGARY), 1972.

CLOVERLEAF REPRESENTATIONS OF SIMPLY CONNECTED 3-MANIFOLDS (SUMMARY)

E.E. MOISE

All manifolds discussed will be piecewise linear. By a *linear graph* we maan either a 1-dimensional complex or a 1-dimensional polyhedron, according to convenience in the context. By a *loop-graph* we mean a linear graph L, containing a point P_0 such that L is the union of a finite collection of 1-spheres $\{J_1, J_2, \ldots, J_n\}$ such that the intersection of any two of them is P_0. These 1-spheres are called the *loops* of L, and P_0 is called the *center* of L. Let L be a loop-graph in a triangulated 3-manifold M. Suppose that there is a collection $\{D_1, D_2, \ldots, D_n\}$ of polyhedral disks, such that $\operatorname{Bd} D_i = J_i$ for each i, and such that the intersection of every two of the disks D_i is P_0. Then L is called a *cloverleaf*. (Here, as usual, if D is an m-manifold with boundary, then $\operatorname{Bd} D$ is the boundary of D. The interior $D - \operatorname{Bd} D$ of D will be denoted by $\operatorname{Int} D$.) The set $\{D_1, D_2, \ldots, D_n\}$ is called *a set of spanning disks* for L.

Now let $K = \cup_{i=1}^n I_i$ and $L = \cup_{i=1}^n J_i$ be cloverleaves, with the same number of loops, in the same triangulated 3-manifold M. Suppose that we can choose the order of the loops in K and L, and choose spanning disks $\{D_1, D_2, \ldots, D_n\}$ for K, and spanning disks $\{E_1, E_2, \ldots \ldots, E_n\}$ for L, in such a way that (1) D_i intersects E_j only if $i = j$, and (2) D_i and E_i intersect in the same way (topologically) as two linked circular regions in Euclidean 3-space $\underline{R}^3$. (Here (2) means that (a) $D_i \cap E_i$ is a polygonal arc, joining a point x_i of I_i to a point y_i of J_i. (b) D_i and E_i are in general position relative to one another, and therefore (c) x_i and y_i are "true crossing points" of I_i with $\operatorname{Int} E_i$ and of J_i with $\operatorname{Int} D_i$ respectively.) Then K and L are *simply linked.*

Let K be a linear graph in a triangulated 3-manifold M. By a slight abuse of language, M/K denotes the space whose points are the components of K and the points of $M - K$ (with the usual topology.)

Theorem 1. *(The cloverleaf theorem.) Let M be a compact, connected, simply connected triangulated 3-manifold. Then there is a linear graph K_1 in the 3-sphere S^3, and a linear graph K_2 in M, such that* (1) *either K_2 is empty or K_2 is the union of two simply linked cloverleaves and* (2) *S^2/K_1 and M/K_2 are homeomorphic.*

Without condition (1) in the conclusion, this theorem would be trivial. It is a fact that if M is a compact, connected, orientable, triangulated 3-manifold, then there are linear graphs K_1 and K_2, in M and S^3 respectively, such that M/K_1 and S^3/K_2 are homeomorphic. But the cloverleaf theorem is not due to this phenomenon, because it has a true converse:

Theorem 2. *Let M be a compact, connected, triangulated 3-manifold. If M satisfies the condition in the conclusion of the cloverleaf theorem, then M is simply connected.*

Thus, if the hypothesis of simple connectivity is replaced by the conclusion of the cloverleaf theorem, in an attempt to prove the Poincaré Conjecture, then nothing is lost; and it may be that something is gained.

The proof of the cloverleaf theorem is based on the main result of the author's *Monotonic Mapping Theorem for Simply Connected 3-manifolds* Illinois J. Math. 12 (1968), 451-474. The proof will appear in full elsewhere. It is elementary but technical, and a legible summary of it would be exessively long.

COLLOQUIA MATHEMATICA SOCIETATIS JÁNOS BOLYAI
8. TOPICS IN TOPOLOGY, KESZTHELY (HUNGARY), 1972.

SOME RESULTS ON *M*-SPACES

K. MORITA

In 1963 we introduced the notion of *M*-spaces in connection with the problem of characterizing a space whose product with any metric space is normal. Since then, many interesting results have been obtained by many mathematicians. A survey on these results was given in my talk at the *Pittsburgh International Conference on General Topology and its Applications* in 1970 (cf. [13]).

In this talk we would like to discuss some of the problems which were unsettled in my previous talk mentioned above.

Throughout this talk let us assume that spaces are T_1, paracompact spaces are Hausdorff, and that maps are continuous and onto; N will refer to the set of positive integers.

I. CLOSED IMAGES OF *M*-SPACES

1. We shall begin with the definition of *M*-spaces.

Definition 1.1 ([10]). A space X is called an *M*-space if there is a normal sequence $\{\mathcal{U}_i\}$ of open coverings of X satisfying condition

(M) below:

(M) If $x_n \in \mathrm{St}(x, \mathscr{U}_n)$ for $n \in N$, then $\{x_n\}$ has a cluster point in X.

A characterization of M-spaces is given by the following.

Theorem 1.2 ([10]). *A space X is an M-space (resp. a paracompact M-space) iff (= if and only if) there is a quasi-perfect map (resp. a perfect map) from X onto a metric space.*

Here a map $f: X \to Y$ is called perfect (resp. quasi-perfect) if f is a closed map and the inverse image of each point is compact (resp. countably compact).

Thus, metric spaces and countably compact spaces are M-spaces, and any paracompact space which is G_δ in $\beta(X)$ is also an M-space (Frolík [4]). For paracompact spaces, M-spaces are identical with p-spaces in the sense of Arhangel'skii [1], although both notions are different from each other for general spaces.

2. The images of metric spaces under various kinds of quotient maps are investigated extensively. Recently the corresponding results have been obtained for M-spaces. They are summarized in Michael [9], including his own contributions.

As for closed maps which are not treated in Michael's paper [9], Lašnev [8] obtained a characterization of closed images of metric spaces. Here we shall give a characterization of closed images of M-spaces, which is obtained in our joint paper with T. Rishel [16].

To state our results we shall need a number of definitions.

Definition 2.1 (Michael [9]). *A sequence $\{A_n\}$ of subsets in a space X is called a q-sequence at a point x of X if*

(a) *$x \in A_n$ for $n \in N$, and*

(b) *any point-sequence $\{x_n\}$ with $x_n \in A_n$ for $n \in N$ has a cluster point.*

Definition 2.2 (Michael [9]). *A space X is called a q-space if each point of X has a q-sequence consisting of neighborhoods of x.*

Definition 2.3 (Morita – Rishel [16]). *X is a CM-space if there is a sequence $\{\mathscr{F}_n\}$ of hereditarily closure-preserving closed coverings of X such that*

(a) *any sequence $\{A_n\}$ of subsets of X with $x \in A_n \in \mathscr{F}_n$ for $n \in N$ is either hereditarily closure-preserving or a q-sequence at a point x of X,*

(b) *every point x of X has a q-sequence $\{A_n\}$ with $x \in \in A_n \in \mathscr{F}_n$ for $n \in N$.*

In Definition 2.1, if we replace (b) by the condition that any neighborhood of x contains some A_n, then $\{A_n\}$ is called a local network at x. In Definition 2.3, if we replace the word "*q*-sequence" by "local network", we obtain the condition (L) originated by Lašnev [8].

Definition 2.4 (Nagata [17].) *A space X is called quasi-k if a subset A of X is closed whenever $A \cap C$ is relatively closed in C for every countably compact subset C of X.*

Definition 2.5 (Michael [9]). *A space X is called singly bi-quasi-k (resp. countably bi-quasi-k) if for any subset F of X (resp. any decreasing sequence $\{F_n : n \in N\}$ of subsets of X) with $x \in \mathrm{Cl}F$ (resp. $x \in \mathrm{Cl}F_n$ for $n \in N$) there is a decreasing q-sequence $\{A_n\}$ at x with $x \in \mathrm{Cl}(F \cap A_n)$ for $n \in N$ (resp. $x \in \mathrm{Cl}(F_n \cap A_n)$ for $n \in N$) such that every sequence $\{x_n\}$ with $x_n \in A_n$ has a cluster point in $\cap A_n$.*

3. We are now in a position to state our characterization of closed images of *M*-spaces.

Theorem 3.1 (Morita – Rishel [16]). *For a regular space X the following statements are equivalent.*

(a) *X is the closed image of an M-space S.*

(b) *X is CM and quasi-k.*

(c) *X is CM and singly bi-quasi-k.*

Furthermore, if case (b) or (c) holds, S in (a) can be chosen to be completely regular or normal according as X is completely regular or normal.

Our proof of Theorem 3.1 is based on the following lemma.

Lemma 3.2. *Let X be a CM-space. Then there exist a metric space T and an M-space S, which is a closed subset of $T \times X$, such that if we put*

$$f = \pi_X | S \colon S \to X, \; g = \pi_T | S \colon S \to T$$

where π_X (resp. π_T) is the projection from $T \times X$ onto X (resp. T),

(1) *f is a closed map, a closed map with $\mathrm{Bd}\, f^{-1}(x)$ countably compact for each point x of X, or a perfect map according as*

(a) *X is quasi-k,*

(b) *X is countably bi-quasi-k (or a q-space), or*

(c) *each covering $\mathscr{F}_n$ is locally finite and $\{\mathrm{St}(x, \mathscr{F}_n) \colon n \in N\}$ is a q-sequence at each point x of X;*

(2) *S is paracompact in case every countably compact closed subset of X is compact.*

Here the construction of S and T will be sketched. Let $\mathscr{F}_i = \{F_{ia} | a \in \Omega_i\}$, $i = 1, 2, \ldots$, be a sequence of hereditarily closure-preserving closed coverings of X with the property described in Definition 2.3.

Let us denote by T all the sequences $\{a_i\}$ with $a_i \in \Omega_i$ for $i \in N$, such that $\{F_{ia_i}\}$ is a q-sequence at some point of X, and consider T as a subspace of the topological product of discrete spaces Ω_i with $i \in N$. Then T is a metric space. Let us put

$$S = \{(\{a_i\}, x) \in T \times X | \{F_{ia_i}\} \text{ is a } q\text{-sequence at a point } x \text{ of } X\}.$$

Then it can be proved that these S and T satisfy the conditions of Lemma 3.2.

The second part of Theorem 3.1 follows readily from Lemma 3.2 if we take into consideration the following

Lemma 3.3 ([14]). *Every CM-space is a P-space in the sense of* [10].

4. As another application of Lemma 3.2 we have a number of characterizations of M^*-spaces.

Definition 4.1 (Ishii [6] (resp. Siwiec – Nagata [22])). *A space X is called an M^*-space (resp. $M^{\#}$-space) if there is a sequence $\{\mathscr{F}_i\}$ of locally finite (resp. closure-preserving) closed coverings of X such that $\{\mathrm{St}(x, \mathscr{F}_i)\}$ is a q-sequence at each point x of X.*

Theorem 4.2. *For a space X the following conditions are equivalent.*

(a) *X is an M^*-space.*

(b) *There is a sequence $\{\mathscr{F}_i\}$ of hereditarily closure-preserving closed coverings of X such that $\{\mathrm{St}(x, \mathscr{F}_i)\}$ is a q-sequence at each point x of X.*

(c) *X is the quasi-perfect image of an M-space*

(d) *X is the perfect image of an M-space S.*

(e) *X is CM and q.*

(f) *X is CM and countably bi-quasi-k.*

The implications (a) ⇒ (b) and (d) ⇒ (c) are obvious. (c) ⇒ (a) is proved by Ishii [6]; (b) ⇒ (c) is proved by Morita – Rishel [16]. Nagata [19] proved the important implication (a) ⇒ (d) by using multi-valued maps. For this implication, however, we have given another proof in Morita – Rishel [16] by utilizing our Lemma 3.2; our proof shows further that S in condition (d) can be chosen to be normal if X is normal. (a) ⇔ (e) and (a) ⇔ (f) are proved by Morita [14].

Problem 4.3 ([16]). Is every $M^{\#}$-space an M^*-space?

As an application of Theorem 4.2 we have a simple proof of the following

Theorem 4.4 (Ishii [7]). *A normal M^*-space is an M-space.*

Indeed, if X is a normal M^*-space, then by Theorem 4.2 X is the quasi-perfect image of an M-space and hence X is an M-space by Ishii [6] and Morita [11].

5. The closed image of a metric space is characterized by Lašnev [8] as was mentioned before.

Theorem 5.1 (Lašnev [8]). *A space X is the closed image of a metric space iff X satisfies condition* (L) *(for the definition, cf. the paragraph following Definition 2.3) and X is a Fréchet space in the sense of Arhangels'kii* (for the definition, cf. [9]).

Another characterization is obtained from our results.

Theorem 5.2 ([14]). *A Hausdorff space X is the closed image of a metric space iff X is CM, quasi-k, and semi-stratifiable in the sense of Michael* (cf. Creede [3]) *(or a σ-space in the sense of Okuyama* [20].

Now, let us consider the following three properties for a space X:

(i) *CM,* (ii) q-space, (iii) semi-stratifiable.

Theorem 5.3 ([14]). *For a space X with the separation axiom indicated in parentheses, we have the following equivalences.*

M^*-space $\Leftrightarrow$ (i) and (ii)

semi-metrizable $\Leftrightarrow$ (ii) and (iii) *(for X regular),*

metrizable $\Leftrightarrow$ (i), (ii) *and* (iii) *(for X Hausdorff).*

However, we do not know what a space is characterized by (i) and (iii).

6. Quite recently R. Telgársky [25] has proved the following theorem.

Theorem 6.1 (Telgársky [25]). *A space X is the closed image of a locally compact, paracompact space iff X has a hereditarily closure-preserving covering by compact sets.*

A locally compact, paracompact space is an M-space. Thus, the theorems concerning closed images of M-spaces are applicable to the present case. For example, as a corollary to Theorem 7.2 below we have

Theorem 6.2. *Let X and Y be spaces. If the product $X \times Y$ has a hereditarily closure-preserving closed covering by compact sets, then either X is locally compact and paracompact or Y is discrete.*

II. PRODUCTS OF CLOSED IMAGES OF M-SPACES

7. As is proved in our previous paper [10], the product of any two paracompact M-spaces is a paracompact M-space. As for the product of closed images of paracompact M-spaces, however, the situation is entirely different. Indeed, we have

Theorem 7.1. *Let X and Y be spaces. If $X \times Y$ is the closed continuous image of a paracompact (resp. normal) M-space, then either X is a paracompact (resp. normal) M-space or Y is discrete.*

Theorem 7.1 follows from a more general theorem below.

Theorem 7.2. *Let X and Y be spaces. If $X \times Y$ is the closed continuous image of an M-space S, then either X is an M^*-space homeomorphic to the quasi-perfect image of a closed subspace of S or Y is discrete.*

As is proved by Ishii [6] and Morita [11], the image of a normal M-space under a quasi-perfect map is a normal M-space, and hence Theorem 7.1 is an immediate consequence of Theorem 7.2.

If we apply Theorem 7.2 to the case where S is a metric space, we have the following theorem of Hyman [5] by virtue of Stone [23] and Morita – Hanai [15].

Theorem 7.3 (Hyman [5]). *Let X and Y be spaces. If $X \times Y$ is the closed continuous image of a metric space, then either X is metrizable or Y is discrete.*

Recently it has been communicated to us by A. Okuyama that he has solved the following question of K. Nagami affirmatively:

"If the product $X \times I$ of a paracompact Hausdorff space X with the closed unit interval I is the closed image of an M-space, is X itself an M-space? "

Since I is not discrete, Okuyama's result is a direct consequence of our Theorem 7.2.

8. Since closed maps are pseudo-open, by Rishel [21] we have the following lemma which was already observed in [16, Remark to Theorem 3.1].

Lemma 8.1 ([21]). *Suppose that f is a closed map from an M-space X onto a space Y. Then Y is singly bi-quasi-k.*

Lemma 8.2 ([14]). *Under the same assumption as in Lemma 8.1, if Y is countably bi-quasi-k, then* $\mathrm{Bd}\, f^{-1}(y)$ *is countably compact for every point y of Y.*

Lemma 8.3. *If Y is a quotient space of an M-space and if Y is not discrete, then there is a sequence $\{y_n \mid n \in N\}$ of distinct points of Y, which has a cluster point y_0 with $y_0 \neq y_n$ for all $n \in N$.*

Proof. Let $f\colon X \to Y$ be a quotient map from an M-space X onto Y. Suppose that $y \in \mathrm{Cl}(Y - y)$. Then we have $\mathrm{Bd}\, f^{-1}(y) \neq \phi$. Pick a point x from $\mathrm{Bd}\, f^{-1}(y)$.

Let $\{U_n \mid n \in N\}$ be a q-sequence consisting of neighborhoods of x. Then there is a point x_1 of X such that

$$x_1 \in U_1 \cap (X - f^{-1}(y)) .$$

In this case $f(x_1) \neq y$. Hence we can find a sequence of points x_n, $n \in$ $\in N$, by induction, such that

$$x_n \in U_n \cap (X - f^{-1}(y)) \cap \left(X - \cup \{ f^{-1} f(x_i) \mid 1 \leqslant i < n \}\right) .$$

Since $\{U_n\}$ is a q-sequence at x, the sequence $\{x_n\}$ has a cluster point x_0. Putting $y_i = f(x_i)$ for $i = 0, 1, 2, \ldots$, we see that y_0 is a cluster point of the sequence $\{y_n\}$ and the y_n, $n \in N$ are all distinct from one another. Here we can assume that $y_0 \neq y_n$ for $n \in N$; in case $y_0 = y_{n_0}$ with some $n_0 \in N$, we have only to consider $\{y_n \mid n > n_0\}$ instead of $\{y_n \mid n \in N\}$. Thus our proof of Lemma 8.3 is completed.

9. **Proof of Theorem 7.2.** Suppose that $X \times Y$ is the closed image of an M-space S and that Y is not discrete. Then by Lemma 8.3 there is a sequence $\{y_n \mid n \in N\}$ consisting of distinct points of Y, such that it has a cluster point y_0 with $y_0 \neq y_n$ for all $n \in N$. Let us put

$$V_n = Y - \{y_i \mid 1 \leqslant i < n\}, \quad \text{for} \quad n \in N.$$

Let $F_n \mid n \in N$ be a decreasing sequence of subsets of X such that $x \in \mathrm{Cl} F_n$ for $n \in N$. Let us put

$$K = \cup \{F_n \times y_n \mid n \in N\},$$

$$z = (x, y_0).$$

Then $z \in \mathrm{Cl} K$.

By Lemma 8.1 $X \times Y$ is singly bi-quasi-k and hence there is a decreasing q-sequence $\{C_n \mid n \in N\}$ at z such that

$$z \in \mathrm{Cl}\,(K \cap C_n) \quad \text{for} \quad n \in N.$$

Let π be the projection from $X \times Y$ onto X. Then $\{\pi(C_n) \mid n \in \in N\}$ is a decreasing q-sequence at x.

On the other hand, for any open neighborhood W of z we have

$$W \cap (X \times V_n) \cap K \cap (\pi(C_n) \times Y) \neq \phi$$

since $C_n \subset \pi(C_n) \times Y$. Hence

$$z \in \mathrm{Cl}\,(K \cap (\pi(C_n) \times V_n)) \quad \text{for} \quad n \in N.$$

Since $\pi(K \cap (\pi(C_n) \times V_n)) = \cup \{F_i \cap \pi(C_n) \mid i \geqslant n\} = F_n \cap \pi(C_n)$, we have

$$x = \pi(z) \in \mathrm{Cl}\,(F_n \cap \pi(C_n)) \quad \text{for} \quad n \in N.$$

Therefore X is countably bi-quasi-k. In view of Lemma 8.2 we see, using the technique in [15], that X is the quasi-perfect image of a closed subspace of S. Since the quasi-perfect image of an M-space is an M^*-space by Ishii [6], this completes our proof of Theorem 7.2.

9. Related results. Since a singly bi-quasi-k space is the image of

an M-space under a pseudo-open map by Rishel [21] and a pseudo-open map is a hereditarily quotient map (cf. Micheal [9]), our proof of Theorem 7.2 gives rise to the following theorem.

Theorem 9.1. *If the product* $X \times Y$ *of two spaces* X *and* Y *is singly bi-quasi-k, then either* X *is countably bi-quasi-k or* Y *is discrete.*

By a similar method we can prove the following theorem; for the terminology used here cf. Michael [9]. Indeed, our results (and proofs) are elaborations of those of Michael [9].

Theorem 9.2. *Let* X *and* Y *be spaces such that* Y *is non-discrete. Then* X *is countably bi-k, countably bi-sequential (=strongly Fréchet) or a strongly* k'*-space according as* $X \times Y$ *is singly bi-k, Fréchet or a* k'*-space.*

For example, let X_0 be a locally compact metric space such that there is a closed subset A_0 with a non-compact boundary. Then the quotient space $X_1 = X_0/A_0$, which is obtained from X_0 by contracting A_0 to a point, is a paracompact k'-space which is not countably bi-quasi-k; this is seen from [14, Lemma 2.1]. Therefore it follows from our results that the product space $X_1 \times Y$ for any non-discrete space Y is not of any type of spaces listed below:

singly bi-quasi-k, singly bi-k, Fréchet,

k'-space, closed image of an M-space.

Thus we have a generalization of Bagley and Weddington [2], Theorem 1].

III. COUNTABLY-COMPACTIFIABLE SPACES

10. In this part by a space we shall always mean a completely regular Hausdorff space.

Definition 10.1. *A space* S *is called a countably compactification of a space* X *if*

(a) S *is countably compact and contains* X *as a dense subset,*

(b) *every countably compact closed subset of* X *is closed also in* S.

In case X has a countably-compactification we shall say that X is countably-compactifiable.

Countably compact spaces and paracompact spaces, or more generally, those spaces for which every countably compact closed subset is compact are trivial examples of countably-compactifiable spaces. Non-trivial examples are provided by the following theorem.

Theorem 10.2. *The product of a countably-compactifiable space with a product of paracompact spaces is countably-compactifiable.*

However, even a locally compact space is not countably-compactifiable in general; indeed, the Tychonoff plank $T = W(w_0 + 1) \times W(w_1 + 1) - (w_0, w_1)$ is not countably-compactifiable.

As is well known, for a locally countably compact space X which is not countably compact we can construct a countably compact space S by adding a new point p_∞ to X; as neighborhoods of p_∞ we take the sets of the form: $p_\infty \cup (X - F)$ with a countably compact closed subset F of X. However, the space S is not necessarily a countably-compactification of X in our sense because S is not necessarily completely regular. In case S is a countably-compactification in our sense, we shall call S a one-point countably-compactification of X.

Theorem 10.3. *A space* X *admits a one-point countably-compactification iff for any countably compact closed subset* A *of* X *there is a real-valued continuous function* f *over* X *such that*

$$f = 0 \text{ on } A \text{ and } f = 1 \text{ on } X - B$$

for some countably compact closed subset B *of* X.

It is pointed out by M. Atsuji that if X is a normal space such that every countably compact closed subset is contained in an open set with a countably compact closure, then X admits a one-point countably-compactification which is normal. This, combined with Theorem 10.3, yields

Theorem 10.4. *Let X be a space which admits a one-point countably-compactification S. If X is normal, so is S.*

As for M-spaces, we have the following theorem.

Theorem 10.5. *An M-space X admits a one-point countably-compactification iff there is a quasi-perfect map from X onto a locally compact metric space.*

In concluding this section, we would like to remark that even a locally compact, countably-compactifiable space does not necessarily admit a one-point countably-compactification.

11. Countably-compactifiable spaces are related to M-spaces as follows.

Theorem 11.1. *An M-space X is homeomorphic to a closed subset of the product of a metric space T with a countably compact space C iff X is countably-compactifiable.*

From our Theorem 11.1 we obtain the following corollaries by virtue of Theorems 10.2 and 10.5.

Corollary 11.2 (Nagata [18]). *A space X is a paracompact M-space iff X is homeomorphic to a closed subspace of the product of a metric space with a compact space.*

Corollary 11.3. *A space X is mapped onto a locally compact metric space under a quasi-perfect map iff X is homeomorphic to a closed subspace of the product of a locally compact metric space with a countably compact space.*

Thus, the problem of Nagata on the embeddability for M-spaces is reduced to

Problem 11.4. Is every M-space countably-compactifiable?

Concerning countably-compactifiable spaces there are many unsolved problems. Here only two of them will be stated.

Problem 11.5. Is every normal space countably-compactifiable?

According to A.K. Steiner [24], there is a product of two countably compact spaces which is an M-space but not countably compact. This product space is countably-compactifiable by our Theorem 10.5.

Problem 11.6. Is the product of two countably compact (or two countably-compactifiable) spaces countably-compactifiable?

REFERENCES

[1] A.V. Arhangels'kii, On a class of spaces containing all metric and all locally compact spaces, *Dokl. Akad. Nauk, SSSR*, 151 (1963), 751-754 (*Soviet Math. Dokl.*, 4 (1963), 1051-1055).

[2] R.W. Bagley – D.D. Weddington, Products of k'-spaces, *Proc. Amer. Math. Soc.*, 22 (1969), 392-394.

[3] G.D. Creede, Conserning semistratifiable spaces, *Pacific J. Math.*, 32 (1970), 47-54.

[4] Z. Frolík, On the topological product of paracompact spaces, *Bull Acad. Polon. Sci.*, 8 (1960), 747-750.

[5] D.M. Hyman, A note on closed maps and metrizability, *Proc. Amer. Math. Soc.*, 21 (1969), 109-112.

[6] T. Ishii, On closed mappings and M-spaces. I, II, *Proc. Japan Acad.*, 43 (1967), 752-756; 757-761.

[7] T. Ishii, On M and M^*-spaces, *Proc. Japan Acad.*, 10 (1968), 1028-1030.

[8] L. Lašnev, Closed image of metric spaces, *Dokl. Akad. Nauk SSSR*, 170 (1966) (*Soviet Math. Dokl.*, 7 (1966), 1219-1221).

[9] E.A. Michael, A quintuple quotient quest, *to appear.*

[10] K. Morita, Products of normal spaces with metric spaces, *Math. Ann.*, 154 (1964), 365-382.

[11] K. Morita, Some properties of M-spaces, *Proc. Japan Acad.*, 43 (1967), 869-872.

[12] K. Morita, Topological completions and *M*-spaces, *Sci. Rep. Tokyo Kyoiku Daigaku*, Sec. A, 10 (1970), 271-288.

[13] K. Morita, A survey of the theory of *M*-spaces, *General Topology and its Applications,* 1 (1971), 49-55.

[14] K. Morita, Results related to closed images of *M*-spaces. III, *Proc. Japan Acad.*, 48 (1972), 16-20.

[15] K. Morita – S. Hanai, Closed mapping and metric spaces, *Proc. Japan Acad.*, 32 (1956), 10-14.

[16] K. Morita – T. Rishel, Results related to closed images of *M*-spaces. I, II, *Proc. Japan Acad.*, *Supplements* to Vol. 47 (1971), 1004-1007; 1008-1011.

[17] J. Nagata, Quotient and bi-quotient spaces of *M*-spaces, *Proc. Japan Acad.*, 45 (1969), 25-29.

[18] J. Nagata, A note on *M*-spaces, *Proc. Japan Acad.*, 45 (1969), 541-543.

[19] J. Nagata, Some theorems on generalized metric spaces, *to appear.*

[20] A. Okuyama, A survey of the theory of σ-spaces, *General Topology and its Applications,* 1 (1971), 57-63.

[21] T. Rishel, A characterization of pseudo-open images of *M*-spaces, *Proc. Japan Acad.*, 45 (1969), 910-912.

[22] F. Siwiec – J. Nagata, A note on nets and metrization, *Proc. Japan Acad.*, 44 (1968), 623-627.

[23] A.H. Stone, Metrizability of decomposition spaces, *Proc. Amer. Math. Soc.*, 7 (1956), 690-700.

[24] A.K. Steiner, On the topological completion of *M*-space products, *Proc. Amer. Math. Soc.*, 29 (1971), 617-620.

[25] R. Telgársky, Concerning two covering properties, submitted to *Proc. Japan Acad.*

[26] K. Morita, Countably-compactifiable spaces, *Sci. Rep. Tokyo Kyoiku Daigaku, Sec. A.*, 1973 (to appear).

COLLOQUIA MATHEMATICA SOCIETATIS JÁNOS BOLYAI
8. TOPICS IN TOPOLOGY, KESZTHELY (HUNGARY), 1972.

THE FOX THEOREM AND THE WHITEHEAD THEOREM IN THE THEORY OF SHAPES

M. MOSZYŃSKA

The well known theorem of J.H.C. Whitehead gives a sufficient condition for a map to be a homotopy equivalence. This is the following

Theorem. [9]. *Let X and Y be two compact connected spaces homotopically dominated by CW-complexes and let* $n_0 = \max(dhX, dhY)$ (where $dhX = \min_{X \underset{h}{\leq} K \in CW} \dim K$).

If $f: X \to Y$ induces isomorphisms $f_n: \Pi_n(X) \to \Pi_n(Y)$ for $n \leq \leq n_0$ then f is a homotopy equivalence.

The proof is based on the following theorem of R.H. Fox

Theorem. [3]. *Two spaces X and Y are of the same homotopy type iff they are both embeddable into a space Z as its deformation retracts.*

We are interested in similar theorems in the theory of shapes for compact metric spaces. As an analogue of Fox's Theorem we obtain the following

Theorem 1. [6]. *Two compact metric spaces X and Y are of the same shape if and only if they are both embeddable into a compactum Z as its fundamental deformation retracts* (in the sense of Borsuk, [1], [2]).

By the results of Mardešić and Segal concerning the *ANR*-system approach to shapes, [4], [5], this theorem can be reduced to a theorem on maps of inverse sequences (Th.1′). Let $f = (\varphi, f_a)\colon X = (X_a, p_a^{a+1}) \to$ $\to (Y_a, q_a^{a+1}) = Y$ be a map of inverse sequences, i.e. $\varphi\colon N \to N$ is an increasing function and all the diagrams

$$\begin{array}{ccc} X_{\varphi(a)} & \xleftarrow{\;p^{\varphi(a+1)}_{\varphi(a)}\;} & X_{\varphi(a+1)} \\ {\scriptstyle f_a}\downarrow & & \downarrow{\scriptstyle f_{a+1}} \\ Y_a & \xleftarrow[\;q_a^{a+1}\;]{} & Y_{a+1} \end{array} \qquad \text{commute up to homotopy.}$$

Let $h_a^{a+1}\colon X_{\varphi(a+1)} \times I \to Y_a$ be a homotopy between the maps $f_a p^{\varphi(a+1)}_{\varphi(a)}$ and $q_a^{a+1} f_{a+1}$. We define a mapping cylinder C_f of the map f (relative to (h_a^{a+1})) as follows:

$$C_f \underset{Df}{=} (C_{f_a}, r_a^{a+1}),$$

where C_{f_a} is the mapping cylinder of f_a and $r_a^{a+1}\colon C_{f_{a+1}} \to C_{f_a}$ is given by the formula:

$$r_a^{a+1}(z) = \begin{cases} [p^{\varphi(a+1)}_{\varphi(a)}(x), 2t] & \text{for } z = [x,t], (x,t) \in X_{\varphi(a+1)} \times \left\langle 0, \frac{1}{2} \right\rangle \\ [h_a^{a+1}(x, 2t-1)] & \text{for } z = [x,t], (x,t) \in X_{\varphi(a+1)} \times \left\langle \frac{1}{2}, 1 \right\rangle \\ [q_a^{a+1}(y)] & \text{for } z = [y], y \in Y_{a+1} \end{cases}$$

The shape of $\varprojlim C_f$ is proved to be independent of the choice of (h_a^{a+1}).

Theorem 1′. [6]. *Let X be an inclusion ANR-sequence, Y – an* *arbitrary inverse sequence and let* $f: X \to Y$ *be a cofinal map (i.e.* $f = (\varphi, f_a)$, *where* $\varphi(N)$ *is cofinal to* N*). Then* f *is a homotopy equivalence if and only if* X *is a deformation retract of* C_f.

The last result is used in the proof of the following theorem which is an analogue of the Whitehead Theorem.

Theorem 2. [8]. *Let* (X, x_0) *and* (Y, y_0) *be two uniformly movable (See* [7]*) pointed continua and let* $n_0 = \max(1 + FdX, FdY)$. *Let* $f_n: \Pi_n(X, x_0) \to \Pi_n(Y, y_0)$ *be the homomorphism of n-th fundamental groups induced by a fundamental sequence* $f = \{f^k, (X, x_0), (Y, y_0)\}$. *If* f_n *is an isomorphism for* $n \leqslant n_0$ *and is an epimorphism for* $n = n_0 + 1$, *then* f *is a fundamental equivalence.*

This theorem can be reduced to the following

Theorem 2′. [8]. *Let* (X, x_0) *and* (Y, y_0) *be two inverse sequences of pointed polyhedra with* $\dim X_a \leqslant n_0 - 1$ *and* $\dim Y_a \leqslant n_0$. *Let* $f_n: \Pi_n(X, x_0) \to \Pi_n(Y, y_0)$ *be the morphism induced by a map* $f = (1_N, f_a): (X, x_0) \to (Y, y_0)$. *If* f_n *is a bimorphism for* $n \leqslant n_0$ *and an epimorphism for* $n = n_0 + 1$, *then* f *is a homotopy equivalence (in the sense of* [4]*).*

(Here the morphisms are considered in some category of inverse sequences of groups.)

REFERENCES

[1] K. Borsuk, Concerning homotopy propetties of compacta, *Fund. Math.*, 62 (1968), 223-254.

[2] K. Borsuk, Fundamental retracts and extensions of fundamental sequences, *Fund. Math.*, 64 (1969), 55-85.

[3] R.H. Fox, On homotopy type and deformation retracts, *Ann. of Math.*, 44 (1943), 40-50.

[4] S. Mardešić – J. Segal, Shapes of compacta and *ANR*-systems, *Fund. Math.*, 72 (1971), 41-59.

[5] S. Mardešić – J. Segal, Equivalence of the Borsuk and the *ANR*-system approach to shapes, *Fund. Math.*, 72 (1971), 61-68.

[6] M. Moszyńska, On shape and fundamental deformation retracts II, *Fund. Math.*, 77 (1972), 235-240.

[7] M. Moszyńska, Uniformly movable compact spaces and their algebraic properties, *Fund. Math.*, 77 (1972), 126-144.

[8] M. Moszyńska, The Whitehead theorem in the theory of shapes, to appear in *Fund. Math.*

[9] J.H.C. Whitehead, Combinatorial homotopy I, *BAMS*, 55 (1949), 313-245.

COLLOQUIA MATHEMATICA SOCIETATIS JÁNOS BOLYAI
8. TOPICS IN TOPOLOGY, KESZTHELY (HUNGARY), 1972.

WHEN ARE THE ISOMETRIES ONTO?

G. MURPHY

Dedicated to W.C. Doyle, S.J. and J.A. McCallin, S.J.

It is well known that a distance preserving mapping of a metric space into itself need not be onto [1, p. 157]. A simple example of such a map is the translation of $\{x: x \geqslant 1\}$ into itself by $f(x) = x + 1$. Throughout this paper a distance-preserving map from one metric space to another will be called an *isometry*. We note that an isometry is not necessarily onto. Lindenbaum [4] called a subset of a metric space *monomorphic* if every isometry of the set into itself is onto. He showed (1) compact metric spaces are monomorphic and (2) if a subset of a compact metric space is both F_σ and G_δ, then it is monomorphic. Freudenthal and Hurewicz [2] gave a different proof of (1). Section 1 contains theorems about monomorphic sets. In section 2 the notion of monomorphism is topologized and that concept investigated.

A CHARACTERIZATION OF MONOMORPHISM

Definition. A subset B of a metric space M is a *complete metric basis* for M if: (1) each point of M is uniquely determined by its distances from the points of B and (2) for each subset B' of M with

$B' \cong B$ (B' isometric with B) and each subset T' of M containing B', a subset T exists such that the isometry of B with B' can be extended to $T \cong T'$.

Theorem 1. *A metric space is monomorphic if and only if it has a complete metric basis.*

Proof. Suppose that M has a complete metric basis B and that f is an isometry of M into M. $B \cong f(B)$ and $f(B) \subset M$ imply the existence of a subset T of M and an isometry g, of T onto M, with $g|_B = f|_B$. If x is an element of M, then $g^{-1}(f(x))$ is an element of T. If y is an element of B, then $g^{-1}(f(y)) = f^{-1}(f(y)) = y$. It follows that $d(x, y) = d(f(x), f(y)) = d(g^{-1}(f(x), g^{-1}(f(y))) = d(g^{-1}(f(x)), y)$ for all y in B. Since B is a complete metric basis, this implies that $g^{-1}(f(x)) = x$. Therefore, g is defined on all of M and $g = f$, so f is onto. If M is monomorphic, then M, itself, satisfies the definition of a complete metric basis.

Theorem 2. *Let A be any subset of a complete metric space M such that $\bar{A}$ is a complete metric basis for M. If A is non-monomorphic, then $M - A$ is non-monomorphic.*

Proof. The proof is accomplished by extending a non-onto isometry f of A to an isometry g of M. Then $g^{-1}|_{M-A}$ is the required non-onto isometry of $M - A$ into itself.

Corollary 2.1. *Suppose that S and $M - S$ are dense subsets of the compact metric space M. S is monomorphic if and only if $M - S$ is monomorphic.*

TOPOLOGICAL MONOMORPHISM

Theorem 3. *Each subset of E_n, the Euclidean n-dimensional space, is homeomorphic to a monomorphic subset of E_n.*

This theorem is proved by showing that the set in question is homeomorphic to a bounded set whose closure has only a finite number of isometries.

Definition. A metric space $M = (X, d)$ is *topologically monomor-*

phic if (X, d') is monomorphic for every metric d' which is topologically equivalent to d.

One method of topologizing metric properties is to assume that some equivalent metric has the property. Theorem 3 indicates that in this case that does not lead to a very interesting class of spaces. On the other hand, with the present definition and the previously mentioned result (1) of Lindenbaum it follows that the class of topologically monomorphic spaces contains the class of compact metric spaces.

An example of de Groot [3] shows that there are non-compact sets which are topologically monomorphic.* One may ask whether Theorem 3 may be extended. In particular, is every metric space homeomorphic to a monomorphic space?

BIBLIOGRAPHY

[1] L.M. Blumenthal, *Theory and applications of distance geometry*, Oxford, 1953.

[2] H. Freudenthal – W. Hurewicz, Dehnungen, Verkürzungen, Isometrien, *Fund. Math.*, 26 (1936), 120-122.

[3] J. de Groot, Groups represented by homeomorphism groups I, *Math. Annalen*, 138 (1959), 80-102.

[4] A. Lindenbaum, Contributions à l'étude de l'espace métrique, *Fund. Math.*, 8 (1926), 209-222.

*The author is indebted to A. Szép for this reference.

COLLOQUIA MATHEMATICA SOCIETATIS JÁNOS BOLYAI
8. TOPICS IN TOPOLOGY, KESZTHELY (HUNGARY), 1972.

A GENERALIZATION OF THE BORSUK-ULAM ANTIPODAL THEOREM

K. NOWINSKI

The following problem is a natural, geometric generalization of the famous Borsuk-Ulam theorem:

Let $f\colon S^n \to E^n$ be a continuous mapping from the standard unit sphere in E^{n+1} into the euclidean n-space E^n and let $d \leqslant 2$ be a positive real number. Do there exist two points $p, q \in S^n$ such that $\rho(p, q) = d$ and $f(p) = f(q)$?

It is clear that in case $d = 2$ we obtain the hypothesis of the classical theorem. The present paper gives a positive answer under the additional assumption that n is even.

Theorem. *Let* n *be an even positive integer. Then for every continuous mapping* $f\colon S^n \to E^n$ *from the standard euclidean unit sphere and for every* $d \in (0, 2]$ *there exist two points* $p, q \in S^n$ *such that* (*) $\rho(p, q) = d$ and (**) $f(p) = f(q)$.

Proof. Let us notice the following

Lemma. *Every antipodity-preserving map* $f\colon S^n \to S^n$ *(that means* $f(-x) = -f(x)$*) is essential (not homotopic to* $*$*).*

This can be proved by considering the following commutative square:

$$\begin{array}{ccc} S^n & \xrightarrow{\ f\ } & S^n \\ p\downarrow & & \downarrow p \\ RP^n & \xrightarrow{\ \tilde{f}\ } & RP^n \end{array}$$

where p is the standard two-point covering map and $\tilde{f}$ is the mapping induced by f in the natural way.

Introducing the standard co-ordinates $x = (x_1, \ldots, x_n)$ in E^n we can represent S^n as the set $\left\{(x_1, \ldots, x_{n+1})\colon \sum_{i=1}^{n+1} x_i^2 = 1\right\}$.

We denote by $h_t\colon S^n \to S^n$, for $t \in [0, 2\pi]$, the following isotopy:

$$h_t(x_1, \ldots, x_{n+1}) = (x_1 \cos t + x_2 \sin t, x_2 \cos t - x_1 \sin t, \ldots, x_{n+1}).$$

It is clear that for every t the mapping h_t is an isometry and that h_0 is the identity map. Now let $u = \arc \sin \frac{d}{2}$ and $g'_t\colon S^{n-1} \to S^n$ be for $t \in [0, 2\pi]$ the homotopy defined as follows:

$$g'_t(x_1, \ldots, x_n) = \begin{cases} (\cos(u+t), x_1 \sin(u+t), \ldots, x_n \sin(n+t))\colon t \in [0, \pi] \\ h_{t-\pi}(g'(x_1, \ldots, x_n))\colon t \in [\pi, 2\pi]. \end{cases}$$

Moreover, define the homotopy $g''_t\colon S^{n-1} \to S^n$ by the formula:

$$g''_t(x_1, \ldots, x_n) = \begin{cases} (\cos(t-u), x_1 \sin(t-u), \ldots, x_n \sin(t-u))\colon t \in [0, \pi] \\ h_{t-\pi}(g''(x_1, \ldots, x_n))\colon t \in [\pi, 2\pi]. \end{cases}$$

By an immediate computation we obtain, that for every $t \in [0, 2\pi]$ and for every $x = (x_1, \ldots, x_n) \in S^{n-1}$ $\rho(g'_t(x), g''_t(x)) = d$. Notice, moreover, that $g'_0(x) = g''_0(-x)$ and $g'_{2\pi}(x) = g''_{2\pi}(-x)$; more precisely if we put $c = \sqrt{1 - \frac{d^2}{4}}$, then

(1) $g_0'(x) = \quad c, \frac{d}{2} x_1, \ldots, \frac{d}{2} x_n$

(2) $g_0''(x) = \quad c, -\frac{d}{2} x_1, \ldots, -\frac{d}{2} x_n$

(3) $g_{2\pi}'(x) = \quad c, \frac{d}{2} x_1, -\frac{d}{2} x_2, \ldots, -\frac{d}{2} x_n$

(4) $g_{2\pi}''(x) = \quad c, -\frac{d}{2} x_1, \frac{d}{2} x_2, \ldots, \frac{d}{2} x_n$.

Now, let $f_t \colon S^{n-1} \to E^n$ be for $t \in [0, 2\pi]$ defined by the formula $f_t(x) = f(g_t'(x)) - f(g_t''(x))$. Notice that if $f_t(x) = 0$ for some x and t then the pair $(p = g_t'(x), q = g_t''(x))$ satisfies conditions (*) and (**) of our theorem. It follows from the equations (1) – (4) that $f_0(x) = -f_0(-x)$ and $f_{2\pi}(x) = -f_{2\pi}(-x)$ and moreover $f_0(x_1, \ldots, x_n) = f_{2\pi}(x_1, -x_2, \ldots$ $\ldots, -x_n)$. Let us assume now, that there exist no x and t such that $f_t(x) = 0$ and notice that putting in this case $\widetilde{f}_t(x) = \dfrac{f_t(x)}{|f_t(x)|}$ we obtain the homotopy $\widetilde{f}_t \colon S^{n-1} \to S^{n-1}$ joining the two antipodity-preserving mappings $\widetilde{f}_0$ and $\widetilde{f}_{2\pi}$. Moreover, similarly as above $f_0(x_1, \ldots, x_n) =$ $= f_2(x_1, -x_2, \ldots, -x_n)$.

It is well-known that if k is even then the antipodal mapping $a(x) = -x$ on S^n into itself is homotopical to the identity map. Therefore it is clear, that the mapping $f_{2\pi}' \colon S^{n-1} \to S^{n-1}$ defined by $\widetilde{f}_{2\pi}(x) =$ $= \widetilde{f}_{2\pi}'(-x)$ is homotopic to $\widetilde{f}_{2\pi}$ and clearly $\widetilde{f}_{2\pi}'(-x_1, x_2, \ldots, x_n) =$ $= f_0(x_1, \ldots, x_n)$, that means, $f_{2\pi} \sim f_0$ and simultaneously $[f_{2\pi}] =$ $= -[f_0]$ in the sense of $\pi_{n-1}(S^{n-1})$. But this group is precisely Z and hence $[f_0]$ must be equal to the zero of this group, that means, f_0 is not essential. But f_0 is antipodity-preserving, so we obtain a contradiction with the hypothesis of the Lemma. Hence there exist x and t such that $f_t(x) = 0$ and the pair $(g_t(x), g_t'(x))$ satisfies conditions (*) and (**) and the proof is completed.

Remark. In the case of an odd n the problem remains open except the trivial case $n = 1$.

COLLOQUIA MATHEMATICA SOCIETATIS JÁNOS BOLYAI
8. TOPICS IN TOPOLOGY, KESZTHELY (HUNGARY), 1972.

THE p-ADIC TOPOLOGIES AND THEIR GENERALIZATIONS

P. NYIKOS

The p-adic topologies on abelian groups, and their various generalizations, are important examples in what might be called topological algebra. Unlike algebraic topology or even the theory of topological groups, the theory presented here consists largely of algebraic results obtained by topological means. In this respect it is more like the Stone duality of Boolean algebras, but it is even more "lopsided" because there are almost no topological results obtained by algebraic means.

A manuscript now in preparation [5] will give proof of the results in this paper, and of many others which had to be omitted due to lack of space.

1. THE p-ADIC TOPOLOGIES ON ABELIAN GROUPS

There is much "folklore" on the p-adic topologies, but very little has actually appeared in print. For some reason, most books and papers on abelian groups generally avoid mentioning topology or use as little as possible [1], [2], [4]). This is surprising because the topological tools used here, except the result that a completion is a projective limit, are extremely elementary facts about pseudometric spaces. Moreover, many of the results in the works cited are awkward to state in algebraic language but have very simple topological characterization.

Definition. Let G be an abelian group and let p be a prime number. The *p-height in* G of an element $x \in G$, denoted $h_p^G(x)$ or simply $h(x)$, is the greatest integer n for which there exists $y \in G$ such that $p^n y = x$. If no such integer exists, $h(x) = \infty$. *The p-adic value of* x in G, denoted $|x|G$ or simply $|x|$ is $1/p^{h(x)}$, with the convention that $1/p^\infty = 0$. Given two elements x, y in G, their *p-adic distance in* G, denoted $d_p^G(x, y)$ or $d_p(x, y)$, is the real number $|x - y|G$.

The p-adic value and distance get their names from the p-adic valuations of algebraic number theory, which are used to define topologies on the field of rational numbers. Given a prime p, one defines the p-adic values of the integers as in the definition above, letting $G = \langle Z, + \rangle$. This value is extended to all the rational numbers by letting $|a/b| = |a| / |b|$. The resulting function, called the p-adic valuation, has many of the properties of the ordinary absolute value, and it even satisfies the strong triangle inequality:

$$|a + b| = \max \{|a|, |b|\} .$$

The importance of the p-adic valuations stems from the topologies and uniformities they give rise to an Q. Similarly, we can use the p-adic distances between elements of an abelian group to produce what we will call the p-adic topology and the p-adic uniformity with respect to a fixed prime p. We pause here to note a discrepancy: if we let $G = \langle Q, + \rangle$ in the definition above, then $|x|Q = 0$ for every rational number x and every prime p. Nevertheless, we have adopted the definition because there will be no further occasion to refer to the classical p-adic valuation on Q in this paper.

For the rest of this section, "group" will mean "abelian group", and topological statements will all refer to the p-adic topology, with respect to a fixed prime p, applied to every abelian group.

It is trivial to show that every homomorphism is contractive (that is, $d_p(f(x), f(y)) \leqslant d_p(x, y)$ for any homomorphism f) and therefore uniformly continuous. Hence every isomorphism and every embedding of a direct summand is an isometry, and summands of metric groups are closed.

There are several formulae reminiscent of the theory of normed vector spaces. Given a family $\{G_a : a \in \mathcal{A}\}$ of groups, and an element $x \in \in \Pi_a G_a$, one has $|x| = \max\{|x_a| : a \in \mathcal{A}\}$. The same "sup norm" formula holds in $\oplus_a G_a$. Given a surjective homomorphism $f: G \to H$, the "quotient norm" formula $|y| = \inf\{|x| : f(x) = y\}$ obtains. Every product of complete groups is complete in the p-adic uniformity (which is finer than the product uniformity). Every product of metric groups is metric, and every homomorphic image of a complete group is complete. The proofs are almost identical with those of the analogous results in the theory of normed vector spaces.

Let C_n denote the cyclic group of order n. The group $G = \prod_{n=1}^{\infty} C_p n$ is a complete metric space. The closure of $H = \bigoplus_{n=1}^{\infty} C_p n$ in G is the set of all elements $\{x_n\}$ of G such that $\{|x_n|\}$ converges to 0. This contains the torsion subgroup T of G properly. *Hence neither* H *nor* T *is a direct summand of* G. The group $K = G/\mathrm{cl}\, H$ is a complete, metric, torsion-free group. If $x \in G$, let x^* be its image in K; then $|x^*| = \limsup |x_n|$.

If we define a "seminorm" on $\mathrm{Hom}(G, H)$ as in the theory of normed vector spaces $\left(\|f\| = \inf\{k : |f(x)| \leqslant k|x| \text{ for all } x \in G\}\right)$ then we have the familiar formula $\|f\| = \sup\{|f(x)| : |x| = 1\}$, and the supremum is actually a maximum. But we can go a step further and show that $\|f\| = \max\{|f(x)| : x \in G\}$. In other words, $\|f\|$ is the p-adic value of f in the group K of all functions from G to H. It is *not* necessarily the p-adic value of f in the group $\mathrm{Hom}(G, H)$. Indeed, the two *topologies* are often distinct, that produced by the seminorm being the coarser of the two. As in the theory of normed vector spaces, $\mathrm{Hom}(G, H)$ is a complete metric space under $\|\ \|$ whenever H is. The same is true of the p-adic metric on $\mathrm{Hom}(G, H)$, but this requires a rather deep theorem:

Theorem 1. *Let* G *be a metric group and let* $(\tilde{G}, \tilde{d})$ be *the completion of* G. *The metric* $\tilde{d}$ *on* $\tilde{G}$ *is the* p-*adic metric. In particular,*

G is dense and isometrically embedded in $\widetilde{G}$ with the p-adic metric.

From this theorem it follows that a closed subgroup G of a complete metric group K is complete in the p-adic uniformity whether or not it is isometrically or even topologically embedded. One need only extend the algebraic embedding to a homomorphism f from $\widetilde{G}$ to K. Clearly $f(\widetilde{G}) = f(G) = G$, and so G is complete.

Most simple topological concepts also have simple algebraic characterizations. Here is a partial listing. For the sake of brevity, we use "iff" to mean "if and only if".

The group G is a metric space iff $\bigcap_{n=1}^{\infty} p^n G = 0$, where $p^n G = \{x \in G\colon x = p^n y \text{ for some } y \in G\}$.

The group G is discrete iff $p^n G = 0$ for some n, and it is indiscrete iff $pG = G$.

The group G is totally bounded iff $G/p^n G$ is finite for each n. This condition may be weakened to G/pG being finite in view of the fact that $G/p^n G$ is a direct sum of cyclic groups. The group G is separable iff G/pG is countable, and in general the cardinality, if infinite, of G/pG is the same as the smallest cardinality a dense subset of G can have.

A subgroup H of a group G is open iff it contains $p^n G$ for some n iff G/H is discrete.

A subset S of a group G is dense iff $S + p^n G = G$ for all n, and S is closed iff the image of each element x of G not in S is outside the image of S in one of the factor groups $G/p^n G$.

More interesting characterizations are obtainable if S is a subgroup. There is what I call the Closed Kernel Theorem: a subgroup H of a group G is closed iff G/H is a metric space. As a corollary, H is dense iff G/H is indiscrete iff $H + pG = G$.

The Closed Kernel Theorem can be proven quickly by using the Weak Open Mapping Theorem: every surjective homomorphism is an open map. More generally, we have the Strong Open Mapping Theorem: a homomorphism $f\colon G \to H$ is open iff $f(G)$ has a nonempty interior.

Definition. A subgroup H of a group G is *p-pure* if $H \cap p^n G = p^n H$ for all n. A metric group G is *conditionally complete* if, given a group G' containing G such that (i) $\bigcap_{n=1}^{\infty} p^n G' = 0$, (ii) G is a p-pure subgroup of G' and (iii) $G \cap pG' = G'$, the equality $G = G'$ holds.

Topologically speaking, H is p-pure in G iff it is isometrically embedded. To say that G is conditionally complete is to say that if G is isometrically embedded and dense in a metric group G', then G is all of G'. The following theorem is a corollary of the first one.

Theorem 2. *A metric group* G *is complete if, and only if, it is conditionally complete.*

From the theory of uniform spaces, the projective limit (lim) of

$$\ldots \to G/p^n G \to \ldots \to G/pG \to 0$$

is the completion of G when endowed with the relative uniformity as a subspace of $\Pi_{n=1}^{\infty} G/p^n G$. The projective limit is a p-pure subgroup of the product. The product uniformity on the projective limit is the p-adic uniformity, even though these do not coincide on the product as a whole.

Using either this characterization of the completion of G, or an elementary analytic proof, one can show the following:

Theorem 3. *Let* H *be a p-pure dense subgroup of the metric group* G. *Let* f *be a homomorphism from* H *to a complete metric group* G', *There is a unique homomorphism* φ *from* G *to* G' *whose restriction to* H *is* f.

In fact, since f factors through $H/H \cap (\bigcap_{n=1}^{\infty} p^n G)$, the condition that G be a metric group may be dropped. Completeness of G' may not be dropped. However, as long as G' is a metric group, the extension φ, if it exists, is unique.

A glance at Theorem 3 should bring out the importance of the following concept:

Definition. Let G be a group and let p be a prime. A *p-basic* subgroup of G is a subgroup B of G which is (i) dense and (ii) p-

pure in G, and is (iii) a metric space in the p-adic topology and (iv) a direct sum of cyclic groups.

One of the most remarkable and useful results of group theory was discovered by Fuchs [2].

Theorem 4. *Every* (abelian) *group* G *has a* p*-basic subgroup.**

An immediate corollary it that every (abelian) group G satisfying $p^n G = 0$ for some integer n, being a discrete space, is a direct sum of cyclic p-groups. By using the fact that every torsion (abelian) group G is the direct sum of its primary components, one deduces that, if $nG = 0$ for any integer n (for example, if G is finite) then G is a direct sum of cyclic groups of prime-power order.

Using elementary isomorphism theorems, Fuchs also showed that any two p-basic subgroups of G are isomorphic. This allows us to completely characterize all groups which are complete metric spaces in each p-adic uniformity. Given a sequence $\alpha = \{\mathfrak{m}_n\}$ of cardinal numbers, we let $B[\alpha] = \oplus_{n=0}^{\infty} B_n$ where B_0 is the direct sum of $\mathfrak{m}_0$ infinite cyclic groups and, for $n \geqslant 1$, B_n is a direct sum of $\mathfrak{m}_n$ cyclic groups of order p_n. *Every complete metric group is the completion of* $B[\alpha]$ *for a unique sequence* α. Trivially, this completion is compact iff α has only finitely many nonzero terms, each of which is finite. It is also a simple exercise to characterize all groups which are locally compact and metric in the p-adic topology.

Fuchs' method of proving Theorem 4 is of independent interest. He defines a subset S of a group G to be p-independent if, for every finite subset $\{x_1, \ldots, x_k\}$ of S and every relation

$$n_1 x_1 + \ldots + n_k x_k \in p^r G \qquad (n_i x_i \neq 0)$$

the integer p^r divides n_i for each i. He then shows that *every maximal* p*-independent subset of* G (such exist, by Zorn's Lemma) *generates* a p*-basic subgroup of* G. This method allows one to extend p-basic subgroups of p-pure subgroups of G to p-basic subgroups of G itself. A quick consequence is this: if one defines a group to be *p-injective* if it is a direct summand of every (abelian) group containing it as a p-pure subgroup, then

*For p-groups this was first shown by Kulikov (cf. [1, p. 98]).

a metric group G is p-injective if, and only if, it is complete.

2. GENERALIZATIONS

The theory outlined in Section 1 remains largely intact if we substitute for "(abelian) group" the phrase, "module over a principal ideal demain R," and make the usual substitutions elsewhere. (cf. [4], Section 12). More generally, one can take the class of modules over an integral domain R having a maximal principal ideal (p). Anything more general than this probably entails the loss of Theorem 4 and its many corollaries. Still, what went before Theorem 4 may itself be worth generalizing. Here we look at there possible generalizations, the third encompassing the first two.

First, we let R be a ring and let $\mathscr{J} = \{J_n\}_{n=0}^{\infty}$ be a descending sequence of left ideals of R intersecting in 0, with $J_0 = R$. Let M be a left R-module. We let the submodules $J_n M$ be a base for the neighbourhoods of 0, defining the $\mathscr{J}$-adic value of $x \in M$ to be e^{-n} where n is the greatest integer such that x is in $J_n M$, or zero if $x \in \bigcap_{n=0}^{\infty} J_n M$. The $\mathscr{J}$-adic distance between x and y is then the $\mathscr{J}$-adic value of $(x - y)$.

Possible "casualities" are the formula $|x| = \max\{|x_a|: a \in \mathscr{A}\}$ on the product (though not the direct sum) of R-modules; the *Strong Open Mapping Theorem,* the formula $\|f\| = \max\{|f|(x): |x| = 1\}$; the fact that the product of complete modules is complete; Theorem 2; and, of course, Theorem 4 and its offshoots. But Theorem 1 and Theorem 3 remain true if we insert the word "conditional(ly)" before "completion" ("complete") [See below]. Every metric R-module M is $\mathscr{J}$-pure in its metric completion $\widetilde{M}$. The "completion metric" $\tilde{d}$ on $\widetilde{M}$ is the $\mathscr{J}$-adic metric if, and only if, M is dense in $\widetilde{M}$ with the $\mathscr{J}$-adic metric. Otherwise $\tilde{d}$ is strictly smaller than $|\ |$ on at least some elements of $\widetilde{M}$.

The second generalization consists of putting a certain topology on every group, abelian or not. Let $\mathscr{K} = \{\mathscr{K}_n\}_{n=0}^{\infty}$ be an ascending sequence of varieties of groups [6], with $\mathscr{K}_0 = \{0\}$. For each group G let G_n be the subgroup of all elements of G which are in the kernel of every homomorphism from G to every group in $\mathscr{K}_n$. The $\mathscr{K}$-adic topology, distance, etc. are defined using G_n in the same way as $J_n M$ in the first generalization. Examples of such topologies are where $\{G_n\}$ is the lower central series or the commutator series of G.

The same results which hold for modules over a general ring, are obtainable for groups with this kind of topology. This is hardly surprising when one considers the following. If $\mathscr{G}$ is the free group on a countably infinite set of generators, then for any group G, the subgroup G_n is the set of all elements of G which are images of some element of $\mathscr{G}_n$ under some homomorphism. Similarly, if we let $\mathscr{M}$ be the direct sum of $\aleph_0$ copies of the ring R, then $J_n M$ for any R-module M is the submodule of all possible homomorphic images of elements in $J_n \mathscr{M}$.

The third generalization involves taking an abstract complete category C with a zero object and letting $\mathscr{K} = \{\mathscr{K}_n\}_{n=0}^{\infty}$ be an ascending sequence of epireflective subcategories [3] of $\mathscr{C}$, with $\mathscr{K}_0 = \{0\}$. Given an object G of $\mathscr{C}$, we let $(r_n^G, r_n G)$ be the reflection of G in $\mathscr{K}_n$. Since elements have no meaning in an abstract categorical setting, we cannot go about defining topologies, but we can translate most of the concepts into categorical terms. For example, letting $\mathscr{D}G$ be the diagram

$$\ldots \to r_n G \to r_{n-1} G \to \ldots \to 0$$

we define G to be *separated* if the induced map into $\varprojlim \mathscr{D}G$ is a monomorphism, and *complete* if it is an extremal epimorphism. Given a subobject $f\colon H \to G$ of G, we say H is *dense* in G if the induced map from $r_n H$ to $r_n G$ is an epimorphism, and $\mathscr{K}$*-pure* if it is an extremal monomorphism. An object G is *conditionally complete* if it is separated, and every monomorphism f from G to a separated object H such that G is $\mathscr{K}$-pure and dense, is an isomorphism.

Despite the extreme generality, we still have some interseting results. Every product of separated objects is separated, every separated object G is $\mathscr{K}$-pure in $\varprojlim \mathscr{D}G$ (indeed, the induced map from $r_n G$ is a retraction) and $\varprojlim \mathscr{D}G$ is conditionally complete and a retract of $\varprojlim \mathscr{D}(\varprojlim \mathscr{D}G)$.

Applying these results to the case where $\mathscr{C}$ is the category of groups or the category of R-modules, one can construct the conditional completion G^* of G as a subgroup [submodule] of $\widetilde{G}$. Let $G(0) = \widetilde{G}$, and for each ordinal a let $G(a+1)$ be the closure of G in $G(a)$. If a is a limit ordinal let $G(a) = \bigcap_{\beta < a} G(C)$. The intersection of this transfinite sequence is G^*. Conditionally complete metric groups [R-modules] from a reflective subcategory of $\mathscr{C}$. In particular, the product of conditionally complete groups or R-modules is conditionally complete.

REFERENCES

[1] L. Fuchs, *Abelian Groups,* Pergamon Press, 1960.

[2] L. Fuchs, Notes on abelian groups, II, *Acta Math. Acad. Sci. Hung.,* 11 (1960), 117-125.

[3] H. Herrlich, *Topologische Reflexionen und Coreflexionen,* Springer-Verlag, 1968.

[4] I. Kaplansky, *Infinite abelian groups* (revised ed.), The University of Michigan Press, 1969.

[5] P. Nyikos, *The p-adic topology on abelian groups.* To appear.

[6] E. Schenkman, *Group theory,* D. Van Nostrand co. 1965.

COLLOQUIA MATHEMATICA SOCIETATIS JÁNOS BOLYAI
8. TOPICS IN TOPOLOGY, KESZTHELY (HUNGARY), 1972.

A CHARACTERIZATION OF COMPACT, SEPARABLE, ORDERED SPACES

A.J. OSTASZEWSKI

We give below a characterization of totally ordered sets whose order topology (see [1] for its definition and properties) is compact and separable – the latter meaning that there is a countable subset which is dense in the topology. We recall that compactness is here equivalent to the existence of a first and last element in the ordering and of a supremum for every subset of the ordered set. The characterization is the following:

Theorem 1. *Let* $(X, <)$ *be a non-empty totally ordered set. The order topology of* X *is compact and separable if and only if* X *is order isomorphic to a subset* Y *of* $[0, 1] \times \{0, 1\}$, *with* Y *ordered lexicographically subject to the condition that*

$$Y \cap [0, 1] \times \{0\}$$

is a closed subset of the order topology of $[0, 1] \times \{0\}$, *and*

$$\langle x, 1\rangle \in Y \to \langle x, 0\rangle \in Y .$$

It is however convenient to prove first a somewhat different theorem from which we can deduce the characterization above. It is:

Theorem 2. *Let* $(X, <)$ *be a non-empty totally ordered set. The order topology on* X *is compact and separable if and only if* $(X, <)$ *is order isomorphic either to a closed countable subset of the unit interval or to a subset* Y *of the unit square* S, *with* Y *ordered lexicographically, such that* Y *enjoys the properties*:

(i) $[0, 1] \times \{0\} \subseteq Y$,

(ii) for each x in $[0, 1]$ the set Y^x defined by $Y^x = Y \cap \cap \{x\} \times [0, 1]$ is finite or countable and closed in S,

(iii) apart from an at most countable set of x in $[0, 1]$, Y^x consists of no more than two points.

A proof of Theorem 1 will appear in the Journal of the London Math. Soc.

REFERENCES

[1] L. Gillman – M. Jerison, Rings of continuous functions, 52-3.

COLLOQUIA MATHEMATICA SOCIETATIS JÁNOS BOLYAI
8. TOPICS IN TOPOLOGY, KESZTHELY (HUNGARY), 1972.

AXIOMATIC HOMOLOGY THEORY AND THE ALEXANDER DUALITY THEOREM

M.M. POSTNIKOV

The aim of this lecture is to prove the Alexander theorem by very simple, standard means of the homotopy theory, using the axiomatic approach. For simplicity, the theorem will be proved under some severe restrictions (fulfilled in the classical polyhedral case).

Let $\{H_q, \partial_q\}$ be a homology theory (in the sense of Eilenberg-Steenrod) and (X, A) be a pair (A is closed) such that.

1) X is *acyclic* (in dimensions $p-1$, p). Then, by the exactness axiom

$$H_{p-1}(A) \approx H_p(X, A) .$$

Let

$$C_i = X \cup CA$$

be the cone of the inclusion map i: $A \to X$. Then by the excision and homotopy axioms

$$H_p(X, A) \approx H_p(C_i, CA).$$

Now let us assume that

2) (X, A) is a *cofibration.* This implies (by well-known arguments) a homotopy equivalence

$$(C_i, CA) \sim (X/A, *).$$

Therefore

$$H_p(C_i, CA) \approx H_p(X/A).$$

If

3) X is *compact*, then $X \setminus A$ is locally compact and let us consider the are-point compactification $(X \setminus A)^\circ$ of $X \setminus A$.

Then

$$H_p(X/A) \approx H_p((X \setminus A)^\circ).$$

So

$$H_{p-1}(A) \approx H_p(X, A) \approx H_p(C_i, CA) \approx H_p(X/A) \approx H_p((X \setminus A)^\circ),$$

that is

(1) $$H_{p-1}(A) \approx \hat{H}_p(X \setminus A),$$

where $\{\hat{H}_p\}$ is the homology theory on the category of locally compact spaces, defined by formula

$$\hat{H}_p(X) = H_p(X^\circ).$$

The isomorphism (1) is in fact the Alexander duality theorem.

REFERENCES

[1] K. Borsuk, *Theory of retracts,* Monografie Matematyczne, Tom 44. Państwowe Wydawnictwo Naukowe, Warsaw, 1967.

[2] S.T. Hu, *Theory of retracts,* Wayne State University Press, Detroit, 1965.

[3] Z.Yu. Lisitsa, to appear in Fund. Math.

COLLOQUIA MATHEMATICA SOCIETATIS JÁNOS BOLYAI
8. TOPICS IN TOPOLOGY, KESZTHELY (HUNGARY), 1972.

A LINDELÖF SPACE X SUCH THAT X^2 IS NORMAL BUT NOT PARACOMPACT

T. PRZYMUSIŃSKI

It is known that the assumption of Martin's Axiom and $2^{\aleph_0} = 2^{\aleph_1}$ is independent from the ZFC axioms of the set theory and that under this assumption there exists a normal Moore space, which is not metrizable (cf. F. D. Tall [2]).

M. Maurice (cf. E. Michael [1], Problem 7.2) and in a special case H. Tamano [3] raised the problem, whether the product of two paracompact spaces, which is normal must be necessarily paracompact. The following theorem solves this problem in the negative under the assumption of Martin's Axiom and $2^{\aleph_0} = 2^{\aleph_1}$.

Theorem. (Martin's Axiom and $2^{\aleph_0} = 2^{\aleph_1}$). There exists a hereditarily Lindelöf space X such, that X^2 is perfectly normal but not collectionwise normal.

Moreover, the space X is hereditarily separable and satisfies the first axiom of countability and the space X^2 is subparacompact.

Added in proof: The proof of the Theorem appeared in *Fundamenta Mathematicae,* 78 (1973.

REFERENCES.

[1] E. Michael, Paracompactness and the Lindelöf property in finite and countable certesian products, *Compositio Mathematica,* 23 (1971), 199-214.

[2] F.D. Tall, Set theoretic consistency results . . . , Thesis, University of Wisconsin 1969.

[3] H. Tamano, Note on paracompactness, *J. Math. Kyoto Univ.*, 3 (1963), 137-143.

COLLOQUIA MATHEMATICA SOCIETATIS JÁNOS BOLYAI
8. TOPICS IN TOPOLOGY, KESZTHELY (HUNGARY), 1972.

REALIZATIONS OF TOPOLOGIES BY SET-SYSTEMS

J. ROSICKÝ — M. SEKANINA

The objective of this paper is to study the connection between categories of topological spaces and the category $\mathscr{S}^-$, objects of which are the pairs $(A, \mathfrak{A})$ where A is a set and $\mathfrak{A} \subseteq \exp A$ and morphisms from $(A, \mathfrak{A})$ to $(B, \mathfrak{B})$ are maps $f: A \to B$ such that for $C \in \mathfrak{B}$ we have $f^{-1}(C) \in \mathfrak{A}$. This category has been studied manytimes (e.g. [3], [4]). It is clear that the category of all topological spaces in the sense of [1] (it will be denoted by $\mathscr{B}$) is a full subcategory of $\mathscr{S}^-$. Given an object $(A, \mathfrak{A})$ we shall often drop A and parentheses and use the symbol $\mathfrak{A}$ for it only. Further, we shall use the following notations:

$$\bar{\mathfrak{A}} = \{A - X: X \in \mathfrak{A}\}.$$

$$\mathfrak{A}^0 = \mathfrak{A} \cap \bar{\mathfrak{A}}.$$

For $C \subseteq A$, $\mathfrak{A} | C = \{X \cap C: X \in \mathfrak{A}\}$.

If $(B, \mathfrak{B})$ is a further object of $\mathscr{S}^-$, $[(A, \mathfrak{A}), (B, \mathfrak{B})]$ (briefly $[\mathfrak{A}, \mathfrak{B}]$) will be the set of all morphisms from $(A, \mathfrak{A})$ in $(B, \mathfrak{B})$.

SOME GENERAL ASSERTIONS

Lemma 1. *Let* $\phi, A \in \mathfrak{A}$. *Then* $[\mathfrak{A}, \mathfrak{B}]$ *contains all constant maps from* A *to* B. *On the contrary, if* $[\mathfrak{A}, \mathfrak{B}]$ *contains all constant maps and there is a* $Y \in \mathfrak{B}$, $\phi \neq Y \neq B$, *then* $\phi, A \in \mathfrak{A}$.

Proof is clear.

Lemma 2. *Let* $\{\phi, A\} \subsetneqq \mathfrak{A} \subsetneqq \exp A$, $\mathfrak{A}_1 \subseteq \exp A$, $[\mathfrak{A}, \mathfrak{A}] = [\mathfrak{A}_1, \mathfrak{A}_1]$. *Then*

a) $\{\phi, A\} \subsetneqq \mathfrak{A}_1 \neq \exp A$

b) $\mathfrak{A}^0 = \mathfrak{A}_1^0$

c) *For every two-point subset* C *of* A *we have* $\mathfrak{A} \mid C = \mathfrak{A}_1 \mid C$ *or* $\mathfrak{A} \mid C = \bar{\mathfrak{A}}_1 \mid C$.

Proof. a) If $\mathfrak{A}_1 = \exp A$ or $\{\phi, A\} \supset \mathfrak{A}_1$, then $[\mathfrak{A}_1, \mathfrak{A}_1] = A^A = [\mathfrak{A}, \mathfrak{A}]$ and therefore $\mathfrak{A} = \exp A$ or $\mathfrak{A} \subset \{\phi, A\}$, a contradiction. Hence Lemma 1 implies a).

b) Let $C \in \mathfrak{A}^0$. Let $C_1 \in \mathfrak{A}_1$, $\phi \neq C_1 \neq A$, $c \in C_1$, $d \in A - C_1$. Let $f\colon A \to A$ with $f(x) = c$ for $x \in C$, $f(x) = d$ otherwise. Then $f \in [\mathfrak{A}, \mathfrak{A}]$, therefore $f \in [\mathfrak{A}_1, \mathfrak{A}_1]$ and $f^{-1}(C_1) = C \in \mathfrak{A}_1$. By symmetry $A - C \in \mathfrak{A}_1$, too, and we have $C \in \mathfrak{A}_1^0$. We have got $\mathfrak{A}^0 \subset \mathfrak{A}_1^0$. Taking $\mathfrak{A}_1$ for $\mathfrak{A}$ we get the opposite inclusion.

c) Let C be a two-point subset of A, $C = \{x, y\}$.

c$_1$) Let $\mathfrak{A} \mid C = \{\phi, C\}$. We have $f\colon A \to A$, $f(A) \subseteq C \Rightarrow f \in [\mathfrak{A}, \mathfrak{A}] = [\mathfrak{A}_1, \mathfrak{A}_1]$. If there is $D \in \mathfrak{A}_1$, $\phi \neq C \cap D \subsetneqq C$ then $\mathfrak{A}_1 = \exp A$, a contradiction. Therefore $\mathfrak{A}_1 \mid C = \{\phi, C\}$.

c$_2$) Let $\mathfrak{A} \mid C = \exp C$. By c$_1$) $\mathfrak{A}_1 \mid C \neq \{\phi, C\}$. Suppose $\mathfrak{A}_1 \mid C = \{\phi, \{x\}, C\}$. Then there exists $X \in \mathfrak{A}_1$ such that $X \in \mathfrak{A}_1 - \mathfrak{A}_1^0$. Let $f\colon A \to A$ be defined as follows:

$$f(z) = x \quad \text{for} \quad z \in X,$$
$$f(z) = y \quad \text{otherwise.}$$

Then $f \in [\mathfrak{A}_1, \mathfrak{A}_1] = [\mathfrak{A}, \mathfrak{A}]$ and $X = f^{-1}(x) \in \mathfrak{A}$. $A - X = f^{-1}(y) \in \mathfrak{A}$, hence $X \in \mathfrak{A}^0$, which contradicts to b). Similarly for $\mathfrak{A}_1 | C = \{\phi, \{y\}, C\}$. The rest of the proof is obvious by using c_1), c_2) and symmetry.

Remark. Warndof [7] calls a topology $\mathfrak{T}$ special if for any topology $\mathfrak{S}$ with $[\mathfrak{T}, \mathfrak{T}] = [\mathfrak{S}, \mathfrak{S}]$ we have $\mathfrak{T} = \mathfrak{S}$. If this is valid in the case when $\mathfrak{S}$ is a T_1-topology, a T_1-topology $\mathfrak{T}$ is called T_1-special. In [7] some sufficient conditions for a non-discrete T_1-special topology to be special have been deduced. By c) of Lemma 2 one sees that every non-discrete T_1-special topology is special.

Namely, $\mathfrak{T} | C = \exp C$ for any two-point subset C of A.

Corollary 1. *Let* $\{\phi, A\} \subsetneqq \mathfrak{A} \subsetneqq \exp A$. *Let there exist a two-point-set* $C \subseteq A$ *such that* $\{\phi, C\} \subsetneqq \mathfrak{A} | C \neq \exp C$. *Let* $\mathfrak{A}_1 \subseteq \exp A$ *with* $[\mathfrak{A}, \mathfrak{A}] = [\mathfrak{A}_1, \mathfrak{A}_1]$. *Then* $\mathfrak{A} = \mathfrak{A}_1$ *or* $\mathfrak{A} = \bar{\mathfrak{A}}_1$.

Proof. Let $X \in \mathfrak{A}$. Let $C = \{x, y\}$ and $\mathfrak{A} | C = \{\phi, \{x\}, C\}$. By c) of Lemma 2 we have $\mathfrak{A}_1 | C = \{\phi, \{x\}, C\}$ or $\mathfrak{A}_1 | C = \{\phi, \{y\}, C\}$. In the first case the map f with $f(z) = x$ for $z \in X$, $f(z) = y$ otherwise is in $[\mathfrak{A}, \mathfrak{A}] = [\mathfrak{A}_1, \mathfrak{A}_1]$, therefore $f^{-1}(x) = X \in \mathfrak{A}_1$. In the second case similarly $A - X \in \mathfrak{A}_1$. Therefore $\mathfrak{A} \subseteq \mathfrak{A}_1$ or $\mathfrak{A} \subseteq \bar{\mathfrak{A}}_1$. By Lemma 2 one can take $\mathfrak{A}_1$ for $\mathfrak{A}$ and get $\mathfrak{A}_1 \subseteq \mathfrak{A}$ or $\mathfrak{A}_1 \subseteq \bar{\mathfrak{A}}_1$.

Let $\{\phi, A\} \subseteq \mathfrak{A} \subseteq \exp A$. The relation $\rho_{\mathfrak{A}}$ on A will be defined by $x, y \in A \Rightarrow \left(x \rho_{\mathfrak{A}} y \equiv \left(X \in \mathfrak{A} \Rightarrow \{x, y\} \subseteq X \text{ or } \{x, y\} \cap X = \phi\right)\right)$. $\rho_{\mathfrak{A}}$ is an equivalence relation. Let $A | \rho_{\mathfrak{A}}$ be the corresponding decomposition of A and $\psi_{\mathfrak{A}}\colon A \to A | \rho_{\mathfrak{A}}$ the canonical map. Let $S(\mathfrak{A}) = \{X\colon X \subseteq A | \rho_{\mathfrak{A}}, \psi_{\mathfrak{A}}^{-1}(X) \in \mathfrak{A}\}$. Let $\mathscr{R} = \{(A, \mathfrak{A})\colon \rho_{\mathfrak{A}}$ has only one-element classes$\}$. Then $\mathscr{R}$ is clearly an epireflective subcategory in $\mathscr{S}^-$, $\psi_{\mathfrak{A}}$ is the corresponding reflection.

Corollary 2. *Let* $\{\phi, A\} \subsetneqq \mathfrak{A} \subsetneqq \exp A$. *The following assertions are equivalent.*

(i) $\mathfrak{A}_1 \subseteq \exp A$, $[\mathfrak{A}, \mathfrak{A}] = [\mathfrak{A}_1, \mathfrak{A}_1] \Rightarrow \mathfrak{A} = \mathfrak{A}_1$ *or* $\mathfrak{A} = \bar{\mathfrak{A}}_1$.

(ii) $\mathfrak{B} \subseteq \exp A | \rho_{\mathfrak{A}}$, $\mathfrak{B} \in \mathscr{R}$, $[S(\mathfrak{A}), S(\mathfrak{A})] = [\mathfrak{B}, \mathfrak{B}] \Rightarrow S(\mathfrak{A}) = \mathfrak{B}$ *or* $S(\mathfrak{A}) = \bar{\mathfrak{B}}$.

Proof. First, we prove *a)*

a) $[\mathfrak{A}, \mathfrak{A}] = \{f \in A^A :$ there exists $h \in [S(\mathfrak{A}), S(\mathfrak{A})]$ such that $\psi_{\mathfrak{A}} f = h\psi_{\mathfrak{A}}\}$.

Suppose f has the property indicated on the right hand side of the equality. Let $X \in \mathfrak{A}$. Then $\psi_{\mathfrak{A}}(X) \in S(\mathfrak{A})$. Therefore $h^{-1}\psi_{\mathfrak{A}}(X) \in$ $\in S(\mathfrak{A})$ and the fact that X is saturated in the canonical decomposition of A induced by $\psi_{\mathfrak{A}}$ implies $\psi_{\mathfrak{A}}^{-1} h^{-1} \psi_{\mathfrak{A}}(X) = f^{-1}(X)$. Necessity of the condition follows from the definition of reflective subcategory.

Suppose now (i), $\mathfrak{B} \subseteq \exp A \mid \rho_{\mathfrak{A}}$, $\mathfrak{B} \in \mathscr{R}$, $[S(\mathfrak{A}), S(\mathfrak{A})] = [\mathfrak{B}, \mathfrak{B}]$. Let $\mathfrak{A}_1 = \{X \subseteq A : \psi_{\mathfrak{A}}(X) \in \mathfrak{B}\}$. Then $S(\mathfrak{A}_1) = \mathfrak{B}$ and $\psi_{\mathfrak{A}}$ (taken as a map from $(A, \mathfrak{A}_1)$ to $(A \mid \rho_{\mathfrak{A}}, \mathfrak{B})$ is the reflection. By *a*) we have $[\mathfrak{A}, \mathfrak{A}] = [\mathfrak{A}_1, \mathfrak{A}_1]$. Therefore $\mathfrak{A} = \mathfrak{A}_1$ or $\bar{\mathfrak{A}}_1$, so $S(\mathfrak{A}) = \mathfrak{B}$ or $\bar{\mathfrak{B}}$.

Let (ii) be valid, $\mathfrak{A}_1 \subseteq \exp A$, $[\mathfrak{A}, \mathfrak{A}] = [\mathfrak{A}_1, \mathfrak{A}_1]$. By c) of Lemma 2 $\psi_{\mathfrak{A}} = \psi_{\mathfrak{A}_1}$. Therefore $[S(\mathfrak{A}), S(\mathfrak{A})] = [S(\mathfrak{A}_1), S(\mathfrak{A}_1)]$ whence $S(\mathfrak{A}) =$ $= S(\mathfrak{A}_1)$ or $S(\mathfrak{A}) = \overline{S(\mathfrak{A}_1)}$, so $\mathfrak{A} = \mathfrak{A}_1$ or $\bar{\mathfrak{A}}_1$.

Corollaries 1 and 2 imply that for finding all special topologies it is sufficient to deal with T_1-special topologies.

If A is a set, $\mathfrak{A} \subseteq \exp A$ and if for every $\mathfrak{A}_1 \subseteq \exp A$, $[\mathfrak{A}_1, \mathfrak{A}_1] =$ $= [\mathfrak{A}, \mathfrak{A}]$ implies $\mathfrak{A} = \mathfrak{A}_1$ or $\mathfrak{A} = \bar{\mathfrak{A}}_1$, we say that $\mathfrak{A}$ is system-special.

ALL TOPOLOGIES AND T_1-TOPOLOGIES

In the sequel, two topologies on the same support will be compared by means of inclusion. In this sense, for a given set A, $\mathscr{B}(A)$, $\mathscr{K}(A)$ will denote the lattice of all topologies or T_1-topologies on A resp. It will be often used, without any special reference, that having a topology $\mathfrak{T}$ and its subbase $\mathfrak{A}$ then $[\mathfrak{A}, \mathfrak{A}] \subseteq [\mathfrak{T}, \mathfrak{T}]$.

First, the immediate consequence of Corollary 1.

Proposition 1. *Atoms in* $\mathscr{B}(A)$ *are system-special.*

Proof. Atoms $(A, \mathfrak{T})$ in $\mathscr{B}(A)$ are of the form $\mathfrak{T} = \{\phi, X, A\}$, where $\phi \neq X \neq A$.

Proposition 2. *Let* A *be infinite. The Fréchet space on* A, *i.e. the minimal topology in* $\mathscr{K}(A)$, *is system-special.*

Proof. Let $(A, \mathfrak{T})$ be the Fréchet space on A, i.e. $\mathfrak{T}$ consists of sets with finite complements and of the empty set. $f \in [\mathfrak{T}, \mathfrak{T}]$ if and only if the preimage of each point of A at f is a finite set or all A. Suppose we have $\mathfrak{A} \subseteq \exp A$ such that $[\mathfrak{A}, \mathfrak{A}] = [\mathfrak{T}, \mathfrak{T}]$. Clearly there exists $T \in \mathfrak{A}$, $\phi \neq T \neq A$. Suppose T is infinite. Then, as all permutations of A are in $[\mathfrak{A}, \mathfrak{A}]$.

(β) each $T_1 \subseteq A$ with $\operatorname{card} T_1 = \operatorname{card} T$, $\operatorname{card}(A - T_1) = \operatorname{card}(A - T)$ is contained in $\mathfrak{A}$, too.

Furthermore

(γ) if $T_1 \subseteq A$, $T_2 \subseteq A$, $\operatorname{card} T_1 = \operatorname{card} T$, $\operatorname{card} T_2 = \operatorname{card}(A - T)$, then $T \cup T_1 \in \mathfrak{A}$, $T - T_2 \in \mathfrak{A}$.

(γ) is a direct consequence of the form of the maps from $[\mathfrak{A}, \mathfrak{A}]$.

Suppose that for each $T_1 \in \mathfrak{A}$ with $\phi \neq T_1 \neq A$ we have $\operatorname{card} A = \operatorname{card}(A - T_1)$. Then there exists an infinite $\mathfrak{m}$ such that for $X \subseteq A$ with $\operatorname{card} X < \mathfrak{m}$ we have $X \in \mathfrak{A}$ but for Y with $\operatorname{card} Y = \mathfrak{m}$, $Y \notin \mathfrak{A}$. We have $\mathfrak{m} > \aleph_0$.

Every mapping $g: A \to A$ in which the preimage of each point of A has the cardinality $\aleph_0$ is in $[\mathfrak{A}, \mathfrak{A}]$, a contradiction.

We have just seen we must have a $T \in \mathfrak{A}$ with $\operatorname{card}(A - T) < \operatorname{card} A$, $T \neq A$. Suppose we have still some $T_1 \in \mathfrak{A}$ with $\operatorname{card}(A - T_1) = \operatorname{card} A$, $\phi \neq T_1$. Then all finite subsets of A are in $\mathfrak{A}$ and all subsets of A with the finite complement as well. Therefore any finite subset of A belongs to $\mathfrak{A}^0$, a contradiction with b) of Lemma 2.

Therefore $\operatorname{card}(A - T) < \operatorname{card} A$ for each $T \in \mathfrak{A}$, $T \neq \phi$. Suppose $\operatorname{card}(A - T_1)$ be infinite for some such T_1. Let $\cup A_i = A$ be a decomposition of A into countable subsets of A. Pick up $a_i \in A_i$. Define $g: A \to A$ as $g(z) = a_i$ for $z \in A_i$. For each $T \in \mathfrak{A}$ we have at most $\operatorname{card}(A - T)$ of a_i's in $A - T$. Therefore $A - g^{-1}(T)$ is countable or of cardinality at most $\operatorname{card}(A - T)$. By ($\beta$) and ($\gamma$) we have $g^{-1}(T) \in \mathfrak{A}$ and $g \in [\mathfrak{A}, \mathfrak{A}]$, a contradiction.

Hence $\operatorname{card}(A - T) < \aleph_0$ for all $T \in \mathfrak{A}$. By (β) and (γ) we have $\mathfrak{A} = \mathfrak{T}$.

If there are only finite $T \in \mathfrak{A}$, we prove by dual considerations $\mathfrak{A} = \bar{\mathfrak{T}}$. The proof is finished.

Let A be a set, $\mathfrak{F}$ a free proper filter (i.e. $\phi \notin \mathfrak{F}$) in A, $a \in A$. Put $\mathfrak{S}(A, \mathfrak{F}, a) = \mathfrak{F} \cup \exp\left(A - \{a\}\right)$. $\mathfrak{S}(X, \mathfrak{F}, x)$ is a T_1-topology. If $\mathfrak{F}$ is an ultrafilter, we get in this way just all dual atoms in $\mathscr{K}(A)$. We call them free ultraspaces. It is known that every T_1-topology is the intersection in $\mathscr{K}(A)$ of all free ultraspaces in which it is contained.

Lemma 3. *Let* $\mathfrak{S}_1 = \mathfrak{S}(A_1, \mathfrak{F}_1, a_1)$, $\mathfrak{S}_2 = \mathfrak{S}(A_2, \mathfrak{F}_2, a_2)$, *where* $\mathfrak{F}_i$ *is a free proper filter on* A_i, $a_i \in A_i$, $(i = 1, 2)$. *Then*

a) $[\mathfrak{S}_1, \mathfrak{S}_2] = [\mathfrak{S}_1, \mathfrak{S}_2^0] = \{f \in A_2^{A_1}: f(a_1) = a_2$ *and* $\{f^{-1}(Z): a_2 \in Z \in \mathfrak{F}_2\} \subseteq \mathfrak{F}_1\} \cup \{f \in A_2^{A_1}: f(a_1) \neq a_2, f^{-1}\{f(a_1)\} \in \mathfrak{F}_1\}$.

b) $[\bar{\mathfrak{S}}_1, \mathfrak{S}_2] = [\mathfrak{S}_1^0, \mathfrak{S}_2] = \{f \in A_2^{A_1}: f^{-1}\{f(a_1)\} \in \mathfrak{F}_1\}$.

Proof. a) Let $f \in [\mathfrak{S}_1, \mathfrak{S}_2]$. If $f(a_1) = a_2$, then from $a_2 \in Z \in \in \mathfrak{F}_2$ we have $a_1 \in f^{-1}(Z) \in \mathfrak{S}_1$, which implies $f^{-1}(Z) \in \mathfrak{F}_1$. If $f(a_1) \neq \neq a_2$, then $\{f(a_1)\} \in \mathfrak{S}_2$, therefore $f^{-1}\{f(a_1)\} \in \mathfrak{F}_1$.

Let $f \in A_2^{A_1}$, $f(a_1) = a_2$, $\{f^{-1}(Z): a_2 \in Z \in \mathfrak{F}_2\} \subseteq \mathfrak{F}_1$. Let $Z \in \mathfrak{S}_2$. When $a_2 \notin Z$ then $a_1 \notin f^{-1}(Z)$ and $f^{-1}(Z) \in \mathfrak{S}_1$. Suppose $a_2 \in Z$, so $Z \in \mathfrak{F}_2$ and by our assumption $f^{-1}(Z) \in \mathfrak{F}_1 \subseteq \mathfrak{S}_1$. Therefore $f \in [\mathfrak{S}_1, \mathfrak{S}_2]$.

Let $f \in A_2^{A_1}$, $f(a_1) \neq a_2$, $f^{-1}\{f(a_1)\} \in \mathfrak{F}_1$. Let $Z \in \mathfrak{S}_2$. When $f(a_1) \notin Z$, $a_1 \notin f^{-1}(Z)$ and $f^{-1}(Z) \in \mathfrak{S}_1$. If $f(a_1) \in Z$, $f^{-1}(Z) \supseteq \supseteq f^{-1}\{f(a_1)\}$ and therefore $f^{-1}(Z) \in \mathfrak{S}_1$ and $f \in [\mathfrak{S}_1, \mathfrak{S}_2]$. We have

$$\mathfrak{S}_2^0 = \{X: X \subseteq A_2, a_2 \in X \in \mathfrak{F}_2\} \cup \{X: X \subseteq A_2, a_2 \in A_2 - X \in \mathfrak{F}_2\}.$$

As $\mathfrak{S}_2^0$ is a base for $\mathfrak{S}_2$ we have $[\mathfrak{S}_1, \mathfrak{S}_2] = [\mathfrak{S}_1, \mathfrak{S}_2^0]$.

b) Let $f \in [\bar{\mathfrak{S}}_1, \mathfrak{S}_2]$. We have $A_2 - \{f(a_1)\} \in \mathfrak{S}_2$, therefore $f^{-1}\left(A_2 - \{f(a_1)\}\right) \in \bar{\mathfrak{S}}_1$. Hence $f^{-1}(f(a_1)) = A_1 - f^{-1}\left(A_2 - \{f(a_1)\}\right) \in \in \mathfrak{S}_1$, i.e. $f^{-1}(f(a_1)) \in \mathfrak{F}_1$.

Let $f \in A_2^{A_1}$, $f^{-1}\{f(a_1)\} \in \mathfrak{F}_1$. Let $Z \in \mathfrak{S}_2$. If $f(a_1) \in Z$, then $f^{-1}(f(a_1)) \subseteq f^{-1}(Z)$ which means $f^{-1}(Z) \in \bar{\mathfrak{S}}_1$ (moreover $f^{-1}(Z) \in \mathfrak{S}_1^0$).

If $f(a_1) \notin Z$, then $f^{-1}(f(a_1)) \subseteq A_1 - f^{-1}(Z)$. Therefore $a_1 \in A_1 - f^{-1}(Z) \in \mathfrak{F}_1$ which means $f^{-1}(Z) \in \mathfrak{S}_1^0 \subseteq \bar{\mathfrak{S}}_1$. Therefore $f \in [\bar{\mathfrak{S}}_1, \mathfrak{S}_2]$.

We have proved even more namely $f \in [\mathfrak{S}_1^0, \mathfrak{S}_2]$. Therefore $[\bar{\mathfrak{S}}_1, \mathfrak{S}_2] \subseteq [\mathfrak{S}_1^0, \mathfrak{S}_2]$. The opposite inclusion is trivial, so $[\bar{\mathfrak{S}}_1, \mathfrak{S}_2] = [\mathfrak{S}_1^0, \mathfrak{S}_2]$.

Lemma 4. *Let $\mathfrak{F}$ be a free proper filter on A, $a \in A$, $\mathfrak{A} \subseteq \exp A$, $[\mathfrak{A}, \mathfrak{A}] = [\mathfrak{S}(A, \mathfrak{F}, a), \mathfrak{S}(A, \mathfrak{F}, a)]$, $A - \{a\} \in \mathfrak{A}$. Then $\mathfrak{A} = \mathfrak{S}(A, \mathfrak{F}, a)$.*

Proof. Denote $\mathfrak{S} = \mathfrak{S}(A, \mathfrak{F}, a)$. Let $X \in \mathfrak{S}$. If $a \in X$, then by b) of Lemma 2 $X \in \mathfrak{A}$. Let $a \notin X$. Define $f: A \to A$ with $f(z) = a$ for $z \in A - X$, $f(z) = z$ otherwise. To prove $f \in [\mathfrak{S}, \mathfrak{S}]$ we verify a) of Lemma 3. Let $a \in Z \in \mathfrak{F}$. Then $f^{-1}(Z) = (Z \cap X) \cup (A - X) \supseteq Z$. This implies $f^{-1}(Z) \in \mathfrak{F}$, what was to be proved. $X = f^{-1}(A - \{a\}) \in \mathfrak{A}$. Therefore $\mathfrak{S} \subseteq \mathfrak{A}$. Suppose $X \in \mathfrak{A} - \mathfrak{S}$. Then $a \in X$, therefore $A - X \in \mathfrak{S}$ and so $A \in \mathfrak{A}^0 = \mathfrak{S}^0$, a contradiction.

Lemma 5. *Let $\mathfrak{S} = \mathfrak{S}(A, \mathfrak{F}, a)$ be a free ultraspace, $\mathfrak{A} \subseteq \exp A$, $[\mathfrak{A}, \mathfrak{A}] = [\mathfrak{S}, \mathfrak{S}]$. Then $\mathfrak{A} = \mathfrak{S}$ or $\mathfrak{A} = \bar{\mathfrak{S}}$ or $\mathfrak{A} = \mathfrak{S}^0$.*

Proof. By b) of Lemma 2 $\mathfrak{A} \supseteq \mathfrak{S}^0$. Let $X \in \mathfrak{A} - \mathfrak{S}^0$. We have $X \in \mathfrak{S}$ or $X \in \bar{\mathfrak{S}}$.

1) $X \in \mathfrak{S}$. Then $a \notin X$ and $A - X \notin \mathfrak{F}$. Then $X \in \mathfrak{F}$. Choose $y \in X$. Put $f(z) = z$ for $z \in X \cup \{a\}$, $f(z) = y$ otherwise. By a) of Lemma 3 we have $f \in [\mathfrak{S}, \mathfrak{S}]$. Then $A - \{a\} = f^{-1}(X)$. By Lemma 4 $\mathfrak{A} = \mathfrak{S}$.

2) $X \in \bar{\mathfrak{S}}$ implies $A - X \in \mathfrak{S}$. We have $A - X \in \bar{\mathfrak{A}}$ as well. Clearly $[\bar{\mathfrak{A}}, \bar{\mathfrak{A}}] = [\mathfrak{A}, \mathfrak{A}] = [\mathfrak{S}, \mathfrak{S}]$. By 1) we have $\bar{\mathfrak{A}} = \mathfrak{S}$, therefore $\mathfrak{A} = \bar{\mathfrak{S}}$.

Let us pass to considerations on categories. Subcategory always means a full subcategory. Categories (taken as subcategories in $\mathscr{B}$) of T_1-, T_2-, regular, completely regular, normal, perfectly normal, compact, metrizable, free ultraspaces are denoted in turn by $\mathscr{K}$, $\mathscr{H}$, $\mathscr{R}$, $\mathscr{CR}$, $\mathscr{N}$, $\mathscr{PN}$, $\mathscr{C}$, $\mathscr{M}$, $\mathscr{FU}$. Realization F of a subcategory $\mathscr{L}$ of $\mathscr{S}^-$ in $\mathscr{S}^-$ is a full embedding preserving supports of objects and maps, i.e. the diagram

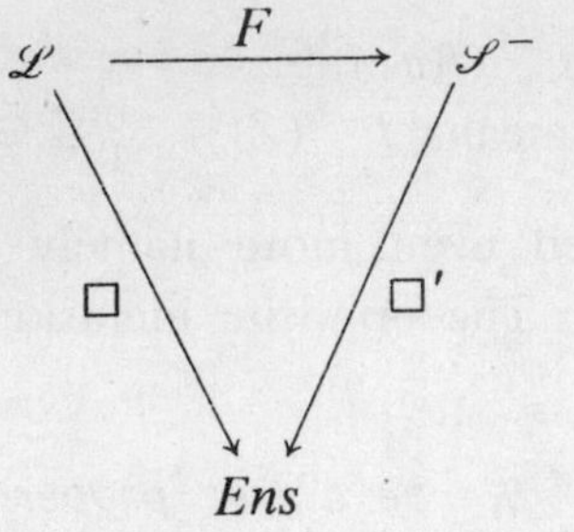

where Ens is the category of all sets and $\square, \square'$ are forgetful functors, i.e. $\square(A, \mathfrak{A}) = A$ $\square(f) = f$, is commutative (see [5].).

Proposition 3. *Let* $\mathscr{L}$ *be a full subcategory of* $\mathscr{S}^-$ *in which every object* $(A, \mathfrak{A})$, *different from* $(A, \exp A)$ *contains a two-point subset* $\{a_1, a_2\}$ *such that* $\mathfrak{A} \mid \{a_1, a_2\}$ *contains* $\{a_1\}$ *but not* $\{a_2\}$. *Furthermore,* $\phi, A \in \mathfrak{A}$ *and in* $\mathscr{L}$ *there exists an object different from objects of the form* $(A, \exp A)$. *Then there are only two realizations of* $\mathscr{L}$, F_1, F_2, $F_1(A, \mathfrak{A}) = (A, \mathfrak{A})$, $F_2(A, \mathfrak{A}) = (A, \bar{\mathfrak{A}})$.

Proof. Let F be a realization of $\mathscr{L}$. By Corollary 1, if $F(A, \mathfrak{A}) = (A, \mathfrak{A}_1)$, $\mathfrak{A} \neq \exp A$, we have $\mathfrak{A} = \mathfrak{A}_1$ or $\bar{\mathfrak{A}} = \mathfrak{A}_1$. Suppose there is an object $(A, \mathfrak{A}) \neq (A, \exp A)$ with $F(A, \mathfrak{A}) = (A, \mathfrak{A})$. Let $a_1, a_2 \in A$ be the two points mentioned in the Proposition. Let $(B, \mathfrak{B}) \in \mathscr{L}$, $F(B, \mathfrak{B}) = (B, \mathfrak{B}_1)$ and $X \in \mathfrak{B}$. Define $f\colon B \to A$ by $f(z) = a_1$ for $z \in X$, $f(z) = a_2$ otherwise. Then $f \in [\mathfrak{B}, \mathfrak{A}]$, therefore $X \in \mathfrak{B}_1$. By our assumption the inequality $\mathfrak{B} \neq \exp B$ implies $\mathfrak{B} \not\subseteq \bar{\mathfrak{B}}$ and so $\mathfrak{B}_1 = \mathfrak{B}$. Therefore $F = F_1$. The rest of the proof can be settled by dual considerations and by observing the evident fact that F_1, F_2 are realizations.

Let $\mathscr{U}$ be the category of all partially ordered sets. By [6], up to equivalence, there are only two full embeddings of $\mathscr{U}$ into $\mathscr{B}$, one given by right topologies, the other one by left topologies. Realization of $\mathscr{U}$ in $\mathscr{S}^-$ is defined similarly as for subcategories of $\mathscr{S}^-$. Proposition 3 implies

Corollary 3. *There are just two realizations of* $\mathscr{U}$ *in* $\mathscr{S}^-$. *These are given by right topologies or left topologies.*

Proposition 4. *Let* F *be a realization of* $\mathscr{F}\mathscr{U}$. *Then* $F(\mathfrak{S}) \subseteq \mathfrak{S}$

for all $\mathfrak{S} \in \mathscr{F}\mathscr{U}$ *or* $F(\mathfrak{S}) \subseteq \bar{\mathfrak{S}}$ *for all* $\mathfrak{S} \in \mathscr{F}\mathscr{U}$.

Proof. By Lemma 5 it is sufficient to prove that there do not exist $\mathfrak{S}_1 = \mathfrak{S}(A_1, \mathfrak{F}_1, a_1)$, $\mathfrak{S}_2 = \mathfrak{S}(A_2, \mathfrak{F}_2, a_2)$ in $\mathscr{F}\mathscr{U}$ with $F(\mathfrak{S}_1) = \bar{\mathfrak{S}}_1$, $F(\mathfrak{S}_2) = \mathfrak{S}_2$.

Suppose we have such spaces. Put $\mathfrak{m}_i = \operatorname{card} A_i$ for $i = 1, 2$ ($\mathfrak{m}_1, \mathfrak{m}_2$ are infinite). Let A_3 be a set with $\operatorname{card} A_3 = \mathfrak{m} = \max(\mathfrak{m}_1, \mathfrak{m}_2)$. There exists a mapping $g: A_3 \to A_1$ such that $\operatorname{card} g^{-1}(a_1) = 1$, $\operatorname{card} g^{-1}(z) = \mathfrak{m}_2$ for each $z \in A_1 - \{a_1\}$. Put $\{a_3\} = g^{-1}(a_1)$. Let $h: A_3 \to A_2$ with the following properties: $h(a_3) = a_2$, $h \mid g^{-1}(z)$ is a bijection of $g^{-1}(z)$ onto $A_2 - \{a_2\}$ for $z \neq a_1$, $z \in A_1$. Let $Z \in \mathfrak{F}_2$. Z is infinite and therefore $h^{-1}(Z) \cap g^{-1}(z)$ is infinite for $z \in A_1 - \{a_1\}$. So $h^{-1}(Z)$ has an infinite intersection with all $g^{-1}(V)$, $V \in \mathfrak{F}_1$. We can find a free ultrafilter $\mathfrak{F}_3$ on A_3 containing all $g^{-1}(V)$, $V \in \mathfrak{F}_1$ and $h^{-1}(Z)$, $Z \in \mathfrak{F}_2$. Put $\mathfrak{S}_3 = \mathfrak{S}(A_3, \mathfrak{F}_3, a_3)$. By construction of $\mathfrak{S}_3$ and by a) from Lemma 3 we have $g \in [\mathfrak{S}_3, \mathfrak{S}_1]$, $h \in [\mathfrak{S}_3, \mathfrak{S}_2]$. Therefore $\{a_3\} = g^{-1}(a_1) \in F(\mathfrak{S}_3)$, $A_3 - \{a_3\} = h^{-1}\left(A_2 - \{a_2\}\right) \in F(\mathfrak{S}_3)$ which contradicts Lemma 5.

Corollary 4. *Let* $\mathscr{L}$ *be a subcategory of* $\mathscr{K}$ *containing* $\mathscr{F}\mathscr{U}$ *and* F *be a realization of* $\mathscr{L}$. *Then* $F(\mathfrak{T}) \subseteq \mathfrak{T}$ *for all* $\mathfrak{T}$ *or* $F(\mathfrak{T}) \subseteq \bar{\mathfrak{T}}$ *for all* $\mathfrak{T} \in \mathscr{L}$.

Proof. If $(A, \mathfrak{T}_1), (A, \mathfrak{T}_2) \in \mathscr{L}$, $\mathfrak{T}_1 \subseteq \mathfrak{T}_2$ then $\mathrm{id}_A \in [\mathfrak{T}_2, \mathfrak{T}_1]$. Therefore $F(\mathfrak{T}_1) \subseteq F(\mathfrak{T}_2)$. Let $(A, \mathfrak{T}) \in \mathscr{L}$, $\mathfrak{T} \neq \exp A$ and $\mathfrak{T} = \cap \mathfrak{S}_i$ where $\mathfrak{S}_i$ are free ultraspaces. Then $F(\mathfrak{T}) \subseteq \cap F(\mathfrak{S}_i)$ and we can use Proposition 4. If $\mathfrak{T} = \exp A$, then $\mathfrak{T} = \bar{\mathfrak{T}} \supseteq F(\mathfrak{T})$.

Remark. Every epireflective subcategory of $\mathscr{B}$ containing a non-indiscrete space contains $\mathscr{F}\mathscr{U}$ because any free ultraspace has a base consisting of clopen sets (see [2]).

Proposition 5. *Let* F *be a realization of* $\mathscr{K}$. *Than* $F(\mathfrak{T}) = \mathfrak{T}$ *for all* $\mathfrak{T} \in \mathscr{K}$ *or* $F(\mathfrak{T}) = \bar{\mathfrak{T}}$ *for all* $\mathfrak{T} \in \mathscr{K}$.

Proof. By Corollary 4 we have $F(\mathfrak{T}) \subseteq \mathfrak{T}$ for all $\mathfrak{T} \in \mathscr{K}$ or $F(\mathfrak{T}) \subseteq \bar{\mathfrak{T}}$ for all $\mathfrak{T} \in \mathscr{K}$. Let e.g. first case occur.

Let $(A, \mathfrak{T}) \in \mathscr{K}$, A infinite, $(A, \mathfrak{T}^{**})$ the Fréchet space on A.

By Proposition 2 $F(\mathfrak{T}^{**}) = \mathfrak{T}^{**}$. Take $X \in \mathfrak{T}$ and $a \in A$. Define f: $(A, \mathfrak{T}) \to (A, \mathfrak{T}^{**})$ such that $f(z) = a$ for all $z \in A - X$, f/X be one-to-one map of X in $A - \{a\}$. $f \in [\mathfrak{T}, \mathfrak{T}^{**}]$ since the preimage of a finite set is finite or $(A - X) \cup Y$ where Y is finite and so again closed. Therefore $X = f^{-1}(A - \{a\}) \in F(\mathfrak{T})$. So $\mathfrak{T} \subseteq F(\mathfrak{T})$ and therefore $F = F(\mathfrak{T})$.

If A is finite, $(A, \mathfrak{T})$ is discrete, therefore $\mathfrak{T} = \exp A$. Since for every $(A_1, \mathfrak{T}_1)$, $[\mathfrak{T}, \mathfrak{T}_1] = A_1^A$ we have $F(\mathfrak{T}) = \exp A$.

Proposition 6. *Let* $\mathscr{L}$ *be a subcategory of* $\mathscr{B}$ *containing* $\mathscr{K}$. *Let* F *be a realization of* $\mathscr{L}$. *Then* $F(\mathfrak{T}) = \mathfrak{T}$ *for all* $\mathfrak{T} \in \mathscr{L}$ *or* $F(\mathfrak{T}) = \bar{\mathfrak{T}}$ *for all* $\mathfrak{T} \in \mathscr{L}$.

Proof. The assertion of Proposition 6 is true for $\mathfrak{T} \in \mathscr{K}$. Let e.g. the first case occur. Let $(A, \mathfrak{T})$ be a space from $\mathscr{L} - \mathscr{K}$ such that there exists a two-point set $C = \{x, y\} \subseteq A$ with $\{\phi, C\} \neq \mathfrak{T} \mid C \neq \exp C$. Then by Corollary 1 $F(\mathfrak{T}) = \mathfrak{T}$ or $F(\mathfrak{T}) = \bar{\mathfrak{T}}$. Let e.g. $\mathfrak{T} \mid C = \{\phi, \{x\}, C\}$. Take $(A_1, \mathfrak{T}_1) \in \mathscr{K}$ such that $\mathfrak{T}_1$ contains an open set X which is not closed. Put $g(z) = x$ for $z \in X$, $g(z) = y$ otherwise. $g \in [\mathfrak{T}_1, \mathfrak{T}]$ and $g \notin [\mathfrak{T}_1, \bar{\mathfrak{T}}]$. Therefore $F(\mathfrak{T}) = \mathfrak{T}$.

Let $(A, \mathfrak{T}) \in \mathscr{L} - \mathscr{K}$, $\mathfrak{T} \neq \{\phi, A\}$ and for every two-point set $C \subseteq A$ it holds that $\{\phi, C\} = \mathfrak{T} \mid C$ or $\exp C = \mathfrak{T} \mid C$. Let $\psi_{\mathfrak{T}}$: $\mathfrak{T} \to S(\mathfrak{T})$ be the reflection defined before Corollary 2. We have $\psi_{\mathfrak{T}} \in [\mathfrak{T}, S(\mathfrak{T})] = [F(\mathfrak{T}), FS(\mathfrak{T})] = [F(\mathfrak{T}), S(\mathfrak{T})]$ for $S(\mathfrak{T}) \in \mathscr{K}$. Therefore $\mathfrak{T} \subseteq F(\mathfrak{T})$. Suppose $\mathfrak{T} \subsetneqq F(\mathfrak{T})$. Let $X \in F(\mathfrak{T}) - \mathfrak{T}$. By Lemma 2 we have $\psi_{\mathfrak{T}} = \psi_{F(\mathfrak{T})}$. Thus $X = \psi_{\mathfrak{T}}^{-1}(Y)$, $Y \subset A \mid \rho_{\mathfrak{T}} = B$. There exists an $f \in [S\mathfrak{T}, \mathfrak{T}] = [F(S\mathfrak{T}), F(\mathfrak{T})]$, $\psi_{\mathfrak{T}} \circ f = \mathrm{id}_B$. $X \in F(\mathfrak{T})$ implies $f^{-1}(X) \in F(S\mathfrak{T})$. $X \notin \mathfrak{T}$ implies $f^{-1}(X) \notin S\mathfrak{T}$. This contradicts to $F(S\mathfrak{T}) = S\mathfrak{T}$. Hence $F(\mathfrak{T}) = \mathfrak{T}$.

Let us consider $(A, \{\phi, A\})$. Put $\mathfrak{T} = \{\phi, A\}$. Then $[\mathfrak{T}, \mathfrak{T}] = A^A$ and there are the following possibilities: $F(\mathfrak{S}) = \phi$ or $\{\phi\}$ or $\{A\}$ or $\{\phi, A\}$ or $\exp A$. As every constant mapping g from $\mathfrak{T}$ in any $(A_1, \mathfrak{T}_1) \in \mathscr{K}$ is a morphism in $\mathscr{L}$ we must have $\phi, A \in F(\mathfrak{T})$. Further, for any $(A_1, \mathfrak{T}_1) \in \mathscr{K}$ any mapping g: $A_1 \to A$ is a morphism from $[(A_1, \mathfrak{T}_1), \mathfrak{T}]$. Therefore for $\operatorname{card} A > 1$, $F(\mathfrak{T}) \neq \exp A$.

Proposition 7. *Let* $(A, \mathfrak{A}) \in \mathcal{K}$, $\operatorname{card} A > 1$. *Let* $\mathcal{L}$ *be the subcategory formed by objects* $(B, \mathfrak{B}) \in \mathcal{B}$ *for which* $\mathfrak{B}^* = \{f^{-1}(X)\colon X \in \bar{\mathfrak{A}},\ f \in [\mathfrak{B}, \mathfrak{A}]\}$ *is a subbase of* $\bar{\mathfrak{B}}$. *Let* $\mathcal{L}_1$ *be a small subcategory of* $\mathcal{K}$. *Than* $\mathcal{L} \cup \mathcal{L}_1$ *has a proper class of realizations.*

Proof. a) Let $(B, \mathfrak{B}) \in \mathcal{L}$. Let $\mathfrak{B}^{**} = \{\bigcup_{i \in I} X_i\colon I$ finite, $X_i \in \mathfrak{B}^*\}$. Then $F_1(\mathfrak{B}) = \mathfrak{B}^{**}$ induces a realization of $\mathcal{L}$. Namely, let $(B, \mathfrak{B}), (B_1, \mathfrak{B}_1) \in \mathcal{L}$. Let $f \in [\mathfrak{B}, \mathfrak{B}_1]$. Let $X \in \mathfrak{B}_1^{**}$. We have $X = \bigcup_{i \in I} X_i$, I finite, $X_i \in \mathfrak{B}_1^*$ and $f^{-1}(X) = f^{-1}\left(\bigcup_{i \in I} X_i\right) = \bigcup_{i \in I} f^{-1}(X_i)$. $f^{-1}(X_i) \in \mathfrak{B}^*$ and so $f^{-1}(X) \in \mathfrak{B}^{**}$ and therefore $f \in [\mathfrak{B}^{**}, \mathfrak{B}_1^{**}]$.

Opposite inclusion is clear as $\mathfrak{B}^{**}, \mathfrak{B}_1^{**}$ are bases for $\bar{\mathfrak{B}}, \bar{\mathfrak{B}}_1$.

b) Let $\mathfrak{m}$ be a cardinal $\geqslant 1$. Let for $(B, \mathfrak{B}) \in \mathcal{L}$, $\mathfrak{B}^{\mathfrak{m}} = \{X\colon X = \bigcap_{i \in I} X_i,\ \operatorname{card} I \leqslant \mathfrak{m},\ X_i \in \mathfrak{B}^{**}\}$. It can be easily proved by means of a) that $F_{\mathfrak{m}}(\mathfrak{B}) = \mathfrak{B}^{\mathfrak{m}}$ induces a realization of $\mathcal{L}$.

c) Let F be a realization of $\mathcal{L}$ with $F(B, \mathfrak{B}) = \bar{\mathfrak{B}}$ for all $(B, \mathfrak{B}) \in \mathcal{L}$. Then $F_{\mathfrak{m}} \neq F$ for every $\mathfrak{m}$. c) will be proved in two steps.

c_1) Let $\mathfrak{n}$ be a cardinal. Because $\mathcal{L}$ is an epireflective subcategory the power $(A, \mathfrak{A})^{\mathfrak{n}}$ is in $\mathcal{L}$.

c_2) Let $\mathfrak{n}, \mathfrak{m}$ be infinite cardinals, $\mathfrak{m} \geqslant \operatorname{card} A$, $\mathfrak{n} > \mathfrak{m}$. Let $a \in A$. Then $\langle \dots, a, a, a, \dots \rangle \notin F_{\mathfrak{m}}((A, \mathfrak{A})^{\mathfrak{n}})$. Put $(B, \mathfrak{B}) = (A, \mathfrak{A})^{\mathfrak{n}}$. Suppose $\langle \dots, a, a, a, \dots \rangle = \bigcap_{j \in J} Y_j$, $\operatorname{card} J \leqslant \mathfrak{m}$, $Y_j \in \mathfrak{B}^{**}$. Let $Y_j = \bigcup_{i \in I_j} Y_{j,i}$, $Y_{j,i} \in \mathfrak{B}^*$, I_j finite. There exists $g_{j,i} \in [\mathfrak{B}, \mathfrak{A}]$ and $X_{j,i} \in \bar{\mathfrak{A}}$ such that $g_{j,i}^{-1}(X_{j,i}) = Y_{j,i}$. Every $X_{j,i}$ is an intersection of at most $\operatorname{card} A$ open sets in $\mathfrak{A}$. Therefore $Y_{j,i}$ is representable as such an intersection in $(B, \mathfrak{B})$, too. Let $Y_{j,i} = \bigcap_{t \in T_{j,i}} Z_t^{j,i}$, $Z_t^{j,i} \in \mathfrak{B}$, $\operatorname{card} T_{j,i} \leqslant \operatorname{card} A$.

For every j we have i_j such that $\langle \dots, a, a, a, \dots \rangle \in Y_{j,i_j}$. Then $\langle \dots, a, a, a, \dots \rangle = \bigcap_{j \in J} Y_{j,i_j} = \bigcap_{j \in J} \bigcap_{t \in T_{j,i_j}} Z_t^{j,i_j}$. This shows $\langle \dots, a, a, a, \dots \rangle$ as an intersection P of at most $\mathfrak{m}$ open sets in $\mathfrak{B}$. It is easily seen that then $\langle \dots, a, a, a, \dots \rangle$ would be an intersection of at most $\mathfrak{m}$ open sets of the form $\prod_{i \in I} X_i$, where all X_i but one are A. But, as $\mathfrak{n} > \mathfrak{m}$ there exists $i_0 \in I$ such that $X_{i_0} = A$ for every such open set and therefore $\prod_{i \in I} V_i \subseteq P$, where $V_i = \{a\}$ for $i \neq i_0$, $V_{i_0} = A$, a contradiction.

c) Clearly implies that there is a proper class of realizations of $\mathscr{L}$. Let us now pass to the assertion about $\mathscr{L}_1$.

Let $(A, \mathfrak{A}) \in \mathscr{L}$. Put $\mathfrak{A}^+ = \{f^{-1}(X) \colon \exists (A_1, \mathfrak{A}_1) \in \mathscr{L}_1 - \mathscr{L}, f \in [\mathfrak{A}, \mathfrak{A}_1], X \in \bar{\mathfrak{A}}_1\}$.

Let $\mathfrak{m} \geqslant 1$ be a cardinal. Put $G_{\mathfrak{m}}(\mathfrak{A}) = \mathfrak{A}^+ \cup F_{\mathfrak{m}}(\mathfrak{A})$. For $(A, \mathfrak{A}) \in \mathscr{L}_1 - \mathscr{L}$ put $G_{\mathfrak{m}}(\mathfrak{A}) = \bar{\mathfrak{A}}$.

d) $G_{\mathfrak{m}}$ induces a realization.

d$_1$) Let $(A_1, \mathfrak{A}_1), (A_2, \mathfrak{A}_2) \in \mathscr{L}$ and $f \in [\mathfrak{A}_1, \mathfrak{A}_2]$. Take $X \in G_{\mathfrak{m}}(\mathfrak{A}_2)$. If $X \in F_{\mathfrak{m}}(\mathfrak{A}_2)$, we have $f^{-1}(X) \in G_{\mathfrak{m}}(\mathfrak{A}_1)$. Let $X = g^{-1}(Y)$ for some $Y \in \bar{\mathfrak{A}}_3$ for some $(A_3, \mathfrak{A}_3) \in \mathscr{L}_1 - \mathscr{L}$ with $g \in [\mathfrak{A}_2, \mathfrak{A}_3]$. Then $f^{-1}(X) = f^{-1}(g^{-1}(Y)) = (gf)^{-1}(Y) \in \mathfrak{A}_1^+ \subseteq G_{\mathfrak{m}}(\mathfrak{A}_1)$. Therefore $f \in [G_{\mathfrak{m}}(\mathfrak{A}_1), G_{\mathfrak{m}}(\mathfrak{A}_2)]$. $[G_{\mathfrak{m}}(\mathfrak{A}_1), G_{\mathfrak{m}}(\mathfrak{A}_2)] \subseteq [\mathfrak{A}_1, \mathfrak{A}_2]$ as $G_{\mathfrak{m}}(\mathfrak{A}_2)$ is a base for closed sets for $\mathfrak{A}_2$.

d$_2$) Let $(A_1, \mathfrak{A}_1) \in \mathscr{L}$, $(A_2, \mathfrak{A}_2) \in \mathscr{L}_1 - \mathscr{L}$ and $f \in [\mathfrak{A}_1, \mathfrak{A}_2]$. Take $X \in G_{\mathfrak{m}}(\mathfrak{A}_2) = \bar{\mathfrak{A}}_2$. Then $f^{-1}(X) \in G_{\mathfrak{m}}(\mathfrak{A}_1)$, whence $f \in [G_{\mathfrak{m}}(\mathfrak{A}_1), G_{\mathfrak{m}}(\mathfrak{A}_2)]$. Let $f \in [G_{\mathfrak{m}}(\mathfrak{A}_1), G_{\mathfrak{m}}(\mathfrak{A}_2)]$ and $X \in \bar{\mathfrak{A}}_2 = G_{\mathfrak{m}}(\mathfrak{A}_2)$. Then $f^{-1}(X) \in G_{\mathfrak{m}}(\mathfrak{A}_1) \subseteq \bar{\mathfrak{A}}_1$ and $f \in [\mathfrak{A}_1, \mathfrak{A}_2]$.

d$_3$) Let $(A_1, \mathfrak{A}_1) \in \mathscr{L}_1 - \mathscr{L}$, $(A_2, \mathfrak{A}_2) \in \mathscr{L}$ and $f \in [\mathfrak{A}_1, \mathfrak{A}_2]$. Take $X \in G_{\mathfrak{m}}(\mathfrak{A}_2)$. Then $f^{-1}(X) \in \bar{\mathfrak{A}}_1 = G_{\mathfrak{m}}(\mathfrak{A}_1)$ and $f \in [G_{\mathfrak{m}}(\mathfrak{A}_1), G_{\mathfrak{m}}(\mathfrak{A}_2)]$.

$[G_{\mathfrak{m}}(\mathfrak{A}_1), G_{\mathfrak{m}}(\mathfrak{A}_2)] \subseteq [\mathfrak{A}_1, \mathfrak{A}_2]$ as $G_{\mathfrak{m}}(\mathfrak{A}_2)$ is a base for closed sets for $\mathfrak{A}_2$.

d$_4$) The case $(A_1, \mathfrak{A}_1)$, $(A_2, \mathfrak{A}_2) \in \mathscr{L}_1 - \mathscr{L}$ is clear.

e) Let $\mathfrak{m}$ be infinite, $\mathfrak{m} \geqslant \operatorname{card} A$ and $\mathfrak{m} \geqslant \operatorname{card} A_i$ for every $(A_i, \mathfrak{A}_i) \in \mathscr{L}_1 - \mathscr{L}$. Then $G_{\mathfrak{m}} \neq F$, where F is a realization of $\mathscr{L} \cup \mathscr{L}_1$ given by $F(A_1, \mathfrak{A}_1) = \bar{\mathfrak{A}}_1$, for all $(A_1, \mathfrak{A}_1) \in \mathscr{L}_1 \cup \mathscr{L}$.

Let $X \in \mathfrak{A}_1^+$ for $(A_1, \mathfrak{A}_1) \in \mathscr{L}$. Then X can be represented as $\bigcap_{i \in I} X_i$ where $X_i \in \mathfrak{A}_1$ and card $I \leqslant \mathfrak{m}$. Therefore, by c_2) $\langle \dots, a, a, a, \dots \rangle$ is not contained in $G_{\mathfrak{m}}((A, \mathfrak{A})^{\mathfrak{n}})$ for $\mathfrak{n} > \mathfrak{m}$.

The proof is ready.

In fact we have proved more. By c) it this clear that

Proposition 7*. *Let* $(A, \mathfrak{A})$ *and* $\mathscr{L}$ *be as in Proposition 7. Let* $\mathscr{L}^*$ *be a full subcategory of* $\mathscr{L}$ *containing all* $(A, \mathfrak{A})^{\mathfrak{n}}$. *Then there exists a proper class of realizations of* $\mathscr{L}^*$.

Corollary 5. *Let* $\mathscr{L}^*$ *be a full subcategory of* $\mathscr{C}\mathscr{R}$ *(category of all completely regular spaces) equal to* $\mathscr{C}\mathscr{H}$ *or consisting by turn of all compact spaces, normal spaces, connected spaces, locally connected spaces, paracompact spaces from* $\mathscr{C}\mathscr{R}$. *Then* $\mathscr{L}^*$ *has a proper class of realizations.*

Proof. It is sufficient to put $(A, \mathfrak{A})$ from Proposition 7* equal to the closed real interval $[0, 1]$.

Corollary 6. *Let* $\mathscr{L}$ *be the subcategory of* $\mathscr{K}$ *consisting of all* T_1*-spaces in which clopen sets form a subbase. Then* $\mathscr{L}$ *has a proper class of realizations.*

Proof. Put the discrete space $\{0, 1\}$ for $(A, \mathfrak{A})$.

Proposition 8. *The category* $\mathscr{F}\mathscr{U}$ *has a proper class of realizations.*

Remark. Proposition 8 is clearly not covered by Proposition 7*.

Proof. Let $\mathfrak{m}$ be an infinite cardinal, $A_{\mathfrak{m}}$ a set with card $A_{\mathfrak{m}} = \mathfrak{m}$, $\mathfrak{F}_{\mathfrak{m}} = \{X\colon X \subseteq A_{\mathfrak{m}},$ card $(A_{\mathfrak{m}} - X) < \mathfrak{m}\}$ a symmetrical filter on $A_{\mathfrak{m}}$. Take $a_{\mathfrak{m}} \in A_{\mathfrak{m}}$ and put $\mathfrak{S}_{\mathfrak{m}} = \mathfrak{S}(A_{\mathfrak{m}}, \mathfrak{F}_{\mathfrak{m}}, a_{\mathfrak{m}})$. Put $F_{\mathfrak{m}}(\mathfrak{S}) = \mathfrak{S}$ if $[\mathfrak{S}, \mathfrak{S}_{\mathfrak{m}}] \neq [\mathfrak{S}^0, \mathfrak{S}_{\mathfrak{m}}]$, $F_{\mathfrak{m}}(\mathfrak{S}) = \mathfrak{S}^0$ otherwise. Let us prove that $F_{\mathfrak{m}}$ gives a realization of $\mathscr{F}\mathscr{U}$.

a) If $\mathfrak{S}_1 = \mathfrak{S}(A_1, \mathfrak{F}_1, a_1)$, $\mathfrak{S}_2 = \mathfrak{S}(A_2, \mathfrak{F}_2, a_2) \in \mathscr{F}\mathscr{U}$, then $[\mathfrak{S}_1, \mathfrak{S}_2] = [\mathfrak{S}_1^0, \mathfrak{S}_2^0]$. Namely, let $f \in [\mathfrak{S}_1, \mathfrak{S}_2]$, take $X \in \mathfrak{S}_2^0$, then $f^{-1}(X) \in \mathfrak{S}_1^0$, therefore $f \in [\mathfrak{S}_1^0, \mathfrak{S}_2^0]$. Opposite inclusion is clear as $\mathfrak{S}_1^0$, $\mathfrak{S}_2^0$ are bases for $\mathfrak{S}_1, \mathfrak{S}_2$.

Further, we prove for $\mathfrak{S}_1, \mathfrak{S}_2 \in \mathcal{FU}$ the implication

b) $F_{\mathfrak{m}}(\mathfrak{S}_1) = \mathfrak{S}_1,\ [\mathfrak{S}_2, \mathfrak{S}_1] \neq [\mathfrak{S}_2^0, \mathfrak{S}_1] \Rightarrow F_{\mathfrak{m}}(\mathfrak{S}_2) = \mathfrak{S}_2$.

Suppose this is not true for some $\mathfrak{S}_1, \mathfrak{S}_2 \in \mathcal{FU}$. Since $[\mathfrak{S}_1, \mathfrak{S}_{\mathfrak{m}}] \neq$ $\neq [\mathfrak{S}_1^0, \mathfrak{S}_{\mathfrak{m}}]$, by Lemma 3 there exists an $f\colon A_1 \to A_{\mathfrak{m}}$ such that $f(a_1) =$ $= a_{\mathfrak{m}}$, $f^{-1}(a_{\mathfrak{m}}) \notin \mathfrak{F}_1$, $\{f^{-1}(Z)\colon a_{\mathfrak{m}} \in Z \in \mathfrak{F}_{\mathfrak{m}}\} \subseteq \mathfrak{F}_1$. From $[\mathfrak{S}_2, \mathfrak{S}_1] \neq$ $\neq [\mathfrak{S}_2^0, \mathfrak{S}_1]$ there follows the existence of such a mapping $g\colon A_2 \to A_1$, for which $g(a_2) = a_1$, $g^{-1}(a_1) \notin \mathfrak{F}_2$, $\{g^{-1}(Z)\colon a_1 \in Z \in \mathfrak{F}_1\} \subseteq \mathfrak{F}_2$. As, by assumption, $F_{\mathfrak{m}}(\mathfrak{S}_2) \notin \mathfrak{S}_2$, we have $[\mathfrak{S}_2, \mathfrak{S}_{\mathfrak{m}}] = [\mathfrak{S}_2^0, \mathfrak{S}_{\mathfrak{m}}]$, i.e. $g^{-1}(f^{-1}(a_{\mathfrak{m}})) = (fg)^{-1}(a_{\mathfrak{m}}) \in \mathfrak{F}_2$. As $\mathfrak{F}_1$ is an ultrafilter, $A_1 -$ $- f^{-1}(a_{\mathfrak{m}}) \in \mathfrak{F}_1$ and so is $\{a_1\} \cup (A_1 - f^{-1}(a_{\mathfrak{m}}))$. Therefore $g^{-1}\big(\{a_1\} \cup (A_1 - f^{-1}(a_{\mathfrak{m}}))\big) \in \mathfrak{F}_2$ and $g^{-1}(a_1) = (g^{-1}(a_1) \cup g^{-1}(A_1 -$ $- f^{-1}(a_{\mathfrak{m}}))) \cap g^{-1}(f^{-1}(a_{\mathfrak{m}})) \in \mathfrak{F}_2$ a contradiction. Therefore $F_{\mathfrak{m}}$ is a realization because by Lemma 3 $[\mathfrak{S}_1, \mathfrak{S}_2] = [\mathfrak{S}_1, \mathfrak{S}_2^0]$. Let $\mathfrak{S} = \mathfrak{S}(A, \mathfrak{F}, a) \in$ $\in \mathcal{FU}$, $\operatorname{card} A < \mathfrak{m}$. Let $f \in [\mathfrak{S}, \mathfrak{S}_{\mathfrak{m}}]$, $f(a) = a_{\mathfrak{m}}$. Then $\operatorname{card} f(A) < \mathfrak{m}$, therefore $(A_{\mathfrak{m}} - f(A)) \cup \{a_{\mathfrak{m}}\} \in \mathfrak{F}_{\mathfrak{m}}$ and so $f^{-1}(a_{\mathfrak{m}}) \in \mathfrak{F}$. Hence $F_{\mathfrak{m}}(\mathfrak{S}) = \mathfrak{S}^0$.

Let $\mathfrak{F} \supseteq \mathfrak{F}_{\mathfrak{m}}$ be an ultrafilter. Put $\mathfrak{S} = \mathfrak{S}(A_{\mathfrak{m}}, \mathfrak{F}, a_{\mathfrak{m}})$, $\mathrm{id}_A \in$ $\in [\mathfrak{S}, \mathfrak{S}_{\mathfrak{m}}]$ but $\mathrm{id}_A \notin [\mathfrak{S}^0, \mathfrak{S}_{\mathfrak{m}}]$ (by b) of Lemma 3). Therefore $F_{\mathfrak{m}}(\mathfrak{S}) =$ $= \mathfrak{S}$. From this and the previous results one gets the assertion about the class of all realizations of $\mathcal{FU}$.

If $\mathcal{L}$ is a subcategory in $\mathcal{K}$ containing $\mathcal{FU}$ and if F is a realization of $\mathcal{L}$, then it can be useful to know the restriction of F to $\mathcal{FU}$.

Lemma 6. *Let* $\mathcal{L}, F$ *be as above. Let* $F(\mathfrak{S}) = \mathfrak{S}$ *for each* $\mathfrak{S} \in$ $\in \mathcal{FU}$. *Then* $F(\mathfrak{T})$ *is a subbase of* $\mathfrak{T}$ *for each* $\mathfrak{T} \in \mathcal{L}$.

Proof. By Corollary 4, $F(\mathfrak{T}) \subseteq \mathfrak{T}$ for each $\mathfrak{T} \in \mathcal{L}$. Suppose there exists $(A, \mathfrak{T}) \in \mathcal{L}$ such that $F(\mathfrak{T})$ is not a subbase in $\mathfrak{T}$. Let $\mathfrak{T}'$ be the topology generated by $F(\mathfrak{T})$. $\mathfrak{T}'$ is a T_1-topology. There exists $a \in A$ and a neighborhood X of a in $(A, \mathfrak{T})$ which is not a neighborhood of a in $(A, \mathfrak{T}')$. Let $\mathfrak{F}(\mathfrak{F}')$ be the filter of all neighborhoods of a in $\mathfrak{T}$ ($\mathfrak{T}'$ resp.). We have $\mathfrak{F} \supsetneqq \mathfrak{F}'$ and therefore there exists an ultrafilter $\mathfrak{G}$, $\mathfrak{G} \supseteq \mathfrak{F}'$, $\mathfrak{G} \not\supseteq \mathfrak{F}$. Put $\mathfrak{S} = \mathfrak{S}(A, \mathfrak{G}, a)$. Then $F(\mathfrak{S}) = \mathfrak{S} \supseteq \mathfrak{T}' \supseteq F(\mathfrak{T})$. Therefore $\mathrm{id}_A \in [F(\mathfrak{S}), F(\mathfrak{T})] = [\mathfrak{S}, \mathfrak{T}]$ which is a contradiction to $\mathfrak{G} \not\supseteq \mathfrak{F}$.

SOME MORE DETAILED THEOREMS ON COMPLETELY REGULAR SPACES

Let $(R, \mathfrak{T}_r)$ be the topological space of the real line.

Lemma 7. *Let* $[\mathfrak{A}, \mathfrak{A}] = [\mathfrak{T}_r, \mathfrak{T}_r]$ *and* $\mathfrak{A} \cap \mathfrak{T}_r \neq \{\phi, R\}$, $\left(\bar{\mathfrak{A}} \cap \cap \mathfrak{T}_r \neq \{\phi, R\}\right)$. *Then* $\mathfrak{T}_r \subseteq \mathfrak{A}(\bar{\mathfrak{T}}_r \subseteq \mathfrak{A})$.

Proof. Let $X \in \mathfrak{A} \cap \mathfrak{T}_r$, $\phi \neq X \neq R$. There is $a \notin X$ such that, for suitable b, the open interval (a, b) or (b, a) is a subset of X. Let the first case occur. Let (c, d) be an open interval. Let $f_{c,d}$: $(c, d) \to \to (a, b)$ for which $f_{c,d}(z) = a + \dfrac{2 \min \left\{d - c, \frac{b-a}{2}\right\}}{d-c}(z - c)$ for $z \in \in \left(c, \frac{c+d}{2}\right)$, $f_{c,d}(z) = a - \dfrac{2 \min \left\{d - c, \frac{b-a}{2}\right\}}{d-c}(z - d)$ otherwise. Clearly $f_{c,d}$ is continuous with $\lim\limits_{z \to c^+} f_{c,d} = a = \lim\limits_{z \to d^-} f_{c,d}$ and $f_{c,d}((c, d)) \subseteq \subseteq (a, b)$.

Similarly for open intervals $(-\infty, c)$, (c, ∞) let $f_{-\infty,c}$, $f_{c,\infty}$ be some continuous mappings: $(-\infty, c) \to (a, b)$ or $(c, \infty) \to (a, b)$ respectively such that $\lim\limits_{z \to c^-} f_{-\infty,c}(z) = a = \lim\limits_{z \to c^+} f_{\infty,c}(z)$. Let now Y be an open set in $\mathfrak{T}_r$, $\phi \neq Y \neq R$ and $Y = \cup\, (c_i, d_i)$ be the decomposition of Y in open intervals (some c_i, d_i can be $-\infty$ or ∞). Let f: $R \to R$ defined as f_{c_i,d_i} on (c_i, d_i) for any i, $f(z) = a$ otherwise. Then f is continuous and $f^{-1}(X) = Y$. Therefore $Y \in \mathfrak{A}$. Quite analogous reasoning applies to the remaining case and the dual one, as well.

Let $\mathfrak{T} \in \mathscr{C}\mathscr{R}$, $\mathfrak{T}^c$ the system of all cozero sets of the topology $\mathfrak{T}$. $\bar{\mathfrak{T}}^c$ is then the set of all zero-sets. $\mathfrak{T}^c$ is a base for $\mathfrak{T}$. Further, one easily sees, $\mathscr{F}\mathscr{U} \subseteq \mathscr{C}\mathscr{R}$ (even $\mathscr{F}\mathscr{U} \subseteq \mathscr{N}$).

Proposition 9. *Let* $F_c\ (\mathfrak{T}) = \mathfrak{T}^c$ *for each* $\mathfrak{T} \in \mathscr{C}\mathscr{R}(\mathscr{N})$. *Then* F_c *is a realization of* $\mathscr{C}\mathscr{R}$ $(\mathscr{N}$ *resp). Let* G *be a realization of* $\mathscr{C}\mathscr{R}(\mathscr{N})$. *Then* $\mathfrak{T}^c \subseteq G(\mathfrak{T}) \subseteq \mathfrak{T}$ *for all* $\mathfrak{T} \in \mathscr{C}\mathscr{R}(\mathscr{N})$ *or* $\overline{\mathfrak{T}^c} \subseteq G(\mathfrak{T}) \subseteq \bar{\mathfrak{T}}$ *for all* $\mathfrak{T} \in \mathscr{C}\mathscr{R}(\mathscr{N})$.

Proof. The first assertion is an immediate consequence of the starting considerations in the proof of Proposition 7. Let G be a realization of $\mathscr{CR}$. By Corollary 4 $G(\mathfrak{T}) \subseteq \mathfrak{T}$ for all $\mathfrak{T} \in \mathscr{CR}$ or $G(\mathfrak{T}) \subseteq \bar{\mathfrak{T}}$ for all $\mathfrak{T} \in \mathscr{CR}$. Consider the first case. Then for $(R, \mathfrak{T}_r)$ we have $G(\mathfrak{T}_r) = = \mathfrak{T}_r$ and so $G(\mathfrak{T}) \supseteq \mathfrak{T}^c$ for each $\mathfrak{T} \in \mathscr{CR}$. Analogously the dual case can be settled.

For $\mathscr{N}$ the reasoning is without any change.

Proposition 10. *The following two assertions are equivalent:*

(i) *For every realization F of $\mathscr{CR}$ we have $F(\mathfrak{S}) = \mathfrak{S}$ for each $\mathfrak{S} \in \mathscr{FU}$ or $F(\mathfrak{S}) = \bar{\mathfrak{S}}$ for each $\mathfrak{S} \in \mathscr{FU}$.*

(ii) *No measurable cardinal exists.*

Proof. Let (i) be valid. Let F_c be as in the previous proposition. So $F_c(\mathfrak{S}) = \mathfrak{S}$ for $\mathfrak{S} \in \mathscr{FU}$ and $\mathfrak{S}^c = \mathfrak{S}$. Let $\mathfrak{S} = \mathfrak{S}(A, \mathfrak{F}, a) \in \mathscr{FU}$. $\{a\} \in \bar{\mathfrak{S}}$, therefore $\{a\} \in \overline{\mathfrak{S}^c}$ and so $\{a\}$ is a G_δ-set. As $\{a\} \notin \mathfrak{F}$, $\mathfrak{F}$ is not closed to countable intersections. Therefore (ii) is valid.

Let (ii) be valid. F a realization of $\mathscr{CR}$ and let the first case from Proposition 9 occur, i.e. $\mathfrak{T}^c \subseteq F(\mathfrak{T}) \subseteq \mathfrak{T}$ for each $\mathfrak{T} \in \mathscr{CR}$. Let $\mathfrak{S} = \mathfrak{S}(A, \mathfrak{F}, a) \in \mathscr{FU}$. By (ii) there exists $X \subseteq A$, $a \in X \notin \mathfrak{F}$ such that X is an intersection of countably many elements of $\mathfrak{F}$. Each element of $\mathfrak{F}$ containing a is a clopen set, therefore a zero set. As an intersection of countably many zero-sets X is a zero set, so $X \in \overline{\mathfrak{S}^c}$. As $X \notin \mathfrak{S}^0$, we have $A - X \in F(\mathfrak{S}) - \mathfrak{S}^0$. Therefore $F(\mathfrak{S}) = \mathfrak{S}$.

Analogously for the second case from Proposition 9.

Proposition 11. *Let us consider the "system" Σ of all realizations F of $\mathscr{CR}$ with $\mathfrak{T}^c \subseteq F(\mathfrak{T}) \subseteq \mathfrak{T}$ for each $\mathfrak{T} \in \mathscr{CR}$. Then Σ is "a complete distributive lattice".*

Proof. Let $F_i \in \Sigma$, $i \in I$, $F(\mathfrak{T}) = \bigcap_{i \in I} F_i(\mathfrak{T})$. Let $f \in [\mathfrak{T}_1, \mathfrak{T}_2]$, $X \in F(\mathfrak{T}_2)$. We have $f^{-1}(X) \in F_i(\mathfrak{T}_1)$ for each i, i.e. $f^{-1}(X) \in F(\mathfrak{T}_1)$. So $[F(\mathfrak{T}_1), F(\mathfrak{T}_2)] \supseteq [\mathfrak{T}_1, \mathfrak{T}_2]$. As $F(\mathfrak{T})$ is a base for $\mathfrak{T}$, the opposite inclusion is valid, too. We can put $F = \wedge F_i$. Put $G(\mathfrak{T}) = \bigcup_{i \in I} F_i(\mathfrak{T})$ and take $f \in [\mathfrak{T}_1, \mathfrak{T}_2]$ $X \in G(\mathfrak{T}_2)$. There exists i so that $f^{-1}(X) \in F_i(\mathfrak{T}_1) \subseteq \subseteq G(\mathfrak{T}_1)$.

The rest of the proof is clear.

Proposition 12. *The greatest subcategory of* $\mathscr{CR}$ *having exactly two realizations is* $\mathscr{PN}$ *(category of perfectly normal spaces).*

Proof. Let $\mathscr{L}$ be a subcategory in $\mathscr{CR}$ having exactly two realizations. As $\mathscr{L}$ has the realizations by means of $\mathfrak{T}, \mathfrak{T}^c, \bar{\mathfrak{T}}, \overline{\mathfrak{T}^c}$ we must have $\mathfrak{T} = \mathfrak{T}^c$ for each $\mathfrak{T} \in \mathscr{L}$, i.e. $\mathfrak{T} \in \mathscr{PN}$ and $\mathscr{L} \subseteq \mathscr{PN}$. Let us prove $\mathscr{PN}$ has exactly two realizations. Since $\mathfrak{T} \in \mathscr{PN}$ exactly if every closed set is a zero-set, $\mathfrak{S} = \mathfrak{S}(A, \mathfrak{F}, a) \in \mathscr{FU}$ is in $\mathscr{PN}$ exactly if $\{a\}$ is a G_δ-set in $\mathfrak{S}$. Therefore if there exists $\mathfrak{T} \in \mathscr{PN}$, $\mathfrak{T} \subseteq \mathfrak{S} \in \mathscr{FU}$, $\mathfrak{S}$ is perfectly normal. One easily verifies that $\mathfrak{S}_3$ from the proof of Proposition 4 is in $\mathscr{PN}$ if $\mathfrak{S}_1, \mathfrak{S}_2$ are. As in Proposition 4 we can see that $F(\mathfrak{S}) \subseteq \bar{\mathfrak{S}}$ for all $\mathfrak{S} \in \mathscr{PN} \cap \mathscr{FU}$ or $F(\mathfrak{S}) \subseteq \mathfrak{S}$ for all $\mathfrak{S} \in \mathscr{PN} \cap \mathscr{FU}$. As in Corollary 4 we see $F(\mathfrak{T}) \subseteq \mathfrak{T}$ for all $\mathfrak{T} \in \mathscr{PN}$ or $F(\mathfrak{T}) \subseteq \bar{\mathfrak{T}}$ for all $\mathfrak{T} \in \mathscr{PN}$. We get $F(R, \mathfrak{T}_r) = \mathfrak{T}_r$ or $\bar{\mathfrak{T}}_r$ respectively and so $\mathfrak{T} = \mathfrak{T}^c \subseteq F(\mathfrak{T}) \subseteq \mathfrak{T}$ for all $\mathfrak{T} \in \mathscr{PN}$ or $\bar{\mathfrak{T}} = \overline{\mathfrak{T}^c} \subseteq F(\mathfrak{T}) \subseteq \bar{\mathfrak{T}}$ for all $\mathfrak{T} \in \mathscr{PN}$.

Similar theorem can be deduced on metrizable spaces.

Proposition 13. $\mathscr{M}$ *has exactly two realizations.*

Proof. Suppose, F is a realization of $\mathscr{M}$ and we have $(A_1, \mathfrak{T}_1)$, $(A_2, \mathfrak{T}_2) \in \mathscr{M}$ and Z_1, Z_2 such that $Z_1 \in F(\mathfrak{T}_1) - \mathfrak{T}_1$, $Z_2 \in F(\mathfrak{T}_2) - \bar{\mathfrak{T}}_2$. There exist $a_1 \in A_1$, $a_1 \in \mathrm{Cl}_{\mathfrak{T}_1}(A_1 - Z_1)$, $a_2 \in A_2 - Z_2$, $a_2 \in \mathrm{Cl}_{\mathfrak{T}_2}(Z_2)$. Therefore there exist a sequence $x_1, x_2, \ldots$ in $A_1 - Z_1$ converging to $a_1, y_1, y_2, \ldots$ a sequence in Z_2 converging to a_2. Let $A = \{t_1, t_2, \ldots, t_\infty\}$ be a countable set turned into a metric space $\mathfrak{A}$ by the one-to-one correspondence $t_n \to \frac{1}{n}$, $t_\infty \to 0$. Let $f\colon A \to A_1$, $g\colon A \to A_2$ such that $f(t_i) = x_i$, $f(t_\infty) = a_1$, $g(t_i) = y_i$, $g(t_\infty) = a_2$. Clearly $f \in [\mathfrak{A}, \mathfrak{T}_1]$, $g \in [\mathfrak{A}, \mathfrak{T}_2]$. We have $\{t_\infty\} = f^{-1}(Z_1) \in F(\mathfrak{A})$, $A - \{t_\infty\} = g^{-1}(Z_2) \in F(\mathfrak{A})$. This is a contradiction to b) of Lemma 2 as $\{t_\infty\}$ is not clopen in $\mathfrak{A}$.

So $F(\mathfrak{T}) \subseteq \mathfrak{T}$ for each $\mathfrak{T} \in \mathscr{M}$ or $F(\mathfrak{T}) \subseteq \bar{\mathfrak{T}}$ for each $\mathfrak{T} \in \mathscr{M}$. Therefore, by Lemma 7, $F((R, \mathfrak{T}_r)) = \mathfrak{T}_r$ or $F((R, \mathfrak{T}_r)) = \bar{\mathfrak{T}}_r$ and therefore for each $\mathfrak{T} \in \mathscr{M}$, $\mathfrak{T}^c \subseteq F(\mathfrak{T})$ or for each $\mathfrak{T} \in \mathscr{M}$, $\overline{\mathfrak{T}^c} \subseteq F(\mathfrak{T})$. As

$\mathfrak{T} = \mathfrak{T}^c$ for metrizable spaces, we have $\mathfrak{T} \subseteq F(\mathfrak{T})$ or $\bar{\mathfrak{T}} \subseteq F(\mathfrak{T})$ simultaneously for all $\mathfrak{T}$ from $\mathscr{M}$.

HAUSDORFF SPACES

According to Corollary 4 $F(\mathfrak{T}) \subseteq \mathfrak{T}$ or $F(\mathfrak{T}) \subseteq \bar{\mathfrak{T}}$ for every realization of $\mathscr{H}$ and every $\mathfrak{T} \in \mathscr{H}$.

Definition. Let $(A, \mathfrak{T}) \in \mathscr{B}$, $a \in A$. Let $(A', \mathfrak{T}')$ be a topological sum of countably many copies of $(A, \mathfrak{T})$. Let $\iota_i: A \to A'$ denote the i-th injection for every $i \in N$ (N is the set of all positive integers). Let $a_i = \iota_i(a)$, $A_i = \iota_i(A)$. Suppose $b \notin A'$. Put $A^* = \{b\} \cup A'$. Let $\mathfrak{T}_a^*$ be the topology on A^* with the system $\mathfrak{T}' \cup \{\{b\} \cup \bigcup_{i \geq k} (A_i - \{a_i\}) : k \in N\}$ as a base for open sets.

Lemma 8. *Let* $(A, \mathfrak{T}) \in \mathscr{B}$, $a \in A$, $\operatorname{card} A > 1$. Then for every $x, y \in A^*$, $x \neq y$, there exists $f \in [\mathfrak{T}_a^*, \mathfrak{T}]$ with $f(x) \neq f(y)$.

Proof. If $x \in A_i$, $y \notin A_i$ then $f \in [\mathfrak{T}_a^*, \mathfrak{T}]$ with $f(x) \neq f(y)$ exists because A_i is a clopen set in $\mathfrak{T}_a^*$. Let $x, y \in A_i$. Put $f(z) = a$ for $z \notin A_i$, $f(z) = \iota_i^{-1}(z)$ for $z \in A_i$. Clearly $f \in [\mathfrak{T}_a^*, \mathfrak{T}]$ and $f(x) \neq f(y)$.

Lemma 9. *Let F be a realization of* $\mathscr{H}$, $F(\mathfrak{T}) \subseteq \mathfrak{T}$, *for each* $(A, \mathfrak{T}) \in \mathscr{H}$. *Then* $A - \{a\} \in F(\mathfrak{T})$ *for every* $(A, \mathfrak{T}) \in \mathscr{H}$, $a \in A$.

Proof. Suppose there exists $(A, \mathfrak{T}) \in \mathscr{H}$, $a \in A$ and $A - \{a\} \notin F(\mathfrak{T})$. Then $\operatorname{card} A > 1$.

By Lemma 8 $(A^*, \mathfrak{T}_a^*) \in \mathscr{H}$.

Let $B \in F(\mathfrak{T}_a^*)$, $A_i - \{a_i\} \subseteq B$. Then $\iota_i^{-1}(B) \in F(\mathfrak{T})$. Since $A - \{a\} \notin F(\mathfrak{T})$, $a_i \in B$.

Let $\mathfrak{F}$ be the filter of all neighborhoods of b in $\mathfrak{T}_a^*$. Let $\mathfrak{F}'$ be the filter in A^* generated by the system $\{B : b \in B \in F(\mathfrak{T}_a^*)\}$. Choose $c_i \in A_i - \{a_i\}$. Put $Y = \{b\} \cup \{c_i : i \in N\} \cup \{a_i : i \in N\}$ and $h: Y \to A$ with $h(t) = t$. Let $\mathfrak{G}$ be the filter generated by $\{h^{-1}(X) : X \in \mathfrak{F}\}$ in Y, $\mathfrak{G}'$ the filter generated by $\{h^{-1}(X) : X \in \mathfrak{F}'\}$. Let us prove $\mathfrak{G} \neq \mathfrak{G}'$. We have $\{b\} \cup \{c_i : i \in N\} = h^{-1}(\{b\} \cup \bigcup_{i \in N} (A_i - \{a_i\})) \in \mathfrak{G}$. If $B \in \mathfrak{F}'$, there exist $B_1, \ldots, B_n \in F(\mathfrak{T}_a^*)$, $b \in \bigcap_{l=1}^{n} B_l$ so that $B \supseteq \left(\bigcap_{l=1}^{n} B_l\right)$. As $B_l \in \mathfrak{T}_a^*$ for any l, the construction of $\mathfrak{T}_a^*$ implies that there exists k_l such that $B_l \supseteq \{b\} \cup \bigcup_{i \geqslant k_l} (A_i - \{a_i\})$ for every l. Let k^* be the greatest number from k_l, $l = 1, \ldots, n$. Since $A_i - \{a_i\} \subseteq B_l$, $i \geqslant k^*$ and $l = 1, \ldots, n$, we have $\{a_i\colon i \geqslant k^*\} \subseteq B_l$ for all $l = 1, \ldots, n$. Hence $B \cap \{a_i\colon i \in N\} \neq \phi$ and then $\{b\} \cup \{c_i\colon i \in N\} \notin \mathfrak{G}'$.

As $\mathfrak{G} \neq \mathfrak{G}'$, there exists an ultrafilter $\mathfrak{G}_1$ on Y such that $\mathfrak{G}' \subseteq \subseteq \mathfrak{G}_1$, $\mathfrak{G} \nsubseteq \mathfrak{G}_1$. Let $\mathfrak{S} = \mathfrak{S}(Y, \mathfrak{G}_1, b)$. As $\aleph_0$ is not a measurable cardinal, then, as in the proof of (ii) ⇒ (i) in Proposition 10, we have $F(\mathfrak{S}) = = \mathfrak{S}$. Therefore $h \in [F(\mathfrak{S}), F(\mathfrak{T}_a^*)]$. As $\mathfrak{G} \nsubseteq \mathfrak{G}_1$, $h \notin [\mathfrak{S}, \mathfrak{T}_a^*]$, a contradiction.

From Lemma 8 it follows that quite analogous theorem holds e.g. for Urysohn spaces or completely Hausdorff spaces. Unfortunately, if a is not isolated, then $\mathfrak{T}_a^*$ is not regular.

A corollary of Lemma 9 is

Proposition 14. *Let F be a realization of $\mathscr{H}$, $F(\mathfrak{T}) \subseteq \mathfrak{T}$ for each $\mathfrak{T} \in \mathscr{H}$. Let X be a closed set in $(A, \mathfrak{T}) \in \mathscr{H}$ such that for every $x \in A - X$ there exist disjoint open sets. O_1, O_2 in $\mathfrak{T}$ such that $X \subseteq O_1$, $x \in O_2$. Then $X \in F(\mathfrak{T})$.*

Proof. $X \cup \{\{x\}\colon x \in A - X\}$ is a decomposition of A and let $\mathfrak{T}_1$ be the factor topology on this decomposition induced by $\mathfrak{T}$. By assumption on X, $\mathfrak{T}_1$ is Hausdorff. Let κ be the canonical map. By Lemma 9 (dual version) $\{X\} \in F(\mathfrak{T}_1)$. So $X = \kappa^{-1}(X) \in F(\mathfrak{T})$.

Corollary 7. *Let F be as in Proposition 14, $(A, \mathfrak{T}) \in \mathscr{H}$. Let X be a compact subset in $(A, \mathfrak{T})$. Then $X \in F(\mathfrak{T})$.*

Corollary 8. *Let F be a realization of $\mathscr{H}$. Then for each $\mathfrak{T} \in \in \mathscr{R} \cap \mathscr{H}$ we have $F(\mathfrak{T}) = \mathfrak{T}$ or for all these topologies $F(\mathfrak{T}) = \overline{\mathfrak{T}}$.*

Proof. In $\mathfrak{T} \in \mathscr{R} \cap \mathscr{H}$ every closed X fulfills the assumptions of Proposition 14.

Corollary 9. *Let F be a realization of $\mathscr{H}$. Then $F(\mathfrak{T})$ is a sub-*

base for $\mathfrak{T}$ *for each* $\mathfrak{T} \in \mathscr{H}$ *or* $F(\mathfrak{T})$ *is a subbase of* $\bar{\mathfrak{T}}$ *for each* $\mathfrak{T} \in \mathscr{H}$.

Proof follows from Corollary 8 and Lemma 6.

Problem. Are the realizations of $\mathscr{H}$ given by open sets or closed sets the only realizations of $\mathscr{H}$?

Proposition 15. *Let* $\mathscr{L}$ *be an epireflective subcategory of* $\mathscr{B}$, $\mathscr{L} \subseteq \mathscr{K}$. *Let for every* $(B, \mathfrak{T}) \in \mathscr{L}$ *there exist* $b \in B$ *with* $(B^*, \mathfrak{T}_b^*) \in \in \mathscr{L}$. *Then* $\mathscr{L}$ *is not simple (i.e.* $\mathscr{L}$ *is not the epireflective hull of a single space).*

Proof. Let F be a realization of $\mathscr{L}$, $F(\mathfrak{T}) \subseteq \bar{\mathfrak{T}}$ for each $\mathfrak{T} \in \mathscr{L}$. Analogously as Lemma 9 it can be proved that for every $(B, \mathfrak{T}) \in \mathscr{L}$ there exists $b \in B$ with $\{b\} \in F(\mathfrak{T})$. Suppose $\mathscr{L}$ is the epireflective hull of a space $(A, \mathfrak{A})$. Therefore $\mathscr{L} \subseteq \{(B, \mathfrak{B}): \{f^{-1}(X): X \in \bar{\mathfrak{A}}, f \in \in [\mathfrak{B}, \mathfrak{A}]\}$ is a subbase for $\bar{\mathfrak{B}}\}$ (see [2]). Analogously as in the proof of Proposition 7 one can construct a realization F_1 of $\mathscr{L} F_1(\mathfrak{T}) \subseteq \bar{\mathfrak{T}}$ such that no one-point set from $(A, \mathfrak{A})^n$, $n > \operatorname{card} A$, belongs to $F_1((A, \mathfrak{A})^n)$; a contradiction.

Let $\mathscr{L}$ be a subcategory of $\mathscr{K}$. Put $\tilde{\mathscr{L}} = \{(A, \mathfrak{T}) \in \mathscr{K}$: for every $x, y \in A$, $x \neq y$, there exists $\mathfrak{A} \in \mathscr{L}$ and $f \in [\mathfrak{T}, \mathfrak{A}]$ such that $f(x) \neq f(y)\}$. Clearly $\tilde{\mathscr{L}}$ is an epireflective subcategory of $\mathscr{B}$.

Corollary 10. *Let* $\mathscr{L} \subseteq \mathscr{K}$. *Then* $\tilde{\mathscr{L}}$ *is not simple.*

Proof follows from Proposition 15 and Lemma 8.

Therefore the category of completely Hausdorff spaces is not simple.

REFERENCES

[1] N. Bourbaki, *Topologie générale,* Russian translation, Moscow, 1968.

[2] H. Herrlich, Topologische Reflexionen und Coreflexionen, *Lecture Notes,* 78 (Berlin, 1968).

[3] J. Chvalina – M. Sekanina, Realizations of closure spaces by set systems, *Proceedings of the 3rd Prague Toposymposium,* 85-87.

[4] L. Kučera, A shrinking of a category of societies is a universal partly ordered class, *Comm. Math. Univ. Carolinae,* 12 (2) (1971), 401-411.

[5] A. Pultr, On selecting of morphisms among all mappings between underlying sets of objects in concrete categories and realizations of these. *Comm. Math. Univ. Carolinae,* 8 (1) (1967), 53-83.

[6] M. Sekanina, Embedding of the category of partially ordered sets into the category of topological spaces, *Fund. Math.,* 66 (1969/70), 95-98.

[7] J.C. Warndof, Topologies uniquely determined by their continuous self maps, *Fund. Math.,* (66) (1969/70), 25-43.

COLLOQUIA MATHEMATICA SOCIETATIS JÁNOS BOLYAI
8. TOPICS IN TOPOLOGY, KESZTHELY (HUNGARY), 1972.

SOUSLIN TREES AND DOWKER SPACES

M.E. RUDIN

A partially ordered set is called a *tree* if every totally ordered subset is well ordered; a chain is a totally ordered subset and an *antichain* a pairwise unordered subset. A partially ordered set is a κ*-Souslin tree* if it is a tree of cardinality κ with no chain or antichain of cardinality κ.

Let Λ be the class of all uncountable cardinals which are neither singular nor the successor of a singular cardinal. If $\kappa \in \Lambda$ [1, 2] it is consistent with the usual axioms of set theory that there exists a κ-Souslin tree; κ-Souslin trees exist trivially for singular κ.

A Dowker space is a normal Hausdorff space which is not countably paracompact [3]. Let us call X a κ-Dowker space provided X is a Dowker space of cardinality κ, but no subspace of X of cardinality less than κ is a Dowker space.

In [4] countably many copies of an $\aleph_1$-Souslin tree are used to construct an $\aleph_1$-Dowker space. In [5] a box product on the chains of an $\aleph_\omega$-Souslin tree is used to construct an $\aleph_\omega^\omega$-Dowker space.

In this paper it will be shown that, by the same technique used in [4], if $\kappa \in \Lambda$, then countably many copies of a κ-Souslin tree may be used to construct a κ-Dowker space.

Observe that there are no $\aleph_\omega$-Dowker spaces.

Suppose $\lambda < \kappa$ is regular. Then, for $\kappa \in \Lambda$, the techniques of [4], and, for the cofinality of κ equal to λ the techniques of [5], yield normal non-λ-paracompact spaces. But for λ uncountable, λ itself with the order topology also has these properties. The interesting case seems to be $\lambda = \aleph_0$.

I feel there is a two way relationship between κ-Dowker spaces and κ-Souslin trees. The basic problems which cry for an answer are:

A. *Find a pure set theoretic translation of the existence of a κ-Dowker space.*

B. *Use an $\aleph_1$-Dowker space to construct an $\aleph_1$-Souslin tree.*

Assume $\kappa \in \Lambda$; then there is a regular cardinal ρ such that κ is the successor of ρ. Assume $(R, \leqslant)$ is a κ-Souslin tree.

If $\beta < \kappa$ let R_β be the set of all terms of R whose predecessors under $\leqslant$ are order isomorphic to β; observe that $R = \bigcup_{\beta < \kappa} R_\beta$.

Let L be the set of all ordinals less than κ which are cofinal with ρ. Without loss of generality assume that, if $x \in R_\lambda$ for some $\lambda \in L$, then there is no other term of R having exactly the same predecessors as x. Also without loss of generality we assume $x \in R_\beta$ and $\beta < a$ imply there is a $y \in R_a$ with $x < y$.

Since ρ is regular, for each $\lambda \in L$ we can define a family $\{f_\beta\}_{\beta < \rho}$ of functions from R_λ into R_λ such that:

(a) $f_\beta(x) = f_a(y)$ only if $\beta = a$ and $x = y$,

(b) $y \in R_\lambda$ and $x < y$ imply there is an a such that $x < f_\beta(y)$ for all $\beta > a$.

CONSTRUCTION OF THE κ-DOWKER SPACE T ASSOCIATED WITH $(R, \leqslant)$

The points of T are the set of all ordered pairs (x, n) where x belongs to R and n belongs to ω.

(1) If $\beta \in \kappa - L$ and $x \in R_\beta$, $\{(x, n)\}$ is a basic open set for (x, n).

(2) If $\lambda \in L$ and $x \in R_\lambda$ and $\beta < \lambda$, then $\{(y, 0) | y \leqslant x$ and $y \in R_\delta$ and $\beta < \delta \leqslant \lambda\}$ is a basic open set for $(x, 0)$.

(3) We now define basic open sets for (x, n) where $\lambda \in L$ and $x \in R_\lambda$ and $n > 0$ by induction on n and λ. Assume that we have defined basic open sets for all points (y, m) of T where $m < n$ or $y \in R_\gamma$ for $\gamma < \lambda$.

For $a < \rho$ define $F_a(x, n) = \{(f_\beta(x), n-1) | \beta > a\}$.

For $\gamma < \lambda$ define $G_\gamma(x, n) = \{(y, n) | y < x$ and $y \in R_\delta$ and $\gamma < \delta < \lambda\}$.

Then N is a basic open set for (x, n) if and only if there exists $a < \rho$ and $\gamma < \lambda$ such that N is the union of $\{(x, n)\}$ and a basic open set for each point of $F_a(x, n) \cup G_\gamma(x, n)$.

The set of all basic open sets described in (1), (2), and (3) form a basis for the topology of T.

For $n \in \omega$, let $C_n = \{(x, m) \in T | m \geqslant n\}$. Then $C_n \supset C_{n+1}$ and C_n is closed. Suppose that for each n, D_n is open and contains C_n; we prove $\bigcap_{n \in \omega} D_n \neq \phi$. Thus by [3], when we prove T is normal and Hausdorff, we know T is a Dowker space. Clearly T has cardinality κ; we prove $Y \subset T$ and the cardinality of Y is less than κ imply Y is paracompact. Thus T is a κ-Dowker space.

Proof that $\bigcap_{n \in \omega} D_n \neq \phi$.

Lemma. *Fix $n \in \omega$. If $x \in R$ there is a $y > x$ such that $z > y$ implies $(z, 0) \in D_n$.*

Proof. Assume not. Let $P = \{z \in R | (z, 0) \notin D_n$ and $z > x\}$. Let Q_0 be a maximal antichain in P. Define an antichain Q_β in R for each $\beta < \rho$ as follows. Suppose Q_γ has been defined for all $\gamma < \beta < \rho$. Select an ordinal $\lambda_\beta < \kappa$ such that $\bigcup_{\gamma<\beta} Q_\gamma \subset \bigcup_{\delta < \lambda_\beta} R_\delta$. Let Q_β be a maximal antichain in $P \cap \bigcup_{\delta > \lambda_\beta} R_\delta$. Let λ be the limit of $\{\lambda_\beta\}_{\beta<\rho}$; $\lambda \in L$. Let $Y = \{y \in R_\lambda | x < y\}$. If $\beta < \rho$ and $y \in Y$, since Q_β is maximal and we are assuming some $z > y$ in P, there is a term $y_\beta < y$ of Q_β. But, by (2), $(y, 0)$ is a limit point of $\{(y_\beta, 0) | \beta < \rho\}$; hence $(y, 0) \notin D_n$. But $w \in Y$ implies $(w, n) \in D_n$ (and $Y \neq \phi$ by assumption). So, by the definition of F_β in (3) and f_β in (b), there is a $y \in Y$ such that $(y, 0) \in$ $\in D_n$. But this is a contradiction.

Now for $n \in \omega$, define $P_n = \{y \in R | z > y$ implies $(z, 0) \in D_n\}$. Choose Q_n to be a maximal antichain in P_n. There is a $\lambda < \kappa$ such that $\bigcup_{n\in\omega} Q_n \subset \bigcup_{\beta<\lambda} R_\beta$. If $x \in R_\lambda$, then $(x, 0) \in D_n$ for all n.

Proof *that* T *is a normal Hausdorff space.* It is clear that every point of T is a closed set.

Let H and K be disjoint closed subsets of T. For $n \in \omega$ define $H_n = \{x \in R | y > x$ implies that there is a $z > y$ such that $(z, n) \in H\}$; define K_n similarly.

Lemma. *If* i *and* j *belong to* ω, $H_i \cap K_j = \phi$.

Proof. Suppose $x \in H_i \cap K_j$. Let $P = \{z \in R | x < z$ and $(z, i) \in$ $\in H\}$ and let $P' = \{z \in R | x < z$ and $(z, j) \in K\}$. Let Q_0 and Q'_0 be maximal antichains in P and P', respectively. For each $\beta < \rho$, choose antichains in R as follows. Assume Q_γ and Q'_γ have been chosen for all $\gamma < \beta < \rho$. Select an ordinal $\lambda_\beta < \kappa$ so that $\bigcup_{\gamma<\beta} (Q_\gamma \cup Q'_\gamma) \subset$ $\subset \bigcup_{\delta<\lambda_\beta} R_\delta$. Then choose Q_β and Q'_β to be maximal antichains in $P \cap$ $\cap \bigcup_{\delta>\lambda_\beta} R_\delta$ and $P' \cap \bigcup_{\delta>\lambda_\beta} R_\delta$ respectively. Let λ be the limit of $\{\lambda_\beta\}_{\beta<\rho}$; $\lambda \in L$. Let $Y = \{y \in R_\lambda | y > x\}$. Clearly, since H and K are closed, by the definition of G_γ in (3), $y \in Y$ implies $(y, i) \in H$ and $(y, j) \in K$. Hence $i \neq j$. Then by the definition of F_β in (3), if $w \in Y$,

(w, j) is a limit point of $\{(y, i) | y \in Y\}$; hence $(w, j) \in H$. But this contradicts H and K being disjoint.

Let $P_1 = \{x \in \bigcup_{n \in \omega} H_n | y > x$ and $i \in \omega$ implies $(y, i) \notin K\}$.

Let $P_2 = \{x \in \bigcup_{n \in \omega} K_n | y > x$ and $i \in \omega$ implies $(y, i) \notin H\}$.

Let $P_3 = \{x \in R | y > x$ and $i \in \omega$ implies $(y, i) \notin H \cup K\}$.

Let Q be a maximal antichain in $P_1 \cup P_2 \cup P_3$. There is a $\lambda < \kappa$ such that $Q \subset \bigcup_{\beta < \lambda} R_\beta$. Observe that $\alpha \geqslant \lambda$ and $y \in R_\alpha$ implies there is a term x of Q such that $x < y$; this follows from the fact that Q is maximal and that there is a $z > y$ such that $z \in P_1 \cup P_2 \cup P_3$.

The set $X = \{(y, n) \in T | y \in R_\delta$ and $\delta > \lambda\}$ is both open and closed. And the sets $U = \{(y, n) \in X|$ there exists an $x \in P_1$ such that $x > y\}$ and $V = \{(y, n) \in X|$ there exists an $x \in P_2 \cup P_3$ such that $x < y\}$ are open and, by the lemma, disjoint. Also $U \supset X \cap K$.

It remains to separate H and K in the open closed set $Y = T - X$. Observe that the cardinality of Y is less than κ, and the uniomof less than κ basic open sets is both open and closed. Index the points of Y as $\{y_\beta\}_{\beta < \rho}$. If $y_\beta \notin H$ choose a basic open set U_β such that $y_\beta \in U_\beta \subset T - H$; if $y_\beta \in H$ choose a basic open set U_β such that $y_\beta \in U_\beta \subset T - K$. Then define $V_\beta = U_\beta - \bigcup_{\alpha < \beta} U_\alpha$. Clearly $V = \bigcup \{V_\beta | y_\beta \in H\}$ and $V' = \bigcup \{V_\beta | y_\beta \in K\}$ are disjoint and open and $(H \cap Y) \subset (V \cap Y)$ and $(K \cap Y) \subset (V' \cap Y)$. Hence T is normal.

Observe that the same argument given in the preceeding paragraph shows *Y is paracompact for any subset Y of T of cardinality less than κ.*

REFERENCES

[1] S. Tennenbaum, *Souslin's problem,* Proc. Nat. Acad. Sci. U.S.A., 59 (1968), 60-63.

[2] R.M. Solovay – S. Tennenbaum, *Iterated Cohen extensions and Souslin's problem* (to appear).

[3] C.H. Dowker, *On countably paracompact spaces,* Can. J. Math., 3 (1951), 219-224.

[4] M.E. Rudin, *Countable paracompactness and Souslin's problem,* Can. J. Math., 7 (1955), 543-547.

[5] M.E. Rudin, *A normal space X for which $X \times I$ is not normal,* Fund. Math., 73 (1971), 179-186.

[6] M. Souslin, Probleme 3, Fund. Math., 1 (1920), 223.

COLLOQUIA MATHEMATICA SOCIETATIS JÁNOS BOLYAI
8. TOPICS IN TOPOLOGY, KESZTHELY (HUNGARY), 1972.

MINIMAL HAUSDORFF SPACES ARE PSEUDO-QUOTIENT IMAGES OF COMPACT HAUSDORFF ONES

L. RUDOLF

Searching for a connection of the kind expressed in the title between the classes of spaces mentioned there follows the idea of P. S. Alexandroff, presented at the *First Prague Topological Symposium*, to characterize topological spaces as continuous images of "nice" spaces under "nice" maps. Clearly, in our case we must resign continuity of maps, still not desisting from require them to be "nice".

The notion of a minimal Hausdorff space as well as much expository material may be found in [4].

Let Q be a decomposition of a topological space X into non-empty disjoint sets and $q: X \to X/Q$ the quotient map. The topology in X/Q generated by a base consisting of sets $q[U] = \{x \in X/Q: q^{-1}(x) \subset U\}$, U running over the family of regularly open sets of X, is called the *pseudo-quotient topology.* This topology in X/Q is, in general, finer than the quotient one and in this case the pseudo-quotient map is not continuous. However, the pseudo-quotient topology has an advantage in comparison with the quotient topology: the pseudo-quotient space X/Q is Hausdorff whenever each pair of distinct elements of Q can be separated by disjoint open sets.

A decomposition Q of X is called *irreducible* whenever each non-empty regularly open set of X contains an element of Q. In particular, for X being a minimal Hausdorff space, irreducibility of Q is equivalent to $q(A) \neq X/Q$ for each closed $A \subsetneq X$, the irreducibility of q.

Theorem 1. *The pseudo-quotient map $q\colon X \to X/Q$ induced by an irreducible decomposition Q of X has the following properties*

(i) *q is Θ-continuous**

(ii) *the basic sets $q[U]$ are regularly open.*

So q is "nice"!

A set $A \subset X$ is *regularly embedded* in X whenever each $x \notin A$ has an open neighbourhood U_x such that $\mathrm{cl}\, U_x \cap A = \phi$. Such sets appear in a natural way: if X/Q is a Hausdorff space, Q being irreducible, then Q consists of regularly embedded sets, by the Θ-continuity of q.

Claim 1. Disjoint regularly embedded sets in a minimal Hausdorff space can be separated by disjoint open sets.

Using Theorem 1 we obtain a

Corollary. *The pseudo-quotient space X/Q of a minimal Hausdorff space X, where Q is an irreducible decomposition consisting of regularly embedded sets, is a minimal Hausdorff space,* which is a half of the theorem announced in the title in the case of X being a compact Hausdorff space. This procedure of generating minimal Hausdorff spaces includes the straightforward constructions of [1] and [7], founded on Urysohn's example [9].

Besides, the Corollary justifies the importance of irreducible pseudo-quotient maps in dealing with minimal Hausdorff spaces.

*A map $f\colon X \to Y$ is Θ-continuous (a notion due to S.W. Fomin [2]) whenever for each $x \in X$ and each open neighbourhood U_y of $y = f(x)$ there exists an open neighbourhood U_x of x such that $f(\mathrm{cl}\, U_x) \subset \mathrm{cl}\, U_y$. This actually generalized continuity notion coincides with the classical continuity, say, when the co-domain is a regular space, and recently proves as very usefull in dealing with non-regular spaces, particularly in the theory of H-closed spaces.

Theorem 2. *A map* $q: X \to Y$ *of a minimal Hausdorff space* X *onto a minimal Hausdorff space* Y *is the pseudo-quotient map induced by an irreducible decomposition iff* q *satisfies any of the following equivalent conditions*

(i) q *is irreducible and* Θ*-continuous*

(ii) q *is irreducible and preserves regularly closed sets and* $q^{-1}(x)$ *is regularly embedded for each* $x \in X$.

Now it is quite immediate that irreducible pseudo-quotient maps between minimal Hausdorff spaces form a category, abbreviated in the sequel by $\mathcal{M}Q$

Claim 2. *Each map of* $\mathcal{M}Q$ *is a monomorphism.*

A curiosity!

The characterization of Theorem 2 yields also a comfortable way in proving

Theorem 3. *A space is minimal Hausdorff iff it is the pseudo-quotient image of a compact Hausdorff space induced by an irreducible decomposition consisting of regularly embedded (=compact, in this case) sets,*

– the central result. The "if" implication follows from the previous Corollary, the proof of the "only if" makes the most of Iliadis results [5], in particular of his construction of the extremally disconnected resolution $a^x: aX \to X$, described there.

Theorem 3 leads to the following definition: a *compact resolution* of a minimal Hausdorff space is a map of $\mathcal{M}Q$ from a compact space onto the given one. In other words, it is a pseudo-quotient irreducible map between a compact and a minimal Hausdorff space. The resolution a^x: $aX \to X$ is not, in general, the only one, – the procedures of forming a minimal space from a compact one, described in [1] and [7] as well as that given by Theorem 3 lead to resolutions of the resultant space which differ from the Iliadis resolution of it whenever the starting-point compact space is not extremally disconnected. This fact raises two questions: whether there exists a maximal resolution and if it is the case, is it functorial?

Theorem 4. *For each compact resolution* $q\colon Y \to X$ *of a minimal Hausdorff space* X *there exists a unique resolutions* $\bar{q}\colon \alpha X \to Y$ *such that the diagram*

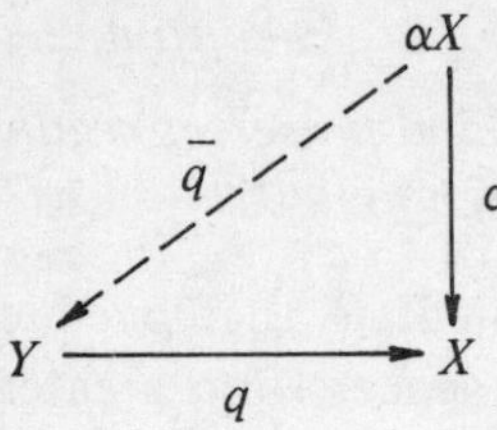

is commutative.

Thus the Iliadis resolution is the greatest one!

Note, that uniqueness of $\bar{q}$ is an immediate consequence of Claim 2. The existence of $\bar{q}$ can be deduced from Theorem 3 of [6]. Observe, that $\bar{q}$ is continuous in view of compactness of Y, thus it is simply a quotient map. Now, given a map $f\colon X \to Y$ from $\mathcal{M}Q$, the resolution $\alpha f = \overline{f \circ \alpha^X}$ completing the diagram

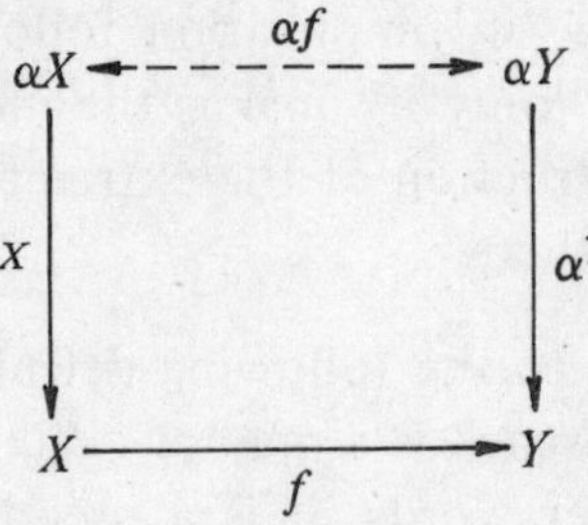

is an irreducible onto map between extremally disconnected compact Hausdorff spaces, thus an isomorphism by Gleason's Theorem [3], hence the resolution $\alpha^X\colon \alpha X \to X$ leads to a functor α and Theorem 4 yields

Corollary. *The functor* α *from the category* $\mathcal{M}Q$ *of minimal Hausdorff spaces and their irreducible pseudo-quotient (= irreducible Θ-continuous onto) maps into their full subcategory* $\mathcal{C}\mathcal{E}Q$ *consisting of*

extremally disconnected compact Hausdorff spaces and their resolutions (= isomorphisms) is a coreflection adjoint to the full embedding $\mu: \mathscr{C}\mathscr{E}Q \to \mathscr{M}Q$.

By Theorem 4, each compact resolution of a minimal Hausdorff space is obtained from a quotient space of the greatest compact resolution of this space. This result is complete with the following.

Theorem 5. *Given a minimal Hausdorff space* X *and a decomposition* $aX \xrightarrow{p} Y \xrightarrow{q} X$ *of* $aX \xrightarrow{a^X} X$ *with* $aX \xrightarrow{p} Y$ *being a quotient map, the factor* $Y \xrightarrow{q} X$ *is a compact resolution of* X, which establishes a one-to-one correspondence between compact resolutions of X and quotient spaces of aX induced by decompositions finer than $\{(a^X)^{-1}(x) \colon x \in X\}$.

A paper containing proofs of the above results will appear in *Bulletin de l'Academie Polonaise des Sciences.*

The functor of compact resolution allows to translate questions on minimal spaces into questions on compact ones. This method has a nice illustration in a categorial proof of the known product theorem for minimal Hausdorff spaces [8]. Also a Baire category theorem for minimal Hausdorff spaces, containing all known results (see [7]), can be obtained by this way.

Theorem 5 enables an insight into the family of all compact resolutions of a given minimal space and it turns out that each subset in this family has a supremum, which leads to a new formal construction of the functor a, and that the space is determined by the family of its resolutions.

Finally, the main results of this paper can be translated to semiregular spaces, using extensions of them to minimal ones, and then even to arbitrary Hausdorff spaces.

The just announced results will be included in papers now in preparation.

REFERENCES

[1] R.F. Dickman – A. Zame, On functionally closed spaces, *Pacific J. Math.*, 2 (1969), 303-311.

[2] S.V. Fomin, On the theory of the extensions of topological spaces, *Mat. Sbornik,* 50 (1940), 285-294.

[3] A.M. Gleason, Projective topological spaces, *Illinois J. Math.*, 2 (1958), 482-489.

[4] H. Herrlich, T_v-Abgeschlossenheit und T_v-Minimalität, *Math. Z.*, 88 (1965), 285-294.

[5] S. Iliadis, Absolutes of Hausdorff spaces, *DAN SSSR,* 149 (1963), 22-25.

[6] S. Iliadis – S.V. Fomin, The method of centered systems in the theory of topological spaces, *Uspehi Mat. Nauk,* 21 (1966), 47-76.

[7] J. Mioduszewski, Remarks on Baire theorem for *H*-closed spaces, *Coll. Math.*, XXIII (1971), 39-41.

[8] L. Rudolf, A categorial proof of the product theorem for minimal Hausdorff spaces, *Coll. Math.*, (to appear).

[9] P.S. Urysohn, Über die Mächtigkeit der zusammenhängenden Mengen, *Math. Ann.*, 94 (1925), 262-295.

COLLOQUIA MATHEMATICA SOCIETATIS JÁNOS BOLYAI
8. TOPICS IN TOPOLOGY, KESZTHELY (HUNGARY), 1972.

EXTENSORS FOR COMPACT AND METRIZABLE SPACES AND APPLICATIONS TO INFINITE-DIMENSIONAL SPACES

YU.M. SMIRNOV*

1. ABSOLUTE NEIGHBOURHOOD RETRACTS AND THE EXTENSION OF ϵ-CLOSE MAPPINGS

Already in the case of separable metrizable spaces** for many purposes it is more convenient to evaluate the closeness between mappings*** not by means of the distance function but by means of coverings:

Definition A. Two mappings $f, g: X \to Y$ are called to be ϵ-close, where ϵ is an open covering of the space Y, if for any point $x \in X$ there exists a set $H \in \epsilon$ such that $fx \in H$ and $gx \in H$.

This relation will be denoted by $d(f, g) < \epsilon$.

*This work has been completed within framework of the scientific cooperation between the Moscow and Sohia universities.

**All spaces considered here are Hausdorff.

***All mappings considered here are continuous.

We recall the following well-known definitions due to Borsuk [1] and Hu [2], respectively.

Definition B. A space Y belonging to the class $\mathcal{K}$ is called an absolute neighbourhood retract for the class $\mathcal{K}$, if for every closed embedding $i: Y \to X$, where $X \in \mathcal{K}$, there exists a retraction $r: U \to iY$ of a certain neighbourhood U of the image iY.

Definition H. The space Y is called an absolute neighbourhood extensor for the class $\mathcal{K}$ if for any space $X \in \mathcal{K}$ and any mapping $f: A \to Y$ with A closed in X, there exists an extension $\operatorname{ext} f: U \to Y$ of the mapping f to a certain neighbourhood U of the set A.

In the first case we shall write $Y \in NR(\mathcal{K})$, and in the second $Y \in NE(\mathcal{K})$.

It is known ([2]) that for the classes $\mathcal{M}$ (of all metrizable spaces), $\mathcal{C}$ (of all compact spaces), and $\mathcal{T}$ (of all Tychonov spaces) the following relations are valid:

$$NR(\mathcal{C}) = \mathcal{C} \cap NE(\mathcal{C}) \supset \mathcal{C} \cap NE(\mathcal{T}),$$

$$NR(\mathcal{M}) = \mathcal{M} \cap NE(\mathcal{M}).$$

Theorem 1. *If $Y \in NR(\mathcal{K})$, where $\mathcal{K} = \mathcal{C}$ or $\mathcal{K} = \mathcal{M}$, then for any covering ϵ there is a covering δ such that whenever $X \in \mathcal{T}$ ($X \in \mathcal{M}$), A is closed in X and $d(f, g: A \to Y) < \delta$ then the existence of an extension $\operatorname{ext} f: X \to Y$, implies the existence of an extension $\operatorname{ext} g: X \to Y$ for which $d(\operatorname{ext} f, \operatorname{ext} g) < \epsilon$.*

This was proved by Borsuk ([1]) for the class $\mathcal{C} \cap \mathcal{M}$ (in an equivalent formulation where ϵ and δ are numbers), and by Dugundji for the class $\mathcal{M}$. However, for the class $\mathcal{C}$ this result seems to be new to me. We remark that, as is shown by very simple examples, Theorem 1, and even its Corollary, to be given below, fail in Borsuk's classical formulation already in the class $\mathcal{M}$ if we do not assume compactness.

Corollary. *If $Y \in NR(\mathcal{K})$, where $\mathcal{K} = \mathcal{C}$ or $\mathcal{K} = \mathcal{M}$, then there exists a covering δ such that whenever $X \in \mathcal{T}$ ($X \in \mathcal{M}$), A is closed in X, and $d(f, g: A \to Y) < \delta$, then the existence of an extension $\operatorname{ext} f$ implies the existence of an extension $\operatorname{ext} g$.*

There are two ways to prove this theorem. Both of them apply, with suitable modifications, to the class $\mathscr{M}$ as well as to the class $\mathscr{C}$. The first method is even simpler for the class $\mathscr{C}$ assuming that X is normal. The second (although it is rather complicated) is interesting because it is needed in proving the following propositions, which are of independent interest.

Theorem 2. *If* $Y \in NE(\mathscr{K})$, *where* $\mathscr{K} = \mathscr{C}$ *or* $\mathscr{K} = \mathscr{M}$, *then for any point* $y \in Y$ *and its arbitrary neighbourhood* U *we can find a neighbourhood* V *of* y *such that* $V \subset U$, *and every mapping* $f\colon A \to V$ *has an extension* $\operatorname{ext} f\colon X \to U$ *for every space* $X \in \mathscr{T}$ $(X \in \mathscr{M})$ *and every closed set* A *in* X.

For $\mathscr{K} = \mathscr{M}$ this theorem was proved by Dugundji ([2]). If $\mathscr{K} = \mathscr{C}$ the statement of our theorem seems to be new.

To prove this we need a lemma which generalizes a construction of K. Kuratowski (which is of countable character and therefore rather simple).

Lemma 1. *Let us be given in the space* Y *a point* y, *its neighbourhood* U, *and an infinite pseudobase* $\{V\}$ *of neighbourhoods* V *of the point* y *satisfying the following condition: for any such neighbourhood* V *there exists a mapping* $f_V\colon A_V \to V$, *where* A_V *is a closed set in a space* X_V, *such that* f_V *is not extendible to a mapping of* X_V *into* U. *Then there exists a space* X *and a mapping* $f\colon A \to Y$ *of some closed subset* A *of* X *which is not extendible to any neighbourhood of the set* A.

Lemma 2. *In lemma* 1, *the space* X *belongs to* $\mathscr{T}$ *if all the spaces* X_V *do.*

The converse of the above corollary is also valid. Hence its assertion (as well as that of theorem 1) provides a necessary and sufficient condition for $Y \in NR(\mathscr{K})$ ($\mathscr{K} = \mathscr{M}$ or $\mathscr{K} = \mathscr{C}$).

Theorem 3. *Suppose that there exists a covering* δ *of the space* $Y \in \mathscr{K}$, *where* $\mathscr{K} = \mathscr{M}$ *or* $\mathscr{K} = \mathscr{N}$ *(the class of all normal spaces), such that for any space* $X \in \mathscr{K}$ *and any closed subset* $A \subset X$ *the relation*

$d(f, g\colon A \to Y) < \delta$ implies that if f is extendible to X then so is g. Then $Y \in NE(\mathscr{K})$.

Curiously, this theorem seems to be new in both cases. Its proof is rather simple, making use of the following additivity property. Let us first call a space Y an extensor for the space X, written $Y \in E(X)$, provided that every mapping $f\colon A \to Y$ from any closed subset A of X is extendible to X.

Theorem 4. *If a paracompact space Y has an open covering ω such that $U \in E(X)$ for all $U \in \omega$, where X is paracompact, then $Y \in E(X)$ as well.*

If only finite coverings are considered then no assumption about the space Y is needed and about the space X instead of paracompactness the assumption of normality suffices. This result is due to Hanner ([2]).

The following problem remains open.

Problem 1. Does the assertion of theorem 1 remain valid if we replace the condition $Y \in NR(\mathscr{K})$ by $Y \in NE(\mathscr{K})$?

2. THE GENERALIZED WHITEHEAD THEOREM ON MATCHING ABSOLUTE NEIGHBOURHOOD RETRACTS

This result we obtain for the class $\mathscr{M}$ only, since for $\mathscr{C}$ it is not valid. In its original form ([1]) it was proved for the class $\mathscr{C} \cap \mathscr{M}$ and asserts that the matching of the absolute neighbourhood retracts X and B by a continuous mapping $f\colon A \to B$, where A is closed and a neighbourhood retract of X, is again an absolute neighbourhood retract.

Theorem 5. *Let us be given a closed mapping $f\colon X \to Y$ onto the space Y and a closed subset B of Y so that f is one-to-one on $X \setminus A$, where $A = f^{-1}B$. Then $Y \in NR(\mathscr{M})$, provided that $X \in NR(\mathscr{M})$, $A \in NR(\mathscr{M})$, and $B \in NR(\mathscr{M})$.*

There is an analogous theorem in Hu's book ([2]). It is easy to construct examples showing that for non-closed mappings the theorem cases to be valid, and other examples showing that the space Y may be not metrizable while X, A and B are. The proof is analogous to that presented in Borsuk's book ([1]) for the case $\mathscr{C} \cap \mathscr{M}$.

Problem 2. Is it true that if in theorem 5 we drop the assumption about the metrizability of Y (preserving the metrizability of the spaces X and B), then $Y \in NE(\mathscr{M})$?

Finally we remark that the conditions $A \in NR(\mathscr{M})$ and $B \in \in NR(\mathscr{M})$ are essential even in the case of $\mathscr{C} \cap \mathscr{M}$.

3. INFINITE-DIMENSIONAL SPACES AND THE EXTENDIBILITY OF MAPPINGS

Definition. A space $X \in \mathscr{M}$ is called locally finite-dimensional if each of its points has a finite-dimensional neighbourhood in the sense of the dimension dim. In this case we write $X \in \mathscr{M}'$.

Theorem 6. *There exists a space* $\Sigma \in \mathscr{M}$ *such that* $X \in \mathscr{M}'$ *if and only if* $\Sigma \in E(X)$.

This theorem was proved by my student Z. Lisitsa. The space Σ is a well-known example of Borsuk ([1]). It is contractible but not locally contractible.

Problem 3. Could the space Σ in theorem 6. be replaced by a locally contractible space?

The following problem arises naturally in connection with the last theorem: How to decompose the class of all – even metrizable – infinite-dimensional compacta into dimensional classes characterizable in the manner indicated in theorem 6, by means of "characterizing" spaces Σ belonging to different, sufficiently good classes. For example by means of spaces $\Sigma \in NR(\mathscr{M})$, or spaces which are metrizable polytopes, or perhaps *CW*-complexes. Concerning this topic I have raised three concrete problems at the problem session of the conference (cf. the list of problems).

COLLOQUIA MATHEMATICA SOCIETATIS JÁNOS BOLYAI
8. TOPICS IN TOPOLOGY, KESZTHELY (HUNGARY), 1972.

A REMARK ON EMBEDDINGS AND DISCRETELY EXPANDABLE SPACES

J.C. SMITH

§1. INTRODUCTION

Recently a number of topological properties have been characterized by various types of subset embeddings. H.L. Shapiro [4] has characterized collectionwise normal spaces by requiring that all closed subsets be *P*-embedded. Alo and Shapiro [1] have characterized strongly normal spaces and normal countably paracompact spaces in a similar way. Nichols and Smith [7] have shown that a number of other type embeddings are equivalent to expandable spaces and almost expandable spaces.

In this paper we continue this investigation of characterization of certain spaces by embeddings. In §2 we show that discretely expandable and almost discretely expandable spaces have similar embedding theorems. In §3 we obtain new embedding results for collectionwise normal and normal spaces, and in §4 we examine special properties of almost discretely expandable space.

The reader is referred to [4] and [7] for the definitions of the various types of expandable spaces and embeddings.

§2. EMBEDDING CHARACTERIZATIONS FOR DISCRETELY EXPANDABLE SPACES

In [6] Krajewski and Smith introduced the notion of discretely expandable spaces and obtained the following characterizations for expandable spaces.

Definition 2.1. A space X is *(almost) discretely expandable iff* for every discrete collection $\{F_a : a \in A\}$ in X there exists a (point finite) locally finite open collection $\{G_a : a \in A\}$ such that $F_a \subseteq G_a$ for each $a \in A$.

Theorem 2.2. (i) X *is expandable iff* X *is discretely expandable and countably paracompact.*

(ii) X *is almost expandable iff* X *is almost discretely expandable and countable metacompact.*

In [7] Nichols and Smith gave characterizations for expandable spaces using embedding notions similar to that of strongly P-embedded subsets due to [1] Alo and Shapiro. The reader is referred to [7] for the various definitions and the following theorem summarizes the characterizations.

Theorem 2.3. *The following are equivalent.*

(1) X *is expandable.*

(2) *Every closed subset of* X *is* E_1*-embedded in* X.

(3) *Every closed subset of* X *is* E_2*-embedded in* X.

(4) *Every closed subset of* X *is* E_3*-embedded in* X.

(5) *Every closed subset of* X *is* E_4*-embedded in* X.

(6) *Every closed subset of* X *is* E_5*-embedded in* X.

It was also shown in [7] that discrete expandability could not be characterized by obvious modifications of Theorem 2.3 above. We now show that discretely expandable spaces enjoy a somewhat nice property and use this property to obtain embedding characterizations for these spaces.

Theorem 2.4. *Let* X *be a discretely expandable space. Then every* *open cover of X with finite order has a locally finite open refinement.*

Proof. Let $\mathfrak{G} = \{G_a : a \in A\}$ be an open cover of X such that order $(\mathfrak{G}) = n$.

Step 1. Define $\mathfrak{B}_1 = \{B \subseteq A : |B| = 1\}$ and let $H(B, 1) = \left[\bigcap_{\beta \in B} G_\beta\right] - \left[\bigcup_{a \in A - B} G_a\right]$ for each $B \in \mathfrak{B}_1$. Since $\mathfrak{H}_1 = \{H(B, 1): B \in \mathfrak{B}_1\}$ is a discrete collection, there exists a locally finite open collection $\mathfrak{U}_1 = \{U(B, 1): B \in \mathfrak{B}_1\}$ such that

(i) $H(B, 1) \subseteq U(B, 1)$ for each $B \in \mathfrak{B}_1$,

(ii) $U(B, 1)$ is contained in some member of $\mathfrak{G}$.

Note that if $\operatorname{ord}(x, \mathfrak{G}) = 1$, then x belongs to some member of $\mathfrak{U}_1$. Let $U_1 = \cup\{U : U \in \mathfrak{U}_1\}$.

Step 2. Define $\mathfrak{B}_2 = \{B \subseteq A : |B| = 2\}$ and let $H(B, 2) = \left[\bigcap_{\beta \in B} G_\beta\right] - \left[\bigcup_{a \in A - B} G_a\right] - U_1$ for each $B \in \mathfrak{B}_2$. We assert that $\mathfrak{H}_2 = \{H(B, 2): B \in \mathfrak{B}_2\}$ is discrete. Indeed, if $x \in U_1$ then U_1 intersects no member of $\mathfrak{H}_2$. If $x \notin U_1$, then $\operatorname{ord}(x, \mathfrak{G}) \geqslant 2$. If $\operatorname{ord}(x, \mathfrak{G}) = 2$, then x belongs to only G_{a_1} and G_{a_2} so that $G_{a_1} \cap G_{a_2}$ intersects only $H(B, 2)$ where $B = \{a_1, a_2\}$. Finally if $\operatorname{ord}(x, \mathfrak{G}) > 2$, then x belongs to exactly G_{a_i}, $i = 1, 2, \ldots, k$; and hence $\bigcap_{i=1}^{k} G_{a_i}$ intersects no member of $\mathfrak{H}_2$.

Therefore there exists a locally finite open collection $\mathfrak{U}_2 = \{U(B, 2): B \in \mathfrak{B}_2\}$ such that

(i) $H(B, 2) \subseteq U(B, 2)$ for each $B \in \mathfrak{B}_2$

(ii) each $U(B, 2)$ is contained in some member of $\mathfrak{G}$.

Define $U_2 = \cup\{U : U \in \mathfrak{U}_2\}$.

Continuing this process we obtain n locally finite collections $\{\mathfrak{U}_i\}_{i=1}^{n}$ so that $\mathfrak{U} = \bigcup_{i=1}^{n} \mathfrak{U}_i$ is the desired locally finite refinement of $\mathfrak{G}$.

Definition 2.5. Let $S \subset X$ and $\mathfrak{G}$ be an open (in S) cover of

S. Then $\mathfrak{G}$ has a *refinement which can be extended to a cover of* X, if there exists a cover $\mathfrak{U}$ of X such that $\mathfrak{U}/S$ refines $\mathfrak{G}$.

Definition 2.6. A subset $S \subseteq X$ is said to be *(weakly) N-embedded in* X iff every open cover of S of finite order has a refinement which can be extended to a (point-finite) locally finite open cover of X.

Theorem 2.7. (1) *X is discretely expandable iff every closed subset of X is N-embedded in X.*

(2) *X is almost discretely expandable iff every closed subset of X is weakly N-embedded in X.*

Proof. We prove only (1) as the proof of (2) is similar.

(i) Let X be discretely expandable, $F = \bar{F} \subseteq X$, and $\mathfrak{G} = \{G_a: a \in A\}$ be an open cover of F of finite order. Let $\mathfrak{G}^* = \{G_a^*: a \in A\}$ be a collection of open sets in X such that $G_a = G_a^* \cap F$. Therefore as in the proof of Theorem 2.4 above there exists locally finite open (in X) collections $\mathfrak{H}_i = \{H(i, a): a \in A\}$ such that $H(i, a) \subseteq G_a^*$ for each $i = 1, 2, \ldots, n = \text{ord}(\mathfrak{G})$. Hence $\mathfrak{H} = \bigcup_{i=1}^{n} \mathfrak{H}_i \cup \{X - F\}$ is the desired locally finite open cover of X.

(ii) Let $\mathfrak{F} = \{F_a: a \in A\}$ be a discrete closed collection in X. Then $\mathfrak{F}$ is an open cover $F = \bigcup_{a \in A} F_a$ of order 1. Thus there exists a locally finite open cover $\mathfrak{G}$ of X such that $\mathfrak{G}|F$ refines $\mathfrak{F}$. Hence X is discretely expandable by Lemma 3.6 of [6].

§3. EMBEDDING CHARACTERIZATIONS FOR NORMAL AND COLLECTIONWISE NORMAL SPACES

The following theorem is well-known. For example see [3].

Theorem 3.1. *A space X is normal iff every locally finite open cover of X is normal.*

We now use this fact to obtain various embedding type characterizations for normal and collectionwise normal spaces.

Definition 3.2. A subset $S \subset X$ is said to be *N_1-embedded in X*

if every open cover of S with finite order has a refinement which can be extended to a normal open cover of X.

Definition 3.3. A subset $S \subseteq X$ is said to be *N_2-embedded in X* if every point finite open cover of S has a refinement which can be extended to a normal cover of X.

Definition 3.4. A cover $\mathfrak{G}$ of X is called a *θ-cover* if $\mathfrak{G}$ has a refinement $\bigcup_{i=1}^{\infty} \mathfrak{G}_i$ satisfying

(i) each $\mathfrak{G}_i$ is an open cover of X,

(ii) for each $x \in X$ there exists an integer $n(x)$ such that x belongs to only finitely many members of $\mathfrak{G}_{n(x)}$.

A space X is called *θ-refinable* if every open cover is a θ-cover. This notion is due to Worrell and Wicke [8].

Definition 3.5. A subset $S \subseteq X$ is said to be *N_3-embedded in X* if every θ-cover of S has a refinement which can be extended to a normal cover of X.

Theorem 3.6. *The following are equivalent:*

(1) *X is collectionwise normal.*

(2) *Every closed subset of X is N_1-embedded in X.*

(3) *Every closed subset of X is N_2-embedded in X.*

(4) *Every closed subset of X is N_3-embedded in X.*

Proof. The fact that $(1) \equiv (2)$ follows from Theorems 2.7 and 3.1 above and Theorem 1.4 of [6]. Clearly $(4) \Rightarrow (3) \Rightarrow (2)$ and hence we show $(1) \Rightarrow (4)$. Let $F = \bar{F} \subseteq X$ and $\mathfrak{G} = \{G_a : a \in A\}$ be a θ-cover of F. Let $\mathfrak{G}^* = \{G_a^* : a \in A\}$ be an open collection in X such that $G_a = G_a^* \cap F$ for each $a \in A$. Since $\mathfrak{G}$ is a θ-cover of X, $\mathfrak{G}$ has a refinement $\bigcup_{i=1}^{\infty} \mathfrak{U}_i$ satisfying

(i) each $\mathfrak{U}_i$ is an open cover of F,

(ii) for each $x \in F$ there exists an integer $n(x)$ such that x belongs to only finitely many members of $\mathfrak{U}_{n(x)}$.

Now for each i we construct a sequence $\{\mathfrak{G}(i,j): j = 0, 1, \ldots\}$ of open collections satisfying;

(1) $\mathfrak{G}(i,j)$ is a discrete collection of cozero subsets of X for each j,

(2) Each member of $\mathfrak{G}(i,j)$ is contained in some member of $\mathfrak{G}^*$,

(3) if $x \in F$ and x belongs to at most m members of $\mathfrak{U}_i$, then $x \in \cup \{G \in \mathfrak{G}(i,j): 0 \leqslant j \leqslant m\}$,

(4) $\cup\{G: G \in \mathfrak{G}(i,j)\}$ is a cozero set in X.

The proof is by induction on j. Define $\mathfrak{G}(i,0) = \phi$ and assume $\mathfrak{G}(i,j)$ has been constructed satisfying (1) – (4) above for $0 \leqslant j \leqslant n$. We now construct $\mathfrak{G}(i, n+1)$. Let $\mathfrak{U}_i = \{U_a: a \in A_i\}$, $\mathfrak{B} = \{B \subseteq A_i: |B| = = n+1\}$ and $G(i,j) = \cup\{G: G \in \mathfrak{G}(i,j)\}$. Define

$$F(B) = \left[X - \bigcup_{j=0}^{n} G(i,j)\right] \cap \left[X - \cup\{U_a: a \in A_i - B\}\right] \text{ for each}$$

$B \in \mathfrak{B}$, and let $\mathfrak{F} = \{F(B) \in \mathfrak{B}\}$.

We assert that $\mathfrak{F}$ is a discrete collection of closed sets in F and hence discrete in X. Let $x \in F$.

Case 1. Suppose x belongs to more than $n+1$ members of $\mathfrak{U}_i$, say $\{U_{a_j}, j = 1, 2, \ldots, k\}$. Then $U(x) = \bigcap_{j=1}^{k} U_{a_j}$ is an open neighborhood of x which intersects no member of $\mathfrak{F}$.

Case 2. Suppose x belongs to exactly $n+1$ members of $\mathfrak{U}_i$, say $\{U_{a_j}: j = 1, 2, \ldots, n+1\}$. Then $U(x) = \bigcap_{j=1}^{n+1} U_{a_j}$ intersects only $F(B)$ where $B = \{a_1, a_2, \ldots, a_{n+1}\}$.

Case 3. Suppose x belongs to less than $n+1$ members of $\mathfrak{U}_i$. Thus from (3) above, $x \in \cup\{G \in \mathfrak{G}(i,j): 0 \leqslant j \leqslant n\}$ which intersects no member of $\mathfrak{F}$.

Since X is collectionwise normal, there exists a discrete cozero collection $\mathfrak{K} = \{K(B): B \in \mathfrak{B}\}$ satisfying

(i) $F(B) \subseteq K(B)$ for each $B \in \mathfrak{B}$,

(ii) each $K(B)$ is contained in some member of $\mathfrak{U}_i$,

(iii) $\bigcup_{B \in \mathfrak{B}} K(B)$ is a cozero set.

Let $\mathfrak{G}(i, n+1) = \mathfrak{K}$. It is easy to verify that $\mathfrak{G}(i, n+1)$ satisfies (1) – (4) above.

Note that $\{G(i,j)\colon i = 1, 2, \ldots ; j = 1, 2, \ldots\}$ is a countable cozero cover of F. By Theorem 2.7 of [3] there exists a locally finite cozero refinement $\{H(i,j)\colon i = 1, 2, \ldots ; j = 1, 2, \ldots\}$ such that $H(i,j) \subseteq G(i,j)$. Let $\mathfrak{H}(i,j) = \{H(i,j) \cap G\colon G \in \mathfrak{G}(i,j)\}$ for each i and j. Then

$$\mathfrak{H} = \left[\bigcup_{\substack{i=1\\j=1}}^{\infty\ \infty} \mathfrak{H}(i,j)\right] \cup [X - F]$$ is a locally finite open cover of X such that $\mathfrak{H} | F$ refines $\mathfrak{G}$. Thus F is N_3-embedded in X.

We now have the following due to H. L. Shapiro [4]. See Theorem 2.1.

Corollary 3.7. *The following are equivalent.*

(1) *X is collectionwise normal.*

(2) *For each closed subset $F \subseteq X$, each normal locally finite cozero cover of F has a refinement which can be extended to a normal open cover of X.*

(3) *For each closed subset $F \subseteq X$, each normal open cover of F has a refinement which can be extended to a normal locally finite cozero cover of X.*

(4) *For each closed subset $F \subseteq X$, each locally finite open cover of F has a refinement which can be extended to a normal open cover of X.*

Remark. If in the above results the covers are assumed to be countable, we obtain analogous characterizations for normal spaces.

§4. NORMAL, ALMOST DISCRETELY EXPANDABLE SPACES

In [5] a space is called PF-*normal* if every point finite open cover is normal. It was shown that collectionwise normal ⇒ Pf-normal ⇒ normal and that neither implication can be reversed. It is also clear that the concepts of paracompactness and metacompactness are equivalent in PF-normal spaces. We now observe that metacompactness has other equivalences in normal, almost discretely expandable spaces.

Theorem 4.1. *Let X be a normal, almost discretely expandable space. The following are equivalent.*

(1) *X is metacompact.*

(2) *Every open cover of X has a point finite cozero refinement.*

(3) *X is θ-refinable.*

Proof. (1) ⇒ (2) Since every point finite open cover $\mathfrak{G} = \{G_a : a \in A\}$ is shrinkable to a closed cover $\mathfrak{F} = \{F_a : a \in A\}$ such that $F_a \subseteq \subseteq G_a$. Then by the normality of X there exists a cozero cover $\mathfrak{W} = \{W_a : a \in A\}$ such that $F_a \subseteq W_a \subseteq G_a$. Clearly (2) ⇒ (3). The fact that (3) ⇒ ⇒ (1) follows from the same argument as (1) ⇒ (4) in Theorem 3.6 above.

Theorem 4.2. *Let X be almost discretely expandable. Then X is strongly normal iff every σ-point finite open cover of X is normal.*

Proof. If every σ-point finite open cover of X is normal then X is PF-normal and countably paracompact. Since X is almost discretely expandable, X is collectionwise normal by Theorem 3.5 of [5]. Therefore X is strongly normal.

REFERENCES

[1] R.A. Alo – H.L. Shapiro, Countably Paracompact, Normal and Collectionwise Normal Spaces, to appear, *Indag, Math.*

[2] K. Morita, Star-finite Coverings and the Star-finite Property, *Math. Japanicae,* 1 (1948), 60-68.

[3] K. Nagami, *Dimension Theory,* Academic Press, 1970.

[4] H.L. Shapiro, Extensions of Pseudometrics, *Can ad. J. Math.*, 19 (1966), 981-998.

[5] J.C. Smith, Properties of Expandable Spaces, *Proceedings of the Third Prague Symposium on General Topology* 1971, Academia Prague.

[6] J.C. Smith – L.L. Krajewski, Expandability and Collection-wise Normality, *Trans. Amer. Math. Soc.*, 160 (1971), 437-451.

[7] J.C. Smith – J.C. Nichols, Embedding Characterizations for Expandable Spaces, to appear, *Duke Math. Journal.*

[8] J.M. Worrell Jr. – H.H. Wicke, Characterizations of Developable Topological Spaces, *Canad. J. Math.*, 17 (1965), 820-830.

COLLOQUIA MATHEMATICA SOCIETATIS JÁNOS BOLYAI
8. TOPICS IN TOPOLOGY, KESZTHELY (HUNGARY), 1972.

ON THE TOPOLOGICAL CHARACTERIZATION OF LIE GROUPS AND TRANSITIVE LIE GROUP ACTIONS

J. SZENTHE

Under the influence of the famous Hilbert problem the characterization of Lie groups and Lie group actions has been for a long time a principal task for the theory of topological groups. Among the various well-konwn results achieved by now in connection of this problem there are two which can be qualified as purely topological. One which has been originally formulated in 1934 by L.S. Pontrjagin for the compact case sais that a finite dimensional locally compact group is a Lie group if it is locally connected (see [4] and [3], pp. 184-185). The other one obtained by D. Montgomery and L. Zippin at first for the compact case in 1940 states that if a σ-compact group G which has a compact factor group by its identity component acts effectively and transitively on a finite dimensional locally compact space X then G is a Lie group provided that X is locally connected (see [2] and [3], pp. 236-246). While the first theorem is obviously a purely topological characterization of Lie groups among the finite dimensional locally compact groups the second one is actually a characterization of transitive Lie group actions among the transitive actions of finite dimensional locally compact groups since by assumptions of the theorem the group G must be finite dimensional. Thus it seems justified to try to find characterizations without dimensionality restrictions among all locally compact groups and among their transitive actions. In fact concerning Lie groups the following theorem can be proved:

A locally compact group is a Lie group if it is locally contractible.

As to the characterization of transitive Lie group actions the following theorem holds:

Let G be a σ-compact group which has a compact factor group by its identity component. If G acts effectively and transitively on a locally compact space X then G is a Lie group provided that X is locally contractible.

Although local contractibility is of course a much more restrictive assumption than local connectedness these theorems materially are not weaker than the above ones since in case of finite dimensional locally compact groups and their coset spaces the two assumptions are equivalent.

In spite of the fact that the new theorems are obtained by simply replaceing local connectedness with local contractibility in the original ones, the proofs of the new theorems require a fundamentally different approach. Actually the basic result from which all the above mentioned can be derived is an adequate description of the local structure of locally compact groups and their coset spaces. To give a precise formulation consider a locally compact group G closed subgroup H and a compact invariant subgroup A of G, let $\pi: G \to G' = G/A$ be the canonical epimorphism then $H' = \pi(H)$ is a closed subgroup of G', let moreover $\chi: G \to$ $\to G/H$ and $\chi': G' \to G'/H'$ be the canonical projections then there is a unique map φ for which the diagram (see next page) commutes. This map φ defines a fiber structure $\varphi: G/H \to G'/H'$ on the coset space G/H. As a matter of fact both the above mentioned two characterizing theorems pertaining to the finite dimensional case are consequences of one fact which in the formulation of E.G. Sklarenko is as follows: Let G be a locally compact group and H a closed subgroup such that the left coset space G/H is finite dimensional, then there is such a compact invariant subgroup A of G that the corresponding fiber structure

$\varphi\colon G/H \to G'/H'$, is locally trivial, has a manifold as base space G'/H' and a generalized Cantor discontinuum as fiber type (see [5]).

$$\begin{array}{ccc} G & \xrightarrow{\ \chi\ } & G/H \\ \Big\downarrow{\scriptstyle \pi} & & \Big\downarrow{\scriptstyle \varphi} \\ G' & \xrightarrow[\ \chi'\ ]{} & G'/H' \end{array}$$

This theorem yields a description of the local structure of finite dimensional coset-spaces of locally compact groups and of finite dimensional locally compact groups in the case when H is the identity element. The above already mentioned basic result is the following generalization of this theorem:

Let G be a locally compact group and H a closed subgroup then there is a compact invariant subgroup A of G such that the corresponding fiber structure $\varphi\colon G/H \to G'/H'$ is locally trivial, has a free union of connected manifolds as base space G'/H' and its fibers are homeomorphic to a coset space A/B.

The proof of this theorem can be reduced to the case when G can be approximated by Lie groups on account of a well-known theorem of H. Yamabe. However even in this special case a rather involved argument must be applied. The basic tool is an other theorem concerning the local structure of groups which can be approximated by Lie groups, this in fact is a complemented version of the well-known theorem of K. Iwasawa concerning groups which can be approximated by Lie groups (see [1]).

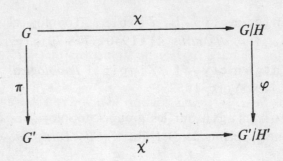

REFERENCES

[1] K. Iwasawa, On some types of topological groups, *Ann. of Math.*, 50 (1949), 507-557.

[2] D. Montgomery – L. Zippin, Topological transformation groups I, *Ann. of Math.*, 41 (1940), 778-791.

[3] D. Montgomery – L. Zippin, *Topological transformation groups,* New York, 1955.

[4] L.S. Pontrjagin, Sur les groupes topologiques compacts et le cinquième problème de D. Hilbert, *C.R. Paris,* 198 (1934), 238-240.

[5] E.G. Sklarenko, On the topological structure of locally compact groups and their quotient spaces, *Mat. Sbor.*, 60 (1963), 68-88.

COLLOQUIA MATHEMATICA SOCIETATIS JÁNOS BOLYAI
8. TOPICS IN TOPOLOGY, KESZTHELY (HUNGARY), 1972.

DIFFERENTIAL TOPOLOGY IN SMOOTH BANACH SPACES

F. SZIGETI

1. Introduction. Let H denote an infinite-dimensional Hilbert space. Cz. Bessaga has proved the following result: There exists a C^∞-smooth diffeomorphism h of H onto $H \setminus \{0\}$, such that $h(x) = x$ for $\|x\| \geqslant 1$ (see Proposition 1 of [2]). Later he raised the question concerning the validity of Proposition 1 for all normed spaces. It is easy to see, that Proposition 1 does not hold for every Banach space B. Suppose, for this purpose, that there exists a C^p-smooth diffeomorphism h of B onto $B \setminus \{0\}$ such that $h(x) = x$ for $\|x\| \geqslant 1$. By the Banach – Hahn theorem there exists a continuous linear functional $\eta: B \to R$ such that $\langle \eta, h(0) \rangle \neq$ $\neq 0$. The support of the C^p-smooth function $\eta \circ (h - id_B)$ is bounded and non-empty. However, R. Bonič and J. Frampton have proved, that there exist separable non-smooth Banach spaces, which have no other C^p-smooth function with bounded support than the identically zero one. For the C^p-smoothness of separable Banach spaces the following definitions have been given:

The separable Banach space B is C^p-smooth, by

Definition 1.1. If there exists a C^p-smooth function $f: B \to R$

with bounded but non-empty support,

Definition 1.2. If each open subset $U \subset B$ admits a C^p-smooth partition of the unity,

Definition 1.3. If for each continuous map Φ of an open subset $U \subset B$ into a Banach space B' and for any $\epsilon > 0$ there exists a C^p-smooth map $\psi \colon U \to B'$ such that

$$\| \Phi(x) - W(x) \|_{B'} < \epsilon \qquad (x \in U) .$$

R. Bonič and J. Frampton have proved the equivalence of these definitions [4], [5]. For example the separable Hilbert space is C^∞-smooth. In fact an adequate C^∞-smooth function is the following:

$$f(x) = \begin{cases} 0 & \text{if} \quad \|x\| \geqslant 1 \\ \exp\{(\|x\|^2 - 1)^{-1}\} & \text{if} \quad \|x\| < 1 . \end{cases}$$

It is easy to see that Banach spaces having C^p-smooth norms are C^p-smooth. It is known, that a separable Banach space has an admissible C^1-smooth norm it and only if the conjugate space B^* is separable too, (M.I. Kad'etz and G. Restrepo [7], [8]), thus a Banach space B with a separable conjugate space B^* is C^1-smooth. But the C^∞-smooth Banach space c_0 has the conjugate space l_∞, so it cannot have an admissible C^1-smooth norm. Finally the spaces $L_p[0, 1]$ are C^∞-smooth for $p = 2q$, where q is a positive integer, and $L_p[0, 1]$ are C^r-smooth but not C^{r+1}-smooth, for $r \leqslant p \leqslant r + 1$([6]).

In this paper we prove theorems, which are more complete characterizations of the C^p-smoothness of a separable Banach space. Further we shall write shortly "Banach space" instead of separable Banach space. One of our most important tools is the concept of the projectible subsets of Banach spaces. We mention, that our definition of projectibility differs from that of R.D. Anderson [1].

Definition 1.4. Let B_1 denote the subspace of a Banach space B spanned by $K \cup \{e_i \mid i \in N\}$ where K is a subset and $\{e_i, \eta_i \mid i \in N\}$ is a biorthonormal system of B. Define the continuous linear map P_n: $B \to R^n$ by

$$x \xrightarrow{P_n} (\langle \eta_1, x \rangle, \langle \eta_2, x \rangle, \ldots, \langle \eta_n, x \rangle)$$

The subset K of the Banach space $\boldsymbol{B}$ is said to be projectible with respect to the biorthogonal system $\{e_i, \eta_i \mid i \in N\}$ if

(i) K is closed,

(ii) for each point $x \in B_1 \setminus K$ there exists an integer $n \in N$ such that $P_n \notin P_n K$

(iii) there exists an infinite subset N_1 of the set N positive integers, such that the subsets $\langle \eta_n, K \rangle \subset \boldsymbol{R}$ have an upper (or lower) bound for each $n \in N_1$.

For example the one-point-set $\{0\} \subset \boldsymbol{B}$ is projectible. This follows from a fact proved by Cz. Bessaga and A. Pelczynski, that every Banach space has a subspace with Schauder bases [3]. The compact subsets of the subspace $\boldsymbol{B}_1 \subset \boldsymbol{B}$ with Schauder bases are projectible too [1].

The first result of this Note is a relatively simple characterization of the C^p-smoothness of a Banach space $\boldsymbol{B}$:

Proposition 1.5. *A Banach space* $\boldsymbol{B}$ *is* C^p*-smooth, if and only if there exists a bounded starlike body with its boundary being* C^p*-smooth submanifold of* $\boldsymbol{B}$.

Our main result is the following theorem.

Theorem 1.6. *Let a subset* K *of a* C^p*-smooth Banach space* $\boldsymbol{B}$ *be projectible with respect to a certain biorthogonal system* $\{e_i, \eta_i \mid i \in N\}$. *Then for any* $\epsilon > 0$ *there exists a* C^p*-diffeomorphism* $f_\epsilon : \boldsymbol{B} \to \boldsymbol{B} \setminus K$, *such that* $f_\epsilon(x) = x$, *if the distance of* x *from the convex hull of* K *is not less, then* ϵ.

Now we mention a few corollaries of 1.6. Theorem.

Corollary 1.7. *(the solution of Bessaga's problem) The Banach space* $\boldsymbol{B}$ *is* C^p*-smooth if and only if, for any* $\epsilon > 0$ *there exists a* C^p*-diffeomorphism* $f_\epsilon : \boldsymbol{B} \to \boldsymbol{B} \setminus \{0\}$, *such that* $f_\epsilon(x) = x$ *if* $\|x\| \geqslant \epsilon$.

Corollary 1.8. *Let* $K \subset \boldsymbol{B}$ *be a bounded projectible subset of a*

Banach space B. *Then* B *is* C^p*-smooth if and only if, for any* $\epsilon > 0$ *there exists a* C^p*-diffeomorphism* $f_\epsilon: B \to B \setminus K$, *such that* $f_\epsilon(x) = x$ *if* x *is off the* ϵ*-neighbourhood of the convex hull of* K.

Corollary 1.9. *A Banach space* B *with Schauder bases is* C^p*-smooth if and only if for any* $\epsilon > 0$ *and compact subset* $K \subset B$ *there exists a* C^p*-diffeomorphism* $f_\epsilon: B \to B \setminus K$, *such that* $f_\epsilon(x) = x$, *for the points* x *which are not in the* ϵ*-neighbourhood of the convex hull of* K.

Further we shall use the notation N_ϵ^Y for the ϵ-neighbourhood of the subset X in Y and $\mathrm{Ch}(K)$ for the convex hull of K.

2. **Definition 2.1.** We say, that the closed subset C of the Banach space B is a starlike body with respect to the point $x_0 \in \mathrm{Int}\, C$ (the interior of C), if for each $x \in C$ the body C contains the interval $[x_0, x] = \{y = x_0 + \lambda(x - x_0) \mid \lambda \in [0, 1]\}$ too.

A convex body $K \subset B$ is also a starlike body with respect to each point contained in $\mathrm{Int}\, K$. The Minkowski functional defined by the starlike body C with respect to $0 \in \mathrm{Int}\, C$ is denoted by ω_c, that is ω_c is the following function

$$\omega_c(x) = \begin{cases} 0 & \text{if} \quad x = 0\ (\in B) \\ \lambda & \text{if} \quad \lambda > 0 \quad \text{and} \quad \frac{1}{\lambda} x \in \mathrm{Bd}\, C \end{cases}$$

where $\mathrm{Bd}\, C$ is the boundary of C. Now we prove two lemmas.

Lemma 2.2. *Let* $C \in B$ *be a bounded starlike body with respect to* $0 \in \mathrm{Int}\, C$. The restriction $\omega_c|_{B\setminus\{0\}}$ of the Minkowski functional ω_c *is* C^p*-smooth function if and only if* $\mathrm{Bd}\, C$ *is a* C^p*-smooth submanifold of the Banach space* B.

Proof. If ω_c is C^p-smooth, then $d\omega_c(x) \neq 0$ in the subset $\omega_c^{-1}(1) = \mathrm{Bd}\, C$, so $\omega_c^{-1}(1)$ is a C^p-submanifold too.

Let us assume, that $\mathrm{Bd}\, C \subset B$ is a C^p-submanifold. Then there exists a C^p-atlas $\{(U_i, \varphi_i) \mid i \in I\}$ of $\mathrm{Bd}\, C$, such that $j \circ \varphi_i^{-1}: \varphi_i(U_i) \to$ $\to B$ is C^p-smooth, where $j: \mathrm{Bd}\, C \to B$ is the inclusion map, and $\varphi_i(U_i)$ is supposed to be contained in the 1-codimensional subspace B_1 of B.

By the homogenity of the Minkowski functional ω_c we prove the C^p-smoothness in the point $x_0 \in \operatorname{Bd} C$. Define the maps $\Phi_i\colon \varphi_i(U_i) \times \times \left(\frac{1}{2}, \frac{3}{2}\right) \to B$ by the formula:

$$\Phi_i(x, t) = t(j \circ \varphi_i^{-1})(x) \qquad (i \in I) \tag{1}$$

These maps are C^p-smooth, and the differential maps $d\Phi_i(x, 1) = (dj \circ \varphi_i^{-1}(x);\ j \circ \varphi_i^{-1}(x))\colon B_1 \times R^1 \to B$ are continuous and linear. The maps $d(j \circ \varphi_i^{-1})(x)\colon B_1 \to T_{\varphi_i^{-1}(x)}(\operatorname{Bd} C)$ are continuous linear isomorphism $(i \in I)$. From the assumption $d\Phi_i(x, 1)(u, t) = d(j \circ \varphi_i^{-1}(x))u + t \cdot j \circ \varphi_i^{-1}(x) = 0$ the relations $t = 0$ and $u = 0$ follow, so $\operatorname{Ker} d\Phi_i = 0$, but $\operatorname{Im} d\Phi_i = T_{\varphi_i^{-1}(x)}(\operatorname{Bd} C) + R^1 j \circ \varphi_i^{-1}(x) = B$, so by the closed graph theorem $d\Phi_i(x, 1)$ is a continuous linear isomorphism int he point $(x, 1)$ for each $x \in \operatorname{Bd} C$. By the inverse-function theorem Φ_i is a local C^p-isomorphism in the point $(x, 1)$, that is there is a neighbourhood $G_1 \times (1 - \delta, 1 + \delta)$ of the point $(x, 1)$, and one G_2 of the point $j \circ \circ \varphi_i^{-1}(0)$, such that the C^p-smooth inverse map $(\psi_i, \omega_i)\colon G_2 \to G_1 \times \times (1 - \delta, 1 + \delta)$ exists. It is obvious, that the component ω_i must be C^p-smooth too. Now we compute $\omega_c(y)$ in the neighbourhood G_2. Suppose, that $\Phi_i(\psi_i(y), \omega_i(y)) = y$. Then $\omega_c(y) = \omega(\Phi_i(\psi_i(y), \omega_i(y)) = \omega_c[\omega_i(y) \cdot j \circ \varphi_i^{-1}(y)] = \omega_i(y)$ in $y \in G_2$, so ω_c is C^p-smooth in x, for each $x \in \operatorname{Bd}(G)$. Thus Lemma 2.2. is proved.

Lemma 2.3. *Let a convex body of a C^p-smooth Banach space B be denoted by K. Suppose, that x_0 is an interior point of K. Then for each number $\epsilon > 0$ there exists a starlike body C the boundary of which is a C^p-smooth submanifold in B, such that* $\operatorname{Bd} C \subset N_{2\epsilon}(K) \setminus N_\epsilon(K)$.

Proof. We can suppose that $X_0 = 0$.

a) It is obvious that $N_\epsilon(N_\epsilon(K)) \subseteq N_{2\epsilon}(K)$

b) Each point x in $\operatorname{Bd}(K)$ has an open neighbourhood U_x in $\operatorname{Bd} K$, such that for each pair of points $y_1, y_2 \in U_x$ the following property holds: Let $y \in B^*$ be an arbitrary continuous functional with the property $\langle \eta, K\rangle \leqslant \langle \eta, y_1\rangle = m$, and let the intersection of $y^{-1}(m)$ and $R^1 y_2$ be denoted by y_2', then $\| y_2 - y_2' \| \leqslant \epsilon$ holds. The proof of assertion b) is very elementary. Let $C(K, x')$ be the minimal cone with the vertex $x' = x + \frac{\epsilon}{2} \cdot \frac{x}{\|x\|}$ containing K. Now we make the intersection U'_x of $\mathrm{Bd}\, N_\epsilon(K)$ with the cone $2x' - C(K, x')$. The cone generated by U'_x with the vertex 0 intersects $\mathrm{Bd}\, K$ in the required neighbourhood U_x.

c) Now, we cover $\mathrm{Bd}\, K$ with the open covering $\{U_i \mid i \in I\}$ given in b). Since $\mathrm{Bd}\, K$ is a paracompact space in the induced topology, there exist locally finite refinements $\{V_i \mid i \in I\}$, $\{W_i \mid i \in I\}$, such that $\bar{V}_i \subset U_i$ and $\bar{W}_i \subset V_i$ hold ($i \in I$). If $W_i = \phi$, than we discard the subsets V_i and U_i too. Choose a point X_i from W_i for each $i \in I$, and a continuous linear functional η_i such that, the following conditions hold;

(i) $\|\eta_i\| = 1$

(ii) $\langle \eta_i, x_i \rangle = \|x_i\|$

(iii) $\langle \eta_i, x \rangle > 0$ if $x \in U_i$.

Where if necessary we divide the subsets U_i, V_i, W_i so as to satisfy condition (iii) too. Now we project the subsets $U_i, V_i, W_i \subset \mathrm{Bd}\, K$ $(i \in I)$ to the linear subvariety $\eta_i^{-1}(\|x_i\|) \subset \boldsymbol{B}$, and the images are denoted respectively by $\tilde{U}_i$, $\tilde{V}_i$, $\tilde{W}$. As the Banach space $\boldsymbol{B}$ and the subvarieties $\eta_i^{-1}(\|x_i\|)$ are C^p-smooth, so there exist C^p-functions $\tilde{\varphi}_i$: $\eta_i^{-1}(\|x_i\|) \subset$ $\subset \boldsymbol{B} \to \boldsymbol{R}$ $(i \in I)$, such that

(iv) the support $\mathrm{Supp}\, \tilde{\varphi}_i$ is contained in $\tilde{V}_i$

(v) $0 \leqslant \tilde{\varphi}_i(x) \leqslant 1$ and

(vi) $\tilde{\varphi}_i(W_i) = 1$.

There exists a number $r > 0$ such that $N_{3r}(0) \subset K$. Then by the C^p-smoothness of the Banach space $\boldsymbol{B}$ there exists a C^p-function ψ: $\boldsymbol{B} \to \boldsymbol{R}$ such that $0 \leqslant \psi(x) \leqslant 1$, $\mathrm{Supp}\, \psi \subset N_{2r}(0)$ and $\psi(x) = 1$ on $N_r(0)$. Now we construct a "C^p-partition of unity" by the formula:

$$(2) \qquad \varphi_i(x) = \begin{cases} \dfrac{(1 - \psi(x))\tilde{\varphi}_i[x\langle \eta_i, x \rangle^{-1}]}{\sum\limits_{i \in I} \tilde{\varphi}_i[x\langle \eta_i, x \rangle^{-1}]} & \text{if } \langle \eta_i, x \rangle > 0 \\ 0 & \text{if } \langle \eta_i, x \rangle \leqslant 0 \end{cases}$$

From the construction we get the following properties: $\varphi_i(B_r(0)) = 0$; $\varphi_i(x) = \varphi_i(\lambda x)$ if $\|x\|$, $\lambda\|x\| \geqslant 2r$; $\lambda\|x\| \geqslant 2r$; $\sum_{i\in I} \varphi_i(x) = 1$ if $\|x\| \geqslant 2r$; $0 \leqslant \varphi_i(x) \leqslant 1$ and finally $\operatorname{Supp}\varphi_i$ is in the cone generated by V_i. So the family of the functions $\{\varphi_i | i \in I\}$ is a so-called homogeneous C^p-partition of unity.

d) Next we define a C^p function $h: B \to R$ by

$$h(x) = 1 - \sum_{i\in I} \langle \eta_i, x \rangle \|x_i\|^{-1} \varphi_i(x) . \tag{3}$$

Firstly we observe that $\langle dh(x), x \rangle = - \sum_{i\in I} \varphi_i(x) \langle \eta_i, x \rangle \frac{1}{\|x_i\|} - \sum_{i\in I} \langle \eta_i, x \rangle \langle d\varphi_i(x), x \rangle \frac{1}{\|x_i\|} = 1 \neq 0$, because $\langle d\varphi_i(x), x \rangle = 0$ by the homogenity of the partition of unity off the neighbourhood $N_{2r}(0)$. Thus $dh(x) \neq 0$ in the boundary $h^{-1}(0)$ of the body $C = h^{-1}[0, 1] \subset B$. But then $h^{-1}(0)$ is a C^p-submanifold of B. If $h(x) = h(\lambda x)$ for x and λx from the submanifold $h^{-1}(0)$, then by the homogenity of the partition of unity $\lambda = 1$ follows. So the closed subset $C \subset B$ is a starlike body with respect to 0. It is clear, that K is contained in C. The covering $\{U_i | i \in I\}$ is chosen by the method of b), so $h^{-1}(0) = \operatorname{Bd} C$ is contained in $N_\epsilon(K)$.

e) Finally we apply our starlike-body construction for the convex body $\overline{N_\epsilon(K)}$. So we get a starlike body C such that $\operatorname{Bd} C$ is a C^p-smooth submanifold of B, and that $N_\epsilon(K) \subset C$, and $\operatorname{Bd} C \subset N_\epsilon(N_\epsilon(K)) \subset N_{2\epsilon}(K)$ hold by a). So we have proved Lemma 2.3.

Now we can easily prove Proposition 1.5. Suppose that C is a bounded starlike body with respect to 0 with C^p-smooth boundary $\operatorname{Bd} C$. By Lemma 2.2. the restriction of the Minkowski functional $\omega_c: B \setminus \{0\} \to R^1$ is C^p-smooth, so the function

$$f(x) = \begin{cases} \zeta_{3/4}^{1/4} \circ \omega_c(x) & (x \in B \setminus \{0\}) \\ 1 & (x = 0) \end{cases}$$

is C^p-smooth with bounded support, where the function $\zeta_x^r: R^1 \to R^1$ is defined as follows: Consider first the function

$$(4)\qquad \xi_x^r(t) = \begin{cases} \exp\{[x-t)^2 - r^2]^{-1}\} & (|x-t| < r) \\ 0 & (|x-t| \geqslant r) \end{cases}$$

and then $\zeta_x^r : R^1 \to R^1$ is given by

$$(5)\qquad \zeta_x^r(t) = \left[\int_{-\infty}^{\infty} \xi_x^r(t)dt \right]^{-1} \cdot \int_t^{\infty} \xi_x^r(\tau)d\tau \, .$$

So the Banach space B is C^p-smooth.

If B is C^p-smooth Banach space then the unit-ball can be approximated uniformly by starlike body with C^p-smooth boundary by Lemma 2.3. So Proposition 1.5. is proved.

An another obvious consequence of Lemmas 2.2, 2.3. is the following proposition.

Proposition 2.4. *In a C^p-smooth Banach space B each open set has a subset U such that U is C^p-diffeomorphic to the whole Banach space.*

Proof. Each open subset contains an open-ball $N_{2r}(x_0)$ for a suitable positive number r. Then $N_r(x_0)$ we approximate by a starlike domain C with C^p-boundary, such that $N_r(x_0) \subset C \subset N_{2r}(x_0)$. So we must prove, that the interior of a starlike body with C^p-smooth boundary is C^p-diffeomorphic to the whole Banach space. We can suppose, that C is starlike body with respect to 0. Then we can define a C^p-diffeomorphism between C and B by the mapping

$$x \to \left[\zeta_{3/4}^{1/4} \circ \omega_c(x) \right]^{-1} \cdot x \qquad (x \in \operatorname{Int} C) \, .$$

So Proposition 2.4 is proved.

Now we shall prove some other lemmas too.

Lemma 2.5. *Let us assume, that K is a closed subset of the infinite-codimensional direct summand B_1 in a C^p-smooth Banach space B. For each positive number ϵ there exists a C^p-diffeomorphism g_ϵ: $B \to B \setminus K$, such that $\|g_\epsilon(x) - x\| < \epsilon$ $(x \in B)$.*

We mention, that this lemma is a special case of a weaker from of

Lemma 2.6. *Let B_1 be an infinite-codimensional subspace of a C^p-smooth Banach space B. Then for each $\epsilon > 0$ there exists a C^p-diffeomorphism $g_\epsilon : B \to B \setminus B_1$, such that $\| g_\epsilon(x) - x \| < \epsilon$ $(x \in B)$.*

Proof of Lemma 2.5. Let us assume, that $B = B_1 \oplus B_2$, where B_2 is an infinite-dimensional Banach subspace. By the result of C. Bessaga and A. Pelczyński [3] there exists an infinite-dimensional subspace $B_3 \supset B_2$ with the bases $\{\widetilde{e}_i\}_{i \in N}$.

Let $\{\widetilde{\eta}_i\}_{i \in N}$ $(\widetilde{\eta}_i \in B_3^*)$ be the coordinate functionals respectively. We can prolong the functionals $\widetilde{\eta}_i$ to the whole subspace B_2 in a norm-preserving manner by the Banach – Hahn theorem. Let these functionals be denoted by $\{\bar{\eta}_i\}_{i \in N}$ respectively. Finally let the functionals $\{\eta'_i\}_{i \in N}$ $(\eta'_i \in B^*)$ be defined by $\langle \eta'_i, x \rangle = \langle \widetilde{\eta}_i, P_2 x \rangle$, where $P_i : B \to B_i$ $(i = 1, 2)$ are the continuous projections defined by the direct sum $B = B_1 \oplus B_2$. It is obvious, that $\langle \eta'_i, x \rangle = 0$ if $x \in B_1$.

a) Now we show that we can choose a bases $\{e_i\}_{i \in N}$ of B_3 so as to satisfy the following condition too

$$\inf_{\sum_{i=1}^{3} \lambda_i = 1} \left\| \sum_{i=1}^{3} \lambda_i e'_{2n+i} \right\| \geqslant \frac{1}{2} \quad (n = 0, 1, \ldots) . \tag{6}$$

Let us consider a three dimensional Banach space B^3 and let us choose a functional $a \in B^{3*}$ such that $\| a \| = 1$. In the plane $a^{-1}\left(\frac{1}{2}\right)$ there exist three linearly independent unit vectors e_1, e_2, e_3. For the vectors e_1, e_2, e_3 satisfy (6). This construction is applied especially to that subspace $B^3 \subset B$ which is spanned by e'_1, e'_2, e'_3. We can find vectors e_1, e_2, e_3 for which (6) holds in such a way that $e_1 = e'_1$. The new coordinate functionals are dentoed by η_1, η_2, η''_3. It is clear, that the new bases $e_1, e_2, e_3, e'_4, e'_5, \ldots$ and coordinate functionals $\eta_1, \eta_2, \eta''_3, \eta'_4, \eta'_5, \ldots$ satisfy the equalities

$$\langle \eta_i^{(k)}, e_j^{(l)} \rangle = \delta_{ij} \quad (i, j \in N,\ k = 0, 1, 2,\ l = 0, 1) . \tag{7}$$

Suppose, that the vectors $e'_1, e'_2, \ldots, e_{2n-1}$ are already substituted by vectors $e_1, e_2, \ldots, e_{2n-1}$ satisfying (6). Similarly the coordinate functionals $\eta'_1, \eta'_2, \ldots, \eta'_{2n-1}$ are canonically transformed to $\eta_1, \ldots$ $\ldots, \eta_{2n-2}, \eta''_{2n-1}, \eta'_{2n}, \eta'_{2n+1}, \ldots$ and (7) holds.

We consider an inductive step to construct a required bases. For this we can choose B^3 to be that subspace of $\boldsymbol{B}$ which is spanned by the linearly independent vectors $e_{2n-1}, e'_{2n}, e'_{2n+1}$. By our construction mentioned above the vectors e'_{2n}, e'_{2n+1} are substituted by e_{2n}, e_{2n+1} and the coordinate functionals $\eta''_{2n-1}, \eta'_{2n}, \eta'_{2n+1}$ are transformed canonically to $\eta_{2n-1}, \eta_{2n}, \eta''_{2n+1}$. It is obvious, that the relations (7) are satisfied. So we get a bases $\{e_i\}$ satisfying (6).

b) Define two families of the C^p-smooth functions $\{\gamma_n\}_{n\in N}$, $\{\delta_n\}_{n\in N}$ by the formulas:

$$\gamma_n(x) = \left| \sum_{i=-1}^{1} \langle \eta_{2n-i}, x\rangle - 1 \right| \qquad (x \in \boldsymbol{B})$$

$$\delta_n(x) = \left(\sum_{i=1}^{2n-2} \langle \eta_i, x\rangle^2 \right)^{1/2} \qquad (x \in \boldsymbol{B})$$

Put $K_j = K + e_j$. The open subsets $U_{n,i}$ $(n \geqslant 2,\ i = 1, 2, 3)$ are defined as the intersection of the following sets:

$$\{x \mid \gamma_n(x), \delta_n(x) < 2^{-3n-i}\}$$

$$\{x \mid P_1 x \in N_{2^{-3n-2}}(K_{2n-2+i})\}$$

$$N_{2^{-3n-i}}(B_1 \oplus B_3)\,.$$

It is obvious, that the relations

$$U_{n,i} \subset U_{n,j} + e_{2n-2+j} - e_{2n-2+j} \qquad (1 \leqslant j < i \leqslant 3)$$

$$K_{2n-2+i} \subset U_{n,i} \qquad (i = 1, 2, 3;\ n \geqslant 2)$$

hold. The subspace $\operatorname{Ker} \eta_{2n} \subset \boldsymbol{B}$ is a C^p-smooth Banach space, so we can find the C^p-smooth functions $\varphi_n\colon \operatorname{Ker} \eta_{2n} \to \boldsymbol{R}^1$ for which

$$\varphi_n(x) = \begin{cases} 1 & \left(x \in \operatorname{Ker} \eta_{2n} \cap \overline{\{U_{n,2} + e_{2n-1} - e_{2n}\}}\right) \\ 0 & (x \notin \operatorname{Ker} \eta_{2n} \setminus U_{n,1}) \end{cases}$$

r.s. and $0 \leqslant \varphi_n \leqslant 1$. Then C^p-diffeomorphisms $\{h_{n,1}\}_{n=1}^{\infty}$ are given by

(8) $h_{n,1}(x) = x + \varphi_n[x - \langle \eta_{2n}, x \rangle (e_{2n} - e_{2n-1})](e_{2n} - e_{2n-1})$

Similarly, there exists C^p-smooth ψ_n: $\operatorname{Ker} \eta_{2n+1} \to R^1$, such that $0 \leqslant \psi_n \leqslant 1$,

$$\psi_n(x) = \begin{cases} 1 & (x \in \operatorname{Ker} \eta_{2n+1} \cap \overline{\{U_{n,3} + e_{2n} - e_{2n+1}\}}) \\ 0 & (x \notin \operatorname{Ker} \eta_{2n+1} \setminus U_{n,2} . \end{cases}$$

define now C^p-diffeomorphisms $h_{n,2}$: $B \to B$ by

(9) $h_{n,2}(x) = x + \psi_n[x - \langle \eta_{2n+1}, x \rangle (e_{2n+1} - e_{2n})](e_{2n+1} - e_{2n})$.

Consider C^p-diffeomorphisms g_n: $B \to B$ $(n \geqslant 2)$ composed of the C^p-diffeomorphisms $h_{n,i}$: $B \to B$ $(i = 1, 2)$ as $g_n = h_{n,2} \circ h_{n,1}$ $(n \geqslant 2)$. It is bovious, that $g_n(x) = x + e_{2n} - e_{2n-1}$ $(x \in K_{2n-1})$ holds. For $n = 1$, the definition of the C^p-smooth diffeomorphism g_1: $B \to B$ is given by $g_1(x) = x + e_1$. The formally infinite composition $g = \ldots$ $\ldots \circ g_n \circ g_{n-1} \circ \ldots \circ g_2 \circ g_1$ of the C^p-diffeomorphisms g_i: $B \to B$. is not defined in K. For the formal composition g we prove two assertion from which follows, that g is a C^p-diffeomorphism of $B \setminus K$ onto B. Then we show, that the C^p-diffeomorphism $f = g^{-1}$; $B \to B \setminus K$ satisfies $\|f(x) - x\| \leqslant M < \infty$ for every point x of B. Thus, if we construct the C^p-diffeomorphism f for the subset $\frac{M}{\epsilon} \cdot K$, then the map

$$f_\epsilon(x) = \frac{\epsilon}{M} f\left(\frac{Mx}{\epsilon}\right)$$

is the required C^p-diffeomorphism which renders K negligible.

c) The first statement is the following: for each point $x \in B \setminus K$ there exist a neighbourhood $U_x \subset B \setminus K$ of the point x and a positive integer n_x, such that $g_n(y) = y$ for each $n \geqslant n_x$ in the subset $g_{n_x} \circ \ldots$ $\ldots \circ g_2 \circ g_1(U_x) \subset B$. The second statement says, that for each $x \in B$ there exist a neighbourhood $V_x \subset B$ of the point x, and a positive integer m_x, such that $g_m(y) = y$ for each point $y \in V_x (m \geqslant m_x)$.

Proof of these statements. If $x \in B_1 \oplus B_2$, then $\rho(x, B_1 \oplus$ $\oplus B_3) > 0$ so there exists a positive integer $n_x = m_x$ such that $2^{-3n_x} <$

$< \rho(x, B_1 \oplus B_3)$. Then we choose the neighbourhood by $U_x = V_x = N_{2^{-3n_x - 1}}(x)$. So the statement holds in this case. If $x \in B_1 \oplus B_3$, we consider two cases. In the first case, when $x \in B_1$ we can choose $m_x = 2$, and $V_x = P_2^{-1}\left(N_{1/4}^{B_2}(0)\right)$ (where P_2 denotes the natural projection of B onto B_2 induced by the direct sum $B = B_1 \oplus B_2$), because the bases $\{e_i\}_{i \in N}$ satisfies (6). If $x \in B_1 \setminus K$, then there exists an integer n_x, such that $x \in B_1 \setminus N_{2^{-3n_x}}(K)$. Thus an open neighbourhood U_x of x with the property $U_x \subset P_2^{-1}\left(N^{B_2}_{2^{-3n_x - 1}}(K)\right)$ satisfies our requirement. The most difficult case is, when $x \in B_1 + B_3 \setminus B_1$ that is when we can find a coordinate functional η_i, such that $\langle \eta_i, x \rangle \neq 0$. Suppose even, that i is already the smallest such integer. If $i = 2j$, then from the construction of the g_i the equality $\langle \eta_i, g_i \circ \ldots \circ g_2 \circ g_1(x) \rangle = \langle \eta_i, x \rangle$ follows. Than we can verify easily that one of the following statements is valid:

$$\langle \eta_i, g_{j+1} \circ \ldots \circ g_2 \circ g_1(x) \rangle \neq 0$$

$$\langle \eta_{i+1}, g_{j+1} \circ \ldots \circ g_2 \circ g_1(x) \rangle > 2 - 2^{3j} .$$

In these cases we choose an integer satisfying the equality $n_x = j + 1$ or the inequality $2^{-3n_x} < |\langle \eta_i, g_{j+1} \circ \ldots \circ g_1(x) \rangle|$, and a neighbourhood $U_x = g_1^{-1} \circ g_2^{-1} \circ \ldots g_{n_x}^{-1}\left(N_{2^{-3n_x - 1}}(g_{n_x} \circ \ldots \circ g_2 \circ g_1(x))\right)$.

For the second assertion $m_x (> j)$ is an integer with the property $2^{-3m_x} < |\langle \eta_i, x \rangle|$, and $V_x = \delta^{-1}_{m_x + 1}\left(2^{-3m_x - 3}\right)$.

Finally, if $i = 2j + 1$, then in the construction of g_{j+1}, we can consider the following three cases:

$$\langle \eta_{2j-1}, g_{j+1} \circ \ldots \circ g_2 \circ g_1(x) \rangle \neq 0$$

$$\langle \eta_{2j}, \quad g_{j+1} \circ \ldots \circ g_2 \circ g_1(x) \rangle \neq 0$$

$$\langle \eta_{2j+1}, g_{j+1} \circ \ldots \circ g_2 \circ g_1(x) \rangle = \langle \eta_{2j+1}, x \rangle + 1 .$$

In the first two cases we apply the method showed above. In the third case we get the following three new cases:

$$\langle \eta_{2j+1}, g_{j+2} \circ g_{j+1} \circ \ldots \circ g_1(x) \rangle \neq 0$$

$$\langle \eta_{2j+2}, g_{j+2} \circ g_{j+1} \circ \ldots \circ g_1(x) \rangle \neq 0$$

$$\langle \eta_{2j+3}, g_{j+2} \circ g_{j+1} \circ \ldots \circ g_1(x) \rangle > 2 - 2^{-3j-3} > \frac{15}{8}.$$

So we prove our assertion, if we apply again the construction as just showed. Thus Lemma 2.5 is proved.

Proof of Lemma 2.6. The infinite-dimensional factor space B/B_1 has an infinite dimensional subspace with a normed bases $\{e'_i\}_{i \in N}$ with the coordinate functionals $\{\eta'_i\}_{i \in N}$ $(\eta'_i \in [B/B_1]^*)$. The coset $e'_i \in B/B_1$ contains an element e_i of B with a norm non greater than 2. With the help of the coordinate functionals $\{\eta'_i\}_{i \in N}$ we may define the functionals $\{\eta_i\}_{i \in N}$ $(\eta_i \in B^*)$ by the formula $\langle \eta_i, x \rangle = \langle \eta'_i, x + B_1 \rangle$. $\operatorname{Ker} \eta_i \subset B$ holds for each η_i, and $\langle \eta_i, e_j \rangle = \langle \eta'_i, e'_j \rangle = \delta_{ij}$ so the vectors $\{e_i\}_{i \in N}$ and the functionals constitute a biorthogonal system. The subspace spanned by B_1 and $\{e_i\}_{i \in N}$ is B_2, because in the subspace B_2/B_1 the $\{e'_i\}_{i \in N}$ is a bases. So the following cases are satisfied for each

(i) $x \in B_1$

(ii) there exists a coordinate functional $\eta_i \in B^*$ such that $\langle \eta_i, x \rangle \neq 0$. We may suppose that the bases $\{e_i\}_{i \in N}$ is of property (6). Let the functions $\gamma_n, \delta_n : B \to R^1$ be defined as in the previous lemma. Naturally, we must reasonably modify the construction of the subsets $U_{n,i}$ $(i = 1, 2, 3)$ as it follows:

$$U_{n,i} = \{x \mid \delta_n(x), \gamma_n(x) < 2^{-3n_x - i + 1}\} \cap N_{2^{-3n-i+1}}(B_2)$$

Then we may follow the proof of Lemma 2.5, and we get the proof of Lemma 2.6.

Remark: Both lemmas may be proved using an additional property of the C^p-diffeomorphism $g_\epsilon : B \to B \setminus K$ $(g_\epsilon : B \to B \setminus B_1)$. With the use of the notations of Lemma 2.5 (and respectively one of Lemma 2.6) we may modify the construction of the map $f_\epsilon : B \to B \setminus K$ (respectively the map $f_\epsilon : B \to B \setminus B_1$), such that $f_\epsilon(x) = x$ off the ϵ-neighbourhood of

$B_1 \oplus B_2 \subset B$ (respectively one of $B_2 \subset B$). For this we can modify the C^p-diffeomorphism $g_1: B \to B$. There exists a C^p-function φ: $\operatorname{Ker} \eta_1 \to R^1$, such that $0 \leqslant \varphi \leqslant 1$, $\varphi(N_{\epsilon/2}(B_1 \oplus B_2)) = 1$, $(\varphi[N_{\epsilon/2}(B_2)] = 1)$ and $\varphi[B \setminus N_\epsilon(B_1 \oplus B_2)] = 0$, $(\varphi[B \setminus N_\epsilon(B_2)] = 0)$ by the C^p-smoothness of the Banach space B, (so $\operatorname{Ker} \eta_1$). Finally the modified C^p-diffeomorphism g_1 is given by the formula

$$g_1(x) = x + \varphi[x - \langle \eta_1, x \rangle e_1] e_1 \,. \tag{10}$$

Definition 2.7. We say, that a subset C of a C^p-smooth Banach space B is a starlike body with respect to a direction x_0 if for each point $x \in C$ the intersection of the line $x + Rx_0$ and C is a half line. Suppose, that this is $\bar{x} + R^+ x_0$, with the only point $\bar{x} \in \operatorname{Bd} C$. Analogously to Lemma 2.3 it is true, that for the convex cone K having an interior point x_0 there exists a starlike domain with respect to the direction x_0 with a C^p-smooth boundary $\operatorname{Bd} C$, such that $N_{2\epsilon}(K) \supset \operatorname{Bd} C \supset N_\epsilon(K)$ holds.

Lemma 2.8. *Let C be a starlike domain of the C^p-smooth Banach space B with respect ot the direction x_0. Suppose, that the boundary $\operatorname{Bd} C$ of this body is a C^p-smooth subvariety, and the projection of $\operatorname{Bd} C$ into has subspace $\operatorname{Ker} \eta_0$ defined by $x_1 \to x - \langle \eta_0, x \rangle x_0$ (where $\langle \eta_0, x_0 \rangle = 1$) is a C^p-diffeomorphic open embedding. Then for each $\epsilon > 0$ there exists a C^p-diffeomorphism f: $B \to B \setminus C$ which renders C negligible, such that f is the identity off $N_\epsilon(C)$.*

Proof. By an elementary argument we can construct an C^p-diffeomorphism h: $R \to (-\infty . \epsilon)$, such that $h(x) = x$ if $x \leqslant 0$. Then we can define a C^p-function $\tilde{h}$ of B in the following manner: $\tilde{h}(x) = h(\lambda)$, where $x - \lambda x_0 \in \operatorname{Bd} C$. It is obvious, that the C^p-diffeomorphism defined by

$$g(x) = x - h^{-1}(\tilde{h}(x))x_0 + \tilde{h}(x)x_0$$

is a required neglecting C^p-diffeomorphism of C. Thus Lemma 2.8 is proved.

Lemma 2.9. *Let us assume, that $B = B' \oplus B_1$ where B' is a finite-dimensional subspace of a C^p-smooth Banach space B. Let us as-*

sume, that $K \subset B$ *is a projectible subset with respect to the biorthogonal system* $\{e_i, \eta_i \mid i \in I\}$. *Then there exists a* C^p*-diffeomorphism* $f\colon B \to \to B \setminus K$, *such that* f *is identity off* $P_1^{-1}(U)$, *where* $P_1\colon B \to B'$ *is a projection defined by the direct sum* $B = B' \oplus B_1$ *and* U *is an* ϵ*-neighbourhood of the image* $P_1(\mathrm{Ch}\, K)$.

Proof. The assertion of this lemma is similar to a result of R.D. Anderson [1]. We also use a technique very similar to that of [1]. Suppose, that we have two complementary subspaces B_1, B_2 of the finite-dimensional subspace B', that is $B = B' \oplus B_1$ and $B = B' \oplus B_2$ are also satisfied. We can define a C^p-diffeomorphism $g\colon B \to B$ such that the image of $N_\epsilon(P^{-1}(U))$ contains the ϵ-neighbourhoods of $P_1^{-1}(U)$, and B_2, and the restriction of the map $g\colon B \to B$ to the set $N_{\epsilon/2}(P_1^{-1}(U))$ is the identity. We can construct such a diffeomorphism by enlargeing the component of the point $x \in B \setminus N_\epsilon(P_1^{-1}(U))$, such that the image of $B \setminus N_\epsilon(P_1^{-1}(K))$ does not contain the ϵ-neighbourhood of B_2. In the $\epsilon/2$-neighbourhood of $P_1^{-1}(U)$ define g to be the identity. Then we can easily extend the map g C^p-smoothly, to the whole Banach space B. Thus we can suppose, that the subspace B_1 contains the vectors $\{e_i\}_{i \in N}$ of the biorthogonal system. Now we apply the basic idea of a construction due to R.D. Anderson [1]. Suppose, that the subset $P_1^{-1}(U)$ contains the ϵ-neighbourhood to the projectible subset K and the vectors of the biorthogonal system $\{e_i, \eta_i \mid i \in N\}$. For the subsequence $\{\eta_i \mid i \in N\}$ of the integers the relations $\eta_i > i$ and $\langle \eta_{n_i}, K\rangle \geq M_i$ $(i \geq 1)$ hold by the definition of the projectible sets. Let U_i and V_i denote respectively the $\epsilon/2^i$- and $\epsilon/2^{i+1}$-neighbourhoods of the subset $P_1(K) \oplus P^{(i)}(K)$ where the projections $P^{(i)}$ are defined by

$$P^{(i)}x = \sum_{j=1}^{i} \langle \eta_j, x\rangle e_j = (\langle \eta_1, x\rangle, \langle \eta_2, x\rangle, \ldots, \langle \eta_i, x\rangle) \quad (\in R^i)\,.$$

We choose a C^p-function $\varphi_i\colon B' \oplus R^i \to R^1$ such that $\varphi_i(x) = 0$ if $x \notin U_i$, $\varphi_i(x) = 2|M_i| + 2^i$ if $x \in \bar{V}_i$ and $0 \leq \varphi_i \leq 2|M_i| + 2^i$. The map $g_i\colon B \to B$ defined by $g_i(x) = x + \varphi_i[P_1 \oplus P^{(i)}(x)]e_{n_i}$ is a C^p-diffeomorphism. The formally infinite composition $g = \ldots \circ g_3 \circ g_2 \circ g_1$ is not defined in the set K. The modification of the construction of g_i in [1] is inessential, then we can prove the C^p-smoothness of the map g:

$B \setminus K \to B$ and of its inverse $g^{-1}: B \to B \setminus K$. From the previous construction follows, that the maps g_i-s are identity off the ϵ-neighbourhood of $P_1^{-1}(U)$. Thus Lemma 2.9 is proved.

Definition 2.10. Let K be a subset of a Banach space B. By the characteristic cone of vertex x of K the following set is meant: $Cc(K, x) = Cc(K) = \{y \mid y \in B \;\; x + [0, \infty)(y - x) \subset \mathrm{Ch}\, K\}$.

(The characteristic cones of K with different vertices are congruent, so very often we may omit the vertex x in our notations: $Cc(K, x) = Cc(K)$). Now we consider three different cases: (i) the characteristic cone of K is contained in an infinite-codimensional subspace, (ii) $Cc(K)$ does not contain any finite-codimensional subspace and is not contained in an infinite-codimensional subspace, (iii) $Cc(K)$ contains a finite-codimensional subspace. In the proof of Theorem 1.6 we must distinguish these three cases.

Proof of Theorem 1.6. In the case (i) we define a C^p-diffeomorphism $k_\epsilon: B \to B \setminus B_1$, where B_1 is an infinite-dimensional subspace of B spanned by $Cc(K)$. If $Cc(K)$ is contained in a subspace B_1 being a direct summand of B, then we can choose a C^p-diffeomorphism k_ϵ- $B \to B \setminus P^{-1}[P(Cc(K))]$ too (where P denotes the projection of B to the kernel of the coordinate functional η_1 defined by $x \xrightarrow{P} \langle \eta_1, x \rangle e_1$. Here we can suppose, that $\langle \eta_1, K \rangle \subset R^1$ is bounded from below, that is there exists a such real number m that $\langle \eta_1, K \rangle \geqslant m$ holds). The map k_ϵ satisfies $\| k_\epsilon(x) - x \| < 3/5$. In the second case we can find a cone with an interior point x_0 containing the characteristic cone, but not containing any finite-codimensional subspace of B. Then we can approximate this cone with a starlike body X with respect to the direction x_0 with C^p-smooth boundary. So we can suppose, that there exists a starlike body X with C^p-smooth boundary containing the characteristic cone of K, but $Cc(X)$ does not contain any finite-codimensional subspace of B. We can construct X also with the property, that the projection of $\mathrm{Bd}\, X$ into the subspace $\mathrm{Ker}\, \eta_0$ defined by $x \to x - \langle \eta_0, x \rangle x_0$ (where $\eta_0 \in B^*$ and $\langle \eta_0, x_0 \rangle = 1$) is a C^p-diffeomorphism. So we can construct a C_p-diffeomorphism $k_\epsilon: B \to B \setminus P^{-1}(P(x))$ such that $k_\epsilon(x) = x$ if $x \in N_{\epsilon/5}(X)$. From now on our argument is valid simultaneously for cases (i), (ii). Then we transform the subset $B \setminus P^{-1}(P(B_1))$ (respectively $B \setminus P^{-1}(P(X))$) with the C^p-diffeomorphism l, such that $l(x) = x$ if $x \in N_{\epsilon/2}(B_1)$ (respectively $N_{\epsilon/2}(x)$), and $N_\epsilon(H) \setminus P^{-1}(P(B_1)) \subset l[N_{4\epsilon/5}(B_1) \setminus P^{-1}(P(B_1))]$ where H is the cylinder $P^{-1}[P(N_\epsilon(B_1))]$. Then in the following step the C^p-diffeomorphism $h: B \to B$ transforms the cylinder H, so that $\eta_1^{-1}(m-1, \infty) \subset h(H)$, and h is the identity in the cylinder $H_1 = P^{-1}(P(N_{\epsilon/2}(B_1))]$. Finally we can find a C^p-diffeomorphism $g: B \to B \setminus K$ with the property $g(x) = x$ if $\langle \eta_1, x \rangle \leqslant m-1$. The required C^p-diffeomorphism $f_\epsilon: B \to B \setminus K$ is the composition:

$$f_\epsilon = k_\epsilon^{-1} \circ l \circ k_\epsilon \circ h^{-1} \circ g \circ h \circ k^{-1} \circ l \circ k_\epsilon . \tag{11}$$

Now, we define the just mentioned C^p-diffeomorphisms.

First we give the map k_ϵ. We can suppose, that $0 \in \operatorname{Int}(\operatorname{Ch} K)$. K is a projectible set with respect to the biorthogonal system $\{e_i, \eta_i \mid i \in N\}$ with its characteristic cone contained in an infinite codimensional subspace B_1 of B. Thus we may suppose, that $e_1 \in Cc(K, 0)$ and that $\langle \eta_1, K \rangle \geqslant m$ holds. P is continuous projection of B onto $\operatorname{Ker} \eta_1$ so by Lemma 2.6 there exists a C^p-diffeomorphism

$$\bar{k}_\epsilon : \operatorname{Ker} \eta_1 \to \operatorname{Ker} \eta_1 \setminus P(B_1) ,$$

such that $\| \bar{k}_\epsilon(x) - x \| \leqslant \epsilon/5$ for each $x \in \operatorname{Ker} \eta_1$. Then we can define the map $k_\epsilon : B \to B \setminus P^{-1}(P(B_1))$ by the formula $k_\epsilon(x) = \bar{k}_\epsilon(Px) + \langle \eta_1, x \rangle e_1$. The map k_ϵ is a C^p-diffeomorphism and satisfies the property $\| k_\epsilon(x) - x \| \leqslant \epsilon/5$. In the second case first we may transform X to a starlike body X' with respect to the origin of B, by a transformation which restricted to $\operatorname{Ch}(K) \cup [B \setminus N_\epsilon(X')]$ is the identity. Then by Lemma 2.8 we give a required C^p-diffeomorphism k_ϵ by an argument similar to that one applied in case (i).

Let H be the convex cylinder $P^{-1}[P(N_\epsilon(B_1))] \subset B$ and respectively $P^{-1}[P(N_\epsilon(X'))]$. For the cylinder H and the convex bodies $N_{\epsilon/5}(B_1)$, $N_{3\epsilon/5}(B_1)$ (and respectively $N_{\epsilon/5}(X')$, $N_{3\epsilon/5}(X')$) we can find the starlike bodies C_1, C_2, C_3, with respect to the origin satisfying the properties:

(ii) $N_{\epsilon/2}(B_1) \subset C_2 \subset N_{2\epsilon/5}(B_1)$ $\quad (N_{\epsilon/5}(X') \subset C_2 \subset N_{2\epsilon/5}(X'))$,

(iii) $N_{3\epsilon/5}(B_1) \subset C_3 \subset N_{4\epsilon/5}(B_1)$ $(N_{3\epsilon/5}(X') \subset C_3 \subset N_{4\epsilon/5}(X'))$,

and the boundaries $\operatorname{Bd} C_1, \operatorname{Bd} C_2, \operatorname{Bd} C_3$ are C^p-smooth submanifolds. The required C^p-diffeomorphism l is denoted by:

$$l(x) = 2X \cdot \frac{\omega_{c_2}(x)}{\omega_{c_1}(x)} \zeta^{1/2(\omega_3^2(x)-1)}_{1/2(\omega_3^2(x)+1)} \circ \omega_{c_2}(x)$$

where $\omega_j^i(x) = \dfrac{\omega_{c_i}(x)}{\omega_{c_j}(x)}$.

Next, we deal with the construction of the C^p-diffeomorphism $h\colon B \to B$. In the $\operatorname{Ker} \eta_1$ we can approximate the B_1 (and resp. $\operatorname{Cc} K$) by the starlike domain C_4, C_5 having a C^p-smooth boundary such that

(i) $P(B_1) \subset C_4 \subset N_{\epsilon/3}(P(B_1))$
$(P[\operatorname{Cc}(K)] \subset C_4 \subset N_{\epsilon/3}[P(\operatorname{Cd}(K))])$

(iii) $N_{2\epsilon/3}[P(B_1)] \subset C_5 \subset N_\epsilon[P(B_1)]$
$(N_{2\epsilon/3}(p[\operatorname{Cc}(K)]) \subset C_5 \subset N_\epsilon[P(\operatorname{Cc}(K))])$

hold. Let the C^p-smooth function $t \to 1 + \exp\{1 + t - m\} \cdot \zeta^{1/2}_{m-1/2}(t)$ be denote by $a(t)$. An another C^p-function $\beta \circ f$ B is defined by

$$x \xrightarrow{\beta} \zeta^{1/2(\omega_5^4(x)-1)}_{1/2(\omega_5^4(x)+1)} \circ \omega_4 \circ P(x)\,.$$

It is clear, that $\beta | P^{-1}[P(B_1)]$ is the everywhere 0-function. The map h is defined by the composition $h_2 = h_1$ of the C^p-diffeomorphisms

$$x \xrightarrow{h_1} [a(\langle \eta_1, x\rangle)\beta(x) + 1]Px + \langle \eta_1, x\rangle e_1\,,$$

$$x \xrightarrow{h_2} x - \beta(x)[\omega_{c_4}(Px) + |m| + 1]e_1\,.$$

Finally by Lemma 2.9 there exists a C^p-diffeomorphism $g\colon B \to B \setminus K$, such that the restriction of g to $\eta^{-1}(-\infty, m-1)$ is the identity. So in the first two cases Theorem 1.6 is proved.

In the third case our theorem follows from Lemma 2.9.

Proof of Corollary 1.7. If there exists a required C^p-diffeomorphism $f_\epsilon: B \to B \setminus \{0\}$ then by the Banach – Hahn theorem we can choose a continuous linear functional $\eta \in B^*$, such that $\langle \eta, f_\epsilon(0) \rangle \neq 0$. Now we may define a C^p-function $f(x) = \langle \eta, x - f_\epsilon(x) \rangle$. The support of f_ϵ is bounded and non-empty, so the Banach space is C^p-smooth.

If B is C^p-smooth, then the assertion follows from the fact, that the one-point-set $\{0\} \subset B$ is projectible. For this by a theorem of C. Bessaga and A, Pelczynski each Banach space B has an infinite dimensional subspace B_1 with a Schauder bases. Since $\{0\} \subset B$ is projectible with respect to this bases and its coordinates.

Proof of Corollary 1.8. This proof is analogous to the previous one.

Proof of Corollary 1.9. In a Banach space with Schauder bases any compact subset K is projectible by [1]. So this statement follows from the previous one.

REFERENCES

[1] R.D. Anderson, On a theorem of Klee, *Proc. Amer. Math. Soc.*, 17 (1966), 1401-1404.

[2] Cz. Bessaga, Every infinite-dimensional Hilbert space is diffeomorpism with its unit sphere, *Bull. Acad. Polon. Sci.*, XIV. 1 (1966), 27-31.

[3] Cz. Bessaga – A. Pelczyński, On bases and unconditional convergence of series in Banach spaces, *Studia Math.*, 17 (1958), 151-164.

[4] R. Bonič – J. Frampton, Differentiable function on certain Banach spaces, *Bull. Amer. Math. Soc.*, 71 (1965), 393-395.

[5] R. Bonič – J. Frampton, Smooth functions on Banach manifolds, *J. Math. Mech.*, 15 (1966), 877-898.

[6] J. Eells, A setting for global analysis, *Bull. Am. Math. Soc.*, 72 (1966), 751-807.

[7] M.I. Kad'etz, Conditions for the differentiability of a norm in a Banach space, *Usp. Math. Nauk,* (3) XX. (1965), 183-187.

[8] G. Restrepo, Differentiable norms in Banach spaces, *Bull. Am. Math. Soc.,* 70 (1964), 413-414.

COLLOQUIA MATHEMATICA SOCIETATIS JÁNOS BOLYAI
8. TOPICS IN TOPOLOGY, KESZTHELY (HUNGARY), 1972.

SOUSLIN'S CONJECTURE REVISITED

F.D. TALL

In this note we use topological methods to derive Souslin's conjecture from Martin's Axiom (stated as an extension of the Baire category theorem) plus $2^{\aleph_0} > \aleph_1$. The result is not new then, but its proof should be of interest to topologists. We also discuss a number of related propositions. This paper is a portion of a work [11] which will appear elsewhere. There we use the same topological methods to prove a number of new results, some of which we list at the end of this note.

Let X be a linearly ordered set with no first or last element, such that every collection of disjoint open intervals is countable. Suppose X is connected in the induced interval topology. *Souslin's conjecture* [8] is that X is separable and (hence) homeomorphic to the real line. An excellent exposition of the conjecture and its history is [6], where additional references may be found.

Martin's Axiom [5] (stated below) is an alternative to the continuum hypothesis which is becoming increasingly useful in general topology. See for example [10] and the many references induced therein. It has many alternate formulations, for example using partial orders, boolean algebras, or topological spaces. Souslin's conjecture also can be formulated in terms of partial orders, more precisely, trees. In [7] Solovay and Tennenbaum derive Souslin's conjecture from Martin's Axiom plus $2^{\aleph_0} > \aleph_1$, using the partial order versions. They then prove the consistency of Martin's Axiom plus $2^{\aleph_0} > \aleph_1$ with the axioms of Zermelo-Fraenkel set theory, thus establishing the consistency of Souslin's conjecture (its independence was known).

Definition 1. A topological space satisfies the countable chain condition (briefly, is a CCC space) if every collection of disjoint open sets is countable. A collection of sets is *centered* if every finite subcollection of it has non-empty intersection. A space has *precaliber* $\aleph_1$ if every uncountable collection of open sets has an uncountable centered subcollection. A space has *caliber* $\aleph_1$ if every uncountable collection of open sets has an uncountable subcollection with non-empty intersection.

Remark. Clearly separable implies caliber $\aleph_1$ implies precaliber $\aleph_1$ implies CCC. It is easy to see that a subspace X is CCC if and only if $\bar{X}$ is, and that X has precaliber $\aleph_1$ if $\bar{X}$ has caliber $\aleph_1$. Also, if X is completely regular and $\bar{X}$ is a compactification of X, then if X has precaliber $\aleph_1$, $\bar{X}$ has caliber $\aleph_1$. Finally, note that a space has caliber $\aleph_1$ if and only if every point-countable open cover of it is countable.

Definition 2. Martin's Axiom is the statement that *no compact Hausdorff CCC space is the union of fewer than* $2^{\aleph_0}$ *nowhere dense sets.* Let "M" stand for "Martin's Axiom plus $2^{\aleph_0} > \aleph_1$". Let "H" stand for "*every CCC compact Hausdorff space has caliber* $\aleph_1$".

Our two fundamental results are:

Theorem 1. *M implies H.*

Theorem 2. *H implies Souslin's conjecture.*

The proof of Theorem 1 utilizes an idea of Arhangel'skiĭ. (For further details, see [11].) Assume M. Let X be a compact Hausdorff CCC space. Suppose $\{U_a\}_{a<\omega_1}$ is an uncountable point-countable open cover of X. We will derive a contradiction. We observe (I. Juhász), that given any uncountable open cover of a CCC space, there is an open set V such that any open subset of V intersects uncountably many members of the cover. In particular then, let $F_\beta = V - \bigcup_{\beta \leq a < \omega_1} U_a$. By point-countability, $V = \bigcup_{\beta < \omega_1} F_\beta$. In the subspace V, each F_β is closed, and moreover has empty interior since it intersects only countably many members of the cover. But a repetition of the proof of a standard corollary to the Baire category theorem yields that, assuming M, no open subset V of a compact Hausdorff CCC space is the union of $\aleph_1$ sets, each nowhere dense in V.

There are many ways to prove Theorem 2. One way is to use

H': Every (completely regular) CCC space has precaliber $\aleph_1$. It follows from the remark following Definition 1 that H' is equivalent to H. (The hypothesis of complete regularity can be removed with some effort, e.g. by going back and forth between the original space and the Stone space of its boolean algebra of regular open sets. Or see [3, chapter 5] for a different approach.) None of the other implications in the remark can be (consistently) reversed.

Given a Souslin space (i.e. a counterexample X to Souslin's conjecture). one violates H' as follows. First remove countably many points from the space. For each countable ordinal, assume countably many points have been removed at each smaller ordinal. X minus the closure of what's been removed is the union of a non-empty collection of disjoint open intervals. Remove a point from each of these intervals. It is not hard to show that the collection of $\aleph_1$ open intervals obtained in this way contradicts H'. (The details of the construction can be found in [6]. It remains only to observe that a centered subcollection of the collection is in fact a chain).

Another way to prove Theorem 2 is to quote

Theorem [3, 5.5]. H' implies the product of an arbitrary collection of CCC spaces is CCC.

Theorem [4]. The product of a Souslin space with itself is not CCC.

We shall sketch yet another way to deduce Souslin's conjecture from H since we think the proposition needed is of independent interest. First a definition.

Definition 3. Let X be a space, $x \in X$. $t(x, X)$ is the least cardinal κ such that for each $A \subset X$, if $x \in \bar{A}$, there is $B \subset A$ such that $x \in \bar{B}$ and $|B| \leqslant \kappa$. $t(X) = \sup\ t(x, X)\colon x \in X$ is the *tightness* of X. $d(X)$, the *density* of X, is the least cardinal of a dense subset of X.

The key step is

Theorem 3. *If* $t(X) \leqslant \aleph_0$ *and each closed subset of* X *has caliber* $\aleph_1$, *then* X *is hereditarily separable.*

From this theorem we easily deduce that if H holds, every CCC linearly ordered topological space X is separable, and in particular, Souslin's conjecture. For consider the Dedekind completion $\bar{X}$ of X. X is a dense subspace of the compact ordered space $\bar{X}$, and the subspace topology on X coincides with its original order topology [3, 0.17]. Since X is CCC, so is $\bar{X}$. Since $\bar{X}$ is CCC and linearly ordered, $\bar{X}$ is hereditarily CCC and also first countable [3, 2.8], so certainly has countable tightness. By H, closed subsets of $\bar{X}$ have caliber $\aleph_1$, so $\bar{X}$ is hereditarily separable by Theorem 3, and therefore X is separable.

To prove Theorem 3, we first observe [1, Lemma 7] that if a space X is not (hereditarily) separable, then it has a non-separable subspace Y of cardinality $\aleph_1$. It suffices to show $\bar{Y}$ is separable to get a contradiction because [3, 2.26] $d(Y) \leqslant d(\bar{Y}) \cdot t(\bar{Y})$ and $t(\bar{Y}) \leqslant t(\bar{X}) \leqslant \aleph_0$. Let $Y = \{y_\alpha\}_{\alpha < \omega_1}$. Y is dense in $\bar{Y}$ which has countable tightness, so $\bar{Y} =$ $= \bigcup_{\beta < \omega_1} \overline{\{y_\alpha\colon \alpha < \beta\}}$. But $\bar{Y}$ has caliber $\aleph_1$, so cannot be the union of a strictly ascending sequence of $\aleph_1$ closed sets. Therefore some countable subset of Y is dense in $\bar{Y}$.

A corollary of Theorem 3 is

Corollary. *If* X *is compact Hausdorff and every closed subset of* X *has caliber* $\aleph_1$, *then* X *is hereditarily separable.*

The point is that X is hereditarily CCC, and therefore by [1] has countable tightness. The next result, due to Arhangel'skiĭ, follows immediately.

Corollary. *M implies that if X is compact Hausdorff and hereditarily CCC, then it is hereditarily separable.*

Before ending with a problem, we shall list without proof some of the results obtained in [11]. All of them may be regarded as generalizations of the fact that M implies Souslin's conjecture. Terms not defined here are defined in [11]. First we note that Theorem 1 holds for absolute G_λ spaces, $\lambda < 2^{\aleph_0}$. As a result, all consequences of M in this area that we know of, can be extended from compact spaces to absolute G_δ spaces, and most to absolute G_λ spaces, $\lambda < 2^{\aleph_0}$. For example we have the following two theorems, where π, χ, $|\ |$ dd respectively stand for π-weight, character, cardinality, hereditary density.

Theorem 4. *Suppose M. Let X be completely regular CCC and absolute G_λ; $\lambda < 2^{\aleph_0}$. Then X is separable if any of the following conditions hold:*

a) $\pi(X) < 2^{\aleph_0}$,

b) $(\chi(X))^+ < 2^{\aleph_0}$,

c) $|X| < 2^{2^{\aleph_0}}$, $dd(X) < 2^{\aleph_0}$, $\lambda \leq \aleph_0$.

Theorem 5. *Suppose M. Let X be completely regular absolute G_δ. Then X is hereditarily separable if either of the following conditions holds:*

a) *X is hereditarily Lindelöf,*

b) *every closed subset of X is CCC.*

For each n in the following list, the n-th statement implies the $(n+1)$-st

1. M.

2. H (or H').

3. Given an uncountable collection $\mathcal{U}$ of open sets in a CCC

space, there is an uncountable subcollection $\mathscr{V}$ such that any two members of $\mathscr{V}$ intersect.

4. The arbitrary product of CCC spaces is CCC.

5. Souslin's conjecture.

Problem. Can any of these implications be reversed?

We conjecture that none of them can. Certainly not all of them can, because R. Jensen has shown Souslin's conjecture to be consistent with the continuum hypothesis.

Remark. (added in proof) B. Šapirovskiĭ achieved results overlapping ours in [9]. In response to the author's question, P. Erdős proved that the Stone space of the usual measure algebra on the unit interval satisfies (3) of the Problem but, if the continuum hypothesis is assumed, does not have precaliber $\aleph_1$.

REFERENCES

[1] A.V. Arhangel'skiĭ, On bicompacta hereditarily satisfying Suslin's condition. Tightness and free sequences, *Soviet Math. Dokl.*, 12 (1971), 1253-1257.

[2] D.D. Booth, *Countably indexed ultrafilters,* thesis, University of Wisconsin, Madison, 1969.

[3] I. Juhász, Cardinal functions in topology, *Mathematical Center Tract,* 34, Mathematisch Centrum, Amsterdam, 1971.

[4] D. Kurepa, La condition de Suslin et une propriété caractéristique des nombres réels, *CR. Acad. Sci. Paris,* 231 (1950), 1113-1114.

[5] D. Martin – R.M. Solovay, Internal Cohen extensions, *Ann. Math. Logic,* 2 (1970), 143-178.

[6] M.E. Rudin, Souslin's conjecture. *Amer. Math. Monthly,* 76 (1969), 1113-1119.

[7] R.M. Solovay – S. Tennenbaum, Iterated Cohen extensions and Souslin's Problem, *Ann. of Math.*, 94 (1971), 201-245.

[8] M. Souslin, Problème 3, *Fund. Math.*, 1 (1920), 223.

[9] B. Šapirovskiĭ, On separability and metrizability of spaces with Souslin's condition, *Dokl. Akad. Nauk SSSR*, 207 (1972), 800-803.

[10] D.F. Tall, An alternative to the continuum hypothesis and its uses in general topology, *preprint*.

[11] F.D. Tall, The countable chain condition vs. separability – applications of Martin's Axiom, *to appear in Gen. Top. Appl.*

COLLOQUIA MATHEMATICA SOCIETATIS JÁNOS BOLYAI
8. TOPICS IN TOPOLOGY, KESZTHELY (HUNGARY), 1972.

ON TOPOLOGICAL PROPERTIES DEFINED BY GAMES

R. TELGÁRSKY

Let K be a class of completely regular spaces such that (i) if $X \in K$ and E is a closed subset of X, then $E \in K$; and (ii) if X has a finite closed cover $\{E_m : m \leq n\}$ such that each $E_m \in K$, then $X \in K$.

Let X be a completely regular space. $G(K, X)$ denotes the following infinite positional game with perfect information. There are two players I and II (the pursuer and the evader). They choose alternately consecutive terms of a sequence $\{E_n : n \in N\}$ (where $N = \{0, 1, 2, \ldots, n, \ldots\}$) of subsets of X so that each player knows $K, E_0, E_1, \ldots, E_n$ when he is choosing E_{n+1}. A sequence $\{E_n : n \in N\}$ of subsets of X is a *play* of $G(K, X)$ if $E_0 = X$ and if for each $n \in N$:

1) E_{2n+1} is the choice of I;
2) E_{2n} is the choice of II;
3) $E_{2n+1} \in K$;
4) $E_n \in 2^X$ (2^X denotes the family of all closed subsets of X);
5) $E_{2n+1} \subseteq E_{2n}$;

6) $E_{2n+2} \subseteq E_{2n}$;

7) $E_{2n+1} \cap E_{2n+2} = 0$.

Player I *wins* if $\cap \{E_{2n} : n \in N\} = 0$. Player II wins if $\cap \{E_{2n} : n \in N\} \neq 0$. A finite sequence $\{E_m : m \leqslant n\}$ of subsets of X is *admissible* for $G(K, X)$ if the sequence $(E_0, E_1, \ldots, E_n, 0, 0, \ldots, 0, \ldots)$ is a play of $G(K, X)$. A function s is a *strategy* for I (II) in $G(K, X)$ if the domain of s consists of admissible sequences $(E_0, \ldots, E_n)$ with n even (odd), $s(E_0, \ldots, E_n) \subseteq X$, and $(E_0, \ldots, E_n, E_{n+1})$ is admissible with $E_{n+1} = s(E_0, \ldots, E_n)$. A strategy s is a *winning strategy* for I (II) in $G(K, X)$ if I (resp. II) using s wins every play of $G(K, X)$. A space X is *K-like* (*anti-K-like*) if there is a winning strategy for I (resp. II) in $G(K, X)$. $G(K, X)$ is *determined* if X is K-like or anti-K-like. A function t: $2^X \to 2^X \cap K$ is a *simple strategy* for I in $G(K, X)$ if $t(E) \subseteq E$ for each $E \in 2^X$. t is used in the play $\{E_n : n \in N\}$ iff $E_{2n+1} = t(E_{2n})$ for each $n \in N$.

The topological terminology is that of [3]. X is *K-scattered* if for each nonvoid closed subset E of X there is a point $x \in E$ and an open nbhd U of x for which $E \cap \operatorname{cl} U \in K$. X is locally K if for each point $x \in X$ there is an open nbhd U for which $\operatorname{cl} U \in K$. We denote the class of all locally K spaces by LK. X is *σ-K* if X has a cover $\{E_n : n \in N\}$ where $E_n \in 2^X \cap K$ for each $n \in N$.

The game-theoretical approach to topological problems has been suggested to me by Professor Czesław Ryll-Nardzewski.

§1. THE GENERAL CASE

1.1. If $X \in K$, then X is K-like.

1.2. If $Y \in 2^X$ and X is K-like, then Y is K-like.

1.3. If $Y \in 2^X$ and Y is anti-K-like, then X is anti-K-like,

1.4. If $X \in \sigma\text{-}K$, then X is K-like.

1.5. If X is a K-scattered Lindelöf space, then X is K-like.

1.6. If X is K-scattered and paracompact, then X is LK-like.

1.7. If X is paracompact, $X \in \sigma\text{-}LK$ and $\dim X = 0$, then $2^X \cap K$ contains a closure-preserving cover of X.

1.8. If X is K-like and each $E \in 2^X$ is a G_δ-set in X, then $X \in \sigma\text{-}K$.

1.9. If X is K-like and hereditarily paracompact, then X has a cover $\{E_n : n \in N\}$ where E_n is a K-scattered closed subset of X for each $n \in N$.

1.10. **Problem.** Let X be a paracompact LK-like space. Does X have a cover $\{E_n : n \in N\}$ where each E_n is a K-scattered closed subset of X?

1.11. If X is K-like and complete in the sense of Čech, then X is K-scattered.

1.12. If Y is complete in the sense of Čech, $X \subseteq Y$, each $E \in$ $\in 2^X \cap K$ is a set in X of first category and $Y - X$ is a set of first category in Y, then X is anti-K-like.

1.13. If X is basiscompact (see [9], p. 24) and each $E \in 2^X \cap K$ is a set of first category in X, then X is anti-K-like.

1.14. If there is a regular Baire measure m on X such that $m(X) > 0$ and $m^*(E) = 0$ for each $E \in 2^X \cap K$, then X is anti-K-like.

1.15. If the class K is perfect, then the class of all K-like spaces is perfect as well.

1.16. If X is a Lindelöf space and complete in the sense of Čech, then $G(K, X)$ is determined.

1.17. If X is paracompact and complete in the sense of Čech, then the game $G(LK, X)$ is determined.

1.18. Let K be the family of all $E \in 2^X$ with the Lindelöf property. Then X is K-like iff $X \in K$.

1.19. Let K be the family of all a-paracompact subsets of X (see [1]). Then X is K-like iff $X \in K$.

§2. FINITE SPACES

F = the class of all finite spaces.

2.1. Let X be a compact space. Then X is F-like iff X is scattered.

2.2. If X is compact, then $G(F, X)$ is determined.

2.3. Let X be a separable metric space. Then X is F-like iff X is countable.

2.4. If X contains a copy of the Cantor discontinuum, then X is anti-F-like.

2.5. If X is an analytic set (see [4], p. 478) in a separable complete metric space, then $G(F, X)$ is determined.

2.6. Let us assume that the axiom (A) of H. Steinhaus and J. Mycielski (see [7], p. 207) holds. Then $G(F, X)$ is determined for each separable metric space.

(A) is inconsistent with the axiom of choice. However (A) implies a weak form of the axiom of choice (see [7], p. 207).

2.7. If X is F-like, then X has the property C'' (see [4], p. 527).

2.8. If X is totally paracompact (see [5]) and $X \subseteq Y$ where Y is a separable complete metric space, then $G(F, Y - X)$ is determined.

§3. COMPACT SPACES

C = the class of all compact spaces.

3.1. If X is σ-compact, then X is C-like.

3.2. Let X be locally compact. Then X is C-like iff X is σ-compact.

3.3. If X is locally compact, then $G(C, X)$ is detemined.

3.4. If X is C-like and is complete in the sense of Čech, then X

is C-scattered.

3.5. If X is C-like, then X has the Hurewicz property (see [6], p. 209).

3.6. The Sorgenfrey line is anti-C-like.

§4. DISCRETE SPACES

D = the class of all discrete spaces.

4.1. If X is σ-discrete, then X is D-like.

4.2. If X contains a compact perfect subset, then X is anti-D-like.

4.3. If X is an A-set in a complete metric space (see [2]), then $G(D, X)$ is determined.

4.4. If X is D-like and collectionwise normal, then X is paracompact.

4.5. If X is a D-like Lindelöf space, then X has the property C''.

4.6. If X is paracompact and σ-discrete, then X has a closure-preserving cover by finite sets.

4.7. If X has a closure-preserving cover by finite sets, then X is D-like.

4.8. If X has a closure-preserving cover by finite sets, then X has a cover $\{E_n : n \in N\}$ where each E_n is a scattered closed subset of X.

4.9. If X is paracompact and scattered, then X is D-like.

4.10. If X is D-like and each $E \in 2^X$ is a G_δ-set in X, then X is σ-discrete.

4.11. If X is metrizable, then the following conditions are equivalent:

a) X is D-like;

b) X is σ-discrete;

c) X has a closure-preserving cover by finite sets.

4.12. If X is paracompact and is has a closure-preserving cover by finite sets, then X is totally paracompact.

§5. LOCALLY COMPACT SPACES

LC = the class of all locally compact spaces.

5.1. If X is σ-locally compact, then X is LC-like.

5.2. If X has a closure-preserving cover by compact sets, then X is LC-like; moreover, the player I has a simple winning strategy in $G(LC, X)$ (cf. [8], Lemma 6.).

5.3. If X is paracompact and σ-locally compact, then X has a closure-preserving cover by compact sets.

5.4. If X is LC-like and each $E \in 2^X$ is a G_δ-set in X, then X is σ-locally compact.

5.5. If X is metrizable, then the following conditions are equivalent:

a) X is LC-like;

b) X is σ-locally compact;

c) X has a closure-preserving cover by compact sets.

5.6. If X is paracompact and C-scattered, then X is LC-like.

5.7. If X is paracompact and LC-like, and Y is paracompact, then the product space $X \times Y$ is paracompact (cf. [10], Theorem 2.3).

5.8. If X is an A-set in a complete metric space (see [2]), then $G(LC, X)$ is determined.

5.9. If X is paracompact and complete in the sense of Čech, then $G(LC, X)$ is determined.

5.10. If X is paracompact and LC-like, then each open base of X contains a σ-locally finite cover of X.

5.11. **Problem.** Let X be a paracompact LC-like space. Is X totally paracompact?

§6. COUNTABLE SPACES

K_0 = the class of all countable spaces.

$G'(K_0, X)$ denotes the following modification of $G(K_0, X)$: the condition (4) is weakend to (4'): $F_{2n} \in 2^X$.

6.1. Let X be a compact metric space. Then X is K_0-like iff X is countable.

6.2. If X is chain concentrated about a countable subset (see [11], p. 154), then X is K_0-like.

6.3. If X is K_0-like, then X has the property C''.

6.4. Let $\{E_n : n \in N\}$ be a cover of X by arbitrary sets. If each E_n is K_0-like, then X is also K_0-like.

REFERENCES

[1] C.E. Aull, Paracompact subsets, *Proc. of the Second Prague Topological Symp.*, 1966, Prague, 1967, 45-51.

[2] A.G. El'kin, *A*-sets in complete metric spaces, *DAN SSSR*, 175 (1967), No. 3, 517-520 (Russian).

[3] R. Engelking, *Outline of General Topology*, Amsterdam, 1968.

[4] K. Kuratowski, *Topology*, Vol. I, New York – London – Warszawa, 1966.

[5] A. Lelek, On totally paracompact metric spaces, *Proc. Amer. Math. Soc.*, 19 (1968), 168-170.

[6] A. Lelek, Some cover properties of spaces, *Fund. Math.*, 64 (1969), 209-218.

[7] J. Mycielski, On the axiom of determinateness, *Fund. Math.*, 53 (1964), 205-224.

[8] H.B. Potoczny, *Closure-preserving families of compact sets, General Topology and its Applications* (in press).

[9] J. van der Slot, Some properties related to compactness, *Mathematical Centre Tracts,* No. 19, Amsterdam, 1968.

[10] R. Telgársky, *C*-scattered and paracompact spaces, *Fund. Math.*, 73 (1971), 59-74.

[11] R. Telgársky, Concerning product of paracompact spaces, *Fund. Math.,* 74 (1972), 153-159.

COLLOQUIA MATHEMATICA SOCIETATIS JÁNOS BOLYAI
8. TOPICS IN TOPOLOGY, KESZTHELY (HUNGARY), 1972.

COHOMOLOGY THEORIES IN SIMPLICIAL TOPOLOGY

W. VOGEL

For the purpose of simplicial topology, A. Dold and D. Puppe outline a theory of derived functors for non-additive functors between abelian categories (see: *Ann. Inst. Fourier* 11, 201-312 (1961)). If in addition the functor is additive, then S. Eilenberg and J.C. Moore conconsider a relative theory (see: *Memoirs Amer. Math. Soc.* 55 (1965)).

In this lecture we outline a theory of simplicial resolutions and derived functors that is a common generalization of the relative theory of Eilenberg – Moore and the theory of Dold – Puppe. To this theory we are able to derive arbitrary functors $E\colon \underline{A} \to \underline{B}$, where $\underline{A}$ is a category with finite limits and a projective class and $\underline{B}$ is abelian. We show a Theorem (*) that the derived functors in this theory are naturally equivalent to the cotriple derived functors of Barr and Beck (see: Homology and standard constructions. *Springer Lecture Notes in Mathematics* No. 80 (1969)) if there is a cotriple G in $\underline{A}$ that realizes the given projective class. We know that the cotriple derived functors of the crossed homomorphism functor yield (with a shift in dimension) the Eilenberg – MacLane cohomology groups of a group. Thus the same statement holds by (*) if we use any exact simplicial resolution by free groups. That these can be chosen to be much smaller and easier to handle than the cotriple is clear. If we take as "models" a projective class, then the derived functors in our theory coincide also with those of M. André (see: Methods simpliciale en algèbre homologique et algèbre commutative. *Springer Lecture Notes in Mathematics* No. 32 (1967)).

Problem: for the purpose of extending (*) to the consideration of homotopy derived functors as by P. Huber "Homotopy theory in general categories", *Math. Ann.* 144, 361-385 (1961), it would be desirable to have a proof that $GKA \xrightarrow{\epsilon A} A$ (functorial G-projective resolution) and $GA \xrightarrow{\epsilon A} A$ (a cotriple G gives rise to a functorial argumented simplicial object $GA \xrightarrow{\epsilon A} A$) are simplicially homotopy equivalent. One might hope, for example, that the simplicial comparison theorem of H. Kleisli (see *C.R. Acad. Sci.* Paris 264, (1967) 11-16) could be applied to obtain this, but the condition of representability required there seems difficult to achieve for $GKA \xrightarrow{\epsilon A} A$ – if at all possible.

Added in proof. In the meanwhile H. Fridow showed in his diploma work (*Über simpliziale Auflösungen und Vergleichsätze in Kategorien mit Multiplikation,* Halle University, Mathematics Department, 1972) that, under weaker assumptions, the simplicial comparison theorem of H. Kleisli can be applied to the above problem.

REFERENCES

M. Tierney – W. Vogel, Simplicial derived functors, *Springer Lecture Notes in Mathematics* No. 86, 167-180 (1969).

COLLOQUIA MATHEMATICA SOCIETATIS JÁNOS BOLYAI
8. TOPICS IN TOPOLOGY, KESZTHELY (HUNGARY), 1972.

COMPACT SUBMANIFOLDS OF HOMOLOGY 3-SPHERES

J. VRABEC

We work in the PL category. Each manifold is supposed to have a fixed PL structure. If M is a manifold, then by a submanifold of M or by a surface, simple closed curve, etc., in M we always mean a respective object contained in M as a subpolyhedron (in the chosen PL structure of M). All maps are assumed to be PL. A *surface* is a compact, connected, orientable 2-manifold. A cube with handles is a 3-manifold M homeomorphic to a "fattened up" connected finite linear graph G in S^3, the 3-sphere; we say that M has n handles if G has Euler characteristic $1 - n$. All homology groups have integer coefficients.

Let S be a closed surface of genus n in S^3. It is easy to see that S must separate S^3. Let U and V be the closures of the two components of $S^3 - S$. We say that S is standardly embedded if there exist disjoint properly embedded disks $D_1, \ldots, D_n$ in U and disjoint properly embedded disks $E_1, \ldots, E_n$ in V such that $D_i \cap E_j = \phi$ if $i \neq j$ and, for each i, $\partial D_i \cap \partial E_i \subset S$ is a single crossing point. It is not difficult to prove that, up to homeomorphism of S^3, there is only one standardly embedded closed surface of genus n in S^3. Of course, there also exist non-standardly embedded surfaces in S^3.

If we speak of a simple closed curve J in a space W, then, from the viewpoint of homology theory, there is no difference between "J bounds a disk in W" and "J bounds a surface in W"; both mean "$J \sim 0$ in W" (we use "$\sim$" for "is homologous to"). Therefore our Theorem 1 below says essentially that, in the homological sense, all embeddings of a surface S in S^3 are standard. It is not strange that this result also holds for *homology 3-spheres*, i.e. 3-manifolds that have homology groups isomorphic to those of S^3. In fact Theorem 1 is proved for an arbitrary connected 3-manifold W with $H_1(W) = 0$. Such a 3-manifold W will be called *1-acyclic*. As in the case of S^3, any closed surface in W separates W. We also remark that W is necessarily orientable.

Theorem 1. *Let W be a 1-acyclic 3-manifold and S a closed surface of genus n in the interior of W. Denote by U and V the closures of the two components of $W - S$. Then there exist oriented simple closed curves J_i, K_i in S $(i = 1, 2, \ldots, n)$ such that*

(1) *J_i and K_i intersect transversely at a single point and $J_i \cap J_j = J_i \cap K_j = K_i \cap K_j = \phi$ if $i \neq j$;*

(2) *$J_i \sim 0$ in U and $K_i \sim 0$ in V;*

(3) *the homology classes of $J_1, \ldots, J_n$ form a free basis of $H_1(V)$ and the homology classes of $K_1, \ldots, K_n$ form a free basis of $H_1(U)$.*

This theorem is similar to the following theorem of Papakyriakopoulos [1; 32.3]: If M is a cube with n handles in a homology 3-sphere Σ and if $D_1, \ldots, D_n$ are disjoint properly embedded disks in M such that they together do not separate M, then there exist disjoint simple closed curves $K_1, \ldots, K_n$ in ∂M such that: (1) each K_i crosses ∂D_i at exactly one point and misses any other ∂D_j, (2) each K_i is homologous to zero in the closure of $\Sigma - M$. Papakyriakopoulos' proof uses the disks D_i and cannot be directly extended to a proof of our Theorem 1. Our proof is more algebraic. Denote by $\alpha\colon S \to U$, $\beta\colon S \to V$ the inclusions and by α_*, β_* the corresponding homomorphisms of first homology groups. Let $A = \operatorname{Ker} \alpha_*$, $B = \operatorname{Ker} \beta_*$. In the first step we prove that $H_1(S) = A \oplus B$ and that $\alpha_* | B\colon B \to H_1(U)$, $\beta_* | A\colon A \to H_1(V)$ are isomorphisms. If W is a homology 3-sphere, this follows directly from the Mayer – Vietoris sequence of $(W; U, V)$; if W has only $H_1(W) = 0$, a small additional argument is necessary. Here is in fact the only difference in the proof between homology 3-spheres and arbitrary 1-acyclic 3-manifolds. In the second step we choose bases $\{a_1, \ldots, a_n\}$ for A and $\{b_1, \ldots, b_n\}$ for B such that each a_i has zero intersection number (in S) with any b_j such that $i \neq j$ and intersection number 1 with b_i. In the third step we find for each i an oriented simple closed curve J_i to represent a_i and an oriented simple closed curve K_i to represent b_i, such that the J_i and K_i satisfy assertion (1) of Theorem 1.

Our proof gives in fact a slightly more general result, which is then also a generalization of Papakyriakopoulos' theorem: if someone has done part of the job by finding "good" curves $J_1, \ldots, J_r$, $K_1, \ldots, K_s$ $(0 \leqslant r \leqslant n,\ 0 \leqslant s \leqslant n)$, then his work can be saved, i.e. we can choose the rest of the J_i's and K_i's so that the whole collection satisfies the conclusions of Theorem 1.

Theorem 1 has the following two easy corollaries.

Corollary 1. *Let* W, S, U, V *be as in Theorem 1. Then we can attach a cube with* n *handles,* V', *to* U *along* S *in such a way that the 3-manifold* $W' = U \cup V'$ *is again 1-acyclic.*

Proof. Find $J_1, \ldots, J_n$, $K_1, \ldots, K_n$ satisfying the conclusions of Theorem 1. Attach a pillbox to U along each K_i and call the resultant 3-manifold W'' (a *pillbox attached to* U *along* K_i is a 3-cell $D \times I$, where D is a disk and I an interval, with the annulus $\partial D \times I$ on its boundary identified with an annulus neighborhood of K_i in S). The boundary component of W'' which intersects S is a 2-sphere. Attach a 3-cell to W'' along this 2-sphere and denote the new 3-manifold by W'. It easily follows from Theorem 1 that $H_1(W') = H_1(W'') = 0$. Clearly, the closure of $W' - U$ is a cube with n handles.

Corollary 2. *Any compact subset of a 1-acyclic 3-manifold can be embedded in a homology 3-sphere.*

Proof. Let C be a compact subset of a 1-acyclic 3-manifold W. By adding an open collar on ∂W, if W has boundary, we can assume that $\partial W = \phi$. Cover C with a compact connected 3-submanifold M of W. Take a boundary component S of M. Denote by U and V the closures of the two components of $W - S$; let U be the one which contains M. By Corollary 1 we can replace V by a cube with handles, V', in such a way that $W' = U \cup V'$ is still 1-acyclic. If we perform a similar surgery along each boundary component of M, we end up with M embedded in a closed 1-acyclic 3-manifold, Σ say. It follows from Poincaré duality that Σ is a homology 3-sphere.

Our second theorem gives a homological characterization of compact 3-manifolds that can be embedded in some homology 3-sphere (or, equivalently, in some 1-acyclic 3-manifold).

Theorem 2. *A compact connected 3-manifold M whose boundary has m components $(m > 0)$ can be embedded in a homology 3-sphere if and only if it satisfies the following conditions* (1), (2), *and either* (3') *or* (3''):

(1) M *is orientable;*

(2) $H_1(M)$ *is free;*

(3') $H_2(M)$ *is free of rank* $m - 1$;

(3'') $H_1(\partial M) \to H_1(M)$ *is onto.*

(By standard homological techniques we can prove that, in presence of conditions (1) and (2), the conditions (3') and (3'') are really equivalent.)

Suppose that $M \subset \Sigma$, where Σ is a homology 3-sphere. It follows immediately from the Mayer – Vietoris sequence of $(\Sigma; M, \Sigma - \operatorname{Int} M)$ that M satisfies (1), (2), and (3'') of Theorem 2.

To prove the other direction of Theorem 2 is more difficult. We first consider the case $m = 2$ (the case $m = 1$ then follows easily). By a careful examination of the homology of M we can show that there exists a simple closed curve K in ∂M which generates a nonzero direct summand of $H_1(M)$. Attaching a pillbox to M along K we obtain a 3-manifold M' which has again two boundary components, again satisfies the conditions of Theorem 2, but whose boundary has smaller total genus than the boundary of M. Theorem 2 for $m = 2$ can thus be proved by induction on the total genus of ∂M. Then we prove the

Lemma. *Let* M' *be a compact connected 3-manifold that can be embedded in a homology 3-sphere,* S *a boundary component of* M', *and* $J \subset S$ *a separating simple closed curve. Let* M *be the 3-manifold obtained by attaching a pillbox to* M' *along* J. *Then* M *can be embedded in a homology 3-sphere.*

We prove this as follows. Suppose that M' lies in a homology 3-sphere Σ. Let U' be the closure of the component of $\Sigma - S$ which contains $M' - S$ and let U be the 3-manifold obtained by attaching a pillbox to U' along J. Then U has two boundary components and satisfies the conditions of Theorem 2. By the already proved part of Theorem 2, U, and hence M, can be embedded in a homology 3-sphere.

Theorem 2 in general is now proved by induction on m. Suppose that $m > 2$. Choose two boundary components of M, say S' and S''. Drill a hole through M from S' to S'' to merge them in a single surface S. Denote the new 3-manifold by M'. Since M is obtained from M' by plugging the hole, which is the same as attaching a pillbox to M' along an appropriate separating simple closed curve in S, it follows from the Lemma that M can be embedded in a homology 3-sphere if M' can be. But M' has fewer boundary components than M and it can easily be shown that M' again satisfies the conditions of Theorem 2. The induction is thus complete.

This paper, with one or two additional results and with complete proofs, will be published in the *Pacific Journal of Mathematics.*

REFERENCES

C.D. Papakyriakopoulos, A reduction of the Poincaré Conjecture to group theoretic conjectures, *Ann. of Math.*, 77 (1963), 250-305.

PROBLEMS

L. Babai – A. Máté

Is there a *locally compact* space which is not Lindelöf and is such that any of its subsets of a cardinality not cofinal to ω has a complete accumulation point? (Without assuming that the space is locally compact the answer is yes).

M. Bognár

1. Let K be a geometrical complex, such that $\widetilde{K}$ (the body of K) is an n-dimensional euclidean manifold. Is it true, that the body of the star of each vertex is homeomorphic to the n-dimensional ball?

2. Let X be a continuum in E^n such that each point $q \in X$ has a base Σ for its neighbourhood system in E^n satisfying the following conditions for any $V \in \Sigma$:

a) V is a domain in E^n,

b) $V \cap X$ is a domain in X,

c) $V \setminus X$ consists of two components $p^1(V)$ and $p^2(V)$,

d) the closure of each $p^i(V)$ contains $V \cap X$.

*Then X is a regular domain boundary in E^n (i.e. $E^n \setminus X$ consists of two components, and the boundary of each of these components is X.)

Now the question is: Does * remain true if we omit *d*)?

Á. Császár

Let E be a T_0-space, and Γ a transitive group of homeomorphisms of E. Γ is said to *separate* $A, B \subset E$ if, for each $p \in E$, there is a neighbourhood V of p such that $a \in \Gamma$, $a(V) \cap A \neq \phi$ imply $a(V) \cap B = \phi$. Consider the following conditions:

(a) Γ separates $\{x\}$ and F if $x \notin F = \bar{F}$,

(b) Γ separates K_1 and K_2 if they are compact and disjoint,

(c) Γ separates $\{x\}$ and K is $x \notin K$ and K is compact.

It can be shown that (a) $\Rightarrow$ (b) $\Rightarrow$ (c). Is (c) $\Rightarrow$ (a) or (b) $\Rightarrow$ (a) valid if E is locally compact?

R.Z. Domiaty

Sei (R, d) ein metrischer Raum und $\mathfrak{T}_d^*$ die gröbste Topologie, für die die abgeschlossenen Kugeln abgeschlossene Mengen sind (d.h. die abgeschlossenen Kugeln bilden ein Subbasis für die abgeschlossenen Mengen).

Wir nennen einen topologischen Raum $(R, \mathfrak{T})$ *$*$-metrisierbar*, wenn es eine Metrik d auf R gibt, so dass

$$\mathfrak{T} = \mathfrak{T}_d^*$$

ist. Man gebe notwendige und hinreichende Bedingungen dafür an, dass $(R, \mathfrak{T})$ $*$-metrisierbar ist.

C.H. Dowker

Approximately how many different (mutually non-homeomorphic) connected T_0-spaces have exactly 100 points?

If a_n is the number of topologies on a set of n points, b_n is the number of homeomorphism classes of spaces with n points and C_n is the number of homeomorphism classes of connected T_0 spaces with n points, then a_n, b_n and c_n behave like $2^{n^2/4}$ when n is large. Let $v_n = \frac{4}{n^2} \log_2 c_n$. Then $v_n \to 1$ and the curve through the points $(1/n, v_n)$ passes through the point $(0, 1)$.

Conjecture: This curve has slope $-\infty$ at $(0, 1)$.

If so, polynomials cannot be used for extrapolation to find v_{100} and c_{100}. What method of extrapolation can be used?

R. Duda

As is well known, the family $T(X)$ of all topologies on a set X is a lattice: the meet of topologies $\{O_s\}_{s\in S}$ is the topology $\bigcap_{s\in S} O_s$ and the join of topologies $\{O_s\}_{s\in S}$ is the smallest topology containing $\bigcup_{s\in S} O_s$. The lattice is quite complicated and has many interesting properties, e. g. it is complemented [4].

In 1970 I have raised [2] the conjecture that each lattice A can be embedded in $T(X)$ for some X, and even more, that each lattice A can be embedded in $T(X)$ as a sublattice consisting of topologies satisfying some restrictive conditions, e.g. of T_1-topologies.

As has been pointed out by Roland E. Larson [3], the affirmative answer in the general case follows from Whitman's result [7] who proved that any lattice is isomorphic to a sublattice of the lattice of all equivalence relations on some set. It remains to consider equivalence classes as the subbasis of the partition topology, what leads to an isomorphism between equivalence relations on X and partition topologies on X. In this way one gets embedding of any lattice into the lattice of all partition topologies on X.

Recently, R i c h a r d V a l e n t has shown [6] that for any infinite set X the lattice $\Lambda(X)$ of all T_1-topologies on X is embeddable in the lattice $T(X)$, thus confirming the conjecture also for T_1-topologies.

Independently (but also using Whitman's result), J . R o s i c k y has proved [4] that the conjecture remains true even for completely Hausdorff topologies (stronger than Hausdorff but weaker than regular). However, it fails for metrisable topologies [4].

In connexion with these results now some new questions arise:

1) Given the cardinality of A, what is the least cardinality of X in each case? Can we have $|A| = 2^{|X|}$ if $|A| > \aleph_0$?

2) Which is the highest T_i separation axiom for which the conjecture holds true $\left(i = 2, 3, 3\frac{1}{2} \text{ or } 4\right)$?

3) Can the embeddings be made in a way to preserve not only order but also all joins and meets? More precisely, does there exist for each lattice A a set X and an embedding $f\colon A \to T(X)$ such that $f(a)$ is a topology satisfying some restrictive condition (e.g., a T_i-separation axiom) for each $a \in A$ and $f\left(\bigvee_{\tau \in T} a_\tau\right) = \bigvee_{\tau \in T} f(a_\tau)$, $f\left(\bigwedge_{\tau \in T} a_\tau\right) = \bigwedge_{\tau \in T} f(a_\tau)$ for any family $\{a_\tau\}_{\tau \in T}$ of elements of A for which $\bigvee_{\tau \in T} a_\tau$ or $\bigwedge_{\tau \in T} a_\tau$ exists.

The third question seems to be the most interesting of the three and is apparently difficult. It requires a new approach in dealing with the embedding question, since none of the embeddings obtained with the help of Whitman's result preserves infinite joins and meets. However, such is the embedding one really wants to have.

Concerning the lecture presented to this Conference by Prof. C . H . D o w k e r [1] the following problem arises:

4) Is it true that every distributive lattice (frame) A can be embedded in the lattice tX of all open sets of some topological space X? If so, can it be done in the lattice of all regularily open sets of some Y? Embeddings should preserve all existing joins and meets.

REFERENCES

[1] C.H. Dowker, The lattice of open sets, *these Proceedings.*

[2] R. Duda, Problem 749, *Colloquium Mathematicum,* 23 (1971), 326.

[3] Roland E. Larson, P 749, R 1, ibidem, 25 (1972), 161-162.

[4] J. Rosicky, Embeddings of lattices in the lattice of topologies, *Archivum Mathematicum* (Brno) (to appear).

[5] A.K. Steiner, The lattice of topologies: structure and complementation, *Transactions of the American Mathematical Society,* 122 (1966), 379-398.

[6] Richard Valent, Every lattice is embeddable in the lattice of T_1-topologies, *Colloquium Mathematicum,* 28 (1973), 27-28.

[7] P. Whitman, Lattices, equivalence relations, and subgroups, *Bulletin of the American Mathematical Society,* 52 (1946), 507-522.

A. Hajnal – I. Juhász

1. Suppose X is a "good" space which does not contain an uncountable discrete subspace. Is it true that X is the union of a hereditarily separable and a hereditarily Lindelöf subspace?

2. Is it true that if a is the number of all open subsets of an infinte T_2 space then $a^\omega = a$?

L. Lovász

Is there a homeomorphism of the plane onto itself which commutes with its powers only (I can construct a homeomorphism φ such that if $\varphi\psi = \psi\varphi$ for a ψ then $\psi | L = \varphi^n | L$ where L is a given line; this implies that $\varphi \neq a^k$ for any a and k; such a φ was constructed by Kerékjártó).

K. Morita

Characterize those completely regular Hausdorff spaces X, which have a compactification by adding a countable number of new points.

$$\text{Necessary condition: } \begin{cases} X \text{ is } G_\delta \text{ in } \beta X, \\ X \text{ is rim-compact.} \end{cases}$$

This condition is sufficient in case X is separable and metric (due to L. Zippin).

X is said to be *rim-compact* if for any neighbourhood U of any point of X there is another neighbourhood W of X such that $W \subset U$, $\mathrm{Cl}\, W \setminus W$ is compact.

K. Nowinski

Problem generalising the Borsuk-Ulam theorem.

Let $f\colon S^n \to R^n$ be a continuous mapping from the standard euclidean unit sphere and let $0 < d \leqslant 2$.

Do there exist two points p and q in S^n such that $\rho(p, q) = d$ and $f(p) = f(q)$?

Remark. The answer is positive if $d = 2$ or if $n = 1$ or $n = 2k$.

J. Pelikán – L. Babai

Give an example of an abelian group of cardinality continuum which is not (algebraically) isomorphic to any compact topological group.

(We remark that the additive group of real numbers is not such an example.)

T. Przymusinski

1. Is the existence of a paracompact, separable, first countable space X such that X^2 is normal but not paracompact independent of the axioms of set theory?

2. Is the existence of a paracompact, first countable space X such that X^2 is normal but not collectionwise normal equivalent to the existence of a normal, non-metrizable Moore space?

Ju.M. Smirnov

a) Let $\mathscr{K}$ be one of the following classes of compact metric spaces:

$\mathscr{K}_a = \{X \mid \operatorname{ind} X = a\}$, where $a = \omega_0, \omega_0 + 1, \omega_0 + 2, \ldots$.

$\mathscr{K}_\# = \{X \mid X$ is weakly infinite dimensional$\}$

$\mathscr{K}_s = \{X \mid X$ is countable-dimensional$\}$

$\mathscr{K}_b = \{X \mid \dim A = 0$ or $\dim A = \infty$ for every closed subset A of $X\}$.

Question. Does there exist for any class $\mathscr{K}$ a space (or *CW*-complex) $Y_{\mathscr{K}}$ such that $\mathscr{K} = \{X \mid$ every mapping $f\colon A \to Y_{\mathscr{K}}$ has $\operatorname{ext} f\colon X \to Y_{\mathscr{K}}$ for every closed set $A \subset X\}$.

b) Does there exist a compact metric space Y_a for every a such that $\mathscr{K}_a = \{X \mid \forall A = \bar{A} \subset X$ and $\forall f\colon A \to Y_a \ \exists \operatorname{ext} f\colon X \to Y_a\}$?

0. Construct an $\aleph_1$ Dowker space.

1. Find a set theoretic translation of "there exists a Dowker space".

2. Does the existence of an $\aleph_1$ Dowker space imply the existence of a Souslin line?

3. Can one prove without set theoretic assumptions (such as the Continuum Hypothesis) that the box product of countably many copies of $\omega_0 + 1$ is paracompact?

4. Is the box product of $\aleph_1$ copies of $\omega_0 + 1$ paracompact (normal)?

5. Is the box product of countably many copies of ω_1 normal?

6. Is the box product of any family of compact spaces paracompact?

7. Is the box product of any family of metric spaces paracompact?

8. Suppose T is an Aronszajn tree (an $\aleph_1$ tree with no uncountable chain and every level countable). Give T the topology induced by the order topology on the chains of T. Prove that it is consistent with the axioms of set theory that T be normal.

9. Construct an Aronszajn tree T such that A is a maximal antichain in T implies $\{t \in T \mid t < a$ for some a in $A\}$ is the union of countably many antichains but $\{t \in T \mid t > a$ for some a in $A\}$ is *not* the union of countably many antichains.

F. Tall

1. Find an example of a (preferably normal) space which is collectionwise normal with respect to points, but is not collectionwise normal.

2. Find an example of a collectionwise normal metalindelöf space which is not paracompact.

3. Is it true that every perfectly normal, locally compact, metalindelöf space is paracompact? (It is consistent with the axioms of set theory. It is true for metacompact instead of metalinedlöf (Arhangel'skii)).

4. Hajnal and Juhász have proved (see Juhász' monograph) that every first countable but not countable space satisfying the countable chain condition has a subspace of cardinality $\aleph_1$, satisfying that condition. Can first countable be weakened to countable tightness?

5. If X is regular and (hereditarily) Lindelöf, is X^2 subparacompact?

6. If X is paralindelöf, countably paracompact and regular, is X paracompact?

R. Telgársky

1. Let $X \neq \phi$ be a compact Hausdorff space without isolated points. Does X contain a closed subspace $X_0 \neq \phi$ such that X_0 is zero-dimensional and without isolated points?

2. Let X be a closed, continuous image of a scattered space. Is X scattered?

3. Let X be a compact, linearly ordered space and let Y be a paracompact, linearly ordered space. Is the lexicographical product $X \times_L Y$ of X and Y a paracompact space?

H. Torunczyk

Let X be a metrisable space and $\mathcal{U}$ a base for its topology. Give some sufficient conditions on $\mathcal{U}$ to contain a σ-locally finite family which is still a base.

Does every $\mathcal{U}$ contain such a family, if $X \in ANR(\mathcal{M})$? What if X is a normed linear space and $\mathcal{U}$ is the family of all open convex sets?

Remarks.

1) Prof. R. Engelking has observed that the family of clopen subsets of Roy's O-dimensional space contains no σ-locally finite base.

2) Every open cover of the normed linear space $c_0(A)$ has a locally finite refinement consisting of open balls; the space $c_0(A)$ cannot here be replaced by a reflexive Banach space (see the theorem of H.H. Corson in Fund. Math.)*

3) The problem has connection with constructing partitions of unity consisting of functions from a given class (see my paper in Studia Math.).

L. Tumarkin

Let $\zeta = \{X\}$ be the class of all topological spaces satisfying the following conditions:

a) If $X \in \zeta$ then every subspace of X belongs to ζ too.

b) $\operatorname{ind} X = \dim X = \operatorname{Ind} X$.

c) If $X_1 \subset X_2$ then $\dim X_1 \leqslant \dim X_2$.

d) If $X_1 \subset X_2$ then $\dim\left(X_1 \cup \{x\}\right) = \dim X_1$, x is a point of $X_2 \setminus X_1$.

*$c_0(A)$ is the space of real-valued functions f on A for which the sets $\left\{a \in A\colon f(a) > \frac{1}{n}\right\}$ are finite for all $n \in N$; the space is regarded under the sup-norm.

e) $\dim (X_1 \times X_2) \leqslant \dim X_1 + \dim X_2$, $X_1 \times X_2$ is the cartesian product.

Problems:

I. Characterise these classes.

I′. The same problem for metric spaces.

II″. The same problem for O-dimensional metric spaces.